ORGANIC CROP PRODUCTION MANAGEMENT

Focus on India, with Global Implications

ORGANIC CROP PRODUCTION MANAGEMENT

Focus on India, with Global Implications

Edited by

D. P. Singh, PhD

H. G. Prakash, PhD

M. Swapna, PhD

S. Solomon, PhD

AAP | APPLE ACADEMIC PRESS

First edition published 2023

Apple Academic Press Inc.
1265 Goldenrod Circle, NE,
Palm Bay, FL 32905 USA

760 Laurentian Drive, Unit 19,
Burlington, ON L7N 0A4, CANADA

CRC Press
6000 Broken Sound Parkway NW,
Suite 300, Boca Raton, FL 33487-2742 USA

4 Park Square, Milton Park,
Abingdon, Oxon, OX14 4RN UK

Library and Archives Canada Cataloguing in Publication

Title: Organic crop production management / edited by D.P. Singh, PhD, H.G. Prakash, PhD, M. Swapna, PhD, S. Solomon, PhD.

Names: Singh, D. P. (Davinder Pal), 1964- editor. | Prakash, H. G., editor. | Swapna, M., editor. | Solomon, S., editor.

Description: First edition. | Includes bibliographical references and index.

Identifiers: Canadiana (print) 20220489998 | Canadiana (ebook) 20220490023 | ISBN 9781774910580 (hardcover) | ISBN 9781774910597 (softcover) | ISBN 9781003283560 (ebook)

Subjects: LCSH: Food crops—Organic farming. | LCSH: Vegetables—Organic farming. | LCSH: Sugarcane—Organic farming.

Classification: LCC S605.5 .O68 2023 | DDC 631.5/84—dc23

Library of Congress Cataloging-in-Publication Data

CIP data on file with US Library of Congress

ISBN: 978-1-77491-058-0 (hbk)
ISBN: 978-1-77491-059-7 (pbk)
ISBN: 978-1-00328-356-0 (ebk)

About the Editors

D. P. Singh, PhD, is Associate Professor/Sr. Scientist (Vegetable Seeds) at Chandra Shekhar Azad University of Agriculture and Technology, Kanpur, U.P., India, as well as the Officer in Charge of the All India Coordinated Research Project on Vegetable Crops. He has also served as Joint Director Research in CSAUAT, Kanpur. Dr. Singh has more than 32 years of experience in teaching, research, and extension activities in the area of seed production technology dissemination and rural extension programs and has developed nine technologies of different vegetable crops for the agro-climatic zone-IV. In addition, he is actively engaged in traditional and organic vegetable production technologies for farmers of north India. He has published over 55 research papers, over 20 papers and review articles, several extended summaries, over 160 research abstracts, 14 books, nine book chapters, proceedings, technical training manuals, popular articles, and more. He is recipient of many prestigious awards such as the CSAU Award of Excellence–2019, Fellow Award–2019, SRDA Award, Azad Krishi Samman–2013, Lifetime Achievement Award–2011, and Vishvidyalaya Ratna–2008. Dr. Singh is a life member of several scientific/social societies and has visited many countries for scientific discussions. Dr. Singh earned his MSc (Ag) in Horticulture from Meerut University (India) and holds a PhD in Horticulture from Kanpur University. He Joined C.S.A. University of Agriculture & Technology, Kanpur, in 1988.

H. G. Prakash, PhD, served as Joint Director of Research and was later on appointed as Director of Research at CSAU A & T, Kanpur, India, where he contributed in the execution, monitoring, and impact assessment of R&D of various research projects, leading to the development of 43 high-yielding cultivars and 105 agro-techniques. Dr. Prakash has 10 technologies, 256 outreach activities, and 218 publications to his credit. He was instrumental in organizing 41 seminars, webinars, advanced trainings for capacity building of postgraduate students and faculty under a World Bank supported project for 4300 postgraduate students across the country to enhance their research competency, entrepreneurship development, and subject competitiveness. He has completed 14 research projects in animal nutrition, forage production, farming system-based research, and diversification and intensification-based operational research, and seven projects are in operation. He is actively engaged in the transfer of agricultural production technologies to all stakeholders. He served as chairman and member of different scientific and study groups, councils, and boards. He was honored with various awards such as Distinguished Scientist Award, Rajiv Gandhi Excellence Award–2016, Award of Excellence–2017, Appreciation Award–2017, Distinguished Scientist Award–2018, Outstanding Scientist Award–2019, International Education Awards–2020 (top 10 directors), and Excellence in Communication Awards–2020 for his distinguished services and outstanding contribution to the nation. He is also a life member of five scientific societies. Professor Prakash was awarded his MSc in Agri-culture from G. B. Pant University of Agriculture & Technology, Pantnagar (India) and PhD from the Indian Veterinary Research Institute, Izatnagar (Bareilly, India).

M. Swapna, PhD, is a Principal Scientist in the Division of Crop Improvement at ICAR-Indian Institute of Sugarcane Research, Lucknow, India. She joined the Agriculture Research Service under the Indian Council of Agricultural Research (ICAR) in 1998 and is working on sugarcane genetic improvement, making use of the conventional as well as nonconventional tools. She is involved in sugarcane improvement programs including prebreeding strategies for improving sugar content and varietal development. She is part of a team that has developed improved sugarcane varieties for the subtropical sugarcane growing areas of the country, which has led to population improvement and prebreeding studies that have yielded high sugar accumulating parental genetic stocks that are being used in commercial varietal development programs. She is also engaged in molecular investigations related to sugar accumulation, especially molecular marker development and its applications in sugarcane improvement. Dr. Swapna is an editorial manager of the journal *Sugar Tech* (Springer Nature) and has organized many national and international conferences.

S. Solomon, PhD, joined the Agricultural Research Service of Indian Council of Agricultural Research (ICAR-Ministry of Agriculture) in 1977. As a Director of the Indian Institute of Sugarcane Research, Lucknow (2012–2014), he was actively involved in the development and transfer of relevant technologies to sugarcane farmers and industry for the sustainable development of the Indian sugar industry. During his career of 36 years of research, he has published over 120 research papers, 22 books, and many technical reports for the benefit of the global sugar industry. Dr. Solomon is the President of the Society for Sugar Research and Promotion and Vice President of the International Association of Professionals in Sugar and Integrated Industries (IAPSIT). He is also on the advisory bodies of many international organizations. As an advisor, he has visited Brazil, Australia, China, Vietnam, Egypt, Iran, Sri Lanka, Cuba, Thailand, Taiwan, and other countries. He is also the editor-in-chief of the international journal *Sugar Tech*, published by Springer Nature. Dr. Solomon has organized many international conferences.

Dr. Solomon is recipient of the most prestigious honors such as Friendship Award and Jin Xiu Qui Award, P. R. China (2005), Award of Excellence–IAPSIT (2006), Sinai University Peace Award, (Egypt, 2008), Global Award of Excellence–IAPSIT (2008), Indira Gandhi Award (2013), Lifetime Achievement Award–IAPSIT (2014), Lifetime Achievement Award–UPAAS (2014), Hari Om Ashram Trust Award–ICAR (2012 and 2013), Dr. Rajendra Prasad Award–ICAR and Noel Deerr Gold Medal–STAI (2014, 2016), and Leadership Excellence Award (2018) from the Thailand Society of Sugarcane & Sugar Technologists. He is a member of many national and international societies and on the board of leading agricultural universities and research institutions in the country. Dr. Solomon was the Vice Chancellor of Chandra Shekhar Azad University of Agriculture & Technology, Kanpur (December 2016–Feb 2020), a premier agricultural university in north India. Acknowledging his relentless contributions to agricultural education and research, he was given Dr. Sampoornan and Rangbharti and ICN awards. Dr. Solomon received his PhD degree from Punjab Agricultural University, Ludhiana (India).

Contents

Contributors

Thomas Abraham
Department of Agronomy, Sam Higginbottom University of Agriculture, Technology and Sciences, Prayagraj (Allahabad) 211007, Uttar Pradesh, India; E-mail: thomas.abraham@shuats.edu.in

Nazeer Ahmed
Sher-e-Kashmir University of Agricultural Sciences and Technology of Kashmir, Shalimar 190025, Jammu and Kashmir, India

S. I. Anwar
ICAR-Indian Institute of Sugarcane Research, Lucknow 226002, Uttar Pradesh, India

R. K. Avasthi
ICAR-National Organic Farming Research Institute, Tadong, Gangtok 737102, Sikkim, India

S. K. Awasthi
Agronomy, ICAR-Indian Institute of Sugarcane Research, Lucknow 226002, Uttar Pradesh, India

Priyanka Babel
Biotechnology & Microbiology, MLS University, Udaipur 313001, Rajasthan, India

Subhash Babu
ICAR Research Complex for NEH Region, Umiam 793103, Meghalaya, India

Prakash M. Bafna
Jay Research and Biotech India Private Limited, Bafna Group, 111 Tower 1,World Trade Centre, Kharadi, Pune 411014, Maharashtra, India; E-mail:prakash@bafnagroup.com

Neelam Bhardwaj
Department of Organic Agriculture and Natural Farming, CSK Himachal Pradesh Agricultural University, Palampur 176062, Himachal Pradesh, India

Santosh B. Chavan
Jay Research and Biotech India Private Limited, Bafna Group, 111 Tower 1, World Trade Centre, Kharadi, Pune 411014, Maharashtra, India

Santosh Choudhary
Agriculture University, Jodhpur 342304, Rajasthan, India

S. Nirmala Devi
Department of Vegetable Science, College of Horticulture, Kerala Agricultural University, KAU P. O., Thrissur 680656, Kerala, India; E-mail: drsnirmala@yahoo.com

Rajesh Kumar Dubey
National Resource Centre- Biotechnology, MLS University, Udaipur 313001, Rajasthan, India

A. Gaur
Agronomy, ICAR-Indian Institute of Sugarcane Research, Lucknow 226002, Uttar Pradesh, India

Shilpi Gupta
Department of Agronomy, College of Agriculture, G.B.P.U.A.&T., Pantnagar, U.S. Nagar 263145, Uttarakhand, India

S. S. Hebbar
Division of Vegetable Crops, ICAR-Indian Institute of Horticultural Research, Hesaraghatta Lake P.O., Bengaluru 560089, Karnataka, India; E-mail: hebbar@iihr.res.in

K. Hussain
Sher-e-Kashmir University of Agricultural Sciences and Technology of Kashmir, Shalimar 190025, Jammu and Kashmir, India

P. Indira
Department of Vegetable Science, College of Horticulture, Kerala Agricultural University, KAU P. O., Thrissur 680656, Kerala, India

V. P. Jaiswal
Agronomy, ICAR-Indian Institute of Sugarcane Research, Lucknow 226002, Uttar Pradesh, India

Gograj Singh Jat
Division of Vegetable Science, ICAR-Indian Agricultural Research Institute, New Delhi 110012, India

A. K. Jha
Soil Science and Agricultural Chemistry, Bihar Agricultural University, Sabour 813210, Bhagalpur, Bihar, India

Minnu Ann Jose
Department of Vegetable Science, College of Horticulture, Kerala Agricultural University, KAU P. O., Thrissur 680656, Kerala, India

R. B. Khandagave
S. Nijalingappa Sugar Institute, Belgaum 591108, Karnataka, India; E-mail: dr.khadagave_r@yahoo.com

Hare Krishna
ICAR-Central Institute for Arid Horticulture, Beechwal, Bikaner 334006, Rajasthan, India; E-mail: kishun@rediffmail.com

Munish Kumar
Soil Conservation & Water Management/Director, Administration & Monitoring C.S. Azad University of Agriculture and Technology, Kanpur 208002, Uttar Pradesh, India; E-mail: munish.csa@gmail.com

M. Senthil Kumar
Division of Vegetable Crops, ICAR-Indian Institute of Horticultural Research, Hesaraghatta Lake P.O., Bengaluru 560089, Karnataka, India

Anitta Judy Kurian
Department of Vegetable Science, College of Horticulture, Kerala Agricultural University, KAU P. O., Thrissur 680656, Kerala, India

Ajaz Ahmed Malik
Sher-e-Kashmir University of Agricultural Sciences and Technology of Kashmir, Shalimar 190025, Jammu and Kashmir, India

A. K. Nair
Division of Vegetable Crops, ICAR-Indian Institute of Horticultural Research, Hesaraghatta Lake P.O., Bengaluru 560089, Karnataka, India

Sumati Narayan
Sher-e-Kashmir University of Agricultural Sciences and Technology of Kashmir, Shalimar 190025, Jammu and Kashmir, India; E-mail: sumatinarayan@gmail.com

Rakesh Pandey
Department of Entomology, BUAT, Banda 210001, Uttar Pradesh, India

R. K. Pandey
Chandra Shekhar Azad University of Agriculture & Technology, Kanpur 208002, Uttar Pradesh, India

Vibha Pandey
Department of Plant Physiology, JNKVV, Jabalpur 482004, Madhya Pradesh, India; E-mail: vibhapandey93@gmail.com

A. S. Panwar
ICAR-Indian Institute of Farming Systems Research, Modipuram, Meerut 250110, Uttar Pradesh, India

K. V. Peter
Kerala Agricultural University, KAU P.O., Thrissur 680656, Kerala, India

M. Prabhakar
Division of Vegetable Crops, ICAR-Indian Institute of Horticultural Research, Hesaraghatta Lake P.O., Bengaluru 560089, Karnataka, India

H. G. Prakash
Chandra Shekhar Azad University of Agriculture & Technology, Kanpur 208002, Uttar Pradesh, India; E-mail: drhp1962@gmail.com

Govind. P. Rao
Department of Plant Pathology, ICAR-Indian Agricultural Research Institute, New Delhi 110012, India; E-mail: gprao_gor@rediffmail.com

N. Ravisankar
ICAR-Indian Institute of Farming Systems Research, Modipuram, Meerut 250110, Uttar Pradesh, India

J. P. Saini
Department of Organic Agriculture and Natural Farming, CSK Himachal Pradesh Agricultural University, Palampur 176062, Himachal Pradesh, India; E-mail: drjpsaini@gmail.com

P. L. Saroj
ICAR-Central Institute for Arid Horticulture, Beechwal, Bikaner 334006, Rajasthan, India

Lalan Sharma
Plant Pathology, ICAR-Indian Institute of Sugarcane Research, Lucknow 226002, Uttar Pradesh, India

Rajiv
Vegetable Research Station, C.S. Azad University of Agriculture and Technology, Kalyanpur, Kanpur 208024, Uttar Pradesh, India; E-mail: rajiv.agro69@gmail.com

R. A. Ram
ICAR-Central Institute for Subtropical Horticulture, Rehmankhera, Lucknow 226101, Uttar Pradesh, India; E-mail: raram_cish@yahoo.co.in

Rajvir Singh Rathore
Former OSD to Governor of Uttar Pradesh, Lucknow 226001, Uttar Pradesh, India;
E-mail: rajvir_ddg@yahoo.co.in

B. L. Sharma
Uttar Pradesh Council of Sugarcane Research, Shahjahanpur 242001, Uttar Pradesh, India

Ritesh Sharma
Basmati Export Development, Modipuram, Meerut 250110, Uttar Pradesh, India

Y. Sharma
Department of Agronomy, College of Agriculture, G.B.P.U.A.&T., Pantnagar, U.S. Nagar 263145, Uttarakhand, India

S. K. Shukla
AICRP on Sugarcane, ICAR-Indian Institute of Sugarcane Research, Lucknow 226002, Uttar Pradesh, India; E-mail:sudhirshukla151@gmail.com

A. K. Singh
ICAR-Indian Agricultural Research Institute, New Delhi 110012, India; E-mail: aksicar@gmail.com
Bihar Agricultural University, Sabour, Bhagalpur 813210, Bihar, India

B. Singh
ICAR-Indian Institute of Vegetable Research, Post Box No. 01, P.O. Jakhani (Shahanshahpur), Varanasi, 221305, Uttar Pradesh, India

Balraj Singh
Agriculture University, Jodhpur 342304, Rajasthan, India; E-mail: drbsingh2000@yahoo.com

B. N. Singh
Centre for Research and Development (CRD), Gorakhpur, Uttar Pradesh, India; E-mail: baijnathsingh08@gmail.com

D. K. Singh
Department of Agronomy, College of Agriculture, G.B.P.U.A.&T., Pantnagar, U.S. Nagar 263145, Uttarakhand, India; E-mail: dhananjayrahul@rediffmail.com

D. P. Singh
Chandra Shekhar Azad University of Agriculture & Technology, Kanpur 208002, Uttar Pradesh, India; Email: dpsinghjdrcsa@gmail.com

Jogendra Singh
Division of Vegetable Science, ICAR-Indian Agricultural Research Institute, New Delhi 110012, India

M. M. Singh
Uttar Pradesh Council of Sugarcane Research, Shahjahanpur 242001, Uttar Pradesh, India

Priyanka Singh
Uttar Pradesh Council of Sugarcane Research, Shahjahanpur 242001, Uttar Pradesh, India; E-mail: priyanka.vishen75@gmail.com

R. A. Singh
Organic Farming, C.S. Azad University of Agriculture & Technology, Kanpur 208002, Uttar Pradesh, India; E-mail: rasinghcsau@gmail.com

Raghavendra Singh
ICAR-National Organic Farming Research Institute, Tadong, Gangtok 737102, Sikkim, India; E-mail: raghavenupc@gmail.com

Suryendra Singh
Guru Angad Dev Veterinary and Animal Science University, Ludhiana 141001, Punjab, India

S. K. Singh
ICAR-Indian Institute of Vegetable Research, Post Box No. 01, P.O. Jakhani (Shahanshahpur), Varanasi, 221305, Uttar Pradesh, India; E-mail: skscprs@gamil.com

S. R. Singh
Deputy Director (Marketing & Skill), CCS National Institute of Agricultural Marketing, Jaipur 302033, Rajasthan, India; E-mail:sattramsingh@gmail.com

S. Sithanantham
Sun Agro Biotech Research Centre, Chennai 600116, Tamil Nadu, India;
E-mail: sabrcchennai@yahoo.co.in

S. Solomon
Chandra Shekhar Azad University of Agriculture & Technology, Kanpur, U.P., India
Email: drsolomonsushil1952@gmail.com

A. Srinivasaraghavan
Plant Pathology, Bihar Agricultural University, Sabour, Bhagalpur 813210, Bihar, India

M. Swapna
Division of Crop Improvement, ICAR–Indian Institute of Sugarcane Research, Lucknow, India,
Email: sugarswapna@gmail.com

B. S. Tomar
Division of Vegetable Science, ICAR-Indian Agricultural Research Institute, New Delhi 110012, India;
E-mail: bst_spu_iari@rediffmail.com

Vivek Yadav
Rice Research Station (SVPUA&T, Meerut), Nagina, Bijnaur 246762, Uttar Pradesh, India

Ashok Kumar Yadav
APEDA, New Delhi 110016, India
NCOF, Ghaziabad 201002, India; E-mail: akyadav@apeda.gov.in

Vijay Kumar Yadav
C. S. Azad University of Agriculture and Technology, Kanpur 208002, Uttar Pradesh, India;
E-mail: vkyadu@gmail.com

Abbreviations

AMC	Arka Microbial Consortium
AOAC	Allahabad Organic Agriculture Cooperative
APEDA	Agricultural & Processed Food Product Export Development Authority
BCAs	biocontrol agents
BLB	bacterial leaf blight
BLQ	below level of quantification
CBR	cost-benefit ratio
CCC	chloromquet chloride
CGMV	cucumber green mottle virus
CL	cue lure
CPP	cow pat pit
CR	crop residues
EIA	Export Inspection Agency
EPM	enriched press mud
ESB	early shoot borer
ET	ethylene transcription
FIRBS	Furrow Irrigated Raised Bed System
FRAP	fluorescence recovery after photo bleaching
FYM	farmyard manure
GAPs	Good Agriculture Practices
GDP	gross domestic product
GI	geographical indication
GHG	greenhouse gas
GM	green manuring
GMOs	genetically modified organisms
HPLC	high pressure liquid chromatography
HYV	high-yielding varieties
IBAs	insect biocontrol agents
ICAR	Indian Council for Agricultural Research
ICM	integrated approach of crop management
ICROFS	International Centre for Research in Organic Food Systems
IGP	Indo-Gangetic Plain
INDOCERT	Indian Organic Certification Agency

INM	Integrated Nutrient Management
ITK	Indigenous Technological Knowledge
IOFS	Integrated Organic Farming Systems
IPCC	Intergovernmental Panel on Global Climate Change
IS	inorganic sugarcane
ISSR	inter simple sequence repeats
ITSS	internal transcribed spacer sequences
ITKs	indigenous and traditional knowledge
JA	jasmonic acid
JAT	jute agrotextile
KTGR	Kalanamak traditional genetic resources
KVK	Krishi Vigyan Kendra
LC-MS/MS	liquid chromatography-tandem mass spectrometry
MDG	millennium development goal
ME	methyl eugenol
MRL	maximum residue level
MSP	minimum support price
NCOF	National Centre of Organic Farming
NGOs	nongovernmental organizations
NLP	neem leaf powder
NMSA	National Mission on Sustainable Agriculture
NPOF	Network Project on Organic Farming
NPOP	National Programme for Organic Production
NPV	nuclear polyhedrosis virus
NSKE	neem seed kernel extract
OPM	organic pest management
OPS	organic production systems
ORCA	Organic Research Centres Alliance
OS	organic sugarcane
PGS	participatory guarantee system
PPP	public–private partnership
PSB	phosphate solubilizing bacteria
PSB	phosphate solubilising bio-fertilizers
PKVY	Paramparagat Krishi Vikas Yojana
RAPD	random amplified polymorphic DNA
RATDS	Regional Agriculture Testing and Demonstration Stations
RDN	recommended dose of N
RFLP	restriction fragment length polymorphism
ROS	reactive oxygen species

SCI	system of crop intensification
SHG	self-help groups
SMW	Standard Meteorological Week
SOC	soil organic carbon
SOM	soil organic matter
SPMC	sulphitation pressmud cake
SSI	State of Sustainability Initiative
SSRs	simple sequence repeats
TFP	total factor productivity
TLCV	tomato leaf curl virus
TS	tensile strength
VC	vermicompost
VCGs	vegetative compatibility groups

Foreword

Intensive agricultural practices from *Rainbow revolution* to *Gene revolution* have resulted in a quantum jump in the production of farm produce. Nonetheless these have had their downsides, like fast depleting natural resources, GHG emission, climate change, and related risks. Getting the maximum out of minimum inputs is the emerging motto, with productivity and profitability going hand in hand with sustainability. Organic farming is a holistic production-management system aiming to maximize production with high quality and minimum adverse effect on resources, including on the environment. It is a biointensive farming method that makes use of organic matter and biological materials, like beneficial microbes and other organisms, for a sustainable eco-friendly production. The positive effects like preserving soil fertility, enhancing carbon sequestration in the soil, and using minimal chemicals go a long way in ensuring eco-friendly cultivation and climate change mitigation, while enhancing soil productivity and sustainability. It also improves the agroecosystem, including biodiversity, biological cycles, and soil biological activity.

Adoption of organic farming is an important step toward an environment-friendly sustainable agriculture. Though estimates show that India has approximately 30% of the total organic farmers of the world, we still lag behind with respect to area under organic farming as well as the production of organic produce. There are several challenges faced by organic growers that need to be addressed so that this practice becomes more profitable. Awareness about improved production and management practices, development of local markets for organic products, better access to international markets with highly competitive produce along with the ability to meet the demands, necessary infrastructure, skill development, easy access to finance and additional financing during the transition phase are some of the aspects that need to be taken care of.

This book entitled *Organic Crop Production Management* encompasses a wide array of topics on the status and challenges of organic farming including production, nutrient management, plant protection aspects, processing methods, policy issues, etc. in various crops. The debatable issues, the risk associated etc. have also been discussed in detail. These can serve to chalk out a blueprint for the holistic improvement of the organic farming sector in the country.

I would like to compliment all the contributors and editorial team for their efforts in bringing out this useful publication. I hope that the compendium will be of immense use to all the stakeholders in promoting and advancing organic crop production in India and other developing countries.

T. Mohapatra
New Delhi

Preface

The world population is expected to touch 9.7 billion by 2050, an increase by 2 billion from the present numbers. With the not-so-rich countries expected to see the lion's share of this population increase, the need to ensure food and nutritional security to this burgeoning population looms large over the whole world. The great strides in the form of the green revolution and the related technology-driven advancements have succeeded in ensuring food security in countries including India, with self-sufficiency in almost all food products. But all these have come with a price, with the depletion of natural resources, declining soil health, resistance of pests/diseases to plant-protection chemicals and others occurring on a large scale. This has prompted mankind to look toward more environment-friendly, sustainable means to meet the demands of food, fuel, and fiber.

Agriculture has been viewed by many as both the cause and the victim of climate change. By 2050, greenhouse gas emissions (GHG) emissions from agriculture have been projected to increase by 30%. An improvement in this situation can be foreseen only through mitigation strategies involving eco-friendly sustainable production and processing systems. Organic farming can be an attractive option toward this end, to address the present day concerns, thereby reducing the impact on the environment to the maximum extent possible. This practice of production and processing of goods with the minimum use of synthetic products and chemicals, bio-engineered products etc., is designed to aid in the rejuvenation of land, water, air, and of the ecosystem as a whole. This can be expected to ensure productivity along with sustainability. A lot of research is still underway to set the limits and boundaries for the various organic practices and related activities, to ensure maximum gains with minimum adverse effect on environment and also on the profitability.

India is home to 30% of total organic producers in the world but accounts for just 2.59% of the total organic cultivation area of 57.8 million ha in the world. Extensive efforts are therefore required in our country with respect to the area under organic cultivation and the quantum of organic produce. Sikkim has been declared as the first organic state in the country. In 2010, the government formulated and launched the Sikkim Organic Mission with clear goals that helped them achieve the all-organic status in 2016. The initiative covered 190,000 acres of farmland and benefitted 66,000 farmer families to switch to organic farming. It has helped in improving the overall human health, soil health, wildlife, and bee population and also leads to an increase in ecotourism. It is imperative that the right mix of organic and inorganic technologies to be put in place; the yield dynamics, especially at the transition stage, appropriate support policies wherever needed, are some of the aspects that still need to be debated and finalized.

This book entitled *Organic Crop Production Management* encompasses a wide array of topics on the status and challenges of organic farming, including production, nutrient management, plant protection aspects, processing methods, and policy issues in various crops like food crops, vegetable crops, and sugarcane including organic jaggery production. The debatable issues and the risk associated have also been discussed in detail. These can serve to chalk out a blueprint for the holistic improvement of the organic farming sector in the country.

We are grateful to our contributors who have provided valuable inputs for this publication. The efforts of Mr. Brahmprakash, ICAR-Indian Institute of Sugarcane Research, Lucknow, and Mr. Prashant Gupta, Stuti Enterprises, Lucknow, are highly appreciated in bringing out this publication in the present form. We thank our sponsors who had helped us make this compilation a reality.

We hope that the compendium will be of immense help to all the stakeholders in promoting and advancing organic crop production, including the processing sector. This will help India to evolve into an organic nation with a productive, profitable, and sustainable agricultural sector.

Editors

Introduction

Organic farming is a sustainable cultivation process that is gaining popularity in many countries. Organic food and food products are increasingly being preferred by many consumers, especially in the wake of mounting concerns for healthy environment and lifestyle. The process of organic farming encompasses all the management systems that promote and enhance biodiversity and biological activity, thereby resulting in a productive and sustainable system. Resource conservation coupled with quality is the main concern that is being aimed to be met through the concept of organic farming. This, in turn, is an ideal strategy for countries to fulfill its commitments toward their sustainability development goals.

In 2018, globally 71.5 million ha was organic agricultural land, that is, around 1.5% of total farmland. The global organic market continues to grow worldwide, exceeding USD 100 billion. In India, approximately 2% of the net sown area is under organic cultivation, and about 30% of the farmers out of the total at the global level are engaged in organic farming. The challenges faced by these farmers at various stages, right from the specialized cultivation practices that strictly adhere to organic norms, processing of produce, availability of markets etc., are still matters of concern.

This compendium presents an overall assessment of organic cultivation in general and also with respect to specific crops and products such as organic jaggery from sugarcane, including their marketing aspects. The authors have reiterated that an environmentally sustainable system of agriculture, like organic farming, will be able to maintain a stable resource balance, avoid over exploitation of renewable resource, and conserve inherent soil nutritional quality, soil health, and biodiversity. It will lead us to sustainable crop production system and create a sustainable lifestyle for generations to come.

Starting with an overview on organic cultivation, the chapters deal with the issues and challenges faced at various stages of organic cultivation, the significance of the practice from a sustainability point of view, and how this can ensure rural livelihood security, especially in the backdrop of the present day vagaries of climate. Organic farming in specialized situations like dry lands, temperate regions, management of soil, and other resources, including nutrients and pest management through organic farming, have also been discussed. The response of horticultural crops like vegetables, grapes, and sugarcane to organic farming and production of by-products like jaggery in an organic way are some other aspects that have been discussed. Crop-based case studies like that of table grapes and *Kalanamak* rice also find a place here. A view of the marketing strategies for organic produce presents the upcoming trends and opportunities for successful marketing, including export-oriented quality management.

A holistic improvement of the organic farming sector in the country could be realized through implementation of the ideas and recommendations in this collection. After almost a century of development, organic agriculture is now being embraced by the mainstream and shows great promise commercially, socially, and environmentally.

CHAPTER 1

Organic Agriculture in India: An Overview

ASHOK KUMAR YADAV

Advisor, APEDA, New Delhi 110016, India
Former Director, NCOF, Ghaziabad 201002, India
E-mail: akyadav@apeda.gov.in

ABSTRACT

Organic farming is in a nascent stage in India and about 2.78 million hectare of farmland was under organic cultivation in 2020. National Programme for Organic Production (NPOP) launched during 2001 was the first such quality assurance initiative by the Government of India and had been the main driving force for the growth of organic agriculture in the country. NPOP certification is a system of process certification wherein an independent organization reviews entire production, processing, handling, storage and transport etc to ensure the compliance of organic standards. Its robust assessment and verification system has earned the name in domestic and international markets. Traceability system, introduced during 2006 in the form of online data management, has also contributed to the integrity and transparency of the system in the national and international trades. In this regard, country and provincial governments have also provided timely support in the form of policies, financial incentives, technology packages, and the availability of quality organic inputs. In recent times, consumer awareness for contamination-free food is driving the markets. The sector has not only attracted the attention of policy planners, research institutions, civil society organizations, and consumers but has also instilled confidence among the growers as an economically viable system for the future.

Organic Crop Production Management: Focus on India, with Global Implications, D. P. Singh, PhD, H. G. Prakash, PhD, M. Swapna, PhD, & S. Solomon, PhD (Eds.)

1.1 INTRODUCTION

India, a country known for its wisdom in traditional agriculture with a large pool of indigenous traditional practices on the best practices in organic agriculture, which were essentially organic, is picking up fast with the modern tenets of standard-based organic agriculture and emerging as the hub for organic food products. National Programme for Organic Production (NPOP), launched during 2001 and revised from time to time (Anonymous, 2014), was the first policy intervention by the Government to lay the foundation stone for the systematic development of organic agriculture. NPOP, which provides an institutional framework for accreditation and certification of various facets of organic agriculture processes in India, has earned international recognition and enjoys recognition agreements with European Union, USDA-NOP, and Switzerland. NPOP is being managed and operated by the Agricultural and Processed Food Products Export Development Authority (APEDA) under the Ministry of Commerce and Industry, Government of India. Starting with just 42,000 ha during 2003–2004 (Anonymous, 2016), it has grown to 1.44 million ha (cultivated area) during 2016–2017 with a CAGR of 26.5%. Almost all types of agricultural, horticultural, and nonfood crops are being grown under the organic certification process. Livestock, aquaculture, animal feed processing and handling, mushroom production, seaweeds, aquatic plants, and greenhouse crop production (Anonymous, 2014) have also been brought under the ambit of organic certification. This chapter presents an overview of the organic agriculture sector in India under the domain of NPOP, as on March 2017.

1.2 OVERALL SCENARIO

1.2.1 AREA

By the end of March 2017, 4.45 million-ha area was under NPOP organic certification process, which consisted of 1.44 million ha (32.35%) arable-cultivated area and 3.0 million ha (67.6%) under forest for nontimber minor forest produce collection. The growth of area under organic certification during the last 14 years is depicted in Figure 1.1, while the major states with their area (in lakh ha) under certification are shown in Figure 1.2.

1.2.2 PRODUCTION

With its varied agroecological conditions, Indian organic farmers are producing almost all types of crops. Top four commodities with a share of about 85% include sugar crops, oilseed crops (mainly soybean), fiber crops, and cereals and millets. Spices, tea, coffee, and medicinal and herbal plants are also being cultivated in a sizeable area. Details on the production of selected crop categories under the organic certification process are given in Table 1.1, while Figure 1.3 shows the share of major states in production (in percentage).

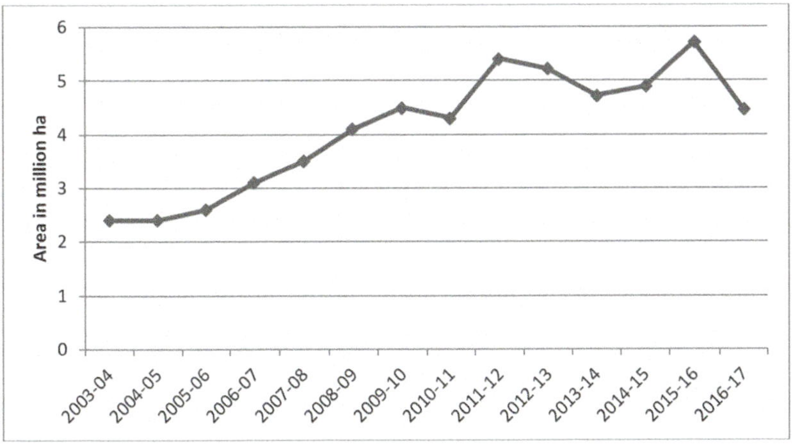

FIGURE 1.1 Growing area under organic certification (cultivated + wild harvest). *Source*: http://apeda.gov.in.

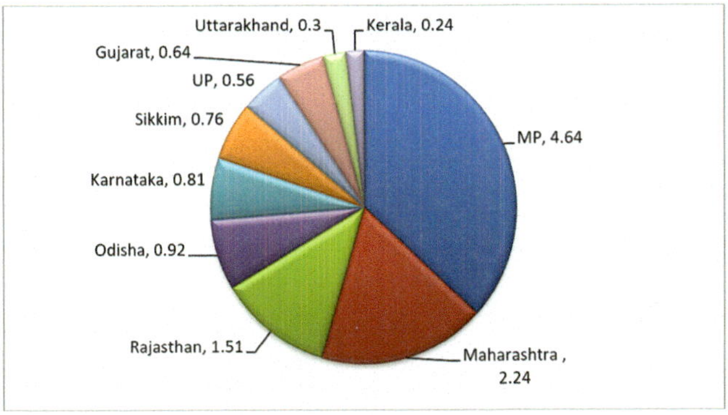

FIGURE 1.2 Major state players with area under organic certification (in lakh ha). *Source*: http://apeda.gov.in.

1.2.3 TRADE

Although export was the main driver for wider acceptance of organic agriculture by the farmers in the initial years, now growing awareness among consumers and economical benefits realized by farmers are driving the domestic demand, which is growing at a CAGR of about 12–15%. However, exports still account for major revenue realization by the growers, processors, and traders. As per the broad estimates, Indian export kitty for organic food products is about US$ 369.8 million (apeda.gov.in). Domestic market accounts for approximately US$ 225 million (estimated).

Important commodities exported during the year 2016–2017 and destination countries are given in Table 1.2 and Figure 1.4, respectively. The year-wise growth in organic food products export during the last 4 years has been depicted in Figure 1.5.

TABLE 1.1 Category-Wise Total Production (in Tons).

S. no.	Category	Production in MT
1.	Cereals and millets	195,874.16
2.	Dry fruits	8241.28
3.	Fiber crops	155,136.88
4.	Fruits	27,851.9
5.	Medicinal (herbal and aromatic) plants	33,477.94
6.	Oilseeds	300,149.23
7.	Pulses, including cluster bean	62,931.03
8.	Plantation crops	47,837.05
9.	Spices and condiments	36,239.07
10.	Sugar crops	281,713.02
11.	Tuber crops	110.93
12.	Vegetables	24,084.98
13.	Fodder crops	869.405
14.	Others such as ornamental plants/flower/*Stevia*/*Henna* etc.	5598.16
Total		1,180,115.035

Source: http://apeda.gov.in.

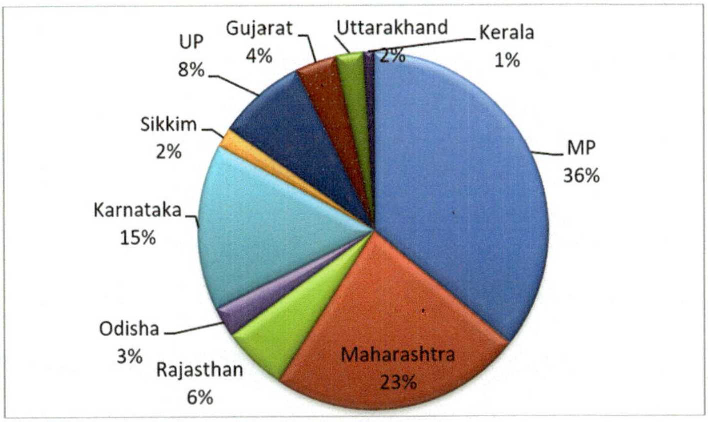

FIGURE 1.3 Share of major states in production (in percentage).
Source: http://apeda.gov.in.

TABLE 1.2 Major Commodities Exported during 2016–2017.

Category	Volume (in MT)	Value in INR (Lakh)
Oilseeds	132,503.93	58,812.34
Processed foods	73,859.65	44,792.97
Cereals and millets	35,356.76	21,493.80
Sugar	31,396.40	12,596.03
Pulses	13,468.09	4708.43
Tea	5918.99	25,208.30
Spices and condiments	4125.68	18,212.56
Medicinal, aromatic, and herbal products	2898.48	12,002.68
Coffee	2224.15	3653.57
Dry fruits	1558.30	24,321.46
Vegetables	497.06	466.46
Essential, aromatic, and other oils	412.35	6582.35
Ornamental plants and flowers	213.36	1693.53
Edible oils	148.02	585.66
Plantation crops other than tea and coffee	30.25	88.06
Honey	23.64	54.91
Fruits	6.46	24.08
Tuber crops	4.75	190.46
Others	5120.62	12,330.01
Total	309,766.94	247,817.67

Source: http://apeda.gov.in.

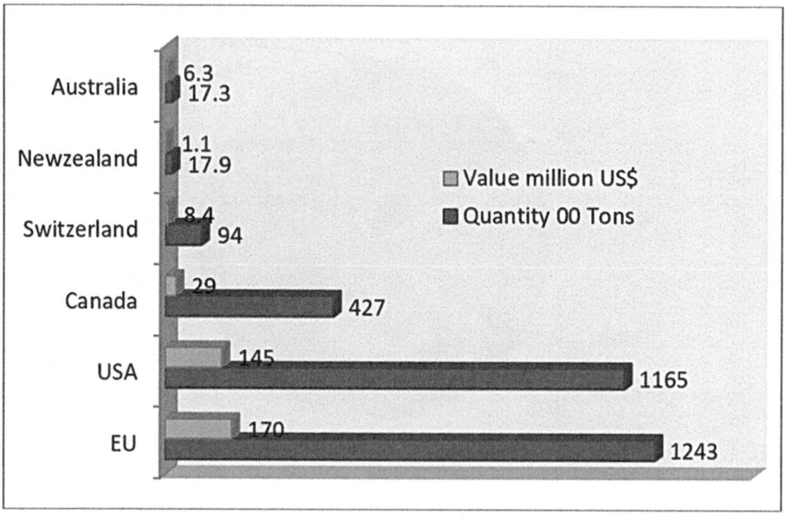

FIGURE 1.4 Major export destinations, quantity exported, and export value.
Source: http://apeda.gov.in.

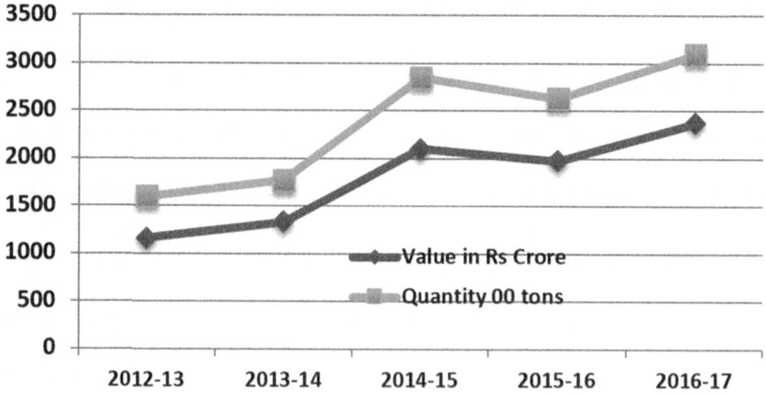

FIGURE 1.5 Growth in organic food products export from India.
Source: http://apeda.gov.in.

1.3 QUALITY ASSURANCE

NPOP has long been the mainstay for quality assurance for organic agriculture products in India and is being operated through 28 accredited certification bodies. A glimpse of the organic agriculture certification scenario is given in Table 1.3.

TABLE 1.3 Organic Certification Process under NPOP.

Parameters	Quantum
Authorized accreditation body	1
Accredited certification bodies	28
Total operators	6674
a. Individual farm producers	1512
b. Grower groups (farm production)	3315
c. Total farmers	1.09 million
d. Processor	885
e. Traders	898
f. Wild harvest projects	64

Source: http://apeda.gov.in.

Participatory Guarantee System under the PGS-India programme is another quality assurance initiative launched by the Ministry of Agriculture and Farmers Welfare (MoA&FW). National Centre for Organic Farming is the nodal implementing agency for PGS-India (Anonymous, 2015).

1.4 PROMOTION POLICY AND INSTITUTIONAL FRAMEWORK

Institutional promotion of organic agriculture started with the launching of the NPOP by the Ministry of Commerce during the year 2001, which defined the NSOP and the procedure for accreditation and certification (Anonymous, 2014). India now has 28 accredited certification agencies for facilitating the certification to growers. For area expansion and technology transfer, the Ministry of Agriculture, Government of India launched a National Project on Promotion of Organic Farming (NPOF-DAC) and earmarked funds for setting up of organic and biological input production units, vermicompost production units, and for organic adoption and certification under various schemes such as National Horticulture Mission (NHM, now Mission for Integrated Development of Horticulture—MIDH), National Mission on Sustainable Agriculture (NMSA), and *Rashtriya Krishi Vikas Yojana* (RKVY-National Agriculture Development Plan). To empower farmers through participation in certification process and to make the certification affordable for domestic and local markets, the Ministry of Agriculture has also launched a farmer group–centric organic guarantee system under the PGS-India programme (Anonymous, 2015). To give domestic organic agriculture a push, the Ministry of Agriculture has recently launched a new scheme under NMSA entitled *Paramparagat Kheti Vikas Yojana* (Traditional Agriculture Development Scheme). Recently under Prime Minister's special initiative for the North Eastern states, a scheme named "Development of Organic Value Chain in North Eastern Region" has been launched with an initial allocation of INR 3000 million (http://agricoop.nic.in/divisiontype/integrated-nutrient-management).

Indian Council for Agricultural Research (ICAR) has launched a Network Project on Organic Farming (NPOF) during 2004 to address research needs with its 13 collaborating centers across the country. This project has already started taking up research activities and now the project is operating from 20 collaborating centers (http://www.iifsr.res.in/npof/index.php). To address the technological needs in horticulture, a Network Project on Organic Horticulture has also been launched by the ICAR during 2014–2015. Three Agricultural Universities, University of Agricultural Sciences, Dharwad; University of Agricultural Sciences, Bangalore in Karnataka; and CSK Himachal Pradesh Agricultural University, Palampur, have set up Centres of Excellence on Organic Farming Research. In 2016, the Government of India has also announced the setting up of the National Organic Farming Research Institute (NOFRI) at Sikkim, India which has now started working.

Many state governments have also put in their efforts for the promotion of organic agriculture. Efforts initiated by the Government of Sikkim in converting the entire state into an organic state and the Government of Uttarakhand of converting their hill districts into organic districts are some noteworthy developments. The area brought under organic certification by these states stands 75,218 and 30,907 ha, respectively (as on March, 2017). The Government of Karnataka's mission of launching organic farming is also an initiative in the right direction through which more than 81,000-ha (as on March 2017) area has been brought under organic certification process.

Twelve state governments, Andhra Pradesh, Karnataka, Kerala, Tamil Nadu, Maharashtra, Madhya Pradesh, Gujarat, Himachal Pradesh, Sikkim, Nagaland, Mizoram, and Uttarakhand, have drafted the policies for the systematic promotion of organic farming. Out of the 12, 3 states, Sikkim, Nagaland, and Mizoram, declared their intention to go 100% organic, while Uttarakhand has declared to convert their all hill districts to organic program (Anonymous, 2016). But it is only the Sikkim that has been able to successfully convert their dream into reality.

1.5 EPILOGUE

As is evident from the details given earlier, NPOP had been the main driving force for the growth of organic agriculture in the country. Its robust assessment and verification system has earned the name in domestic and international markets. Traceability system, introduced during 2006 in the form of online data management, has also contributed to the integrity and transparency of the system in the national and international trades.

Country and provincial governments have also provided timely support in the form of policies, financial incentives, technology packages, and the availability of quality organic inputs. Growing awareness among consumers for contamination-free food is driving the markets. The sector has not only attracted the attention of policy planners, research institutions, civil society organizations, and consumers but has also instilled confidence among the growers as an economically viable system for the future.

KEYWORDS

- **NPOP**
- **trade**
- **organic agriculture**
- **institutional promotion**
- **organic inputs**

REFERENCES

Anonymous. *National Programme for Organic Production*, Published by Agricultural and Processed Food Products Export Development Authority (APEDA), Ministry of Commerce, Government of India, 2014; p 221. http://apeda.gov.in/apedawebsite/organic/ORGANIC_CONTENTS/National_Programme_for_Organic_Production.htm

Anonymous. *Participatory Guarantee System for India—Operational Manual for Domestic Organic Certification*, National Centre of Organic Farming, Government of India, Ghaziabad, 2015; p 68. https://pgsindia-ncof.gov.in/pdf_file/PGS-India%20Operational%20Manual.pdf

Anonymous. *India Organic Sector, Vision 2025*, YES Bank, APEDA and Ingenus Strategy and Creative Research, 2016; p 53.

CHAPTER 2

Organic Farming: Issue and Challenges in India

RAGHAVENDRA SINGH[1*], R. K. AVASTHI[1], and SUBHASH BABU[2]

[1]*ICAR-National Organic Farming Research Institute, Tadong 737102, Gangtok, Sikkim, India*

[2]*ICAR Research Complex for NEH Region, Umiam 793103, Meghalaya, India*

[*]*Corresponding author. E-mail: raghavenupc@gmail.com*

ABSTRACT

Organic farming is found to be superior to conventional farming because of increased human labor employment, lower cost of cultivation, higher profits, better input use efficiency and reduced risk leading to increased income, enhanced self-reliance and livelihood security of the farmers and maintaining soil health and environment. Indian agriculture for long remained sustainable only because of the low external input factors. It provides the least negative impact on the environment. The well-being of the environment and living things are less affected by organic farming. Hence, the promotion of organic farming should be concentrated on niche crops and areas. Similarly, cluster-based market-driven strategies have to be developed through policy interventions. The main issue emerging in organic farming include yield reduction in organic farm is certification, marketing and policy support.

Organic Crop Production Management: Focus on India, with Global Implications, D. P. Singh, PhD, H. G. Prakash, PhD, M. Swapna, PhD, & S. Solomon, PhD (Eds.)

2.1 INTRODUCTION

Self-sufficiency in food grains is the result of judicious management of inputs which was brought through the Green Revolution in India. The indiscriminate and nonjudicious use of chemical fertilizers, herbicides, and pesticides along with high-yielding crop varieties has now shown its ill effects on soil and environment. The response of the per-unit inputs application has also shown decreasing trends over the years due to a decline in the factor productivity (Nweke and Sanders, 2009; Sarkar et al., 2012). Similarly, the human population has been also exposed toward noncurable diseases in recent years. Hence, an alternative method of farming is the need of hours to overcome all these ill effects and also to provide a safe environment. Eco-friendly farming is needed for maintaining crop productivity along with maintaining sustainability of the ecosystems (Singh et al., 2018).

Recent evidence of international researches showed that organic farming not only provides a sustainable soil environment but also creates social, economic, and environmental changes. Organic farming has shown adaptive ability by sequestering more carbon under changing climatic scenarios. Under conservation agriculture, it also performs well in minimum and no-till conditions in different rice- and maize-based cropping systems. Organic agriculture system is also expected to perform a key role in helping against desertification, conserving biodiversity, contributing to sustainable development and supporting plant and animal health (Bonilla et al., 2012; Kilcher, 2007; Vaarst and Alroe, 2012). The demands of organic foods show increasing trends over the years. Hence, new trade opportunities have increased for traders and also growers of the developing countries.

The abovementioned ill effects of the conventional farming systems have only been alleviated by the adoption of organic farming and integrated organic farming systems. Further, it has the ability to reduce the vulnerability of farming with its diverse integration of all the input in a farming system mode. Organic farming has the ability to enhance farm profitability through diverse income sources and provide the strength of flexibility to withstand the adverse climatic effects (Roychowdhury et al., 2013). It provides higher income and lower risk as mostly on-farm inputs are used on rather than off-farm. The cost of inputs has been reduced by using local farm-available resources (recycling) which also makes them more efficient. The extreme climatic variations have reflected loss of production extremely under changing climate, but organic farming can also reduce them at a low level (El-Hage Scialabba and Hattam, 2002).

An organic production system (OPS) has the most sustainable production that not only caters to the agroecosystems but at the same time also offers food security without affecting the environment. It has the ability for sound rural development and also to provide healthy foods with greater employment opportunities (Egelyng and Hogh-Jensen, 2006; Elena, 2009). Moreover, in the recent years, it has witnessed increasing demand of organic products which has fetched two to three times higher returns as compared with conventional products. Additionally, soil organic carbon content increases with the addition of organic matter into the soil. Hence, it provides the porosity by reducing the bulk density, resulting in more soil moisture content compared to the conventional systems (Das et al., 2017). It has been observed from the studies that the local landraces of crops have performed well under organic agriculture rather than high-yielding crop varieties. Organic agriculture can be used as adaptive measures for improving the livelihood security of resource-poor small and marginal farmers who are prone to especially higher vulnerable to climatic variations (El-Hage Scialabba and Hattam, 2002). It is estimated that organic agriculture can have the mitigation potential of about 3.5–4.8 Gt CO_2 through carbon sequestration (55–80% of total global GHG from agriculture) and also two-thirds reduction in N_2O (Niggli et al., 2009). Hence, organic agriculture is designed for the following:

- enhancing the biological diversity of the ecosystem;
- increasing the microbial activities of soil for releasing unavailable nutrients to plants;
- developing ability to maintain long-term soil fertility by putting higher organic biomass;
- integrating ability for recycling of the plant- and animal-based waste materials;
- Depending more on locally available (on-farm) materials in production systems;
- minimizing all kinds of pollution generated from conventional productions systems by the judicious uses of soil, water, and air; and
- carefully handling the processing of raw organic produce and maintaining all the guidelines (NOP, National Programme for Organic Production, NPOP, and Codex) strictly for getting the saleable organic products with higher organic integrity.

2.2 WHAT IS ORGANIC FARMING?

Organic agriculture is a complex production system that promotes the use of on-farm recyclable produce for enhancing the sustainability of the agroecosystems, maintaining all kinds of biodiversity and improving the soil productivity maintaining the soil microbial diversity. OPSs followed the standard specific protocols of production given by certain agencies within the country and outside. The overall aims of the OPS are to make agro-ecosystems more sustainable, provide social acceptance, an ecologically safe environment, and economically sound products. The International Federation of Organic Agriculture Movements (IFOAM) defines as, "a production system that sustains the health of soils, ecosystems and the people. It relies on ecological processes, biodiversity and cycles adapted to local conditions, rather than the use of inputs with adverse effects. Organic agriculture combines traditions, innovation and science to benefit the shared environment and promote fair relationships and a good quality of life for all involved."

The four basic principles that have been described by the IFOAM for organic farming are given in Figure 2.1.

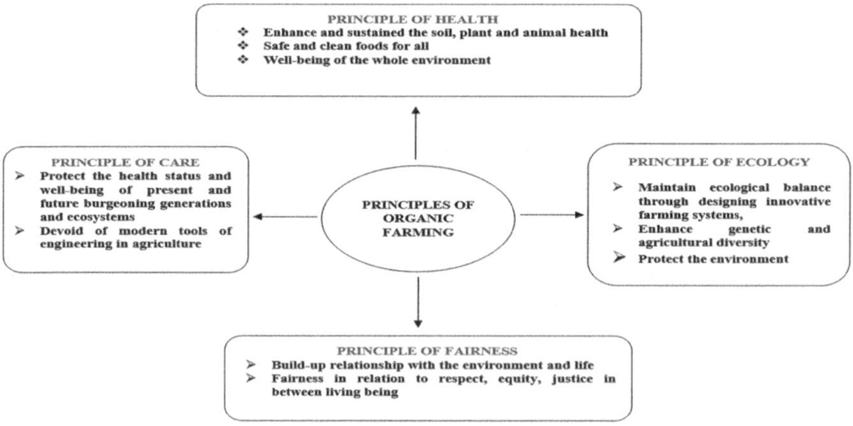

FIGURE 2.1 Principles of organic farming.

2.3 STATUS OF ORGANIC FARMING IN THE WORLD

The recent growth trends in organic agriculture depict an increasing rate of 20.0% at the end of 2017 over the previous years, that is, 2016 (Fibl, 2019). The cultivated area of organic farming is spread to 69.8 million ha which is 1.4% of total agricultural land and 2.9 million growers in 181 countries. Globally, maximum organic producers are in Asia (35%), followed by Africa (30%), and Europe (14%), while the maximum number of producers are in India (835,000) followed by Uganda (210,532), and Mexico (210,000), (IFOAM and FiBL, 2019). The largest areas in terms of land are in Australia (35.6 million ha) followed by Argentina (3.4 million ha) and China (3.0 million ha). Further, apart from the total cultivated organic land, it is being grown as wild collections, including forests, aquaculture, and nonagricultural grazing land. Wild collections have 42.4 million ha area.

The market of organic agriculture has gained importance worldwide in the recent time and now it has reached USD 90.0 billion. The main consumers of organic products are the US and European countries (IFOAM and FiBL, 2019). Among the consumers, the US has the leading market with 40.0 billion euros, followed by Germany (10 billion euros), France (7.9 billion euros), and China (7.6 billion euros).

2.3.1 STATUS OF ORGANIC FARMING IN INDIA

Indian agriculture stands at a crossroads after exploiting the potential of the most fertile zones in the country. The response of the inputs is also showing decreasing trends over the years. However, the production levels should be maintained at the self-sufficiency level to feed the burgeoning population. The organic agriculture history in India is very old and since time immemorial, it is being adopted by the farming community. Still in many parts of the country, the consumption level of inorganic inputs in terms of synthetic fertilizers and chemical is very meager. For the very first time in the year 1900, Sir Albert Howard (British agronomist) started organic farming in northern India. He developed Indore Aerobic Compost Preparation method (Howard, 1929) and later, Bangalore Anaerobic Compost Preparation method (Acharya, 1934). Nowadays, DADEP method is very popular for the preparation of compost. The different agroclimatic conditions make a conducive environment for the preparation of various organic products through locally available resources. That is the reason that makes India the largest producer of organic products.

The cultivable area of organic agriculture in India has been continuously increasing and now, it ranks eighth in the cultivable area of organic agriculture as per the statistics of World's Organic Agriculture (2017 data). The total cultivable area is 1.78 million ha, while 1.80 million ha is under wild harvest, including forest area for wild collections. The ratio of a cultivable area over wild harvest has been reduced over the years. In the year

2016–2017, India added 0.3 million–ha cultivable area that was the highest among other countries during 2017–2018. The highest number of organic producers was recorded in India (835,000) during 2016–2017.

The total organic export potential of India has increased over the years and in 2016–2017, it has touched the value of €582 million (INR 4476.2 crores). At the same time, the domestic market also showed an increasing trend and the annual growth rate was at 15–25%. Cotton and tea are gaining popularity in the European countries which are being the most common organic products in India. The demand for organic spices and condiments in the international market is very high which are being exported by the country. The perishable nature of fruits and vegetables makes lesser interest to the exporters; hence, these are mainly consumed by the domestic customers.

The climatic variability of Indian conditions provides opportunities for producing a range of organic products. The organic growers from Maharashtra, Gujarat, Madhya Pradesh, and some parts of Karnataka are coming forward for taking the training at ICAR-NOFRI after the declaration of Sikkim as a fully organic state in 2016. They seek information on the export potential of different crops growing in their states. The state governments and the Government of India have now changed their focus on agriculture and made it more remunerative and sustainable. Hence, they are now promoting organic farming in all the states across the country through various Government. of India schemes, namely, RKVY, PKVY, and MOVCD-NER. The Government of India has focused on enhancing the area of organic farming and reached at least 10% of the total cultivable area of the country in the future. The accreditation processes were mainly promoted by APEDA and followed the guidelines of the NPOP, particularly for certification and standards of organic farming. Numbers of organic products that include tea, coffee, sugarcane, cotton, basmati rice, oilseeds, spices, pulses, fruits, vegetables, and their value-added products are now being exported to the international market. Madhya Pradesh has the largest area of organic agriculture followed by Himachal Pradesh and Rajasthan. The details are given in Tables 2.1 and 2.2.

TABLE 2.1 Area under Organic Certification Process (2016–2017).

S. no.	State	Organic area (ha)	In conversion area (ha)	Total farm area (ha)	Wild harvest area (WH) (ha)	Total WH+ farm
1	Andhra Pradesh	9812	7871	17,683	155,099	172,783
2	Arunachal Pradesh	21	3989	4011	68,300	72,311
3	Assam	2544	21,326	23,870	60	23,930

TABLE 2.1 *(Continued)*

S. no.	State	Organic area (ha)	In conversion area (ha)	Total farm area (ha)	Wild harvest area (WH) (ha)	Total WH+ farm
4	Bihar	0	1	1	678	679
5	Chhattisgarh	2339	10,372	12,712	167,040	179,752
6	Goa	14,116	1645	15,762	0.0	15,762
7	Gujarat	36,034	28,206	64,241	6253	70,495
8	Haryana	4482	528	5011	20	5031
9	Himachal Pradesh	5903	6473	12,376	2000	14,376
10	J & K	9550	13,058	22,608	159,000	181,608
11	Jharkhand	88	26,725	26,813	10,000	36,813
12	Karnataka	22,478	58,610	81,089	859	81,948
13	Kerala	13,809	11,003	24,812	18,889	43,701
14	Lakshadweep	895	0	895	0	895
15	Madhya Pradesh	213,968	250,891	464,859	1,827,837	2,292,697
16	Maharashtra	84,338	139,668	224,007	68,384	292,391
17	Manipur	0	241	241	0	241
18	Meghalaya	1414	8214	9629	0	9629
19	Mizoram	0	210	210	0	210
20	Nagaland	1508	3191	4699	0	4699
21	New Delhi	9	0	9	0	9
22	Odisha	36,710	55,479	92,190	7546	99,736
23	Pondicherry	3	0	3	0	3
24	Punjab	434	598	1032	16,616	17,648
25	Rajasthan	46,088	105,521	151,609	387,912	539,522
26	Sikkim	72,145	3072	75,218	0.0	75,218
27	Tamil Nadu	2058	3654	5712	5062	10,775
28	Telangana	4457	5230	9687	0	9687
29	Tripura	203	0	203	0	203
30	Uttar Pradesh	39,929	16,319	56,249	45210	101,459
31	Uttarakhand	18,510	12,397	30,907	62,679	93,586
32	West Bengal	4759	416	5176	0	5176
	Total	648,617	794,920	1,443,538	3,009,449	445,2987

Source: Modified Yadav et al. (2017).

TABLE 2.2 State-Wise Production for the Year 2016–2017 (Farm Production).

S. No.	State	Organic production (Mt)	In conversion production (Mt)	Total production (Mt)
1	Andhra Pradesh	8367	5	8372
2	Arunachal Pradesh	38	0	38
3	Assam	32,395	0	32,395
4	Chhattisgarh	3108	0	3108
5	Goa	4759	0	4759
6	Gujarat	41,511	100	41,611
7	Haryana	8193	0	8193
8	Himachal Pradesh	1837	0	1837
9	Jammu & Kashmir	9408	0	9408
10	Karnataka	164,048	117	164,165
11	Kerala	14623	2	14625
12	Madhya Pradesh	382,800	8798	391,598
13	Maharashtra	254,310	1625	255,935
14	Meghalaya	1111	0	1111
15	Nagaland	879	0	879
16	Odisha	26,538	4124	30,661
17	Pondicherry	3	0	3
18	Punjab	711	0	711
19	Rajasthan	64,245	0	64,245
20	Sikkim	184	1	185
21	Tamil Nadu	11,019	66	11,084
22	Telangana	468	53	521
23	Tripura	339	0	339
24	Uttar Pradesh	87,760	423	88,183
25	Uttarakhand	28,155	6	28,161
26	West Bengal	17,977	0	17,977
	Total	1,164,787	15,320	1,180,107

Source: Modified from Yadav et al. (2017).

2.3.1.1 EXPORTED ORGANIC PRODUCTS

During 2017–2018, India exported 458,339-MT organic products to the international market. A total of INR 3453.5 crores (USD 515 million) were realized through the export of organic products during 2017–2018. Among the export products, the maximum share (47.6%) was contributed by oilseeds followed by cereals and millets (10.4%). The share of tea and coffee was 8.96% followed by spices and condiments (7.76%). The major portion of materials was exported to the US, European countries, and Canada.

2.3.1.2 TOTAL ORGANIC EXPORT SHARE UNDER DIFFERENT COMMODITIES

Among the oilseeds, soybean (70%) is the leading crop for export followed by basmati rice (6%), sugar (3%), tea (2%), and pulses (1%) (www.apeda.org).

The implementation of organic farming was started in 2001 by the Government of India agency through NPOP. The government has focused on the promotion of organic farming across all the states in the country through various sponsored schemes. Sikkim is the alone state in the country which has declared itself fully organic in 2016. Similarly, Madhya Pradesh, Maharashtra, Gujarat, Rajasthan, Tamil Nadu, Kerala, all the north eastern states have extensively promoted organic farming in their states. The crop-wise promotion in different states has been given in Table 2.3.

TABLE 2.3 Major Crops Grown State-Wise under Organic Farming in India.

Arunachal Pradesh	Pulses, oilseeds, maize, and kiwifruit tea/coffee and herbal/ medicinal plants
Andhra Pradesh	Cotton, maize, pulses, oilseeds, fruits, and vegetables
Assam	Tea, fruits, and vegetables
Chhattisgarh	Rice, wheat, and vegetables
Goa	Fruits and vegetables
Gujarat	Cotton, pulses, oilseeds, and vegetables
Haryana	Basmati rice, wheat, maize, and vegetables
Himachal Pradesh	Wheat, fruits, and vegetables
Jammu and Kashmir	Spices, fruits, and vegetables
Karnataka	Cotton, rainfed, wheat, maize, and sorghum pulses, oilseeds, and vegetables

TABLE 2.3 *(Continued)*

Kerala	Spices, vegetables, and herbals
Manipur	Spices, vegetables, and herbals
Maharashtra	Cotton, rice, wheat, pulses, oilseeds, spices, and vegetables
Madhya Pradesh	Soybean, wheat, and vegetables
Meghalaya	Spices and vegetables
Punjab	Basmati rice, wheat, and vegetables
Sikkim	Maize, rice, buckwheat, ginger, turmeric, large cardamom, and tea
Rajasthan	Cotton, wheat, seed spices, and vegetables
Tamil Nadu	Tea, herbs, and spices
Uttar Pradesh	Rice, wheat, maize, and vegetables
Uttarakhand	Basmati rice, vegetables, maize, sorghum, herbs, and spices
West Bengal	Tea and vegetables

2.4 CHALLENGES IN ORGANIC AGRICULTURE

2.4.1 LACK OF AWARENESS

- Lack of skill on improved and efficient methods of composting
- Lack of awareness about the concentration, time, and the method of manure and biofertilizer application
- Lack of proper training on organic farming
- Inadequate knowledge of field functionaries about organic farming
- Lack of knowledge among farmers about different biopesticides
- Farmers think that chemical fertilizers are more effective than manures and biofertilizers
- Low credibility of source for purchasing compost and biofertilizers

2.4.2 MARKETING PROBLEMS

In organic agriculture, the marketability of organic produce with premium price has to be assured prior to the adoption of organic farming. The organic produce with proper certification has to be assured premium prices; otherwise, the initial loss in the yield of organic agriculture may not be compensated with conventional agriculture. Once the productivity level reached conventional, it may be waved off.

2.4.3 NONAVAILABILITY OF ORGANIC INPUTS

The organic agriculture needs huge quantities of organic bulky inputs especially in terms of FYM as compared to chemical fertilizers. The small and marginal farmers are not being able to manage huge quantities of the abovesaid plant and animal biomass for proper nutrition. Hence, locally available plant biomass is required for the preparation of compost for plant nutrition. Similarly, wastelands are also gradually decreasing over the years by putting population pressure (Narayanan, 2005).

2.4.4 LIMITED SUPPORTING INFRASTRUCTURE

The nonexistence of green market, inadequate certifying agencies, and low applicability are some of the major issues to be addressed for the proper implementation of organic agriculture in the country. The new trade channels are needed to be formed with adequate infrastructure facilities for value addition and postharvest processing of raw materials.

2.4.5 ABSENCE OF AN APPROPRIATE ORGANIC AGRICULTURE POLICY

Although the country has self-sufficient food grain production and surplus in many crops and commodities, the promotion of organic agriculture throughout the country is still a matter of concern and needs assessment for domestic and export. The niche area and niche crop may be identified before preparing a policy plan on organic agriculture.

2.4.6 GAP IN EXPORT DEMAND

The demand for organic products has rapidly increased in the international market. While a huge deficit was observed in the supply and demand of organic products in the world market, vast diversity and agroclimatic conditions and traditional cultivation practices make India a huge potential for producing organic products rapidly. The urgent need is to promote cluster-based organic farming that should be commodity centric in the country. Basmati rice, Indian tea, coffee, cotton sugar, spices, oilseeds, and condiments have enormous potential organic avenues for the growers in the country.

2.5 DEBATED ISSUES ON ORGANIC FARMING

2.5.1 CAN ORGANIC FARMING PROVIDE ENOUGH FOOD FOR ALL?

Organic farming requires a conversion period of 3 years from conventional farming in which the productivity was low compared with conventional farming. However, studies suggested that organic farming can produce the same product as compared with conventional farming. The demand for food is increasing day by day as the population increases. Hence, it is necessary to maintain the level of production and adopt those areas that are lesser productive. Similarly, several parts of the country are still not using synthetic fertilizers and chemicals in higher amounts. Hence, such areas may also be promoted under organic farming. Hilly regions and particularly the north east region are also still untouched for conventional farming and may be promoted for organic agriculture.

2.5.2 ORGANIC AGRICULTURE VERSUS CONVENTIONAL FARMING ON LABOR INTENSITY

Organic farming often requires more labor as compared with conventional farming. For instance, herbicides are not allowed in organic farming; hence, they require more labor for weeding through manual tools and implements. Similarly, bulky organic inputs are also labor-intensive as compared to synthetic fertilizers. In developing countries, most of the work is done by farm machinery that makes them cost-effective. Similarly, many countries have cheaper labor cost that maintains the cost-effectiveness of organic farming. Many technologies have been developed by researchers for the cost-effectiveness of organic farming compared to conventional farming. One of the examples of weed management is through cover crops. Similarly, reduced tillage is also effective under organic farming. The application of mulches for managing the weeds has also made a profitable venture to organic farming.

2.5.3 IS REQUIREMENT OF NUTRIENTS TO CROPS FROM ORGANIC SOURCES POSSIBLE?

It may not possible to compete for the requirement of nutrients under organic farming where the organic carbon level is lower. It is recommended by the researchers that at least four organic nutrient sources may be used as a source of nutrients for organic crop production. Alternative options are available for the supplementation of organic nutrient sources under organic farming. Plenty of biomass is available in the northeastern hill region which may be utilized as a source for the preparation of compost.

2.5.4 IS ORGANIC FARMING ECOLOGICALLY BENEFICIAL?

Organic farming maintains all kinds of diversity based on the principle of care which means under organic farming, nothing is going to damage if maintained at a level which may not cause economic injury. The products obtained from organic farming are more healthy and tasty and also maintain the agroecosystem.

2.5.5 IS THE PRODUCE SUPERIOR IN QUALITY?

Several studies conducted in the European Union (EU) suggested that organic produce was more beneficial and had several compounds as compared to conventional produce. Organic produce is more qualitative and rich in minerals, vitamins, and tasty. Antioxidants are also higher in organically grown foods.

2.5.6 IS ORGANIC FARMING ECONOMICALLY SOUND?

Organic farming is economically sound as most of the on-farm resources are used for production. Similarly, organic products also fetch a premium price. The premium price should be at least 25% higher to make them economically sound. Integration of all the resources maintained under organic farming provides benefits.

2.5.7 ARE INSECT-PEST AND DISEASES MANAGED EASILY IN ORGANIC FARMING?

The major concern for organic farming is to manage insect pests and diseases. Some plant extracts have the ability to control the pests and diseases that may be used for organic farming. However, preventive measure is the best solution for the management of insect pests and diseases under organic farming. Similarly, the selection of resistant crop varieties, agronomic management, and the use of recommended chemicals may be used for the management. It should be effectively promoted under organic farming.

2.6 RISK ASSOCIATED WITH ORGANIC FARMING

Some of the risks associated with organic farming are as follows:

1. Avoids genetically modified organisms (GMOs) as they may act as contaminants.
2. Timely unavailability of certified organic seeds, inputs, and biological pesticides.
3. Unavailability of capital, because banks are sometimes unfamiliar with organic agriculture.
4. Lack of green market/traders which are unstable for getting premium prices.
5. Some crops in organic rotations do not benefit from USDA commodity program price and income protection (Hansona et al., 2004).

2.7 PROSPECTS OF ORGANIC FARMING

The attention toward organic farming in developing countries is increasing because of its sustainability and requirement of less financial inputs. It is indicated from the study that organic agriculture offers a reasonable benefit in areas with less rainfall and comparatively low natural soil fertility levels. Laborers realize a good return and this is very important where paid laborers are almost nonexistent. Agricultural policies should revise their food supply strategies and valorize local production. Organic agriculture does not need costly investments but rather substantial investments in capacity building through research and training.

2.8 POLICY SUPPORT TO ORGANIC AGRICULTURE

- EU has supported organic agriculture for a long time.

- In 2016, Sri Lanka launched the "Toxin Free Nation Programme," a 3-year plan that lays down 10 areas of action to phase out toxic chemicals from Sri Lankan agriculture through a step-by-step process.
- The Government of India launched the *Paramparagat Krishi Vikas Yojana* (PKVY) program that allocates Euros 40 million in organic support measures.
- Third-party certification is also important which needs the government support in terms of funds allocation to various state governments for setting up public organic certification bodies for accreditation.
- Organic agriculture requires support for subsidies of certification, organic input development, research, and capacity-building programs.
- In Armenia, the government started the "Organic Agriculture Support Initiative" with the support of an EU-funded project which combines a range of supports to boost national capacities and policies in favor of organic growers.
- China has also focused on a capacity-development program in organic agriculture. The government plans to invest around €187 million in 2016–2020 in new farmers' training, with a focus on organic and sustainable agriculture.
- Sao Paulo decided in April 2016 to provide 100% organic meals for 2 million school children in the city every year by 2026 .

The importance of organic agriculture and its rising demands worldwide have made the Indian government think and develop a national strategy (policy framework) to encourage the sector growth with a view to use natural resources sustainably and improve the well-being of farmers and consumers.

2.9 CONCLUSION

Organic farming is the solution to most of the problems faced by the farming community, especially in developing countries. It offers qualitative healthy foods without deteriorating the agroecosystems. It provides the least negative impact on the environment. The well-being of the environment and living things are less affected by organic farming. Hence, the promotion of organic farming should be concentrated on niche crops and areas. Similarly, cluster-based market-driven strategies have to be developed through policy interventions.

KEYWORDS

- **organic farming**
- **income**
- **promotion**
- **certification**
- **organic products**
- **export**
- **challenges**

REFERENCES

Acharya, C. N. Comparisons of Different Methods of Composting Waste Materials. *Indian J. Agric. Sci.* **1934,** *9*, 741–744.

Agricultural and Processed Food Products Export Development Authority. National Programme for Organic Production (NPOP). http://apeda.gov.in/apedawebsite/organic/index.htm (accessed Nov 20, 2017).

Bonilla, N.; Gutierrez-Barranquero, J. A.; de Vicente, A.; Cazorla, F. M. Enhancing Soil Quality and Plant Health through Suppressive Organic Amendments. *Diversity* **2012,** *4*, 475–491. DOI: 10.3390/d4040475.

Egelyng, H.; Hogh-Jensen, H. Towards a Global Research Programme For Organic Food and Farming. In *Global Development of Organic Agriculture: Challenges and Prospects*; Halberg, N., Alroe, H. F., Knudsen, M. T., Kristensen, E. S., Eds.; CABI publishing: Wallingford, 2006. DOI 10.1079/9781845930783.0000.

Elena, S. Impact of Organic Farming Promotion upon the Sustainable Rural Development. *Agric. Econ. Rural Dev.* **2009,** *6* (2), 217–233.

El-Hage Scialabba, N.; Hattam, C. Eds. *Organic Agriculture, Environment, and Food Security*. [online], Environment and Natural Resources Service, Sustainable Development Department, Food and Agriculture Organization of the United Nations (FAO), 2002. http://www.fao.org/docrep/005/y4137e/y4137e00.htm (accessed Feb 12, 2009).

Howard, A. *The Application of Science to Crop-Production; an Experiment Carried Out at the Institute of Plant Industry*; Oxford University Press: Indore, India, 1929; p 81.

Kilcher, L. *How Organic Agriculture Contributes to Sustainable Development*. University of Kassel at Witzenhausen, *Supplement* **2007,** *89*, 31–49.

Narayanan, S. *Organic Farming in India: Relevance, Problems and Constraints.* Occasional Paper. 38, Published by the Department of Economic Analysis and Research. National Bank for Agriculture and Rural Development, Mumbai, 2005; p 73.

Niggli, U.; Fliessbach, A.; Hepperly, P.; Scialabba, N. *Low Greenhouse Gas Agriculture: Mitigation and Adaptation Potential of Sustainable Farming Systems*; FAO, April 2009, Rev. 2–2009.

Nweke, O. C.; Sanders, W. H. Modern Environmental Health Hazards: A Public Health Issue of Increasing Significance in Africa. *Environ. Health Persp.* **2009,** *117* (6), 863–870.

Roychowdhury, R.; Banerjee, U.; Sofkova, S.; Tah, J. Organic Farming for Crop Improvement and Sustainable Agriculture in the Era of Climate Change. *Online J. Biol. Sci.* **2013,** *13* (2), 50–56.

Sarkar, A.; Aronson, K. J.; Patil, S.; Hugar, L. B., Vanloon, G. W. Emerging Health Risks Associated with Modern Agriculture Practices: A Comprehensive Study in India. *Environ. Res.* **2012,** *115*, 37–50.

Vaarst, M.; Alroe, H. F. Concepts of Animal Health and Welfare in Organic Livestock Systems. *J. Agric. Environ. Ethics* **2012,** *25*, 333–347. DOI 10.1007/s10806-011-9314-6.

Willer, H.; Lernoud, J. *The World of Organic Agriculture, Statistics and Emerging Trends*; FiBL and IFOAM- Organics International, 2017; p 340.

Yadav, A. K. Organic Agriculture in India: An Overview. In *Proceedings of 'National Conference on Organic Farming for Sustainable Agriculture and Livelihood Security under Changing Climatic Condition*', Dec 12–13, 2017; pp 1–9.

CHAPTER 3

Organic Farming for Sustainable Agriculture in India

J. P. SAINI* and NEELAM BHARDWAJ

Department of Organic Agriculture and Natural Farming, CSK Himachal Pradesh Agricultural University, Palampur 176062, Himachal Pradesh, India

*Corresponding author. E-mail: drjpsaini@gmail.com

ABSTRACT

Organic farming one of the most frequently accepted alternative farming systems than the traditional once, there is an urgent got to improve the sustainability of farming systems, which may cause better livelihoods of the farmers. It is an exclusive production management system, helpful for improving the agro-system health, including biodiversity, biological cycles, and soil biological activity, making the use of on-farm technologies where no chemical off-farm inputs are used. Organic agriculture has a large potential in India due to its large geographical and arable area with a huge range of agroclimatic zones. Two-thirds of the arable area is rainfed. Organic produce can provide premium prices due to the rapidly growing demand compared with its production, and India would like to trap this opportunity to harness the cash flow. In past organic agriculture was neglected in the agricultural policy, and therefore there was less government assistance for the promotion of organic farming, as it exists for the conventional agriculture in the form of subsidies, agricultural extension services and official research. Given proper encouragement, organic farming will progress tremendously in India.

Organic Crop Production Management: Focus on India, with Global Implications, D. P. Singh, PhD, H. G. Prakash, PhD, M. Swapna, PhD, & S. Solomon, PhD (Eds.)

3.1 INTRODUCTION

Organic farming is an exclusive production management system, helpful for improving the agro-system health, including biodiversity, biological cycles, and soil biological activity, making the use of on-farm technologies where no chemical off-farm inputs are used. A significant emphasis is placed on maintaining soil fertility by returning all the waste to it chiefly through compost to minimize the gap between addition and the depletion of major nutrients like nitrogen, phosphorus, and potassium from the soil (Chhonkar, 2002). Food and Agriculture Organization (FAO) promotes this understanding and perception about organic agriculture, through its own recommendations and programs as well as among the member nations. Countries like the United States, where organic agriculture is an established technology, recognizes, "organic farming as a system which avoids or largely excludes the utilization of synthetic inputs (such as fertilizers, pesticides, hormones, feed additives, etc.), and to the maximum extent feasible depend upon crop rotations, crop residues, animal residues, animal manures, off-farm organic waste, minerals grade rock additives and biological system of nutrient mobilization and plant protection."

Organic farming is a complete production management system that prohibits the use of any type of chemicals. It depends upon ecology, biodiversity, and related cycles that are adapted to the local conditions and ultimately sustains the ecosystem, including soil health and mankind. Traditional farming, the original type of farming, is in the practice since time immemorial. Organic agriculture has a large potential in India due to its large geographical and arable area with a huge range of agroclimatic zones. Two-thirds of the arable area is rainfed. Organic produce can provide premium prices due to the rapidly growing demand compared with its production, and India would like to trap this opportunity to harness the cash flow. In spite of that, there are certain limitations of organic farming, and to overcome such issues and limitations, the Ministry of Agriculture and Farmers Welfare should introduce favorable government policies and strategies like the provision of assistance to farmers to shift to organic farming, increasing the investment on research in organic agriculture, strengthening of links between the government, private sectors, and non-governmental organization (NGO)s, awareness campaigns, support structures for small farmers' group certification, the establishment of monthly information bulletin on local and international prices, and so on. Taking into consideration the principles of health, ecology, fairness, and care and to market the technology in India, we will have to identify niche areas along with the crops which should be selected in a demand-driven manner. There is a need of strengthening organic agriculture as a sustainable alternative production system approach, and long-term studies in different agroecological regions with a strong data support system in terms of productivity, soil health, and quality are urgently required which will be advantageous for the farmers.

3.2 DEFINITION OF ORGANIC FARMING

In the last decades, irrational use of some of the farm technologies has given rise to various hazardous outcomes like degradation of soil health, habitat imbalance, soil erosion, and salinization, lowering groundwater levels, genetic erosion, environmental pollution, degradation of food quality, and increased cost of cultivation, leading to a reduction in farmers' income (Ram, 2003). Now, farmers do not find farming a viable proposition (Deshpande, 2002), maybe due to higher cost of inputs, a shift to commercial farming with purchased inputs, and limited market intervention by the government (Reddy, 2010), leading to an unviable and unsustainable agricultural enterprise. In this context, viable alternative farm technologies are very important. Organic farming, a productive and sustainable technology (Mader et al., 2002), which promotes biological processes is emerging as an attractive option to farmers. The definition of organic farming given by FAO, 1999:

Organic agriculture is a holistic production management system which promotes and enhances agro-ecosystem health, including biodiversity, biological cycles, and soil biological activity. It emphasizes the use of management practices in preference to the use of off-farm inputs, taking into account that regional conditions require locally adapted systems. This is accomplished by using wherever possible, agronomic, biological and mechanical methods, as opposed to using synthetic materials, to fulfill any specific function within the system.

There is a full range of pre- and postplant tillage practices under conventional farming with a high degree of crop specialization in contrast to organic farming which is characterized by crop diversification.

3.2.1 IFOAM DEFINITION

"Organic agriculture is a whole system approach based upon a set of process resulting in a sustainable ecosystem, safe food, good nutrition, animal welfare and social justice" (IFOAM, 2002 Basic Standards).

As per IFOAM, 2002, organic farming has been developed as a holistic tool that has started from the inputs with the following major principles:

1. *Principle of health*: Should sustain and enhance soil, plant, animal, and human health in the entire planet as a whole.
2. *Principle of ecology*: Should be based on living ecological systems and cycles and should sustain them.
3. *Principle of fairness*: Should be based on relationships for ensuring fairness with respect to the common environment and life opportunities.
4. *Principle of care*: Should be implemented in a precautionary fashion to maintain the health and well-being of present and future generations and the environment.

3.3 OBJECTIVES AND IMPORTANCE OF ORGANIC FARMING

India has a strong traditional farming system and a widely diverse climate, with vast drylands and a treasure of natural organic nutrients that are most appropriate for organic farming. Innovative farmers make use of technologies that encourage minimum use of chemicals, like subsistence farming, which are organic by default being implemented since time immemorial by the tribals of the northeast and hilly regions of the country. The dry/marginal soils do not respond well to intensive/conventional farming practices. These respond better to low-input farming systems that exploit the ample biodiversity present, thereby making these lands well suited for organic farming (Pionetti and Reddy, 2002). With its focus on improving soil health, avoidance of pollutants, and maximum utilization of local resources, including labor, organic farming systems aid in advancing the economic and ecological health of the land and the people. The semiarid and arid dryland soils with shallow depth and poor water-holding capacity get improved in their physical properties and their ability to supply balanced plant nutrients, through organic practices like the addition of organic matter to soils. This can be a solution to the problem of overexploitation of natural resources due to the inappropriate use of technologies in these soils (Dhir, 1997). The basic objectives of organic agriculture are to:

- Mitigate the increased use of chemical fertilizers and their residual effects in food products,
- Promote rainfed farming in the hills,

- Meet the demand for safe food without chemical residues,
- Create opportunities for sustainable employment for youth,
- Enhance the productivity sustainably, and
- Maintain soil health.

3.4 POTENTIAL

Tremendous progress in fostering organic agriculture in India has been made during the last 15 years. Eleventh Plan Document on Organic sector and National Commission on Farmers have recommended organic farming as a tool for the Second Green Revolution. Organic farming has been recommended by Food and Agricultural Organization, IFOAM, and other international agencies as a tool to address Millennium Development Goal (MDG) 1 and 7. The first MDG deals with the eradication of extreme poverty and hunger, whereas ensuring environmental sustainability is dealt with in the seventh MDG. Organic agriculture in India has a large potential to meet both of these goals due to the following factors:

- It has an enormous geographical and arable area, with a large variation in different agroclimatic zones. Two-thirds of India's total arable area is rainfed which still await the Green Revolution, as the expected yield is modest and it was not considered to be worth making new investments in agriculture.
- There is not much scope for enhancing the irrigation potential in the country. All the Green Revolution technologies were suitable for irrigated areas of the country like Haryana, Punjab, and western Uttar Pradesh. Due to excessive use of water and chemicals and deficiency of organic matter, soils have been degraded in these areas and yield levels are only being maintained with a major increase in inputs and increased cost of production.
- India has an ancient culture, in which the earth is referred to as a mother, where "organic" techniques are practiced to improve the soil, plant, animal, and human health. Organic farming has the potential to use this traditional culture.
- Increased global demand can provide premium prices, and with the supplies still to catch up, India can trap this opportunity to earn the maximum profit under these conditions.

Even with all the possibilities mentioned above, there are several limitations in implementing organic agriculture.

3.5 LIMITATIONS

The main limitations of organic farming are as follows:

- There is a low level of sufficient knowledge for benefits and products to peasants and consumers.
- The information for Indian and global market information on suppliers, prices, and quality are unavailable.
- There is a lack of training, organic farmers' field schools, and the nonavailability of an adequate extension system, with only the exception of the areas covered by NGOs and private traders promoting the contract farming.
- Storage facilities are not sufficient in the country.
- There is a lack of a guarantee system in the domestic market. Therefore, consumers often feel confused about the genuine nature of organic products before buying any organic products.
- A stagnant local market, owing to a lack of consumer awareness and the minimal number of processed products offered.
- Lack of adequate support by the government.
- Higher costs of certification, particularly for marginal and small farmers.
- Unavailability of socioeconomic and scientific data for organic farming.

3.6 CHALLENGES FOR FUTURE DEVELOPMENTS

Ram (2003) has elaborated that modern agricultural farming practice that involves the use of chemicals in excess has created an imbalance in natural habitat causing the production of low-quality food with enhanced cultivation costs. Yield emerges as a matter of concern when we talk of organic agriculture (Trewavas, 2004), even though conventional agriculture may also not be fully capable of successfully feeding the world. Problems related to proper distribution, social organization, poverty, racism, and gender are emerging as the major reasons for the failure of the present high-yielding systems in completely meeting the world food demands. With a shift from inorganic to organic farming, the availability of food for the poor might still decrease, especially during the initial stage, due to a reduction in yield during the conversion period and a lack of access/affordability, with an increase in the cost of organic food. This gives rise to equity issues. In spite of several challenges, organic agriculture is productive and sustainable (Mader et al., 2002) in the long run. The excessive use of chemicals and inorganic fertilizers, which is the mainstay for farmers in Punjab, Haryana, and western Uttar Pradesh for higher yields (Suresh Reddy, 2010a, 2010b), is a major reason for the farmers to shift from conventional to the organic farming system. In nonirrigated farms, even though there seems to be a yield reduction during the conversion period from a conventional input-intensive system of farming to organic farming, yields after conversion have been found to be comparable to that in conventional farming. In rainfed farming, the difference in the two systems with respect to yield will be narrower. In the long run, the organic management efficacy along with the quality of applied organic manures has a great role to play in the yield restoration in organic farming systems (Kasturi, 2007). Against the prevailing myth, organic agriculture is quite competitive economically with conventional agriculture by adopting suitable approaches during the conversion period along with a well-tailored knowledge delivery system (Cacek and Langner, 1986).

The major challenges of organic farming are as follows:

- Introducing Lu facilities in agricultural universities like well-trained faculty;
- Introducing organic extension services and capacity building for farmers' field schools;
- Creating required infrastructure for processing, storage, transport, and market facilities;
- Creating a guarantee system for the Indian market;
- Enhancing awareness of customers about the organic food production system;
- Giving additional information about organic products to the existing reports for the foreign markets;
- Promoting production and supply of organically cultivated seeds, organic manures, and bio-based products;
- Providing an additional budget for undertaking scientific studies on generating incomes of individual and household levels and ensuring food security, productivity level, and soil enrichment from organic farming.

3.7 RECOMMENDATIONS FOR PROMOTION OF ORGANIC FARMING

Recommendations include formulation and implementation of favorable government policies and strategies by the Ministry of Agriculture and Farmers Welfare for the promotion of organic agriculture.

These should include:

- Launching of a mega-policy for providing assistance to peasants to convert their lands from conventional agriculture to organic agriculture.
- Increasing investment in research and development in organic agriculture: There should be an emphasis on comparative studies on the cost of production, productivity, and other benefits from organic farming as compared to conventional agriculture. This can be accomplished through the establishment of suitable facilities, including trained faculty in the NARS.
- Developing and strengthening of links between the government, the private sector, and NGOs at the national level.
- Government assistance in microcredit and microenterprises to self-help groups of women and landless agricultural families, for bio-intensive agriculture (usufruct rights to common property resources, such as wasteland, exclusively for the resource-poor).
- Awareness campaigns on the advantages of organic agriculture and products.
- Development of a support system for certification of farm produce of small farmers' groups.
- Monthly information bulletin on local and international prices of the most common organic food items.
- Implementation of the IFOAM Accreditation system to reduce overlapping certification work, bureaucracy, and costs.

3.8 CERTIFICATION

In India, the organizational support systems for smallholder organic agriculture fall into four groups. These include the company supporting the farmers, NGO-supported farmers' groups, farmers organized or facilitated by the government, and organizations like cooperatives, associations, and self-help groups formed by the farmers themselves. Organic farming has thrived well with several institutional arrangements. Therefore, it would not be easy to suggest any specific framework for its success (Kasturi, 2007). A support system for the resource-poor small farmers is necessary to overcome various constraints faced by the farmers in the initial "switchover" phase from nonorganic to organic farming. The time between starting of organic management and getting certification of the crop is basically known as the conversion period, which ensures the neutralization of the chemical residues from the previous agricultural practices. Unlike conventional agriculture, organic farm management does not rely on a common strategy. The standard duration of the conversion period is 2 years for annual crops, and 3 years for the perennials. However, depending on the ecological conditions of the farm and the practices during a preconversion phase, the certification authority can extend or reduce the duration of the conversion period.

3.8.1 NEED FOR ORGANIC CERTIFICATION

Consumers demand healthy and environmentally sound farm products even at higher prices, while the farmer produces according to well-specified standards of organic farming for obtaining higher prices. Any agricultural produce can be designated as organic only after it is certified as organic. Organic certification is a process by which organic food producers and processors are certified to have produced various products by following fixed standards and under fixed regulations. Organic certification is basically an auditing process that assures that organic growers and product manufacturers abide by the fixed standards and regulations. In general, any vocation that is directly involved in the production and marketing of organic produce can be certified and this includes seed suppliers, organic food producers, food processors, retailers, and even restaurants that serve organic food to its clients.

3.8.1.1 THIRD-PARTY CERTIFICATION

- For maintaining standards and trust of certified organic farm products, there are well-structured policies and mechanisms in India. Various programs like National Programme on Organic Production under FTDR Act and APGMC Act have specified requirements for quality assurance in organic agriculture for export, import, and domestic markets. Even though there are 16 certifying agencies spread all over the country, keeping in view the Mission 2020 goals, their number will prove inadequate. Therefore, Mission 2020 should include a program to increase the tally of certifying agencies in the country.

- Financial assistance should be allocated for establishing new certification bodies, with at least one agency in each state.
- The group certification system supports ICS management and certification through service providers and data management for thorough traceability. Support may be provided for setting up of national/regional web-hosted database system to develop a databank of farmers involved in organic cultivation and organic produce.
- Funding establishment of 20 new residual testing laboratories across the country, preferably under PPP mode.

3.8.1.2 FARMERS' PARTICIPATORY GUARANTEE SYSTEM

Farmers guarantee an alternative quality assurance system for ensuring the low cost of production. An urgent need is being felt to promote a participatory guarantee system (PGS) or any other form of system that will be helpful in encouraging farmers to have a better understanding of the benefits of quality assurance which will ultimately motivate them to use it as a precursor to certification by the third party. The PGS should be based on National Standards of Organic Production. Further, the PGS system developed for the country should ideally be based on a peer evaluation system rather than individual farmer's record maintenance.

3.9 INFRASTRUCTURE FOR ORGANIC SUPPLY CHAINS

For promoting organic farming, a well-developed market for organic produce is a prerequisite factor, highlighting the efforts required to ensure the strengthening of the supply chains. The mission strategies should, therefore, include provision for providing financial support to develop exclusive organic supply chain components like grading, cleaning, primary processing units for *dal* and flour making,and so on, storage units with organically compatible fumigation/protection facility, and packaging units. Organic marketing strategies are quite different from that of regular marketing. Large markets and distribution channels characteristics to organic supply chains (Reddy, 2010a, 2010b) demand additional costs and specialized skills, know-how, and experience, all of which are mostly beyond the reach of unorganized individual farmers (Kasturi, 2007). As more than 85% of the total organic farm products of India are exported, it results in an underdeveloped domestic market for organics in our country. The lack of well-developed domestic marketing channels adds to the challenges. Market access for small producers depends on (1) a thorough understanding of the markets, (2) a well-organized firm or operations, (3) access to information and communication, and (4) a suitable policy environment. In this changing scenario, the major need of the small farmers is better access to capital and information, along with a proper management capacity. There are the latest products and process standard needs from the supply chain and this demands collective action to handle the requirements. There is an urgent need for professional training in marketing for marginal and small farmers for improving productivity. Cooperatives or associations can be of help to a large extent, to trade at a better price and also to prepare small farmers for higher levels of competition and to new marketing patterns. Producer companies by farmers' groups need to be encouraged to assist in value addition, storage infrastructure development, and direct marketing (Reddy, 2010a, 2010b) and for this, Organic Farming Mission can play a very significant role.

KEYWORDS

- **alternate farming system**
- **definition**
- **potential**
- **limitation**
- **challenges**
- **certification**
- **supply chain**

REFERENCES

Cacek, T.; Linda, L. L. The Economic Implications of Organic Farming. *Am. J. Altern. Agric.* **1986,** *1* (1), 25–29.

Chhonkar, P. K. Organic Farming Science and Belief. *J. Indian Soc. Soil Sci.* **2002,** *51* (4), 365–377.

Deshpande, R. S. Suicides by Farmers in Karnataka-Agrarian Distress and Possible Alleviatory Steps. *Econ. Political Weekly* **2002,** *XXXVII* (26), 2601–2610.

Dhir, R. P. Problems of Desertification in the Arid Zones of Rajasthan, India. *Desertif. Contr. Bull.* **1997,** *27*, 45–52.

FAO. *Organic Agriculture*; Food and Agriculture Organization of the United Nations: Rome, 1999.

Kasturi, D. Towards a Smoother Transition to Organic Farming. *Econ. Political Weekly* June 16, 2007.

Mader, P.; Fliefback, A.; Dubois, D.; Gunst, L.; Fried, P.; Niggili, U. Soil Fertility and Biodiversity in Organic Farming. *Science* **2002,** *296* (5573), 1694–1697.

Ram, B. Impact of Human Activities on Land Use Changes in Arid Rajasthan: Retrospect and Prospects. In *Human Impact on Desert Environments*; Narain, P., Kathaju, S., Kar, A., Singh, M. P., Kumar, P., Kumar, P., Eds.; Scientific Publishers: Jodhpur, 2003; pp 44–59.

Suresh Reddy, B. *Assessment of Economic and Ecological Returns from Millet-based Bio-diverse Organic Farms vis-à-vis Conventional Farms*. CESS Monograph Series No.8, Centre for Economic and Social Studies: Hyderabad, 2010a.

Suresh Reddy, B. Organic Farming: Status, Issues and Prospects—A Review. *Agric. Econ. Res. Rev.* **2010b,** *23*, 343–358.

CHAPTER 4

Organic Farming for Adaptation and Mitigating Impact of Climate Change for Ensuring Rural Livelihood Security

A. K. SINGH

ICAR-Indian Agricultural Research Institute, New Delhi 110012, India
E-mail: aksicar@gmail.com

ABSTRACT

Organic farming has a lot of opportunities for sustainable livelihood security and employment opportunity. In spite of its immense potential, organic farming in India is still at a crossroads. Farmers are not convinced about the availability of biofertilizers and organic supplements/amendments in requisite quantity and the local market for organic produce. In addition, small and marginal farmers are discouraged from adopting organic farming for commercial purposes due to the lack of knowledge and information access, proper guidelines, certification, and higher cost coupled with capital-driven regulation. Thus, there is an urgent need to develop a comprehensive framework that integrates strategy for opting organic farming on a commercial scale that accumulated bottom-up approach, transfer of technology with reciprocal knowledge flow from farmers' associations, and their local resources and innovation will generate large-scale acceptance by the farmers to address not only ecological concerns with respect to climate change but also address the issues of health and livelihood security of large rural masses of the country.

Organic Crop Production Management: Focus on India, with Global Implications, D. P. Singh, PhD, H. G. Prakash, PhD, M. Swapna, PhD, & S. Solomon, PhD (Eds.)

4.1 INTRODUCTION: AGRARIAN SITUATION, FOOD SECURITY, AND CLIMATE CHANGE

Almost 18% of the world's human population and 15% of the livestock residing in India share about 2.3% of the world's geographical area and 4.2% of its water resources of the country. Agricultural sector that was a major contributor to the Indian economy till independence still plays an important role by contributing 18% share in the gross domestic product (GDP) at 2011–2012 prices, 11% of exports, and 53.3% share in total employment or body of workers in 2013–2014. Among the necessary cereals, India is the second largest producer of rice and wheat globally and has the distinction of being the largest producer of pulses and milk. India ranks second in the world in groundnut, cotton, sugarcane, fruits, and vegetables. Long-term trends of household-level consumption patterns reveal that per capita direct consumption of food grains has been reduced and animal products, fruits, and vegetables have been witnessing an upward trend since quite a long time (Kumar et al., 2007; Mittal, 2007; Chand, 2009). The consumption pattern of food grains, which makes contributions to essential dietary intake, is witnessing a declining trend. The consumption pattern that was recorded as 64% at some stage in the base year 2000 may decline to 57 and 48% by 2025 and 2050, respectively (Amarasinghe and Shah, 2007). As per the estimates, there will be total food grain demand of 291 Mt by way of 2025 and 377 Mt via 2050 against the total production of 292 Mt by 2025 and 385 Mt by 2050. However, a deficit is expected in the other cereals, oilseeds, and pulses production. The other cereals may record a deficit of 33 and 43% by 2025 and 2050, respectively, while pulses are expected to record a deficit of 3 and 7% by the abovesaid period.

Even if physical and economic access to food is assured, ecological factors will decide the long-term sustainability of food protection systems. In India, agricultural production depends on the overall performance of the summer season monsoon (June–September) contributing nearly three-fourths of the annual precipitation. Apart from the annual variability in summertime monsoon rainfall, the prevalence of many of the hydrometeorological activities is found to affect Indian agriculture at exclusive spatial scales. India also receives about 15% of annual rainfall during the period of December–March, in addition to the summer monsoon rainfall which is very essential for winter crops. Thus, agricultural production and food security in India are influenced by the climate.

4.2 AGRICULTURE AS BOTH REASON AND SUFFERER OF CLIMATE CHANGE

In general, the climate is one of the most significant determinants of agricultural production, and climate alteration may cause variability in agricultural production. As climate pattern shifts, changes within the distribution of plant diseases and pests may additionally have adverse effects on agriculture. At the same point of time, farming is proved to be the foremost human activity to varied climate conditions. Climate change is anticipated to vary agricultural production dramatically. This amendment in international climate could be a development that's mostly due to the burning of resources of fossil energy such as coal, oil, and natural gas, and to the mineralization of organic matter attributable to land use. These processes are caused by human beings by clearing of natural vegetation by cutting trees, overexploitation of fossil resources, and use of those soils for cultivatable cropping.

Since the industrial revolution, international food yields have increased considerably, chiefly because of new cultivars and a rise within the use of agrochemicals, like pesticides and artificial chemicals, and additionally to a lesser extent because of additional space underneath agriculture (IAASTD, 2009; FAO, 2010). However, as international production has enlarged, there has additionally been a rise within the awareness of the negative effects caused by agriculture. As an example, the widespread use of huge monoculture systems could increase environmental risks, for example, reduction in various levels can be terribly sensitive to global climate change (IAASTD, 2009). Wearing away and loss of soil organic matter (SOM) is another major shortcoming. Terming this as a complication, Gomiero et al. (2008) claimed this as one of the most vital and most studied consequences of agriculture, poignant the longer term prospects of the world crop production. Moreover, Lal (2010) stated that improving the quality of soil and particularly soil organic carbon (SOC) is critical to handling food security. Another risk is the dependence on high energy inputs, particularly within the variety of fossil fuels and artificial chemicals (FAO, 2011). These activities have primarily led to a measurable increase within the carbon dioxide (CO2) content of the atmosphere, an increase of which ends up in heating, as greenhouse gas (GHG) hinders the reflection of daylight into house, and therefore, additional of it's cornered within the Earth's atmosphere.. It's a tangle that must be resolved if we would like to eliminate our GHG emissions. Agriculture contributes considerably to global climate change via emissions of GHG like alkane and azotic oxides.

The Intergovernmental Panel on Global Climate Change (IPCC) has suggested that the annual amount of GHGs emitted by the farm sector is calculable between 5.1 and 6.1 giga tons GHG equivalents in 2005 (Barker et al., 2007). In accordance with UN agency (2014), throughout the period 2001–2010, agriculture, biology, and different land uses created up 21% of the world's GHG emissions, of that Martinmas was directly from agriculture.

Within the same report, the agricultural emissions are projected to rise half an hour by the year 2050. In keeping with current projections, total greenhouse emissions from agriculture are anticipated to succeed in 8.3 Gt GHG equivalents per annum by 2030, in comparison to this level of nearly 6.0 Gt GHG equivalents annually (Smith et al., 2007). Emissions of inhalation general anesthetic originate chiefly from high, soluble element levels within the soil from inorganic fertilizers and organic nitrogen sources, animal housing, and manure management. Most of the sources of alkane emissions are enteric fermentation by ruminants, handling of manure, anaerobic turnover in rice paddies, and soil compaction due to the use of agricultural machinery, biomass burning, for example, from slash-and-burn agriculture, that emits each alkane and inhalation general anesthetic. Molecules of alkane (CH_4) and inhalation general anesthetic (N_2O) have an identical, however, way bigger effect: the worldwide warming potential of alkane is 20 times that of GHG, whereas that of azotic oxide is the maximum amount as 300 times bigger. GHGs emitted from agriculture have been projected to extend significantly unless action is taken.

In addition to being a major contributor to GHG emissions, agriculture is additionally among the most sectors to suffer from the impact of climate change: several farmers, particularly smallholders and people preponderantly within the hemisphere, have already been affected because of their harvests being destroyed or broken by the dynamic environmental conditions; extreme weather events, heat waves, and droughts are progressively frequent within the future and can additionally impact farmers within the hemisphere. In keeping with studies administrated by the IPCC, the common international temperature has enlarged by 0.74°C between 1906 and 2005 and an additional increase of 0.2°C–0.4°C is anticipated within the next 20 years. Regional GHGs are projected to be enhanced from the present level of concentration of 360 ppm to 400–750 ppm by 2100. Water level can rise from 15 to 95 cm by 2100 (IPCC, 2014). Thus, increase in temperature, variation in rainfall patterns, rise in ocean levels, excess actinic radiation, and better incidence of utmost weather events like floods and droughts are already inflicting vital agricultural yield losses and can become even additional prevailing within the returning decades because of the consequences of world amendment (Battisti and Naylor, 2009; Lobell et al., 2008). At the same time, farm production is the basis of the worldwide food providers for the world's voters. Therefore, it's important to scrutinize, but farming will be helpful in reducing GHG emissions and similarly in increasing food and biological process security. The advance will only return by variations within the whole food production and consumption system. So, there is a desperate need to concentrate on completely different effective variations and mitigation ways to attenuate the environmental effect on food production and to preserve the long-run property to create the assembly system additional resilient to global climate change.

The IPPC has instructed a spread of measures for mitigating GHG emissions from agricultural ecosystems (Smith et al., 2007). In accordance with Barker et al. (2007), sink sweetening (carbon sequestration) can contribute most to mitigation during this context. The major mitigation choices in farming in keeping with IPPC (Barker et al., 2007; Smith et al., 2007) embrace improved crop production through efficient nutrient management, residue/tillage management, and water management, improved grassland management through integrated nutrient management and grazing intensity, and the restoration of degraded soils. Agriculture will facilitate to mitigate global climate change by reducing emissions of GHGs and by sequestering GHGs from the atmosphere within the soil.

These challenges can be met through organic agriculture because it has the power to continue functioning even in the face of surprising events of global climate change (Borron, 2006). Predictions regarding the longer term international trends for GHG emissions from agriculture mostly depend upon physical and economic parameters that have direct effects on total emissions. Value of fuel, economic development, the evolution of placental numbers, yield enhancement, new technology, handiness of water, deforestation, shopper attitudes, and diet are some of the important major parameters that should be taken care of (Smith et al., 2007).

4.3 ORGANIC FARMING AS MITIGATION AND ADAPTATION STRATEGY

Organic farming plays an important role in addressing two of the global largest and most challenging concerns of climate change and food security. Unsustainability, vulnerability, and social inequity of agriculture and foods production are the major issues of concern highlighted by climate change and the international crisis for food. There is a general assumption among masses that policies and practices have not succeeded to feed the world's most needy people, failed to adapt to always changing environmental scenarios due to climate change, and failed to protect the sustainability of the ecosystems. The major issues of concern like "soil carbon," "soil organic matter," "ecosystem services," and "holistic" approaches are established core pillars of organic farming (IFOAM, 2006).

Organic agriculture avoids or restricts the application of inorganic chemical fertilizers, resulting thereby in increasing the carbon sequestration in the soil. The general practices of organic farming are helpful in preserving soil fertility and enhancing crop yield under the adverse local weather extremes (ITC and FiBL, 2007; Niggli et al., 2009). The practices of organic farming additionally promote and beautify agroecosystem health, consisting of biodiversity, organic cycles, and soil organic activity. Research experiments conducted in Europe and North America exhibit that natural farming systems are around 30% greater environment friendly in the use of fertilizer nitrogen than traditional cropping systems, leaving a small quantity of nitrogen on the farms in the form of GHGs or as nitrate to pollute the aquatic systems (Drinkwater et al., 1998; Mader et al., 2002). Studies exhibit that fantastic practiced organic agriculture emits less GHGs than conventional agriculture (Mader et al., 2002; Pimentel et al., 2005; Reganold et al., 2001). Higher soil natural count beneath natural farming system improved carbon sequestration and ameliorated some of the fundamental motives of climate alternate that can end result in a net reduction of GHGs. Two long-term assessment trials (21 and 22 years) of conventional and organic systems have discovered that the organic structures use much fewer fossil fuels and, therefore, emit drastically lower degrees of (around 30% less) GHGs (Mader et al., 2002; Pimentel et al., 2005).

Farmers need to intelligently adapt to the altering climate change in order to sustain crop yields and farm income. Enhancing the resilience of agriculture to climate change chance is of paramount importance for protecting livelihoods of small and marginal farmers. Traditionally, science switch in agriculture has aimed at bettering farm productivity. But considering the scenario of climate change and variability, farmers need to adapt quickly to beautify their resilience to face the emerging challenges of climatic variability like floods, droughts, hailstorms, and other extreme climatic events. Over the years, an array of practices and technologies has been developed by way of researchers toward fostering steadiness in agricultural production against the onslaught of seasonal variations. The adoption of such resilient practices and technologies through farmers seems to be essential rather than treating it as an alternative in the modern state of affairs of increasing frequency of occurrence of climate aberrations.

4.3.1 HIGH SEQUESTRATION

The practices adopted under organic farming can definitely increase a part of the primary driver of climate change. Carbon sequestration in soils is the most noteworthy relief capability of organic agriculture, as better carbon sequestration can neutralize up to 40% of the total GHG output of the world. Soil carbon losses caused by agriculture record for 10% of the total carbon dioxide emissions inferable from human activity since 1850. In the case of the carbon released from fossil fuels, the soil carbon store can possibly be reproduced to a generous degree by adopting suitable practices. Organic agriculture provides a suitable framework that can be helpful in reasonably recovering carbon from the air and successfully restoring it in the soil. Niggli et al. (2009) assessed that conversion to organic agriculture would extensively upgrade the sequestration of carbon dioxide using methods that develop SOM, as well as lessen nitrous oxide outflows.

Organic farming helps in sequestering more carbon dioxide in comparison to conventional farms, while practices of reducing soil erosion convert carbon losses into gains (Bellarby et al., 2008; ITC and FiBL, 2007; Niggli et al., 2009). Utilization of compost, reusing of biomass squander for preparation of the compost, and storage and handling of manures enhance the potential for producing carbon credits. Organic agriculture is self-sufficient in nitrogen because of the reapplication of manure from animals and crop residues *via* composting as well as the planting of pulse crops that have the inherent ability to fix atmospheric nitrogen in the soil through symbiosis with *Rhizobium* bacteria present in root nodules (ITC and FiBL, 2007). The soil is managed to depend upon organic agriculture. Organic farming rehearses that accomplish ideal carbon sequestration additionally empower farmers to adjust to climate change and build resilient systems.

4.3.2 CONSERVATION OF SOILS

Fertile, healthy soils are a key resource for long-term agricultural production. Organic agriculture has a strong focus on enhancing and maintaining the fertility and quality of soils, and a number of its core practices support that goal. The soils that are organically managed are enriched by the SOC that enhances the water-holding capacity in comparison to conventionally managed soils. Thus, organic agriculture provides sustainable production even under the adverse conditions of water stress or drought. The soils managed organically are biologically alive and naturally more fertile. The crop diversification followed in an organic system, planted at different periods of time during the year, makes organic farming more stable in unfavorable and aberrant weather scenarios. The practices adopted under organic farming build soil, resulting in enhancing fertile land availability while controlling soil erosion and improving land degradation. The small-scale and poor farmers, largely dependent on biodiversity, soil health, and locally-available resources for agricultural production, can easily adopt approaches of organic farming.

Practices such as covering crops, mulching, and intercropping protect soils against erosion from both run-off water and wind. Organic fertilizers and optimized crop rotations help the accumulation of SOM (Gattinger et al., 2012), which, in turn, improves soil characteristics, such as its water infiltration and holding capacities. In a comprehensive global literature review of studies, Lori et al. (in press) have identified a greater abundance of soil microorganisms in organically managed soils, along with more carbon and nitrogen transformation activities than in conventionally managed soils. This shows that on an average, SOC sequestration tends to be higher in organic than conventional agriculture. Moreover, the higher organic matter shapes the soil as a habitat for soil life. A living soil, in turn, provides a good basis for coping with weather extremities such as droughts and floods, while the good soil structure of organically managed soils effectively reduces the risk of waterlogging and soil erosion (Lal, 2004).

4.3.3 REDUCTION OF EUTROPHICATION AND WATER POLLUTION

The problems of eutrophication and water pollution are caused by the application of nitrogenous fertilizers and associated nitrate leaching. The Nitrate Directive (EC 1991) and Drinking Water Directive (EC 1980) set a maximum permissible concentration of 50 mg/L for nitrate in surface freshwater or in groundwater. Several studies indicate that this maximum value is often exceeded in areas dominated by conventional farming, but less often near organic farms (Mondelaers et al., 2009). Studies show that much higher rates of nitrate leaching occur in conventional farming systems than in organic. Similarly, conventional farming is associated with higher levels of pollution. This can be attributed to the application of a lower dose of nitrogen in organic farming systems and the correspondingly better plant uptake helpful in checking the rate of nitrogen leaching.

4.3.4 RESILIENT CROPS

Building healthy soils is the prerequisite for cultivating healthy resilient plants that are better enough to face environmental pressures like enhanced water stress and insect-pests and disease pressure. Organic farming improves the immune system of the plants by strengthening the defense and self-healing mechanism of crops against insect-pests and diseases. Plants receiving their nutrients through natural biological processes are more resilient to environmental stress than crops receiving their primary nutrition artificially through highly soluble synthetic/inorganic chemical fertilizers. This could be possible through optimal soil and water management, the building of soil structure and fertility, and the selection of location-specific robust crop varieties. Apart from it, organic crops tend to have longer and denser roots that make them able to extract water reserves deeper in the soil profile.

4.3.5 ECO-FUNCTIONAL INTENSIFICATION

Conventional agriculture intensifies production by enhancing the use of external inputs like chemical fertilizers, pesticides, water, hybrid, and genetically modified seeds in the case of plants and feed concentrates, and pharmaceutical drugs in the case of animal production. More food is produced without compromising the quality of the environment and our food under eco-friendly production systems. Eco-functional intensification in organic agriculture is achieved by higher inputs of knowledge, observation skills, and agroecological methods to intensify the advantageous effects of ecosystem functions, including biodiversity and soil fertility, reducing losses from material cycles, and utilizing the self-regulating mechanisms of biological systems for sustainable farming systems.

4.3.6 FOOD SECURITY

Organic agriculture is gaining more recognition due to its immense potential to ensure food security and improve food accessibility. UNEP, UNCTAD, and the IAASTD reports concluded during 2018 that organic farming can enhance farm yield and increase incomes, resulting thereby in improving food security. The International Fund for Agricultural Development also concluded in 2005 that organic farming is specifically beneficial in harsh environments. The IAASTD report (2008) advocated for adopting agroecological and organic principles, emphasizing the strong need for sustainability through better land, crop, and livestock management along with more support to smallholder farmers. It also advises for a participatory process to ensure that science and technology have been designed to help small-scale farmers with particular emphasis on women farmers. Kathleen Merrigan, US Agriculture Deputy Secretary, called for the importance of organic agriculture and its role in agroecology to be elevated within the FAO scope of work.

Underwater stress conditions, organic agriculture performs better than conventional agriculture which highlights the importance of its role in a changing climate (Hepperly et al., 2006). The strong linkages between organic agriculture and natural element cycling increase long-term sustained crop yield in comparison to various intensive conventional farming systems, where there is reduced productivity in spite of the use of high inputs (DFID, 2004; Matson et al., 1997). For making sustainable, affordable, and locally adapted farming systems, the local communities are empowered by organic agriculture to take control of their food production requirements. About 400 million small farms having less than 2-ha area are easily adopting organic farming as it is affordable and dependent on locally available and renewable inputs. These small farms will be transformed for higher crop yield under harsh environments by improving soil fertility, optimizing the use of water, crop diversification, climate-resilient agriculture, and creating new local markets. Organic agriculture has emerged as an affordable low-risk strategy for small farmers after getting the requisite information through extension services. Hunger prevailing all over the world can be attributed to people-oriented agricultural policies, lack of food sovereignty, climate change, degraded farming systems, and destruction of ecosystems. The lack of knowledge for having sustainable, economically viable, climate-resilient, and high yielding farming systems which was being missed for quite a long time got an answer in the form of organic farming that has the strength and power of breaking the vicious circle of poverty. As small farms constitute

90% of the global farms, they can play a crucial role in achieving food security.

Out of the 1.5 billion total global smallholder farmers, most do not adopt easily affordable, highly productive, and climate-resilient farming systems. In the developing world, about 400 million small (having less than 2-ha area) and marginal farmers (having less than 1-ha area) are the foundation of local food security. The majority of these farms are unproductive and vulnerable to climate change. For ensuring food security and easy accessibility to needy people, it is important to optimize the yield and make the farm climate-resilient. The food production in the smallholder farms can be enhanced by using local resources available in abundance, using the location-specific agro-technology for improving soil fertility and the management of insect-pests and diseases. Eco-functional intensification may play an important role in enhancing crop yield and ensuring food security, particularly in harsh environments like drought.

4.3.7 KNOWLEDGE-BASED INCOME GENERATION FOR A GREEN ECONOMY

Only local conditions and opportunities such as topography, climate, biodiversity, ecosystem health, local and traditional farming system knowledge, and the entrepreneurial and innovative spirit of local communities are exploited under organic agriculture. It does not depend upon externally imposed and controlled one-size-fits-all formulae. Rather, organic agriculture provides power to local communities to enable them to develop their own food and farming systems for generating income through crop production and value addition. The realization of the optimization of organic agriculture, including sequestration, depends upon the dissemination of the research, technology, and knowledge of the location-specific ecosystem.

For the adoption of organic farming on a larger scale, there will be a need for a new team of dedicated agronomists who are capable of unlocking the innovative and entrepreneurial spirit of local peasants, increasing local food accessibility, developing climate-resilient farming systems, protecting the ecosystems from the adverse impact of global warming and climate change, and assisting the farmers and communities to have access of financial facilities to boost value-adding activities by organizing self-help groups, cooperatives, or FPOs. The local persons having expert knowledge of local conditions and skills to encourage the local farmers and communities to participate in the training programs specially designed for participatory seed and animal breeding programs should be assigned the duties for conducting research, extension, and training. Organic agriculture is capable of converting the present economy to a green economy ensuring easy access to food for all, protecting the ecosystems by mitigating the impact of global warming, and ensuring all the people a sustainable growth and better quality of life.

4.3.8 SOCIAL INNOVATION AND TRAINING

Organic agriculture has become instrumental in expanding a large number of innovative local production and marketing systems inclusive of community-supported agriculture, periurban or urban farming, rooftop gardening, and other direct or indirect marketing initiatives. Contrary to conventional agriculture that is capital and resource-intensive, organic agriculture is basically knowledge-intensive which exploits indigenous traditional knowledge and believes in exchanging the knowledge and technology among the farmers. The large number of farmers having small landholdings can also sell directly to the community through Participatory Guarantee Systems (PGS) programs as they can guarantee the authenticity of each other's organic production systems and develop efficient markets to sell their organic produce. This model has been proved very successful in Latin America and India as the initiatives taken by local people eliminate the involvement of third-party certification costs and help in promoting sustainable, affordable, and climate-resilient farming systems by empowering them for ensuring food and livelihood security.

4.3.9 REDUCING PRODUCTION AND MARKETING COSTS TO AVOID FINANCIAL RISKS

The people's right to food has been officially recognized by the 2009 World Summit on Food Security. Three-fourths of the global 1 billion hungry people reside in rural areas of underdeveloped and developing nations who deserve access to food. But most of these underprivileged people are not lucky enough to have the essential means for resilient productive farming. The financial and resource barriers are reduced by organic farming, which helps in enhancing people's access to local food. As costly inputs like inorganic fertilizers, pesticides, and diesel are not used in organic farming, the cost of production under organic farming remains cheaper. Low costs incurred on the production of organic produce mitigate financial risks as there is no need for credit which can later be changed to indebtedness. With the rise in prices of fossil fuel, the cost of chemical inputs will increase further making conventional farming a risky affair. Due to crop diversification involved in organic agriculture, the risks involved with the failure of a particular crop are reduced to a great extent with higher ecological and economic stability. Optimum carbon sequestration accrued by the organic agricultural practices is helpful to the farmers to adapt to climate change and build resilient systems. The soils of organic farms being rich in SOC retain more water than conventionally managed soils. This clearly reveals that organic agriculture provides higher yields under the conditions of drought or water stress. The various crops grown in the biologically alive and naturally fertile soil of an organic system, planted at various points of time during the year make organic farming more stable in the conditions of aberrant weather. Organic agricultural practices help in reversing land degradation and soil erosion. The approaches of organic farming are most suitable for small-scale and poor farmers who mainly depend on biodiversity, soil health, and locally available resources for agricultural production.

4.3.10 HIGHER PRODUCTIVITY IN HARSH ENVIRONMENT

In most of the underdeveloped and developing nations, organic farms provide higher productivity even under adverse weather conditions. The water-holding capacity of the farms adopting conventional agriculture remains low with low productivity of the crops in comparison to organic farms. Locally available sources like local varieties of the crops are efficiently used under organic agriculture, which is often limited in areas with extreme poverty and food insecurity. The enhanced crop productivity and food security particularly in the harsh environment is possible only through the intensification of organic agriculture systems. In a UNs' research study of 114 projects undertaken in 24 African countries, it has been revealed that crop productivity on organic farms or in the farms using near organic practices was more than doubled with the 128% enhancement in the yield recorded in East Africa.

4.3.11 SUSTAINABILITY

Unsustainable supply of chemical nitrogen and restricted supply of chemical phosphorous fertilizer are major problems in supplying essential nutrients. At present, chemical nitrogen is manufactured from fossil fuels but its availability is diminishing rapidly. There is heavy dependence on phosphorus in conventional agriculture as it is obtained directly from rapidly reducing nonrenewable deposits. In spite of their unsustainable nature in obtaining good production of crops, chemical nitrogen and phosphorus are the most important ingredients for conventional agriculture. A consortium of research institutes has been proposed by the Food and Agriculture Organization of the United Nations (FAO), the Swiss Research Institute of Organic Agriculture (FiBL), and the Danish International Centre for Research in Organic Food Systems (ICROFS) with the title of the Organic Research Centres Alliance (ORCA) to work in a harmonious manner across the developing world. There is a strong network of nonorganic research institutions as well as core organic institutions at each and every center for undertaking research work on low-input systems and ecology, farmers' organizations, and technology development. The social, economic, and environmental benefits accrued from organic research are shared worldwide by the abovementioned alliance for poverty alleviation and sustainable crop production. Each and every center will emerge as an institution without walls, formed through alliances between producers and scientists, as well as partnerships between institutions in developed and developing nations. The purpose of this alliance is to develop and strengthen the network among the existing organizations so that they can develop as centers of excellence in transdisciplinary and participatory organic agriculture research.

4.3.12 RESILIENT TO DROUGHT AND FLOODS AND ENHANCING WATER USE EFFICIENCY

Prediction of rainfall is not an easy task and is becoming very difficult day by day. The prevailing farming systems can hold and store water for its use in the future. Soils of a farm adopting organic agriculture systems, having a better structure of the soil and higher humus levels and other organic matter compounds, have a better capacity to hold water like a sponge. Humus of the soil has the water-holding capacity to store 30 times of its weight and, thus, plays an important role in ensuring that water received through rains or irrigation is not wasted through the leaching and evaporation processes. Drainage in soils is also increased significantly due to organic matter of the soil and, thus, saves the crops from the menace of surface-water flooding and waterlogging (International Trade Centre UNCTAD, 2007; IFPRI, 2009).

4.3.13 AGRO-GENETIC BIODIVERSITY

Genetic resources of different varieties of various crop species and farm animal breeds are protected by adopting organic farming as it promotes varieties suitable for local situations and decentralizes participatory breeding programs, particularly on-farm-based conservation, breeding, and production. Various varieties of crops may be efficiently maintained for the future requirements by adopting in situ approaches and adaptation to climate change. The critical resource of agriculture genetic diversity can be effectively managed through well adaptation to climate change. Conservation of natural resources as well as the biodiversity of flora and fauna adapted under organic farming systems is helpful for crop production through the most efficient use of nutrients and water (Anonymous, 2007).

Precisely 30% more biodiversity can be ensured by adopting organic agriculture in comparison to conventional farming. Biodiversity can be better maintained in landscapes with a large share of arable crops. Apart from it, prevalent climate, farm management practices, and the landscape along with types and species of the crops also play an important role in maintaining biodiversity under organic agriculture (Hole et al., 2005). However, the benefits to biodiversity from adopting organic agriculture may not be the same at different locations as it is influenced by several factors like climate, geographic location, types, and species of the crops and crop management practices. Crop management practices adopted under organic agriculture are helpful in promoting biodiversity as the management of field margin helps in increasing the population of natural predators for the management of insect-pests without using any chemicals. Not at all or restricted application of chemical insecticides and inorganic fertilizers, sympathetic management of noncrop habitats and field margins, and preservation of mixed farming, the three major options of crop management are helpful in maintaining biodiversity (Hole et al., 2005).

4.3.14 DIVERSIFICATION

To address the concern of farmers, science based on uniformity has not remained appropriate now. The application of science based on diversity can manage risks and uncertainties associated with climate change and increasing vulnerability to environmental and economic variability. Diversification is the foremost important aspect of organic agriculture. Climate resilient is directly associated with farm biodiversity. Organic farms having vast biological diversity with optimized ecological functionality reduce the buildup of insect-pest and diseases levels and are more resilient to various environmental stresses. Temporal as well as spatial crop diversity provides a wide range of rooting depths for increasing soil structure and its stability, improving nutrient and water use efficiencies, and plays an important role in stabilizing microclimate. Farmers adopting organic farming with diversified farming systems have to suffer less in comparison to the farmers adopting conventional farming with monoculture. There is enhanced diversity of landscapes, farming activities, and crops in organic agriculture resulting in the development of climate-resilient varieties suitable to face the adverse effects of climate change. The farming systems of organic agriculture also facilitate protecting sequestrated carbon from climatic disturbances and enhancing its permanency. More soil fauna diversity increases soil biological activities. The adoption of crop diversification prevents the incidence of insect-pests and diseases with better utilization of nutrients and soil water with enhanced efficiencies. The diversification of farming systems also makes more efficient use of available nutrients, with enhanced yield and economic performance, even in the case of limited nutrients and financial constraints. Organic farming involves the diversification of different crops, rotations, landscapes, and farming practices (ITC, 2007; Bengtsson et al., 2005). Rather than preferring monoculture and mono breeds, organic farmers opt for more robust traditional varieties and species, which they tend to conserve and develop. Apart from it, adopting diversified agriculture over time and space helps in enhancing the agroecosystem resilience to external shocks such as aberrant weather situations or price variation. These risks are more likely to increase with climate change.

4.3.15 LOCAL PRODUCTION BY THE CONSUMPTION OF LOCALLY AVAILABLE RESOURCES

Organic farming promotes local production by consuming locally available farm inputs and outputs, leading to the maximization of efficiencies and synergies, reducing emissions from transportation, and enhanced access to local food production and, therefore, enhancing food security. Most of the organic production, distribution, and marketing systems are based on the objective of reducing emissions irrespective of place of production and consumption. The exporting countries can earn more economic returns in the global international farm markets but there is a risk of long-term negative effects on local food security due to reduced accessibility of local food accessibility, particularly for poor people who are highly vulnerable to price fluctuations. For enhancing food accessibility and for reducing emissions, it is important to promote local food production in food-insecure regions.

4.3.16 LOCAL FARMER KNOWLEDGE

Organic agriculture promotes the application of local and indigenous traditional knowledge by duly recognizing the important role played by women throughout the entire food chain, as farmers, consumers, and mothers. Very little attention of research and development organizations has been received by these farming systems and they are generally treated as inherently unproductive. Farmers, having small landholding across the globe, have developed a series of practices and a number of innovations that are the basis for any realistic development, including productivity improvements.

Indigenous and traditional knowledge (ITK) is the major source of information on adaptive capacity which revolves around selective, experimental, and resilient capabilities of peasants (IFPRI, 2009; Niggli et al., 2009; Bolwig and Odeke, 2007; Pimentel et al., 2005; Pfiffner and Luka, 2003). The majority of the global farmers, including women, are small landholders. The vast ecological knowledge gained on organic agriculture by the global smallholder farmers results in increased crop yield with higher resource efficiency, more resilience to global warming, and other climate change and improved access to food and income.

4.3.17 NUTRITION

Several studies (Tinttunen and Lehtonen, 2001; Caris-Veyrat et al., 2004) have revealed that there are much more contents of advantages, health-promoting secondary plant compounds in organic produce in comparison to conventional produce. Phytochemicals like vitamins and antioxidants (e.g., carotenoids and flavonoids) produced by plants prevent human beings from infections of various diseases. Crops under conventional agriculture are more "pushed" or "forced" than in organic agriculture. The growth under organic farming is generally slower, offering the plants sufficient time to synthesize the important components. The dilution effect has been noticed in the produce produced with high water and chemical inorganic fertilizers and pesticide inputs (Mitchell et al., 2007) which is stimulated by the use of a high quantity of nitrogen and rapid plant growth. For example, tomatoes continuously grown in organic farms show higher flavonoid concentrations than the tomatoes cultivated in conventional crops. It has been observed in several studies conducted on organic and conventionally managed farms that agricultural produce grown in farms managed by following organic practices revealed higher content of advantageous, health-promoting secondary plant compounds in organic produce as compared to the produce grown in conventional farms. It is applicable for animal products too. Growth is hastened by including hormones in the feed of conventionally raised livestock in several countries of the world. The use of hormones in the feed results in enhancement of meat weight produced per calorie of food ingested, basically through the retention of water in the flesh. The level of enhanced secondary plant compounds results in improving community resilience to disease and minimizes dependence on health-care interventions. The poor farmers are likely to be hit more due to increasing the incidence of a number of diseases caused by global warming.

4.3.18 HUMAN HEALTH

Health of human beings would also benefit from an increase in organic production. Organically produced food is different from the food produced adopting the conventional practices in respect of the concentration of antioxidants, pesticide residues, and cadmium (Cd). Of particular interest are the concentrations of antioxidants in agricultural products, due to the fact these are related with high-quality influences on human health, along with safety against persistent diseases, positive most cancers types, such as prostate cancer, and neurodegenerative diseases. Concentrations of antioxidants are 20–70% higher in organic crops. In conventional crops, pesticide residues occurred four times more frequently than in organic crops. The variations in pesticide illness are mainly the end results from the reality as synthetically produced plant protection chemicals are banned under organic farming. Finally, the incidence of poisonous metals is drastically higher in conventional than organic products, the concentration of cadmium for instance being twice as high. Cadmium, a toxic metal, is accumulated in the human body and leads to adverse impacts on health. As such, the lower incidence of such components in organic products can extend food security and grant robust advantages to the health of farmers. There is proof that the differences in antioxidant and cadmium content material derive from the particular characteristics of the organic production systems as the chemical fertilizers like mineral nitrogen, potassium chloride, and superphosphate are not used in organic farming.

There are very low levels of pesticides and residues of veterinary drugs in organically produced foods and lower nitrate content in most cases. There is a restriction on the use of chemical pesticides and herbicides in natural farming. Peasants and their family members, rural communities, farm animals, native animals and insects, water bodies, foods, and ecosystems are generally exposed to such very toxic compounds. According to the World Bank, about 5 million agricultural laborers are projected to suffer from pesticide poisoning each year with at least 20,000 deaths annually due to exposure. Most of the dead people belong to developing countries where protection tips and equipment are less accessible and the chemicals banned in most of the developed nations are still manufactured, marketed, and used. In addition to the human health advantages from the reduced use of agrochemicals, organic farming can also assist to limit the air pollution related with farming practices. Organic farming reduces soil erosion and emissions of particulate matter, oxides of nitrogen, carbon, and sulfur, as well as volatile organic compounds and pathogens. These substances have unfavorable implications on human health, being a motive of respiratory diseases, allergic reactions, and other problems.

4.4 BOTTLENECKS OF ORGANIC FARMING AND RESEARCH NEEDS

There are a number of shortcomings in adopting organic farming due to low yield and yield decline in few crops and their production zones. In developing countries where carbon dioxide mitigation will be most advantageous, greater assistance needs to be furnished to organic agriculture projects using low inputs for improving the overall performance of organic farming. There is a need to focus on soil health management, crop growth, incorporation of leguminous or pulse crops in various cropping systems, habitat management with the maintenance of diversity at various levels, plant breeding programs focusing on adaptability of plants to low-input conditions under environmental harsh conditions in organic livestock production and reduced tillage organic systems, crop-weed competition, insect-pests, and disease resistance and plant protection technology. In spite of these bottlenecks, the most promising method for mitigation and adaptation to climate change is only adopting organic agriculture that depicts an advantageous example of mitigating climate change by the farmers and adapting to its predictable and unpredictable impacts. It can serve as a barometer for resource allocation to climate change adaptation, or to measure growth in implementing climate-related multilateral environmental agreements.

4.5 INSTITUTIONAL AND FINANCIAL ASPECTS

The significance of sufficient institutional systems and financial management for adaptation has every now and again been brought up by different investigations. Organic agriculture can expand on the existing general agricultural organizations present in any nation and globally. In any case, a primary obstruction is the way that organic farming is not yet extensively perceived for its potential as an advanced technique and even less as an adaptation or strategy. Its ability to create exceptional returns by supplanting conventional agriculture to a significant amount is often addressed. In developing countries, yields are not really lower, as ongoing research points out. In organic agriculture, prospects for long-term sustained productivity are given and are unique in relation to numerous intensive traditional cultivating systems, where, after certain decades, diminishing yields are observed (see, e.g., Matson et al., 1997; DFID, 2004). Specific establishments like IFOAM (International Federation of Organic Agriculture Movements) also promote organic agriculture. Similarly, larger association such as FAO (Food and Agriculture Organization of the United Nations) plays a significant role in spreading the knowledge about organic farming.

Organic farming can be spread in larger areas as a well-structured adaptation and mitigation strategy could be achieved on the off chance that it turns into a part of national and international agriculture policy discourse. As an adaptation and mitigation strategy, organic farming does not rely on huge extra financing for organic agriculture itself. However, it is essential to approach worldwide markets and to create local markets for the produce. In the transition phase to organic agriculture, extra financing for the firms might be important: Training and extension services should be given and lower yields for the 2–3 years of the transition period may be necessitated. At that point in time, it is reasonable to underline knowledge transfer and infrastructure development rather than direct monetary transfer. The financial suitability of organic farming is additionally prone to increment with increasing prices of energy, which makes conventional farming more costly affair due to expenses on the costs for the production of fertilizer and pesticides and with decreasing levels of subsidies for conventional agriculture. A few choices to meet the monetary prerequisites exist on a basic level. A few choices to address the difficulties of environmental change impacts on farming in the future must be obliged in arrangement (IPCC, 2014).

4.6 STRATEGIES

Given the degree of the risk worldwide temperature rise poses to the total populace and condition, it is of utmost significance to coordinate organic agriculture into both national and international climate change agreements, relief systems, and action plans. It is additionally suggested that as there are streamlined IPCC rules for the transformation of land from one use to another, a rearranged set of guidelines ought to be created for the change of traditional to organic land use. A few proposals and procedures for further application and scaling-up of advantages are, more emphasis on healthy soils, utilizing biological nitrogen fixation and promoting it in programs and activities that are planned for improving farming, combining water harvesting with organic fertilization, using the available biodiversity, empowering low capital-input sustenance crops, and so on. Self-sufficient food crop systems can be successfully intensified without capital inputs. This requires knowledge and labor-intensive organic farming frameworks, making the utilizing a healthy soil, biodiversity, and organic fertilization. Apart from it, the involvement of the farmers in the adaptation of farm practices is another prerequisite. Farmers' situation is profoundly various, and the effect of climate change is continuous. This implies that there will be a requirement for a wide range of adapted proposals for improved farm practices. Research and extension should involve the farmers to ensure that the farmers are liable to continued adaptation and improvement.

4.7 CONCLUSIONS

Organic farming has a lot of opportunities for sustainable livelihood security and employment opportunity. In spite of its immense potential, organic farming in India is still at a crossroads. Farmers are not convinced about the availability of biofertilizers and organic supplements/amendments in requisite quantity and the local market for organic produce. In addition, small and marginal farmers are discouraged from adopting organic farming for commercial purposes due to the lack of knowledge and information access, proper guidelines, certification, and higher cost coupled with capital-driven regulation. Thus, there is an urgent need to develop a comprehensive framework that integrates strategy for opting organic farming on a commercial scale that accumulated bottom-up approach, transfer of technology with reciprocal knowledge flow from farmers' associations, and their local resources and innovation will generate large-scale acceptance by the farmers to address not only ecological concerns with respect to climate change but also address the issues of health and livelihood security of large rural masses of the country.

KEYWORDS

- **food security**
- **climate change**
- **adaptation strategy**
- **carbon sequestration**
- **conservation of soils**
- **water pollution**
- **green economy**

REFERENCES

Amarasinghe, U. A.; Shah, T. J. E. 2007. http://www.iwmi.cgiar.org/NRLP%20Proceeding-2%20Paper%202.pdf (accessed date?).

Anonymous. Organic Farming and Climate Change. International Trade Centre UNCTAD/WTO: ICT, Geneva, Switzerland, 2007; p 31.

Barker T.; Bashmakov, I.; Bernstein, L.; Bogner, J. E.; Bosch, P. R.; Dave, R.; Davidson, O. R.; Fisher, B. S.; Gupta, S.; Halsnæs, K.; Heij, G. J.; Kahn Ribeiro, S.; Kobayashi, S.; Levine, M. D.; Martino, D. L.; Masera, O.; Metz, B.; Meyer, L. A.; Nabuurs, G.-J.; Najam, A.; Nakicenovic, N.; Rogner, H. H.; Roy, J.; Sathaye, J.; Schock, R.; Shukla, P.; Sims, R. E. H.; Smith, P.; Tirpak, D. A.; Urge-Vorsatz, D.; Zhou, D. Technical Summary. In *Climate Change 2007: Mitigation. Contribution of Working Group III to the Fourth Assessment Report of the Intergovernmental Panel on Climate Change*; Metz, B., Davidson, O. R., Bosch, P. R., Dave, R., Meyer, L. A., Eds.; Cambridge University Press: Cambridge, 2007. http://www.mnp.nl/ipcc/pages_media/ FAR4docs/ final_pdfs_ar4/TS.pdf

Battisti, D. S.; Naylor, R. L. Historical Warnings of Future Food Insecurity with Unprecedented Seasonal Heat. *Science* 2009, *323* (5911), 240–244.

Bengtsson, J.; Ahnström, J.; Weibull, A. C. The Effects of Organic Agriculture on Biodiversity and Abundance: A Meta-Analysis. *J. Appl. Ecol.* **2005,** *42*, 261–269.

Bolwig, S.; Odeke, M. *Household Food Security Effects of Certified Organic Export Production in Tropical Africa—A Gendered Analysis*, 2007. http://www.grolink.se/epopa/Publications/EPOPA Report on Food Security impact of organic production.pdf

Caris-Veyrat C.; Amiot, M. J.; Tyssandier, V.; Grasselly, D.; Buret, M.; Mikolajczak, M.; Guilland, J. C.; Bouteloup-Demange, C.; Borel, P. Influence of Organic *Versus* Conventional Agricultural Practice on the Antioxidant Microconstituent Content of Tomatoes and Derived Purees; Consequences on Antioxidant Plasma Status in Humans. *J. Agric. Food Chem.* **2004,** *52* (21), 6503–6509.

Chand, R. *Demand for Food Grains during the 11th Plan and Towards 2020.* Policy Brief No. 28. National Centre for Agricultural Economics and Policy Research: New Delhi, 2009; pp 1–4.

DFID (UK Department for International Development). Agricultural Sustainability. Working Paper, No. 12. DFID: London, 2004. http://dfid-agriculture-consultation.nri.org/summaries/wp12.pdf (accessed Feb 12, 2009).

Drinkwater, L. E.; Wagoner, P.; Sarrantonio, M. Legume-Based Cropping Systems Have Reduced Carbon and Nitrogen Losses. *Nature* **1998,** *396*, 262–65.

FAO (Food and Agriculture Organization). Online Website and Documents. International Conference on Organic Agriculture and Food Security, Rome, May 3–5, 2007. http://www.fao.org/organicag/ofs/index_en.htm. (accessed Oct 13, 2017).

FAO. *"Climate-Smart" Agriculture*; FAO: Rome, 2010.

FAO. *"Energy-Smart" Food for People and Climate*; FAO: Rome, 2011.

FAO. *Agriculture, Forestry and Other Land Use Emissions by Sources and Removals by Sinks. Report ESS/14-02*, 2014.

Gattinger, A.; Muller, A.; Haeni, M.; Skinner, C.; Fliessbach, A.; Buchmann, N.; Mader, P.; Stolze, M. et al. Enhanced Top Soil Carbon Stocks under Organic Farming. *Proc. Natl. Acad. Sci. USA* **2012,** *109*, 18226–18231. DOI: 10.1073/pnas.1209429109

Gomiero, T.; Paoletti, M. G.; Pimentel, D. Energy and Environmental Issues in Organic and Conventional Agriculture. *Crit. Rev. Plant Sci.* **2008,** *27*, 239–254.

Hepperly, P.; Douds, Jr., D.; Seidel, R. The Rodale Farming Systems Trial 1981 to 2005: Long-Term Analysis of Organic and Conventional Maize and Soybean Cropping Systems. In *Long-term Field Experiments in Organic Farming*; Raupp, J., Pekrun, C., Oltmanns, M., Köpke, U., Eds.; International Society of Organic Agriculture Research (ISOFAR): Bonn, 2006; pp 15–32.

Hole, D. G.; Perkins, A. J.; Wilson, J. D.; Alexander, I. H.; Grice, P. V.; Evans, A. D. Does Organic Farming Benefit Biodiversity? *Biol. Conserv.* **2005,** *122*, 113–130.

IAASTD. *Agriculture at a Crossroads International: Assessment of Agricultural Knowledge, Science and Technology for Development*; IAASTD Report, 2009.

IPCC (Intergovernmental Panel on Climate Change).*Climate Change 2014: Synthesis Report*; IPCC: Geneva, 2014.

ITC (International Trade Center). Organic Farming and Climate Change, 2014. www.intracen.org/Organics/documents/Organic_Farming_and_Climate_Change.pdf (accessed Nov 30, 2017).

ITC (International Trade Centre UNCTAD/WTO) and FiBL (Research Institute of Organic Agriculture). *Organic Farming and Climate Change*. ITC: Geneva, 2007.

International Federation of Organic Agriculture Movements (IFOAM). *The IFOAM Basic Standards for Organic Production and Processing Version 2005*; IFOAM: Bonn, Germany, 2006.

International Food Policy Research Institute/Asian Development Bank. *Building Climate Resilience in the Agriculture Sector of Asia and the Pacific* 2009; IFPRI/ADB: Jakarta.

International Trade Centre UNCTAD/WTO, Research Institute of Organic Agriculture (FiBL) Organic Farming and Climate Change; ITC: Geneva, 2007; 7 p.

Kumar, P.; Mruthyunjaya, D.; Dey, M. M. Long-term Changes in Food Basket and Nutrition in India. *Econ. Polit. Wkly.* **2007,** *42* (385), 3567–3572.

Lal, R. Soil Carbon Sequestration Impacts on Global Climate Change and Food Security. *Science* **2004,** *304* (5677), 1623–1627.

Lal, R. Enhancing Eco-Efficiency in Agro-Ecosystems through Soil Carbon Sequestration. *Crop Science* **2010,** *50*, S-120–S-131. DOI: 10.2135/cropsci2010.01.0012

Lobell, D. B.; Burke, M. B.; Tebaldi, C.; Mastrandrea, M. D.; Falcon, W. P.; Naylon, R. L. Prioritizing Climate Change Adaptation Needs for Food Security in 2030. *Science* **2008,** *319*, 607–610.

Mader, P.; Fließbach, A.; Dubois, D.; Gunst, L.; Fried, P.; Niggli, U. Soil Fertility and Biodiversity in Organic Farming. *Science* **2002,** *296*, 1694–1697.

Matson, P. A.; Parton, W. J.; Power, A. G.; Swift, M. J. Agricultural Intensification and Ecosystem Properties. *Science* **1997,** *277*, 504–509.

Mitchell, A. E.; Hong, Y. J.; Koh, E.; Barrett, D. M.; Bryant, D. E.; Denison, R. F.; Kaffka, S. Ten-Year Comparison of the Influences of Organic and Conventional Crop Management Practices on the Content of Flavonoids in Tomatoes. *J. Food Agric. Chem.* **2007,** *55* (15), 6154–6159.

Mittal, S. What Affect Changes in Cereal Consumption? *Econ. Polit. Wkly*. Feb 2007, pp 444–447.

Mondelaers, K.; Aertsens, J.; Van Huylenbroeck, G. A. Meta-Analysis of the Differences in Environmental Impacts between Organic and Conventional Farming. *Br. Food J.* **2009,** *111*, 1098–1119.

Niggli, U.; Fließbach, A.; Hepperly, P.; Scialabba, N. *Low Greenhouse Gas Agriculture: Mitigation and Adaptation Potential of Sustainable Farming Systems*; FAO, April 2009. ftp://ftp.fao.org/docrep/fao/010/ai781e/ai781e00.pdf (accessed Jan 6, 2019).

Pfiffner, L.; Luka, H. Effects of Low-Input Farming Systems on Carabids and Epigeal Spiders—A Paired Farm Approach. *Basic Appl. Ecol.* **2003,** *4*, 117–127.

Pimentel, D.; Hepperly, P.; Hanson, J.; Douds, D.; Seidel, R. Environmental, Energetic, and Economic Comparisons of Organic and Conventional Farming Systems. *BioScience* **2005,** *55* (7), 573–582.

Hepperly, P.; Douds, D. Jr.; Seidel, R. The Rodale Farming Systems Trial, 1981 to 2005: Long-Term Analysis of Organic and Conventional Maize and Soybean Cropping Systems. In *Long-Term Field Experiments in Organic Farming*; Raupp, J., Pekrun, C., Oltmanns, M., Köpke, U., Eds.; ISOFAR: Bonn, Germany, 2006; pp 15–32.

Reganold, J. P.; Glover, J. D.; Andrews, P. K.; Hinman, H. R. Sustainability of Three Apple Production Systems. *Nature* **2001,** *410*, 926–930.

Reganold, J. P.; Wachter, J. M. Organic Agriculture in the Twenty-First Century. *Nat. Plants* **2016,** *2*, 15221. DOI: 10.1038/nplants.2015.221

Smith, B.; Skinner, M. Adaptation Options in Agriculture to Climate Change: A Typology. *Mitig. Adapt. Strat. Global Change* **2002,** *7* (1), 85–114.

Smith, P.; Martino, D.; Cai, Z.; Gwary, D.; Janzen, H.; Kumar, P.; McCarl, B.; Ogle, S.; O'Mara, F.; Rice, C.; Scholes, B.; Sirotenko, O. Agriculture. In *Climate Change 2007: Mitigation. Contribution of Working Group III to the Fourth Assessment Report of the Intergovernmental Panel on Climate Change*; Metz, B., Davidson, O. R., Bosch, P. R., Dave, R., Meyer, L. A., Eds.; Cambridge University Press: Cambridge, 2007. http://www.mnp.nl/ipcc/pages_media/FAR4docs/ final_pdfs_ar4/Chapter08.pdf (accessed Jan 6, 2019).

Tinttunen, S.; Lehtonen, P. Distinguishing Organic Wines from Normal Wines on the Basis of Concentrations of Phenolic Compounds and Spectral Data. *Eur. Food Res. Technol.* **2001,** *212*, 390–394.

CHAPTER 5

Organic Vegetable Production under Protected Conditions

BALRAJ SINGH* and SANTOSH CHOUDHARY

Agriculture University, Jodhpur 342304, Rajasthan, India

**Corresponding author. E-mail: drbsingh2000@yahoo.com*

ABSTRACT

In view of growing awareness of health and environment issues, organic farming especially of vegetables is gaining momentum in India. Organic products are increasingly preferred in developed countries and in major urban centers in India. There is high demand for organic food in domestic and international market which is growing around 20–25 percent annually; as a result the area under organic farming has been increasing consistently. India with its varied climate and variety of soils has an enormous potential for organic vegetable production. The wide product base, high volume of production round the year, strategic geographic location, high international demand, abundant sunlight and availability of labor at comparatively low cost make India an apt location for organic vegetable production. Although India occupies the second position in the world in terms of vegetable production, productivity and the quality of vegetables in the country are low compared with several other countries. In open-field cultivation, several biotic and abiotic factors cause degradation in the productivity and quality of vegetables in our country. Biotic stresses in the form of viruses, insects, and diseases are the main hindrances under open-field conditions

Organic Crop Production Management: Focus on India, with Global Implications, D. P. Singh, PhD, H. G. Prakash, PhD, M. Swapna, PhD, & S. Solomon, PhD (Eds.)

5.1 INTRODUCTION

During the last two decades, "organic agriculture" has emerged as an "alternative farming system" in India. It has emerged in response to resource degradation in pursuit to achieve higher yields to meet the increasing food demand of the escalating population of the country. Although this pursuit is met with success, it has resulted in a change in the soil quality in terms of soil nutrient imbalance and reduction in the diversity of beneficial micro-organisms, reduction in organic matter, and increased salinity and sodicity with other ill effects on the environment. Now, organic farming regulations implemented in India and acreage under organic farming practices are increasing in the country. Some states like Uttarakhand and Sikkim have been declared organic states.

Although India occupies the second position in the world in terms of vegetable production, productivity and the quality of vegetables in the country are low compared with several other countries. In open-field cultivation, several biotic and abiotic factors cause degradation in the productivity and quality of vegetables in our country. Biotic stresses in the form of viruses, insects, and diseases are the main hindrances under open-field conditions. In open-field cultivation, pesticides are mainly used against two significant problems, that is, borers and vectors (mainly whitefly).

In vegetables like okra, brinjal, tomato, chili, and sweet pepper, the pest and diseases incidence is high during the rainy and postrainy seasons. For effective control of these pests and diseases, application of several pesticides is needed. As a result, strong resistance against insecticides has developed in several insect-pests of vegetables besides carrying toxic pesticide residues as potential health hazards. However, it is difficult to grow such vegetables without pesticides' application in open-field conditions. So, considering the pest and disease incidences and pesticide use, organic cultivation is the need of the hour in vegetable cultivation. Besides, organic vegetables are accepted very well in the international markets with a price premium of 20–30% over conventional produce. The organic products, generally used as plant-protection measures, are not so useful in controlling target insect-pests of vegetable crops in open-field cultivation. So protected cultivation provides a strong base for organic vegetable cultivation and by growing vegetables under different protected structures, one can successfully grow vegetables organically by the management of fungi and nematodes in the soil and vectors of viruses infesting plant foliage. Under protected conditions, the vegetables cannot only be raised organically but the fruiting period of the crop can also be extended as compared to open-field cultivation.

5.2 SUITABLE PROTECTED STRUCTURES FOR ORGANIC VEGETABLE CULTIVATION

5.2.1 *GREENHOUSE*

Mostly high-value vegetables, namely, cherry tomatoes, colored capsicum, and parthenocarpic cucumbers, are grown both under climate and semiclimate controlled greenhouses or naturally ventilated greenhouses in several developed countries. It is also possible under the northern plains of India. Under greenhouse, vegetables can be grown for a more extended period, that is, tomatoes for 8–12 months and sweet pepper for 8–11 months as compared with open-field cultivation.

5.2.2 *INSECT-PROOF NET HOUSE*

For organic vegetable cultivation, insect-proof net houses can be fabricated using 40 mesh UV-stabilized insect-proof nylon net. Many vegetables, namely, tomatoes, brinjal, okra, chili, sweet pepper, and some cucurbits, can be grown successfully by minimizing the use of pesticides. Most of the pesticides in open-field vegetable cultivation are used against borers and vectors for viruses, for example, whitefly, which can be avoided in insect-proof net houses through clean cultivation. Unlike greenhouses, in insect-proof net houses, vegetables cannot be grown for a longer duration, and the vegetables cannot be protected from adverse weather, although the crop duration during summers can be extended up to 30–50 days by covering the roof area of the insect-proof net through shade nets. Moreover, the insect-proof net houses are low-cost structures compared to greenhouses; therefore, the cost of cultivation of vegetables under insect-proof net houses is always much lower than greenhouse vegetable cultivation.

5.3 TECHNOLOGIES AND MANAGEMENT PRACTICES FOR ORGANIC VEGETABLE CULTIVATION IN PROTECTED CONDITIONS

5.3.1 *SEEDLING PRODUCTION*

Before going for organic vegetable cultivation under protected conditions, it is a prerequisite to grow vegetable seedlings organically. Seedlings are raised in multicelled plastic plug trays using soilless media consisting of a mixture of good-quality vermicompost, biogas slurry, FYM, and organic cakes in a suitable proportion. The organic medium filled in plug trays supplies the nutrients to seedlings. Besides, the organic material filled in plug trays following material can also be added to supplement the nutrient requirement of the seedlings:

- *Nitrogen sources*: Alfalfa meal, cottonseed meal, soybean meal, blood meal, and feather meal.
- *Phosphorus sources*: Oak leaves, bonemeal, wastage of sugar factory, and rock phosphate
- *Potassium sources*: Soybean meal, ash of fruits and vegetable peels, wood ashes, and others.
- *Material for foliar application*: Spray the aqueous solutions of organic materials like filtered solutions of manures, vermiwash, and others, on seedlings as a supplementary source of the nutrient.

5.3.2 NUTRIENT MANAGEMENT

The use of organic manures is indispensable in organic vegetable production, as they maintain soil fertility status by increasing soil organic matter content, and supplying essential nutrients to crop plants. Use of organic manures like FYM, poultry manures, fish manures, and sheep composts, at 25–38 t/ha and use of organic cakes of neem, groundnut, *Pongamia*, and castor as basal dose becomes imperative in the protected structures. Application of crop residues is essential in organic cultivation and for that, raising and incorporating green manure crops like *Sesbania* or *Dhanicha* into the soil can fulfill the requirement. Always include legume crops like beans, peas, and cowpea in the crop rotation as they not only improve the soil fertility by fixing atmospheric nitrogen but also increase the yield up from 13 to 30–35%. Inoculation of legume crop-specific *Rhizobium* strains can further improve their N-fixing ability.

In organic vegetable cultivation, the main emphasis is laid on the management aspects that promote and enhance ecological harmony. In organic cultivation, no synthetic substances or chemicals and hormones, or genetically modified crops are permitted. Instead, only natural inputs, biofertilizers, botanicals, and bioagents may be used.

The application of biofertilizer is of considerable significance in organic farming as they play a nutritional stimulatory role in improving the yield and quality of vegetable crops. Applications of different biofertilizers to vegetable crops have depicted an encouraging response both to increase the yield and quality of vegetables and improve soil fertility. The application of *Azotobacter* and *Azospirillum* depicted a significant influence on vegetable crops, resulting in the nitrogen economy of 25–50% and an increase in yield from 1 to 42% (Singh, 2017). Similarly, phosphorus solubilizers can also save 40% phosphorus fertilizers and can enhance crop yields from 4.7 to 51%. Nutrients from plant and animal waste may be recycled after making the compost. The technology for converting waste into compost through rapid composting using the microbial consortia has to be upscaled. This would help in organic farming, reducing the cost of cultivation and improving soil health. Where nutrient sources of low concentrations and slow acting are required, certain organic fertilizers are used (Table 5.1).

TABLE 5.1 Some Organic Nutrient Sources for Application in Protected Cultivation.

Material nutrient content (%)	Material nutrient content (%)		
	N	P_2O_5	K_2O
Organic fertilizer			
Blood meal	10–12	1–2	0.5
Feather meal	13	0	0.5
Bonemeal	5	16	–
Chicken pellets	2.2	0.8	1.2
Guano	0.4–9	12–26	–
Organic manures and compost			
Farmyard manure	0.5	0.15	0.5
Poultry manure	2.87	2.9	2.35
Rural waste compost	0.5	0.2	0.5
Urban waste compost	1.5	1.0	1.5

Source: FAO (2013).

5.3.3 PLANT PROTECTION

Plant protection measures under protected cultivation depend on the local climatic conditions; outside pest and disease pressure; the design of protected structures; the extent of climate control inside the structure; and the level of management of protected cultivation. For controlling the insect pest inside the protected structures, insect-proof screens are used to cover the ventilation openings. By keeping away the vectors (insect) from protected structures, viral diseases can be controlled. Under poly-houses and shade net houses (35%), a few pests like the aphid and whiteflies may enter, but these do not cause any severe infestation (Singh et al., 2017).

5.3.4 INSECT PEST MANAGEMENT

5.3.4.1 PEST EXCLUSION

- *Sanitation and cleanliness:* It is essential to keep the inside and outside areas of the protected structures free from weeds and other pest-harboring plants. As per their relative risk to the harbor, high-risk pest plants should be removed.
- *Airlock doorway:* Entrance doors of protected structures provide easy entry to many pests; hence, the growers need to check the likelihood of pest entrance. To prevent pest entry, a safe and sound entry room is imperative to regulate pest entry in the production area. In greenhouses, a provision of airlock entrance must be provided outside the greenhouse.
- *Insect-proof screens: A* fine mesh screen can be used effectively to keep insects out from both fan and pad and passively ventilated greenhouses.
- *Reflective or metalized mulches:* The entry of insects like whitefly and thrips can be reduced effectively by the use of reflective or metalized plastic mulches. The combined application of insect screens and metalized mulch can result in a substantial reduction in the entry of whitefly into greenhouses.
- *Inspection and monitoring of insects*: Even with the use of all sorts of exclusion techniques, some insects still find an entry in the greenhouse. For early detection, the adoption of specific pest-monitoring tools and techniques is essential, for example, the use of yellow sticky traps and the inspection of plants with a hand lens.

5.3.4.2 CONTROL STRATEGIES

When the pest populations increased over a threshold level, control measures are required. Growers have to apply only bio-pesticides in an organic greenhouse as a control measure. Various promising biological control agents have been identified against common pests of vegetables (Tables 5.2 and

5.3). Among the available biopesticides, a safer and effective material has to be applied.

TABLE 5.2 Botanical Pesticides Permitted for Organic Cultivation.

S. no.	Pesticide	Source	Nature of the product	Mode of action
1	Allicin	Garlic	Broad-spectrum pesticide	Acts as an antibacterial and antifungal biopesticide
2	Nicotine sulfate	Tobacco	Insecticides	Aphids, thrips, spider, mites, and other sucking insects
3	Sabadilla	Sabadilla lily	Insecticides	Caterpillars, leaf hoppers, thrips, stink bug, and squash bugs
4	Nematicide	Neem tree	Insecticides	Potato beetle, grasshopper, and moth
5	Pyrethrum	Chrysanthemum	Insecticides	Aphids and ectoparasites of live stocks

TABLE 5.3 Different Bioagents for Plant Protection in Organic Cultivation.

Mode of control	Category	Bioagent	Target pest
Predators	Insects	*Chrysoperla carnea*	Soft-bodied insects, including aphids
		Ladybird beetles	Aphids and mealybugs
		Carabid and staphylinid beetles	Wide range of insect
Parasitoids		*Trichogramma* sp.	Lepidopteran insects
		Apanteles sp.	Lepidopteran larvae
		Trichospilus pupivora	Caterpillars
Pathogens	Bacteria	*Bacillus thuringiensis* (Bt)	Diamondback moth
	Fungi	*Beauveria bassiana*	Various crop pests
		Metarhizium anisopliae	*Helicoverpa armigera*
		Nomuraea rileyi	
	Virus	*Nuclear polyhedrosis virus* (NPV)	*Helicoverpa armigera* *Spodoptera litura*
	Nematode	*Steinernema glaseri*	Soil insects

5.3.4.3 PROPER PLANT SELECTION

The following plant selection measures must be adopted for growing organic vegetables in protected structures:

- Grow only disease pest-free plants raised by the grower himself or buy only disease- and insect-free plants.
- The optimum plant stand has to be maintained in the protected structures. Higher plant stand resulting in increased pest problems due to weak plant growth and check of air movement inside the protected structures.
- Protected structures used for growing organic vegetables should be kept free of host plants like weeds of pests and diseases similar to crop plants.
- To avoid infection of pathogens, avoid any mechanical injury to plants.
- Remove the diseased and infected plants and plant parts from the protected structures to avoid further contamination in healthy plants.
- Various resistances to insects and diseases have to be selected.

5.3.5 DISEASE MANAGEMENT

Diseases of vegetables must be properly identified for adequate control. The organic approaches to disease management involve sanitary measures, healthy cultural practices, the growing of resistant varieties, and timely use of permitted fungicides and antibiotics in adequate quantity.

Selection of resistant cultivars: The resistant cultivars should be selected judiciously as they have to be acceptable to consumers besides minimizing the pest problem.

Treatment of growing media: Soil and soilless medium are to be treated with bio-fungicides as preventive applications against various soil-borne plant pathogens.

Humidity control inside the structure: Continuously high relative humidity is the triggering factor for *Botrytis* blight of vegetables grown into the greenhouse and other protected structures. For humidity control, the greenhouse is to be ventilated in regular intervals by operating exhausts. Humidity within the canopy of plants can be controlled by adopting suitable irrigation practices and maintaining proper plant spacing. Proper canopy height management, optimum planting time, adequate plant nutrition, and suitable irrigation practices are some of the measures to avoid humidity accumulation within the plant canopy.

Use of biofungicides: Biofungicides are living organisms that help to control plant diseases by infecting and controlling the growth of plant pathogens. Biofungicides are applied as soil, seed, and seedling treatment as a preventive measure. As biofungicides cannot cure the diseased plants, hence, they must be applied as a preventive measure with other cultural measures for disease prevention. Several products identified as biofungicides are available for vegetables (Table 5.4).

TABLE 5.4 Permitted Fungicides for Organic Vegetable Cultivation.

Fungicide	Targeted disease/pathogen	Crop
Inorganic salts and minerals		
Basic copper sulfate	Diseases include angular leaf spot, downy mildew, *Alternaria blight*, anthracnose, bacterial blight, and bacterial spot	Tomato, cucumber, brinjal, and peppers
Copper hydroxide	Leaf spots, *Anthracnose*, bacterial spots, etc.	Wide range of vegetables
Cuprous oxide	*Anthracnose, Phomopsis, Botrytis*, various leaf spots and blights	Tomatoes, peppers, and brinjal
Kaolin	Powdery mildew	Cucurbits
Sulfur	Powdery mildew	Crucifers, cucurbits, peppers, tomato, and brinjal
Synthetic compounds and oils		
Horticultural oil, paraffin oil	Powdery mildew	Cucurbits, melons, squash, and others
Insecticidal soap, potassium salts of fatty acids	Powdery mildew	Greenhouse cucumber
Microbes		
Bacillus pumilus	Downy mildew, powdery mildew	Cole crops, cucurbits, fruiting, and leafy vegetables
Bacillus subtilis	Downy mildew, powdery, mildew, bacterial spot, and early blight	Broccoli, leafy vegetables, cucurbits, peppers, and tomatoes
Streptomyces griseoviridis	*Fusarium, Alternaria, Botrytis, Pythium, Phytophthora*, and *Rhizoctonia* in the greenhouse	Lettuce, cole crops, cucumber, melons, pepper, and tomato

TABLE 5.4 *(Continued)*

Fungicide	Targeted disease/pathogen	Crop
Streptomyces lydicus	Downy mildew, powdery mildew, *Botrytis*, *Pythium*, *Phytophthora*, and *Rhizoctonia*	All greenhouse vegetables
Trichoderma harzianum	*Pythium*, *Rhizoctonia*, *Fusarium*, *Cylindrocladium*, and *Thielaviopsis*	Fruiting vegetables, leafy vegetables, and cole crops

Source: FAO (2013).

Soil solarization: Solarization is a practice of disinfecting the soil by solar energy trapped underneath a transparent polythene sheet of 25–30 μm thickness spread over the soil surface. With the phasing out of methyl bromide in many countries, it is a very promising technique for effective control of soil-borne insects, pathogens, and weeds. Soil solarization is found to be effective against many diseases causing soil-borne pathogens, nematodes especially *Meloidogyne* spp., and insect pupa and larva in the soil. For higher efficiency, the solarization of the soil is to be carried out for 4–6 weeks during peak summer months of May to July. The effect of soil solarization can be improved by applying in combination with permitted chemicals, bio-fumigation agents, or the use of grafted seedlings (Tuzel and Özcelik, 2004).

5.3.6 ADOPTION OF GOOD AGRICULTURAL PRACTICES FOR PLANT PROTECTION IN PROTECTED CULTIVATION

- Workers should maintain personal hygiene by hand washing, cleaning toilet facilities, etc.
- The site of protected structures should be disease- and pest-free. It should be away from existing agriculture production areas.
- The greenhouse should be fabricated with the provision of double doors, insect-proof screens, and restriction of entry for unwanted visitors inside.
- Maintain structures clean by removal of weeds and crop wastes like diseased plant parts.
- Regular pests monitoring should be done by sticky traps and pheromone traps to adopt control measures timely.

5.4 STRATEGIES FOR SCALING UP OF ORGANIC VEGETABLE CULTIVATION UNDER PROTECTED CULTIVATION

The following strategies should be adopted for scaling up vegetable cultivation under protected cultivation:

- Training of educated youths in protected cultivation of organic vegetables.
- Government support must be extended for the self-fabrication of temporary low-cost structures like insect-proof net houses, shade net houses, walk-in-tunnels, and plastic low tunnels for the production of vegetables.
- Promotion of protected cultivation in a cluster approach, especially in periurban areas of the country through government incentives, is the need of the hour.
- The government should promote to the development of organic input hubs for protected cultivation at different locations in PPP mode in various regions of the country.
- Marketing is the key to the success of protected cultivation; hence, e-auctioning of organic produce through cluster representatives can be a boon.
- A particular cluster club of protected cultivation growers may be established along with protected cultivation blocks at the state level.
- All the protected cultivation clusters must be mandatorily clubbed with rainwater harvesting infrastructure and facilities.
- Practical demonstration units at each state level may be established for demonstrations of all low-cost protected structures along with the production and pest-management strategies.
- Large-scale promotion of low-pressure drip irrigation system for low-cost, small-scale protected cultivation in various parts of the country.
- Selection and promotion of location-specific designs of protected structures for different regions of the country.
- Promotion of large-scale mechanization in vegetable cultivation by using raised bed makers, plastic laying machines, plastic low tunnel making machines, pipe-bending machines for making walk-in tunnels, and drip lateral laying and binding machines.
- Convergence and synergy among various ongoing government programs in the field of protected cultivation.

- Ensuring adequate links in production under protected conditions, postproduction on-farm value addition, processing, and consumption chain should be adopted.
- Promotion of the use of solar energy for running the drip system and up to some extent for heating and cooling devices of the protected structures.
- Under the new era of FDI (Foreign Direct Investment) in retail, marketing models for high-value vegetables by opting for quality and off-season vegetable cultivation through protected cultivation is to be promoted.

KEYWORDS

- **greenhouse**
- **organic vegetable cultivation**
- **nutrient management**
- **pesticides**
- **bioagent**
- **disease management**
- **fungicides**
- **good agricultural practices**

REFERENCES

FAO. *Good Agricultural Practices for Greenhouse Vegetable Crops: Principles for Mediterranean Climate Areas*; FAO: Rome, 2013. ISBN 978-92-5-107649-1.

Singh, S.K.; Yadav, R.B.; Singh, J.; Singh, B. Organic Farming in Vegetables. IIVR, Technical Bulletin No. 77, ICAR-IIVR, Varanasi, 47 p.

Tuzel, Y.; Ozçelik, A. Recent Trends and Developments in Protected Cultivation of Turkey. Intl Workshop La Produzione in serra dopo l'era del bromuro di metile, April 1–3, 2004, Catania, Italy.

CHAPTER 6

Organic Farming: Issues and Strategies for Improved Production

N. RAVISANKAR* and A. S. PANWAR

ICAR-Indian Institute of Farming Systems Research, Modipuram, Meerut-250110, Uttar Pradesh, India

Corresponding author. E-mail: ifsofr@gmail.com

ABSTRACT

Adoption of integrated crop management practices in large scale leading to toward organic approach is expected to contribute toward food security, economic security, environmental safety, and climate resilience. This will also positively contribute toward health of human beings, livestock along with ecosystem, fulfilling the primary aim of organic agriculture. There is a need to promote organic farming scientifically in the different cropped areas for improving the yields of crops and to make the yields comparable with that in conventional agriculture. The success of organic farming depends to a great extent on the efficiency of agronomic management adopted to stimulate and augment the underlying productivity of the soil resources. Organic agriculture should sustain and enhance the health of soil, plant, animal, human and planet.

6.1 INTRODUCTION

National Institute of Agricultural Economics and Policy Research prepared the total factor productivity growth score which clearly revealed that the

Organic Crop Production Management: Focus on India, with Global Implications, D. P. Singh, PhD, H. G. Prakash, PhD, M. Swapna, PhD, & S. Solomon, PhD (Eds.)

highest technology-driven growth was recorded in Punjab, while the lowest was observed in Himachal Pradesh. It means that few states like Uttarakhand, Himachal Pradesh, Rajasthan, Madhya Pradesh, Jharkhand, and the north-eastern region of India comprising states of Assam, Arunachal Pradesh, Manipur, Mizoram, Meghalaya, Nagaland, Tripura, Sikkim have not been much influenced by the chemical fertilizers and pesticides. The mean fertilizer and pesticide consumption of India still stands at 128.3 kg/ha and 0.31 kg a.i./ha, respectively. In spite of such a technological advancement, the nutrient use efficiency remains low for all the nutrients (33% for N, 15% for P, and 20% for K and micronutrients). But at the same time, it has been approved that the nutrient use efficiencies improve by the integrated use of organic manures with chemical fertilizers due to improvement in physical, chemical, and biological properties of the soil. Regular use of organic manures improves the water holding capacity of the soil.

In industrialized or countries witnessing the green revolution, organic farming systems provide solutions to the number of problems of the farm sector. The biggest benefit of adopting organic agriculture is its dependence on fossil fuel independent cheaper locally available resources causing minimal stress to the agroecology (Scialabba, 2007). Organic agriculture is a "neo-traditional food system" that takes advantage of ancient indigenous knowledge as well as modern advanced science.

6.2 GROWTH OF ORGANIC FARMING

At present, 179 countries across the globe and 2.4 million farm households are practicing organic farming in 50.9 m ha area. Oceania (12.1 million ha or 33% of the organic farmland of the world), Europe (10.6 million ha or 29% of the global organic farmland), and Latin America (6.8 million ha or 23%) are the most important regions having the largest area of organically managed farmland. The organic agricultural land area has witnessed an increase of 14.7% over the area in 2014. Australia (22.7 million ha), Argentina (3.1 million ha), and the United States (>2.0 million ha) are the major countries having the largest area under organic farming. India, Ethiopia, and Mexico have the distinction of having the largest number of organic growers. At the global level, the number of organic producing farmers has recorded an increase of 7.2% compared with that in 2014.

During 2003–2004, there was about 42,000 ha area under certified organic farming in India that has recorded almost 29-fold growth during the past 5 years. More than 5.71 million ha area has been covered under the organic certification process in India during 2015–2016. Out of this, 1.49 million ha area is being cultivated area, while the remaining 4.22 million ha is wild forest harvest collection area. Madhya Pradesh has the maximum area under organic farming followed by Maharashtra and Rajasthan. Sikkim has the distinction of being declared the first organic state of the country (Anonymous, 2017). Meghalaya also has committed 2 lakh ha of land toward organic certification by 2020.

6.3 APPROACHES

Organic farming is the name given to the farm practices adopted for cultivating a crop. Organic farming implies combinations of tradition, science, and innovation. There are two approaches of organic farming, namely, integrated crop management (50% nutrients through inorganic fertilizers and remaining through organic manures without using any synthetics for insect-pests management) and organic management practices as per National Programme of Organic Production (NPOP) standards were evaluated in crops grown in cropping systems across India under a Network Project on Organic Farming (NPOF).

6.4 TOWARD ORGANIC FARMING

It has been established from the scientific studies that the productivity of crops significantly declined by about 10–15% during the conversion of highly intensive agriculture areas to organic systems, particularly during the initial 3–4 years. After that, enhanced crop productivity reaches a comparable level due to regaining by the soil system. The percent variation in the productivity of major crops during the initial 1–3 years of conversion period and afterward across the various locations of high intensive, rainfed, and hilly areas revealed a decline in productivity of crops (except soybean) during the initial 2 years and the mean reduction was observed as 6.5% during the first year and 1.7% during the second year (Table 6.1). Productivity of *desi* cotton (*Gossypium arboretum*) and soybean was enhanced under organic management by 8 and 1%, respectively. From the third year onward under organic management, the productivity of most of the crops except wheat starts enhancing or is at par with chemical management. Thus, national food security may get jeopardized, in case the gross cultivatable areas are covered under organic production systems, highlighting the need of developing a phased strategy. Considering this fact and taking the stock situation of global organic agriculture scenario, the working group on horticulture, plantation crops, and organic farming for XI Five-Year Plan (GoI, 2007) recommended that only 1–5% area may be covered under organic farming each year in the high productive zones. However, in the less exploited areas like rainfed and hilly areas, the more area may be covered under organic farming. The Intergovernmental Panel on Climate Change (IPCC) found that the present-day chemical agriculture accounts for nearly 20% of the anthropogenic greenhouse effect, producing about 50 and 70% of the overall anthropogenic methane and nitrogen oxides emissions, respectively (Charyulu and Biswas, 2010). Soil carbon sequestration potential was assessed after a decade of continuous organic cultivation which concluded 27% higher soil carbon stock under integrated crop management than inorganic management, clearly displaying environmental benefit. Further, chemical-free safe food is also beneficial for human beings. According to Pimentel (1995), only 0.1% of pesticides actually reach the target pests while 99.9% go to nontarget sectors. Therefore, there is a need to develop an integrated approach of crop management (ICM) in which half-dose of fertilizers without using any synthetics/chemicals for pest management may be used which may be considered as a "toward organic" approach. This approach has been found very effective to enhance the nutrients and water use efficiencies. Therefore, an integrated approach of crop management should be adopted particularly in the highly intensive areas, which contributes a lion's share in the food basket of the country. In addition, this approach will also be helpful in achieving "more crop per drop" and "less land, less resource/time and more production" strategies.

In various states of the country, there is a sizable cropped area, which is more prone to weather vagaries; especially situated in rainfed, dryland, and hilly areas. Increasing the productivity in the field and the income of the farmers, along with sustaining soil health, is the major challenge for farm scientists. In these areas, the fertilizers and pesticides are being used at a level, far below the national average. Initially, the proper strategies should be devised for increasing the income of the farmers in the region after the identification of niche crops to be taken under organic production systems that have adequate market demand. As majority (78%) of the Indian consumers of organic products prefer Indian brands, therefore, there is a need of exploitation of domestic and export markets. There is a large demand of diversified organic foods of Indian tropical fruits, vegetables, spices, herbs, essential oils, flowers, and organic cotton

TABLE 6.1 Mean Yield of Crops Recorded in Cropping Systems under Organic Management and Productivity Trends over the Years.

Crop	**No. of observations**	**Mean productivity (kg/ha) under organic management**	**Productivity trend under organic system over the years (% increase (+) or decrease (−) over inorganic input management)**						
			First-year	**Second-year**	**Third-year**	**Fourth-year**	**Fifth-year**	**Sixth-year**	**Seventh-year**
Wheat	56	2952	−15	−9	−7	−3	−7	−3	−4
Rice	56	3639	−12	−13	5	2	1	2	1
Basmati rice	67	3099	−13	−14	−3	2	2	8	7
Maize	55	4541	−5	9	4	0	3	10	16
Chickpea	25	1269	−10	5	9	3	0	1	5
Soybean	58	1697	1	1	5	0	3	0	12
Cotton *(desi)*	29	1243	8	9	11	12	11	14	12
Mean			*−6.5*	*−1.7*	*3.4*	*2.3*	*1.9*	*3.1*	*7.0*

Source: https://www.icar.org.in/files/Base-paper-organic-Farming-16-03-2015.pdf.

from number of countries of the world. Apart from it, the adoption of organic agriculture at such a large scale in these regions will not only help in conserving the environmentally fragile ecosystems but will also play an important role in supplementing overall food production of India. Sikkim which is an agriculturally low productive state located in the north-eastern hill region of India is a good example of the same. Before the Sikkim Organic Mission, the highest consumption of fertilizer (21.5 kg/ha) was recorded along with an average yield of 1.43 t/ha of rice during the year 2002–2003, which has increased to 1.81 t/ha during the year 2013–2014, after initiation of organic agriculture. Thus, it is interesting to note that no yield reduction was recorded during the conversion period. The yield of other crops like maize, finger millet, and buckwheat also recorded an increase to the tune of 11, 17, and 24%, respectively. In addition, carbon stock build-up under the organic management also recorded an increase of 28.2 and 63.1% than integrated and chemical management, respectively. The soil C sequestration rate in 0–60 cm soil depth was recorded to be 1.46 and 2.57 t/ha/year in comparison with integrated and chemical application of soil nutrients, respectively, in the area where organic farming was taken. It clearly reveals that there is immense potential to sequester the carbon in soil by adopting the farm practices of organic farming, resulting in offsetting the C emissions in the atmosphere (Yadav et al., 2015).

6.5 NICHE AREA AND CROPS FOR ORGANIC FARMING

A place or area suitable for organic production system is known as niche area for organic farming. Generally, the following criteria are used for the identification of niche area for organic farming:

- Soil characteristic (organic carbon).
- Resource use level (seeds, fertilizer, pesticides).
- Biomass generation level (only 50% of total production is to be taken).
- Yield level of widely cultivated crops.
- Category of field (irrigated/rainfed/mixed/hills/plains).
- Suitability of location to maintain crop diversity.

6.6 MAJOR ISSUES AND STRATEGIES FOR ORGANIC FARMING

6.6.1 MAJOR ISSUES

Although there are several challenges for organic growers, but practically only three major constraints adversely influence the yield of crops under organic farming in comparison with conventional farming. These major issues have been discussed below in detail:

A. *Supply of nutrients in sufficient quantity through organic management:* Crop needs nitrogen, phosphorus, potassium, and several other secondary and micronutrients for assimilation and better biomass output. These nutrients need to be supplied in a form that does not have synthetics and does not cause environmental degradation. How to meet the nutrient requirement of crops through organic manures and its availability is the starting point of discussion at the time of pursuing organic farming.
B. *Insect and disease management:* This is another important issue that directly relates to crop productivity and environment and the feasibility and effectiveness of diseases and pest management without using synthetics is another major concern among the growers.
C. *Weed management:* It is the major issue for many of the organic growers as it has been observed that under organic management, weeds grow intensively if manures from outside the farm are used.

6.6.2 MAJOR STRATEGIES

To cope with the above-mentioned issues, the following strategies could be adopted:

A. *For supply of sufficient nutrients through organic management:* There is immense scope for the production of sufficient organic inputs in India, which has been calculated as 7 m in terms of nutrients. Among various sources, the major share (nearly 40%) is contributed by livestock followed by crop residues (30%) and other sources (15%). Rural compost, vermicompost, and agricultural wastes are the major constituents of other sources. In addition, organic farming should not be promoted for an individual crop. Rather than that, it should be practiced in cropping/farming systems. The challenge of ensuring smooth supply of nutrients in sufficient quantity under organic systems can be addressed through the following measures:

Practice through farming system: Livestock alone contributes about 40% to total organic manures in India, and it is not easy to imagine organic farming without livestock. Since time immemorial, the Indian farmers are traditionally practicing an integrated farming system comprising of crops and animals/dairy ASA predominant farming system. The existence of 38 types of farming systems was found in the analysis of farming systems practiced by 732 marginal households across the 30 NARP zones. Out of these 38 types of farming systems, 47% of households had a system comprising of crop + dairy, 11% practiced crop + dairy + goat system, and crop + dairy + poultry systems were adopted by 9% households. Hence, the traditional farming practices already existing in the country naturally provide a major boost for promoting organic agriculture in India. Net returns could be improved from 3 to 7 times in comparison with prevailing farming systems from the integrated organic farming systems (IOFS) models established at Coimbatore (Tamil Nadu) and Umiam (Meghalaya) under NPOF (Table 6.2) and meet up to 90% of seeds/planting materials, nutrients, biopesticides, and other inputs within the farm in the 2 years of establishment.

Multiple cropping and crop rotation: Mixed cropping is the hallmark of organic farming in which a number of crops are cultivated at one time or at various points of time on the same land. Growing legume crops in at least 40% of the area needs to be ensured every year. In selecting crop combinations, the compatibility among the crops needs to be kept in mind. For example, maize crop grows well along with cucumber and beans; tomatoes get well with marigold and onions. But on the contrary, beans and onionsdo not grow well when grown together. At any given time, the entire farm should have a minimum of 8–10 types of crops.

Green manures: Green manure crops, the principal source of organic matter, supply fixes additional nitrogen in the soils due to its in-built ability to biological nitrogen fixation through symbiosis of *Rhizobium* bacteria present in nodules of the roots of these crops, particularly if it is a leguminous or pulse crop. Besides it, these crops gradually economize the use of nitrogen to the soil for the succeeding crop and also protect the soil from erosion and leaching. Since these crops are cultivated for their green leafy material, rich in nutrients, the crop needs to be incorporated into the soil before the flowering

TABLE 6.2 Costs and Net Returns in IOFS Models.

Components	Area (ha)	Total cost (Rs/year)	Net returns (Rs/year)				
			Crop	Livestock	Others	Total	Existing system
Coimbatore (Tamil Nadu)							
Crop (Okra, cotton, desmanthus) + dairy (1 milch animal, 1 heifer & 1 bull calf) + vermicompost + boundary plantation	0.40	110,109	64,500 (87%)	8216 (11%)	1600 (2%)	74,316	27,200[a]
Umiam (Meghalaya)							
Crops (Cereals + pulses + vegetables +fruits + fodder) + Dairy (1 cow + 1 calf) + Fishery + Vermicompost	0.43	68,255	33,531 (57%)	13,252 (22 %)	11,538 (21%)	58,321	8618[b]

[a]Finger millet; cotton; sorghum; [b]rice-fallow.

Source: https://www.icar.org.in/files/Base-paper-organic-Farming-16-03-2015.pdf.

stage. Green manure crops can also be intercropped and incorporated which will have dual advantage of managing weeds and soil fertility. Popularly grown green manures are *Sesbania aculeate* (*Dhaincha*), *Sesbania rostrata*, sunhemp, and others.

Integration of multiple sources of organic nutrient: Combining multiple, that is, more than one organic source for the supply of the nutrients to crops has been found quite effective as meeting the nutrient requirement from a single source may not always be possible. For example, rice–wheat system requires around 30 t farm yard manure (FYM)/year to meet its nutrient demand. By adopting strategies of cropping systems involving green manures, legumes, and combined application of FYM + vermicompost and *neem* cake, this can be managed more efficiently in an easy way. Soil incorporation of *neem* cake has been found quite effective. This type of management also helps in reducing the insect/disease incidences. FYM which is commonly used organic manure in India constitutes about 5–6 kg nitrogen, 1.2–2.0 kg phosphorus, and 5–6 kg potash per tonne, but generally, the proper conservation and efficient use of the resource is not given adequate importance. The techniques for preparing the best quality FYM also differ with various factors, like the amount of precipitation received. In general, it is recommended that pit method of manure preparation is the most suitable for areas with less than 1000 mm rainfall, while heap method can be adopted in other places. Few nonedible oil cakes like castor and *neem* cakes also have insecticidal properties. Among the edible oil cakes, coconut, groundnut, niger, rapeseed, and sesame cakes have higher nutrients (N ranging from 3 to 7.3%; P_2O_5 ranging from 1.5 to 2%, and K_2O ranging from 1.2 to 1.8%). In the case of nonedible oil cakes such as castor, cotton, *karanj*, *mahua*, *neem*, and safflower, *neem* cake has higher N (5.2%), while castor and *mahua* cake have higher P_2O_5 (1.8%) and K_2O (1.8%), respectively. Farmers can select any one or combination of composting methods such as Indore method, NADEP compost, NADEP phospho compost, IBS rapid compost, coirpith, sugarcane trash, press mud composts, poultry waste compost using paddy straw, vermi compost, pitcher *khad*, and biogas slurry depending upon the nature and quantity available with them for making compost within the farm. A microbial consortium of various efficient microbes present in nature will be another option. The consortium can include P-solubilizers, N_2-fixers, plant growth promoting rhizobacteria, photosynthetic microorganisms, lactic acid bacteria, yeasts, and various fungi and actinomycetes. Each and every microorganism has its own advantageous role to play in nutrient cycling, plant protection, and soil health and fertility enrichment in this microbial consortium. The various nutrient management packages for different cropping systems are given in Table 6.3.

TABLE 6.3 Identified Nutrient Management Packages for Major Cropping Systems at Different Places of India.

Place (State)	Cropping system (s)	Sources to meet nutrients
Coimbatore (Tamil Nadu)	Cotton–maize–green manure (GM)	Farmyard manure (FYM) + nonedible oil cakes (NEOC) + *Panchagavya* (PG)
	Chillies–sunflower–green manure	
Raipur (Chhattisgarh)	Rice–chickpea	Enriched compost (EC) + FYM + NEOC + Bio dynamic (BD)+PG
Dharwad (Karnataka)	Groundnut–sorghum	EC + VC + Green leaf manure (GLM) + biodynamic and PG spray
	Maize–chickpea	
Ludhiana (Punjab)	Maize–wheat–summer mung bean	FYM + PG + BD in maize, FYM +PG in wheat and FYM alone in mung bean
Bhopal (Madhya Pradesh)	Soybean–wheat	FYM+PG + BD
	Soybean–chickpea	
	Soybean–maize	
Pantnagar (Uttarakhand)	Basmati rice–wheat–green manure	FYM + VC + NC + EC + BD + PG
	Basmati rice–chickpea	
	Basmati rice–vegetable pea	
Ranchi (Jharkhand)	Rice–wheat–green manure	VC+ *Karanj*cake + BD+ PG

Source: https://www.icar.org.in/files/Base-paper-organic-Farming-16-03-2015.pdf.

B. *For insect and disease management:* Generally, the incidence of pests and diseases under organic production systems is comparatively lower than inorganic systems. This can be attributed to the number of factors like use of oil cakes having insecticidal properties and green leaf manures such as *Calotropis* and the slightly higher content of phenols in plant parts from plants under organic management. Further, this form of agriculture also facilitates the survival and growth of the natural enemies on the farm. The population of Syrphids, Coccinellids, spiders, *Chrysopa, Micromus*, and *Campoletis* which are natural enemies of crop pests and pathogens remains high under organic management in comparison with inorganic and integrated management (organic and inorganic both). In the areas where groundnut, cotton, potato, soybean, and maize crops are being cultivated organically, Coccinellids are the predators of leaf folders and hoppers were observed to be two to three times higher (Table 6.4). Similarly, the population of spiders, another bioagent was observed as double under organic farming in comparison with inorganic management. Arthropod population diversity in soil under organic management in respect of Diplura, Collembola, Cryptostigmatids, Pseudoscorpians, and other mite population recorded higher in comparison with inorganic management.

TABLE 6.4 Variations in Population of *Coccinelids* and Other Natural Enemies in Different Crops under Organic and Chemical Management Practices.

Crop	*Coccinelids*		Other natural enemies (*Micromus, Syrphids, Chrysopa, spiders*)		Cumulative % reduction of natural enemies/year under chemical management
	Chemical	Organic	Chemical	Organic	
Maize (nos/m)	0.80	2.65	0.50	1.53	68
Groundnut (nos/m)	0.69	2.58	0.76	2.15	69
Soybean (nos/m)	0.35	1.35	–	–	74
Cotton (nos/plant)	1.60	4.15	0.88	2.67	63
Potato (nos/m)	0.30	1.25	0.09	0.30	74

For the management of insect-pests and diseases, various inputs are available or prepared at the local farm, animals, plants, and microorganisms. *Neem*-based preparations like *neem* oil, *neem* seed kernel extract (NSKE), traps like pheromone traps, mechanical traps, plant-based repellants, soft soap, and clay are the major products that can be used for controlling of insect-pests and diseases. A natural pest repellent paste mixture formulated and prepared by farmers in Tamil Nadu is gaining popularity. This contains 1 kg each of the leaves of *Vitex nigunda, Agave cantala, Datura metha, Calotropis*, and *neem* seeds dissolved in 5 L of cow urine. This is kept in plastic or earthenware. After 15 days of fermentation, 100 L of water are added and the filtrate is sprayed in the field. The farmers informed that majority of the insect-pests were repelled from the treated area.

- *Weed management:* Weeds are a major problem under organic management according to almost 43% of organic growers; for which cost-effective management techniques need to be identified. Slash weeding is also required among the plants. The weeds can also be used as mulch in the field for reducing evaporation. Stale seedbeds, hand, and mechanical weeding are also other options. Crop rotation, mixed and intercropping can be adopted as weed management strategies.

The other major problems in the field are the incidence of termites and rats. Some of the indigenous technical knowledge (ITKs) practiced for termite management include application of a mixture of dye from *Noni* (*Morinda citrifolia*) and garlic extract on trees, tank silt in sandy wetlands, application of filtrate of *Alotropis* (8–10 kg) soaked in water for 24 h on the soil infested with termite-infested soil and use of sheared human hair collected from barber's shop, on live mounds and along the infested pathways. Small packets were made from small pieces of cotton or thermocol, dipped in jaggery solution, and spread in field/orchard. Small packets of packed cement or white cement coated partly cooked sorghum grains spread in the farm also help in the management of rodents.

6.7 RELATIVE PRODUCTIVITY UNDER ORGANIC AND TOWARD ORGANIC MANAGEMENT

Nearly two-thirds of the total energy requirement in human diet is contributed by three crops of wheat (*Triticum aestivum* L.), rice (*Oryza sativa* L.), and maize (*Zea mays* L.). The four major cropping systems, namely, rice–rice, rice–wheat, rice–green gram, and maize–wheat occupy more than 16 m ha area influencing the livelihood and health of the Indian population. Some of the crops like rice, *basmati* rice, corn, *mung* bean, soybean, and peanut give higher relative yield with organic sources over inorganic sources. About 7% of yield in the case of wheat, a major source of providing food security has been recorded under organic over inorganic. However, wheat grown under integrated crop management recorded higher productivity to the tune of 2% in comparison with chemical management reveals that wheat is a suitable crop for turning organic way. Percent variation in crop productivity under organic approaches and organic management in irrigated and rain-fed/hilly region indicated that maize and basmati rice recorded higher yield in irrigated highly intensive areas under both toward organic and organic approaches in comparison with chemical management. But wheat and coarse rice recorded better yield under toward organic approach compared to organic practices (no change and + 1.9% yield of coarse rice and wheat under toward organic practice over inorganic practice). The productivity of coarse rice and wheat was reduced by 1.6 and 12.5%, respectively, in irrigated regions under organic management (Table 6.5). The performance of maize was observed better under organic management in rainfed/hilly regions. While higher yield was recorded in the case of wheat over inorganic management. These data clearly bring out the potential of organic agriculture in rain-fed and hilly regions for improving the food security.

TABLE 6.5 Percentage Changes in the Yield of Major Food Crops at Different Locations as Influenced by Organic and Organic Management.

Location	% Change in yield under toward organic (ICM) over inorganic				% Change in yield under organic over inorganic			
	Basmati rice	Coarse rice	Wheat	Maize	Basmati rice	Coarse rice	Wheat	Maize
Irrigated region								
Chhattisgarh	–	–	8.7	–	–	–	−13.8	–
Jharkhand	8.1	–	−2.5	–	14.0	–	−12.6	–
Madhya Pradesh	−3.7	–	−3.6	–	−7.4	–	−15.7	–
Maharashtra	–	−6.0	–	–	–	−7.7	–	–
Meghalaya	–	−5.9	–	–	–	−0.2	–	–
Punjab	6.4	9.7	−0.8	17.5	2.3	−3.6	−8.6	17.4
Tamil Nadu	–	–	–	6.8	–	–	–	−1.9
Uttarakhand	7.0	–	−2.1	–	2.0	–	−14.7	–
Uttar Pradesh	1.9	1.9	11.9	21.3	5.3	5.3	−9.3	15.1
Mean	**3.9**	**−0.1**	**1.9**	**15.2**	**3.2**	**−1.6**	**−12.5**	**10.2**
Rainfed/hilly region								
Himachal Pradesh	–	–	–	38.0	–	–	–	13.5
Karnataka	–	–	13.4	18.5	–	–	15.0	19.9
Madhya Pradesh	–	–	5.5	–	–	–	4.0	–
Mean	**–**	**–**	**9.5**	**28.3**	**–**	**–**	**9.5**	16.7

6.8 CONCLUSION

It can be well summarized that adoption of integrated crop management practices in large scale leading to "toward organic" approach for intensive agricultural areas and "certified organic farming" with the unique combination of tradition, innovation, and science in the existing organic cultivation areas and rainfed/dryland regions will contribute toward food security, economic security, environmental safety, and climate resilience. This will also positively contribute toward health of human beings, livestock along with ecosystem, fulfilling the primary aim of organic agriculture. There is a need to promote organic farming scientifically in the different cropped areas for improving the yields of crops and to make the yields comparable with that in conventional agriculture.

KEYWORDS

- **organic farming**
- **cropping systems**
- **supply of nutrients**
- **natural enemies**

REFERENCES

Anonymous. FiBL, *Organic Agriculture Worldwide 2017: Current Statistics. The World of Organic Agriculture*, 2017. www.organic-world.net (assessed Jan 6, 2019).

Charyulu, K.; Biswas, S. *Organic Input Production and Marketing in India: Efficiency, Issues and Policies*, CMA Publication 239; Centre for Management in Agriculture: Ahmadabad, 2010.

GOI. *Report of the Working Group on Horticulture, Plantation Crops, and Organic Farming for XI Five Year Plan (2007–12)*; Planning Commission, Government of India: New Delhi, 2007.

Pimentel, D. Amounts of Pesticides Reaching Target Pests: Environmental Impacts and Ethics. *J. Environ. Ethics* **1995,** *8*, 17–29.

Scialabba, N. E. H. FAO Report Presented in International Conference on 'Organic Agriculture and Food Security', Rome, Italy May 3–5, 2007.

Yadav, R. S.; Parmar, D. K.; Arya, M. P. S.; Choudhary, V. P.; Subash, N.; Dutta, D.; Kumar, S.; Ravishankar, N.; Meena, A. L. Soil C Sequestration Potential of Organic Crop Production System in the Western Himalaya Region (Abstracts). *IIFSR Newsletter*, January–June 2015 issue, ICAR-IIFSR, Modipuram: Uttar Pradesh, India, 2015; pp 1–16.

CHAPTER 7

Status and Scope of Organic Vegetable Farming Under the Temperate Himalayan Region of Jammu and Kashmir

NAZEER AHMED, SUMATI NARAYAN*, AJAZ AHMED MALIK, and K. HUSSAIN

Sher-e-Kashmir University of Agricultural Sciences and Technology of Kashmir, Shalimar-190025, J&K, India

Corresponding author. E-mail: sumatinarayan@gmail.com

ABSTRACT

India is bestowed with enormous potential to produce all varieties of organic products due to its various agro-climatic regions. India is endowed with various types of naturally available organic form of nutrients in different parts of the country and it will help for organic cultivation of crops substantially. There is no dearth of purchasers in India if quality and genuineness are assured. The domestic market of the country is quite large, and there is a large demand for organic foods in different parts of the country, especially in the metros. Jammu and Kashmir should explore the opportunity to emerge as a promising state for organic production. There is immense scope for establishing organoagri units with the technological support for national as well as international markets. There is a need to disseminate the requisite information through various extension programs focused on women workers at the rural farm. There is an urgent need for marketing research for organically produce for export potential. There should be proper planning for marketing of organically grown fruits, vegetables and food grains that should help

Organic Crop Production Management: Focus on India, with Global Implications, D. P. Singh, PhD, H. G. Prakash, PhD, M. Swapna, PhD, & S. Solomon, PhD (Eds.)

farmers to get a better price for their produce. This, in turn, should motivate them to invest more in organic farming.

7.1 INTRODUCTION

Agriculture is the mainstay of Jammu and Kashmir state, which provides employment to approximately 70% of its population. The state is heavily dependent on agriculture, being 65% contributor to the state revenue. The average size of landholding in the state is 0.73 ha. The worsening Environment Sustainability Index of Jammu and Kashmir has emerged as a matter of grave concern for the planners, policymakers, scientists, and general public. A number of factors are responsible for the state environmental catastrophe. The increasing use of fertilizers and insecticide–pesticides in the state is leading to soil health deterioration. The green revolution has led to self-sufficiency in food in India but the need for sustainable agricultural production in the backdrop of dwindling natural resources has resulted in a shift from "resource degrading" farming to a "resource protective" organic farming. Advanced technologies like increased application of synthetic agrochemicals, nutrient-responsive high-yielding crop varieties, unrestricted exploitation of irrigation sources, and others have boosted the output in most of the cases. However, indiscriminate application of these technologies has adversely affected the production and yield of various crops, in addition to the deterioration of soil health and environment. New concepts, like organic farming, biodynamic agriculture, natural or eco-farming, and do-nothing agriculture, have emerged to face the number of present challenges that are the need of the hour for sustaining crop productivity, maintaining soil health, and preserving the ecosystem The crux of all these concepts calls for "back to nature," which is based upon the maintaining of the soil health by returning the nutrients to the soil, which are taken from the soil for the crop production rather than to feed the soil (Funtilana, 1990).

The production of vegetable crops, rich in minerals, vitamins, fiber, proteins, and carbohydrates, can solve the problem of food and nutritional security. These can cater to the domestic demand as well as export market. We should develop a proper strategy to produce more vegetables from limited resources without causing any harm to the soil and environment. Vegetable cultivation under organic conditions offers one of the most sustainable cropping systems with continuous benefits not only with respect to long-term soil health but toward sustainable production, by incorporating moderately or highly resistance against various biotic and abiotic stresses, apart from fetching a premium price of 10–50% over the conventional produce. The market for organic vegetables has been recording a growth of about 20%, against the conventional ones (~5%), especially in countries like Japan, the United States, Australia, and European Union. Jammu and Kashmir, a state that has been carrying out organic farming since time immemorial, has a great scope to exploit the high-demand export market for such produce.

7.2 ORGANIC ADVANTAGE: JAMMU AND KASHMIR

Jammu and Kashmir state is still witnessing lower consumption of fertilizer and other agrochemicals (38.3 kg/ha) against 170 kg/ha being consumed in Punjab. It is estimated that the quantity of nutrients mined by crops in Kashmir is 48 kg/ha, causing better prospects for boosting organic production in the state especially for horticultural products. More than 22,316 ha area has already been covered under organically certified cultivation in the state. The diverse agroclimatic conditions coupled with the existing organic farming system provide an ample opportunity for adopting organic farming. Nearly 1180 ha area has already been allocated for adopting organic cultivation in Jammu and Kashmir, with 300 ha for vegetable under urban cluster scheme, 380 ha under horticulture mission, and 500 ha under *Krishi Vikas Yojana*. Gurez, Kishtwar, and Ramban have been identified for organic production of *Rajmash* while Machil for potato. Budgam, Noorbagh Takenwari, Guzerbaland, and Bangidhara districts have been identified for the production of other vegetables. More than 3400, peasants have been trained for organic farming. According to the study titled, "*Organic Jammu and Kashmir: Avenues of Job Creation & Capital Formation* organic farming should be promoted for wealth accumulation worth over Rs 10,600 crore and can generate exports worth Rs 600 crore in Jammu and Kashmir in the next 5 years (Assocham, 2013). The adoption of organic farming has the immense potential to generate more than 80 lakhs jobs in the state, with over 60 lakhs jobs directly in organic agricultural activities and about 20 lakhs additional jobs in related areas like value addition, packaging, storage, and marketing. This can lead to two and a half time increase in the net per capita income of peasants, from a current value of Rs 7050–17,625 in the coming 5 years (Assocham, 2013).

7.3 ORGANIC FARMING: BASIC TENETS

The basic tenets of organic farming are as follows:

- Improving the biological fertility of the soil for the steady uptake of nutrients, by the crops in a need-based manner from the soil.
- Development of a balanced eco-system by the optimal use of biopesticides with the implementation of cultural practices like crop rotation and mixed cropping.
- Recycling all farm wastes and manures in the farm itself to avoid the drain of nutrients.
- Recycling of industrial and urban wastes for returning to agriculture, for conservation of energy and resources, for the sustainability of soil fertility.

7.4 IMPORTANCE OF PRODUCTION OF VEGETABLE CROPS UNDER ORGANIC SITUATIONS IN JAMMU AND KASHMIR

- Generally, vegetables are cultivated by most of the poor, small, and marginal farmers of the state. Organic farming can play an important role in ensuring economic security.
- Increased application of chemical fertilizers is responsible for deteriorating soil health and land productivity.
- The consumption of vegetable crops in fresh form also enhances the risk of chemical contamination leading to health hazards.
- The cost of production in chemical farming is continuously increasing, against very low costs in organic farming.
- High environment pollution.
- There is a need to compete with the best and to meet the global demands of globalization, which highlights the importance of production of vegetable crops under organic farming.
- Organic farming has an immense potential to generate additional income through international exports or by saving production costs. It can also help to secure a respectable position for the state in the international markets.
- There is an urgent need to address issues of threats to the environment quality, ecological stability, and sustainability of production associated with conventional farming methods.

- There is an immense potential for appropriate managed organic farming systems for increasing the crop yield and restore the natural base, esp., under low input situations prevalent in many areas of the country.
- Financial and environmental concerns.

7.5 TECHNOLOGIES FOR ENSURING ORGANIC AND LOW-COST AGRICULTURE

Low-cost organic farming has a positive impact on the global trade for India. Organic farming technology should ensure resource conservation with minimum input demand, without sacrificing the productivity. The steps to be taken are as follows:

- Growing varieties of crops having resistance to biotic and abiotic stresses performing well under limited resources is the need of the hour. Further, such varieties should meet the prescribed standards of organic farming, along with a reduction in the cost of cultivation. The varieties with early vigor are generally found to hamper the growth of weeds.
- The use of biofertilizers and bioagents will aid in the control of environmental pollution and soil health improvement, apart from reduction in the use of chemical inputs. Inoculation by improved *Azotobacter* strains and use of inorganics like biogas slurry, along with other package of practices enhances the nutrient uptake, reduce chemical use, and improve the productivity of crops significantly.
- The use of recyclable nutrients (N, P, K, Zn, Mn, Fe, and Cu) from plant as well as animals along with a well-standardized and efficient technology for converting waste into compost should be adopted. Proper refinement and large-scale verification of these technologies with adequate funding mechanism is a must. There is a need to adopt integrated nutrient management (INM) with varieties that respond well to the integrated nutrient application. For maintaining soil health and enhancing production, green manure is another option that needs to be adopted.
- For efficient weed management, the minimal application of weedicide through the use of integrated techniques should be ensured. This can be done to a large extent through mechanical weeding, changing the

crop dynamics, and by timely sowing. Cultivation practices should favor quicker nutrient uptake and vigorous crop growth to enable the crop to successfully compete with the weeds. Refinement of biocontrol of weeds is necessary.

- Popularization of machinery like tractor-drawn bed former-cum-seeder for furrow irrigated raised bed system (FIRBS) and cell type metering system for sowing to conserve resources without affecting crop production needs to be taken up. There is an urgent need to enhance the water and fertilizer use efficiency by the development of input efficient varieties and agronomic practices. Apart from it, development of efficient agricultural tools for small and marginal farmers and gender-friendly farm equipment is the need of the hour (Table 7.1).

TABLE 7.1 Nitrogen Content and Productivity of Biomass of Different Crops.

Crop	Productivity (t/ha)	Nitrogen (%)
Subabul	09–11	0.80
Sunhemp	12–13	0.43
Dhaincha	20–22	0.43
Cowpea	15.16	0.49
Clusterbean	20–22	0.34
Berseem	15–16	0.43
Mungbean	08–09	0.53

7.6 USE OF MICROBES: ONE OF THE BEST OPTIONS FOR ORGANIC FARMING

There are 104–105 microbes per gram of soil which:

- Act as plant growth promoter.
- Enhance nutrient availability.
- Help in value addition.
- Aid in bioremediation/removal of toxicants.
- Act as biocontrol agents.

7.7 CONVERSION TO ORGANIC PRODUCTION SYSTEMS

The time taken during the initiation of organic management and certification of crops and or animal husbandry is known as the conversion period. There is no need of a conversion period in case, existing cultivation package of practices fulfills the standard principles of the organic cultivation standards or when virgin lands are put in use for organic farming. If required on a field by field basis, the full farm, including livestock, should be converted according to required standards over a period of time. All aspects relevant to the established standards should be covered under the conversion plan.

7.7.1 CONVERSION REQUIREMENTS

There should be a clear plan, with the past history of crops taken, present crops situation, management techniques, and a time frame for conversion. Generally, the criterion of date of application for applying to the certification agencies or the date of last application of unapproved agricultural inputs is considered for calculating the conversion period. There should be a clear demarcation between the conventional and organic farmed plots with easy accessibility to the organic plots for frequent inspections. Anybody cannot go back from organic farming to traditional farming. The sale should be started only after the certificate is received as an organic produce after meeting the standard requirements for a maximum duration of 1 year before the initiation of the production process. For the perennial plants, there is a provision that produce will be certified as organic produce only at the first harvest after at least one and a half years of management under organic conditions. If anyone is into organic and conventional production simultaneously, the use of genetically modified organisms is not permissible on the conventional part., the processing units can be permitted for the certification provided there is clear documented evidence for a well-differentiated processing chain for organic and conventional production.

7.8 CROP PRODUCTION STRATEGIES FOR ORGANIC SYSTEM

7.8.1 SOIL AND WATER CONSERVATION

- Prevention of soil erosion, salinization of water and ground and surface water pollution, ensuring water conservation, and proper use of water using suitable technologies.
- Restricting the burning of organic matter to a minimum.

7.8.2 SELECTION OF SEED AND PLANTING MATERIAL

- Seeds and planting materials produced under traditional/certified organic procedures with permissible seed treatment techniques should be the norm. In the absence of organically produced source, untreated conventional seed materials may be used. Genetically engineered plants or transgenic plants should not be used as a source of seed.
- Seed and plant material of new crop treated with synthetic pesticides or chemicals can only be allowed in certain regions. If organic farming is in its early stage, use of planting material treated with chemicals/ pesticides may be permitted.

7.8.3 CROP ROTATIONS

Taking into account the nature of the crop, presence of weeds, and local conditions, the minimum standards for crop rotations on arable land should be developed and followed.

7.8.4 MANURIAL POLICY

- Manuring may comprise green manure, leaf litter and vermicomposting, organic wastes in their natural composition, inputs sourced from microbes, plants, or animals, excluding any synthetic and chemical fertilizers, and other nutrients, with no adverse effects on environment.
- Manure containing human feces or untreated sewage should not be used for cultivation of vegetables for human consumption.
- All the materials should be in accordance with the standards.

7.8.5 MANAGEMENT OF BIOTIC STRESSES (PESTS, DISEASES, AND WEED MANAGEMENT)

Traditional products prepared from local sources at the farm itself should be used. Need-based thermal sterilization of soil is permitted for pest and disease control.

7.9 TECHNOLOGY PACKAGE FOR ORGANIC VEGETABLES

- Soil should be timely prepared to a fine tilth with minimum tillage. All debris, stubbles, stones, etc., should be removed to avoid infestation of ants and termites.
- Organic manures should be applied as basal dose @ 25–38 t/ha through and /or organic cakes from *neem* and groundnut.
- Incorporation of *Sesbania, dhanicha,* or biomass from other suitable plants in the soil. The use of crop residues is also advocated to improve the soil health and to increase the yield.
- Crop rotation especially with legume crops and *Rhizobium* inoculation of leguminous crop with crop-specific strains not only to improve atmospheric nitrogen fixation, which can increase the yield up to 30%–35% can further improve their N-fixing ability (Table 7.2).
- The varieties of vegetable should be selected on the basis of climate and market preference. Emphasis should be laid on adopting optimum plant density and timely planting, raising plants/seedlings with sufficient organic manures, and biofertilizers with vigorous seedlings for better establishment, growth, and yield should be emphasized.
- The growth of weeds may be minimized by reducing moisture loss and mulching with locally available materials or polythene sheets.
- Disease-resistant varieties suitable for the eco-system, weed control, crop rotation, ensuring field sanitation, and raising trap plant to attract insects must be used (Table 7.3).

TABLE 7.2 Yield of Vegetable Crops as Influenced by Biofertilizer Inoculations.

Biofertilizer	Crop	Increase in yield (%)	Nitrogen economy (%)	References
Azotobacter	Cabbage	24.30	25	Verma et al. (1997)
	Garlic	14.23	25	Anonymous (2003)
	Knol khol	9.60	25	Chatto et al. (1997)
Azospirillum	Onion	21.68	25	Anonymous (2002)
	Knolkhol	14.90	25	Chatto et al. (1997)

TABLE 7.3 Suggested Varieties of Vegetable Crops Tolerance/Resistance to Disease and Pests.

Crop	Pest/disease	Variety
Chilli	Leaf curl virus	Pusa Jwala, Pusa Sadabahar
	Leaf curl virus CMV, TMV, and leaf curl	Punjab Lal
Cabbage	Black rot	Pusa Mukta
Cauliflower	Black rot	Pusa Shubra
Brinjal	Bacterial wilt	BWR 12, Arka Nidhi, Utkal Tarini, Utkal Madhuri, Annamalai

7.10 CERTIFICATION OF ORGANIC FOOD

The procedure of giving assurance in black and white that the product, process, or service meets the specific requirements by the third party between the producer and consumer is known as the organic certification. For ensuring quality and authenticity before certification of organic production, there is a provision of strict supervision and inspection of the farming unit at regular intervals and at various stages of production. At present, Agricultural & Processed Food Product Export Development Authority (APEDA), Tea Board, Coffee Board, Spices Board, Coconut Development Board, and Cocoa & Cashew nut Board are the six authorized accreditation agencies in India that has been approved by the Ministry of Commerce, Government of India. Indian Organic Certification Agency (INDOCERT) is India's first-ever local Organic Certification Body who has been entrusted the authority to provide certification to both domestic as well as export markets. INDOCERT also provides a platform for capacity building, creating general awareness, information dissemination, and networking in the sphere of organic farming. The labeling of organic certification followed by INDOCERT has been summarized in Table 7.4 as per the year of production.

TABLE 7.4 INDOCERT Certification Label (Year of Production).

Crop	First year	Second year	Third year	Fourth year
Annual	No label	In conversion to organic agriculture	Certified organic	Certified organic
Perennial	No label	In conversion to organic agriculture	In conversion to organic agriculture	Certified organic

7.11 MAJOR ADVISORIES FOR PROMOTING ORGANIC FARMING IN INDIA

For the success of organic farming in the country, the following suggestions must be taken care of:

- A Center of Excellence and nationwide network for research on organic farming must be established.
- All the current available Indigenous Technological Knowledge (ITK) along with other technologies developed by various public sector research institutions/NGO/individuals on different aspects of organic farming in the country must be well documented.
- The concepts and practices in organic farming may be included as the core courses in the curriculum of undergraduate and post-graduate degree programs in different agriculture universities and other agricultural institutes.
- Standardization of mechanisms or methods for certifying of organic farming practices.
- A comprehensive package of practices for the production of various crops under organic farming systems must be developed.
- Eminent agriculture as well as social scientists and progressive farmers should be an integral part of the export group for visiting farms of the peasants who have succeeded in adopting scientific organic farming practices.
- *Krishi Vigyan Kendra* (KVK) should disseminate information on organic farming by conducting field demonstrations, programs broadcast on *Akashvani* and *Doordarshan*, and other suitable mass media.

7.12 CONCLUSION

Since it is the beginning of organic farming in our country, there is an urgent need to address the issues associated with production as well as marketing. There is no dearth of purchasers in India if quality and genuineness are assured. The domestic market of the country is quite large, and there is a large demand for organic foods in different parts of the country, especially in the metros. Jammu and Kashmir should explore the opportunity to emerge as a promising state for organic production. There is immense scope for establishing organoagri units with the technological support for national as well as international markets. There is a need to disseminate the requisite information through various extension programs focused on women workers at the rural farm.

KEYWORDS

- **low cost agriculture**
- **microbes**
- **organic production systems**
- **soil and water conservation**
- **biotic stresses**
- **organic vegetables**
- **certification**

REFERENCES

Anonymous. *Annual Report (Rabi)*, Division of Olericulture; SKUAST (K): Shalimar, Srinagar, 2002.

Anonymous. *Annual Report (Rabi)*, Division of Olericulture; SKUAST (K): Shalimar, Srinagar, 2003.

Assocham. *Organic Jammu and Kashmir: Avenues of Job Creation & Capital Formation*; Assocham: New Delhi, 2013.

Chatto, M. A.; Gandorio, M. Y.; Zargar, M. Y. Effect of *Azospirillum* and *Azotobacter* on Growth and Quality of Knolkhol (*Brassica oleracea* L. Var. gongylodes). *Veg.Sci.* **1997,** *24*, 16–19.

Funtilana, S. Safe, Inexpensive, Profitable, and Sensible. *Int. Agric. Dev.* **1990,** March–April, 24.

Verma, T. S.; Thakur, P. C.; Singh, S. Effect of Biofertilizers on Vegetable and Seed Yield of Cabbage. *Veg. Sci.* **1997,** *24*, 1–3.

CHAPTER 8

Organic Farming: Perspective in Sustainable Development Under Changing Climatic Conditions

RAJVIR SINGH RATHORE

Former OSD to Governor of Uttar Pradesh, Lucknow 226001, Uttar Pradesh, India
E-mail: rajvir_ddg@yahoo.co.in

ABSTRACT

Vegetables are generally sensitive to environmental extremes. Under climate change scenario, increasing temperatures, reduced irrigation water availability, flooding, and salinity will be major limiting factors in sustaining and increasing vegetable productivity. The tropical vegetable production environment is a mixture of conditions that varies with season and region. Farmers in our country are usually small-holders, have fewer options and must rely heavily on resources available in their farms or within their communities. Thus, technologies that are simple, affordable, and accessible must be used to increase the resilience of farmers under climate change.

8.1 INTRODUCTION

Organic farming systems exist throughout the world in diverse climates with wide range of management practices. The soil's organic matter is likely to be affected due to changes in the earth's atmosphere. The changes in temperatures and rainfall patterns will further influence the soil properties, such as organic matter content, water-holding capacity, cation-exchange

Organic Crop Production Management: Focus on India, with Global Implications, D. P. Singh, PhD, H. G. Prakash, PhD, M. Swapna, PhD, & S. Solomon, PhD (Eds.)

capacity, and nutrient content in the soil. The research conducted shows that the carbon sequestration in soils is influenced by limitations in major nutrients like nitrogen and phosphorus in changing climate under different agroecological situations. At present, very little information is available on the impact of climate change on soil organisms, which plays an important role in driving the C and N cycles taking place in the soil.

Indian economy is basically agrarian in nature as more than two-thirds of its total population directly or indirectly engaged in agricultural activities. Before the green revolution, only traditional agriculture was in use with very little intervention of high-yielding varieties (HYV) and hybrids, chemical fertilizers, and pesticides. The results of the green revolution were seen in the form of the substantial increase in production and productivity and India could become the self-sufficient in food grains to the large extent. In spite of these significant achievements, the country's present performance in the agriculture sector is highly unsatisfactory. With an increased cost of inputs and declined factor productivity, the agriculture has become an unviable venture. Consequently, a large percentage of farmers and agricultural laborers have been compelled to migrate from the farm sector.

As farm and related activities fulfill almost all the needs of India's growing population, nearly 67% of our population and 54% of the total labor force are dependent on this important sector. It has been estimated that if India has to achieve a double-digit GDP growth rate, agricultural growth of around 4% or more would be essential. Despite having such a vast potential for striving to the needs of the ever-growing population, agriculture is facing various challenges such as small and fragmented farmlands, deteriorating soil health, declining water resources, change in rainfall pattern, increasing temperature, inadequate storage facilities, lack of processing and packaging facilities, and marketing network at the local level.

In modern agriculture, imbalance use of chemical fertilizers, illegitimate use of pesticides and herbicides over the past four decades have resulted in the loss of natural habitat balance and soil health. Apart from this, hazards like pollution due to chemical fertilizers and pesticides, genetic erosion, ill effects on environment, and reduced food quality and increased cost of cultivation are the other serious manifestations that are associated with the fallacious use of chemicals in new-age farming. As a result, farmers do not find farming a viable proposition anymore and those who are still practicing it are in debt and committing suicides in face of any natural calamity adding to these woes. Apart from this, the other factors aiding to this crisis are the significantly high price of seeds especially hybrids and genetically modified seeds, fertilizers, pesticides, other inputs, and the government's slow withdrawal of investment in food processing as well as market interventions and more significantly, shifting of subsistence farming based primarily on locally available inputs to external inputs resulting in an unviable and unsustainable farm enterprise.

8.2 IMPACT OF CLIMATE CHANGE ON CROP YIELDS, SOIL HEALTH, AND FOOD SECURITY

It is projected that due to climate change, the productivity of the major crops at 2035, 2065, and 2100 at maximum and minimum variations in rainfall and temperature will be adversely affected and pulse crops will be the worst sufferer than the other field crops (Table 8.1).

TABLE 8.1 Projected Changes (%) for the Year 2035, 2065, and 2100 in the Productivity of Major Crops at Maximum Variations in Temperature and Rainfall.

Crop	2035	2065	2100
***Kharif* crops**			
Paddy	−7.1	−11.5	−15.4
Maize	−1.2	−3.7	−4.2
Sorghum	−3.3	−5.3	−7.1
Pigeonpea	−10.1	−17.7	−23.3
Groundnut	−5.6	−8.6	−11.8
***Rabi* crops**			
Wheat	−8.3	−15.4	−22.0
Barley	−2.5	−4.7	−6.8
Chickpea	−10.0	−18.6	−26.2
Rapeseed-mustard	0.3	0.7	0.5

Source: Adapted from Birthal et al. (2014).

Although, there will be small impacts of climate change on *Kharif* season crops but these crops will become more susceptible to increased incidence of weather extremes, such as changes in rainfall pattern and its intensity, duration, and frequency of drought and floods, diurnal asymmetry of temperature, change in humidity, and pest incidence and virulence. Similarly, winter crop production will become comparatively more vulnerable due to an increase in temperature, asymmetry of day and night temperature, and higher uncertainties in rainfall during the crop growing period. The climate change does not influence the food production only but will also hamper food security through its direct or indirect effect on other components of the production systems, particularly; livestock production as it is very closely linked with the crop production.

Indian agriculture is key sector for food and nutritional security. Since inception of Green Revolution, agriculture sector has gained tremendous progress but presently this sector is facing numerous challenges viz; deterioration of soil health, stagnation in yield and net sown area, reduction in size of holding and weather vagaries. Agriculture is the most vulnerable to climate change, owing to its huge size and sensitivity to weather parameters, thereby causing huge economic impacts.

India possesses more than 60% rainfed agriculture causing very challenging task to maintain the production and productivity level and also expose to biotic and abiotic stresses arising from weather vagaries and climate change. More than 85% of Indian farmers are small and marginal farmers with poor economic status. Therefore, to maintain the production level of agriculture and its sustainability is a matter of great concern. The climate variability and weather are likely to aggravate the problem of future food and nutritional security and livelihood also.

The rise in global temperature may lead to melting glaciers, causing rising in sea level and increase in greenhouse gases viz; carbon dioxide (CO2), methane (CH4) and nitrous oxide (N2O). These greenhouse gases trap the outing radiation from earth resulting enhanced atmospheric temperature. Perusal of 50 years weather data, it is observed increase in temperature, occurrence of hot days, hot night and heat waves, increasing frequency of heavy precipitant events, increased melting glaciers and rise in sea level.

Perusal of IPCCs report published in 2014, it envisaged the warming of climate system is unequivocal. Climate change and weather vagaries threatens to enhance the productivity losses resulting increase in malnutrition among weaker sections of society, changing the pattern of plant pathogens, insect and pest and reduce productivity of rainfed eco system which more than 60% of net cropped area of our country. Therefore, to sustain and enhanced the agricultural productivity in rainfed eco system of semiarid tropicsmay be the matter of great concern. It can be revived by developing suitable climate smart plants, crop production technologies, climate smart plant pathogens and insects & pests management technologies and cropping patterns. Current reports envisaged that an increase in global annual mean temperature of 10C by 2025 and 30C by the 2100. Precipitation pattern of rains and intensity of rains may also varies. The concentration of greenhouse gases (CO2& O3) emission lead to enhance in global precipitation of 20.5

per 10C temperature warming. Indian agriculture is also highly prone to weather vagaries especially to drought and flood which is depend upon mansoon rains; frost in north west; heat waves in central and northern part and cyclones in eastern coastal region which is major cause of agricultural production and productivity.

Climate variability has direct effect on crop production, soil health, livestock and fisheries productivity and incidence of plant pathogens and insects. Increase in atmospheric greenhouse gases especially carbon dioxide has a fertilization effect on crop with C3 plant photosynthesis pathway and thus promotes the growth and productivity of C3 plants. Increase in ambient level of carbon dioxide in atmosphere leads to increased photosynthesis mechanism in several plants (C3) viz; wheat, rice, maize etc. and also decreased evaporation losses. The crop productivity of cereal crops especially wheat would likely to be reduced because of decrease in crop duration and increase in photo respiration and evaporation rates will be enhanced due to rise in atmospheric temperature. It is also evident from climate change reports that increase in temperature may reduce the crop duration, increase crop respiration rate and alter the photosynthesis process and also affect the plant pathogen and insect and pest population.In rainfed eco system reduction in yield will also be observed due to increase in crop water demand and change in rainfall pattern during mansoon season. Horticultural crops, tea, coffee, aromatic and medicinal plants quality will also be decreased. As temperature will rise, irrigation demand of crops will be increased causing heavy pumping of ground water at crop sown areas and glaciers will melt in Himalayas regions may lead to increase water availability in Ganga, Brahmaputra and their tributaries.Soil health will be reduced due to low quality and quantity of organic matter in the soil. As carbon dioxide concentration in atmosphere increase, resulting higher carbon nitrogen ratio in crop residue which may reduce their rate of decomposition and nutrient supply. As soil temperature may increase, the mineralization of nitrogen will increase but its availability may decrease because of increased in gaseous losses through processes of volatilization and denitrification. Plant pathogens, insects & pests population will be increased due to climate change resulting emergence of new plant pathogens and insect and pest problems.

Weather vagaries have direct effect on feed and foliage production resulting inadequate livestock nutrition. As temperature increases resulting more lignification of plant tissues causes poor digestibility and palatability. Increase water scarcity may also decrease the food and fodder production and productivity. Climate vagaries also effect the incidence of epidemic in livestock by increasing vector population. Change in rainfall pattern may also influence the vector population during wetter season leading to large outbreak of epidemics in animals causing the great losses in production and productivity of animals. Rise in temperature has also create the heat stresses in mulching animals causing their poor fertility and productivity also. The global warming may also enhanced sea and river water bodies causing adverse effect on breeding of fish, migration and their production. Impact of increased temperature and cyclonic activity leads to poor fishing, loss of production and increase fishing and marketing cost of marine fish production.

According to the estimates of IPCC, average temperature of the world is likely to be increased between 1.1°C and 6.4°C during the 21st century and rainfall patterns will also be distorted. Through the cycles of carbon, hydrologic, and nitrogen, soils are also intricately linked to the atmospheric changes. These changes will have an impact on soil formation processes and soil properties. As carbon and nitrogen are the principal ingredients of soil organic matter, it affects the physical as well as chemical properties of soils, namely, water holding and cation-exchange capacities, soil structure formation, and availability of nutrients and their supply to the plants from the soil ecosystem. Soils with enough quantity of organic matter are more productive than soils having low organic content. Thus, how potential changes in the carbon and nitrogen cycles will influence soils is an important question in respect of climate change and its effects on soil processes and properties? How nitrogen cycle will be influenced by the climate change will have an impact on C storage and availability of nitrogen and phosphorus in soils and influence on plants due to enhanced atmospheric carbon dioxide need to be studied in-depth under varied agroecological regions. Therefore, in-depth investigation of soil–climate interactions in a changing scenario is critical for addressing the future food security concerns.

Food security is defined as a situation that prevails when all the people, at all times, have physical, social, and economic access to sufficient, safe, and nutritious food that meets their nutritional requirements essential for an active and healthy life. In terms of calories consumed, the soil only provides more than 97.5% of the food requirement, while the aquatic systems supply less than 2.5% food. Thus, most of our food supply will have to come from the terrestrial environment, which will make soils more important and pertinent for the food security. Variations in temperature and rainfall patterns, soil formation processes and soil properties will definitely be there in the future also, as in the past due to climate change. This has emerged as a matter of great concern that the food security will be compromised by the climate change which will result in deteriorating human health. Soil erosion also has an adverse impact on crop productivity and production. Studies conducted in Spain on a semiarid Mediterranean ecosystem revealed strong correlations between climate change and soil erosion and negative relation on bulk density, aggregate stability, water-holding capacity, organic matter content, pH, total N, and soluble P in the soil, and all other properties important for healthy crop growth. Therefore, it can be concluded that further soil erosion caused by climate change will degrade the soil properties that are important for production of more food and fiber resources required for the growing population.

During 2010, at least 1 billion people were surviving in food insecurity positions. As per the targets for the year 2050, overcoming this insecurity along with feeding 2.3 billion additional people will require about 70% increase in cereal production at the global level. According to the IPCC, climate change will make an impact on all food availability, stability of food supplies, access to food, and food utilization—all the four dimensions of food security. Global food production is expected to rise, if the increase in local average temperatures averages 1°C–3°C, but will decrease, if mean temperature surge goes beyond 3°C. However, the expected variations in distributions of food production will not be uniform.

In most of the developing countries of South Asia and Africa, which are also facing risk of food insecurity and deteriorating soil health, low soil fertility level is the major constraint for the food security. Therefore, proper soil management is the only way that has the potential to reduce the food security issue of these regions to the large extent. For example, with additional use of 1 tonne of soil organic matter per hectare resulted in an increased maize yields by 17 and 10 kg/ha in Thailand and Nigeria, grain yield by 40 kg/ha in Argentina, and cowpea yield by 1 kg/ha in Nigeria. Research studies conducted in West Africa revealed that soil organic carbon (SOC) losses could be restored with the applicable of 30–60 kg N/ha by increasing biomass production and incorporating more crop residues (CRs) to the soil. Enhancing SOC in deteriorated soils will be a welcome step toward enhancing food security in developing countries. But enhanced decomposition of soil organic matter caused by global warming could be a serious threat toward these efforts, in case C sequestration practices are initiated in these nations, having an adverse effect on food security. In the case of noninitiation of C sequestration practices, the already degraded soils will be more vulnerable to the adverse impact of climate change than comparatively healthier soils. Thus, we can say that climate change will have an adverse impact on food security issues through its effects on soil health.

8.3 RELEVANCE AND STATUS OF ORGANIC FARMING

It is apparent that the indiscriminate use of fertilizers and other chemicals is causing an adverse effect on the human health and also on the environment. Besides, the pesticides and fertilizers persisting in the soil are harmful to the beneficial soil microorganisms and earthworms, resulting in deterioration of soil health. In the name of growing more to feed the increasing population, we have taken the wrong path of unsustainability. Several challenges associated with the conventional farming have compelled majority of the developed and some developing countries to adopt organic farming. Organic farming is the most important practice for protecting the environment. Agriculture on which the present and future generations of the globe will be dependent for their survival is still a vital sector for ensuring food security, alleviating poverty, conserving the vital natural resources, and overall economic development for majority of the developing countries. The practices used in organic farming are based on "give back to nature" with the philosophy of feeding the soil rather than maintaining the soil health.

Therefore, the organic farming has to play a vital role in Indian agriculture in view of the conversion of agricultural lands to nonagricultural purposes, fragmentation of small holdings, deteriorating soil health, and declining response to the applied plant nutrients in changing scenario. In spite of that, still there is no consensus of opinion between the farmers and the experts in respect of organic farming. Disputes about the economic viability and yield increases under organic farming are perceptive, but both the groups agree that organic farming is an eco-friendly exercise with inherent ability to protect human health. Strong views have been expressed for and against organic farming on the grounds of the practicability of feeding 135 crore Indian population and economic viability, availability of organic inputs in required quantities at the local level and the technical knowledge and skills. The people advocating conventional farming ignore its ill effects. Large number of supporters of organic farming want to have an integration of both the conventional as well as organic farming or advocate a cautious approach by slowing the process of the conversion of the conventional farms into organic farms. The farmers have their own doubts about the crop productivity as well as economic viability but their queries remain unanswered up to a large extent.

The procedure of organic farming involves the use of locally available and decomposable matter and providing resistance to different crops in a direct or an indirect manner against different pathogens. Organic manures are not only a source of nutrients but also increase microbial population and their activity in soil as the physicochemical and biological processes are influenced significantly with their use.

In India, since January 1994 when "Sevagram Declaration" for promotion of organic agriculture in India was adopted, organic farming has grown many folds. Various initiatives have already been taken at government and nongovernment level for giving it as major boost. Regulated framework of organic farming has already been defined by National Programme on Organic Production (NPOP), while the promotional strategy has been chalked down by NPOF for providing the support to expand the area under certified organic farming. The present status of certified organic farming in India can be understood from the figures given in Table 8.2.

TABLE 8.2 Current Status of Organic Farming in India.

Sl no.	Particulars	Magnitude
	Organic Stakeholders under NPOP Certification Bodies	26
	Number of certified operators	4346
	Number of individual operators	2109
	Number of producers	971
	Number of processors	682
	Number of trader/exporters	693
	Number of grower group	2237
	Number of farmers (lakh)	5.91
	Number of wild operators	63
	Total number of wild collectors	143,610

Source: APEDA (2015).

8.4 SCOPE OF ORGANIC AGRICULTURE

A misconception still persists among masses that organic food is just a superficial concept and it is meant only for the rich people as the emphasis is being given more on their export. In general, organic food is priced over 25% more than conventional food in the domestic market. But now, several

Indians prefer organic food even on the higher prices in view of its health benefits. The consumption of organic food has increased which is evident from the fact that several organic food stores have come up in Indian big cities. Further, organic food has become an essential part of many retail food stores and restaurants. The consumption pattern of organic food in India is much different from what is consumed in the developed countries. Most of the consumers are not able to differentiate between natural and organic food, as they consider products that are labeled "Natural" as organic. Therefore, there is an urgent need to create awareness among the consumers regarding organic food.

India is bestowed with an immense potential to produce various varieties of organic food products due to its wider agroecological zones. Several parts of India has the inherited tradition of organic farming that holds promise for the organic producers to tap the untapped potential of the market, which has been growing slowly and steadily in the domestic market as well as the export market. India ranks 15th in terms of world's organic agricultural land according to the Year Book 2015. There was 5.71 million hectare area under organic certification during 2015–2016, inclusive of 26% cultivable area with 1.49 million hectare and rest 74% (4.22 million hectare) under forest and wild area for collection of minor forest produce.

India produced around 1.35 million MT (2015–2016) of certified organic products including all the food products including cereals and millets, oilseeds, pulses, cotton, sugarcane, medicinal plants, tea, coffee, vegetables, fruits, dry fruits, spices, and others. Among all the states, the largest area under organic certification has been covered by Madhya Pradesh followed by Himachal Pradesh and Rajasthan.

The total volume of 263,687 MT was exported during the year 2015–2016. Around USD 298 million was realized from the export of organic food. European Union, United States, Switzerland, Canada, Australia, New Zealand, Korea, South East Asian countries, Middle East, South Africa, and others. are the leading export markets of Indian organic products. Oilseeds (50%) have the distinction of being leader among the products exported followed by processed food products (25%), cereals and millets (17%), tea (2%), pulses (2%), spices (1%), dry fruits (1%), and others. Although share of India in the international food market is only about 1% but there is immense scope to explore the new arena.

Farmers across the globe are attracted toward the organic farming due to number of benefits over the modern agricultural practices. Basically, it is a farming system with supportive biological processes without the use of inorganic chemicals or biotechnological interventions like genetically modified organisms. As a result, many state-supported agencies, nongovernmental organizations, and individuals are practicing methodologies with organic methods of food production. Nowadays, stress is being laid on the appearance and the quantity of food rather than the intrinsic quality and vivacity of the food grains. Residues of chemicals and fertilizers that are being used at the time of crop production were found in the food. In addition to this, the reduced quality of food has led into an increase in various diseases, mainly various forms of cancer and diseases related to weakened body immunity.

To have a better, efficient, and holistic approach toward the global food security, the concept of biosafety should be implemented at different levels. The marketing of food is also a matter of consideration as the production of food is determined by the vagaries of climate as well as the various marketing constraints at the national and international levels.

8.5 BASIC CONCEPTS OF ORGANIC FARMING

Sustainable use of local resources to maintain the long-term fertility of soils, production of high nutritional quality in sufficient quantity, and minimizing the use of fossil energy for higher production are the basic principles of organic farming. Reduction in farm productivity in some parts of the country due to excessive and indiscriminate use of chemical inputs, decreased soil fertility, and a concern toward the environment has renewed the interest of Indian farmers in the adoption of organic farming.

Health of the soils is very important as it provides nutrients to the crops grown. But in case of nonavailability of the nutrient in the soil or unavailable to the plants due to being immobilized in the soil or through antagonistic affects from other ions, the plants are unable to uptake the plant nutrient and pass it up to the food chain. Unproductive and poor soils generally have poor status of the nutrients. Poor nutrient status in soils not only reduces the availability of food for consumption but also makes the crops less nutrient-rich. Generally, low-nutrient soils become more susceptible to diseases.

8.6 SOIL ORGANIC MATTER AND ITS IMPACT ON MICROBIAL POPULATION

Residues as a source of energy and nutrients are used by soil organisms, thereby releasing carbon dioxide, inorganic compounds, and recalcitrant molecules, contributing to the formation of soil humus. CRs after decomposition release about 55–70% of the carbon to the atmosphere as CO_2, 5–15% is incorporated into microbial biomass, and the remaining 15–40% carbon is partially stabilized in soil as new humus. The results of a specific management system may be visible on the soil carbon after a span of several years due to the very slow process. Numerous calculations have been made to assess the amount of residues required to maintain organic matter at a particular level. Experiments conducted reveal that soil regained its health with enhanced production and the reduced use of inputs as compared to the control after 3 years of switching over to natural cultivation. It is also evident that the yield of the crops can be increased or maintained through the application of recommended dose of N–P–K or N–P–K partially substituted with organics. Therefore, it is essential to use the recommended dose of fertilizer either through inorganic fertilizer alone or in combination with manures, CR, and green manuring (GM) for sustaining yield for the years to come. Apart from it, a quantitative assessment of the carbon sequestration potential of agricultural soils of the Indo-Gangetic Plain may be made under different management practices for different agroecological regions.

Incorporation of residues increased organic carbon as well as total nitrogen in comparison to burning or removal of straw. Incorporation of only wheat or rice straw did not influence organic carbon content significantly from the removal or burning of straw. The straw of rice was found better in enhancing total nitrogen content in soil than wheat straw. Studies have revealed that in rice–rice cropping system, incorporation of rice straw to supply 25% of the recommended N dose for rainy season crop for 6 years significantly increased organic carbon content from 0.98% in straw removal treatment to 1.29%. Burning or removal of CRs has almost similar effects on organic carbon content.

Bacterial and fungal activities were enhanced due to residue incorporation into the soil. Further, the population of protein decomposing microorganisms was increased during the early stages of incubation of rice straw under waterlogged conditions, which was followed by population increase in the case of cellulose-decomposing microorganisms. Sulfate-reducing microorganisms then increased after a lag phase. In a long-term study, it was observed that simultaneous application of rice straw and NH^{+}_{4}–N to the soil under upland conditions enhanced the population of denitrifiers but depressed the nitrogen fixation activity, significantly.

The crop yield under rice-based cropping systems can be increased by the intelligent management and utilization of CRs along with improved soil quality. Retaining CRs in the field is a viable option and its burning should be avoided at any cost. Chopping and spreading of straw in the field or the use of disc-type trash drills should be included as a strategy (Table 8.3).

TABLE 8.3 Organic Carbon and Biological Activity as Influenced by Different Tillage Practices at the End of Four Cropping Cycles in North-East India.

Treatment	Organic carbon (%)	SMBC (μg/g soil)	Earthworm population	Dehydrogenase activity (μg TPF/g/24 h)
Conventional tillage	1.47	91.3	60,000	29.5
Zero tillage	2.23	128.5	1,60,000	131.5
Double no-till	2.51	134.1	3,80,000	166.6
Minimum tillage	2.17	121.3	1,00,000	127.5
CD (p=0.05)	0.78	12.1	–	27.5

OC, organic carbon; SMBC, soil microbial biomass carbon.

Source: Ghosh et al. (2010).

Microbial immobilization of soil and nitrogenous fertilizer results in the decomposition of poor-quality residues with low N contents, high C: N ratios, and high lignin and polyphenol contents. For sustainable productive agricultural enterprise, nutrient cycling in the soil–plant ecosystem is an integral component. Although practices of fertilizer application have played a dominant role in the rice-based cropping systems during the past 30 years, CRs still play an important role in the cycling of nutrients. Incorporation of crop stubbles influences the soil environment, which, in turn, effects the microbial population and their activity in the soil and subsequent nutrient transformations.

8.7 CROP RESIDUE AND CARBON SEQUESTRATION

CRs which are generally treated as agricultural wastes and a problematic material can be easily used for improving soil organic matter dynamics and nutrient cycling and developing a much conducive environment for plant growth, if managed properly. A fundamental property of aggregates known as tensile strength (TS) is a measure of the resistance of the aggregates against breaking forces and therefore, it is highly sensitive for the management of the soil. High TS has been found very helpful for the maintenance of soil tilth and provides a stable traction for farm implements with limited intra-aggregate root growth. On the contrary, the friability of the soil having the tendency of a soil body to break into smaller pieces under an applied stress or load is considered as an important physical property of the farm soils, as it is desirable for better tillage and growing of plant. Although under the intensive rice–wheat cropping system, data on TS and friability under different combinations of organic and inorganic inputs and their interrelationships with management-induced changes in SOC are very meager. Water stability of aggregates reflects soil resistance against disintegration, while aggregates size reveals the effect of management on soil structural stability. For the development and stability of macroaggregates (>0.25 mm), organic binding agents are mostly amenable highlighting the important role of organic matter in aggregate stability. The intensive agriculture reduces SOC content which corresponds to a reduction in aggregate stability by changing its structure.

8.8 FARMING SYSTEM FOR SUSTAINABILITY

The food requirements for 2025 and 2050 are estimated at 590 and 904 lakh metric tonnes, respectively, for Uttar Pradesh. The skewed distribution of small holdings calls for a need to evolve appropriate farming modules that are profitable to small and marginal farmers. Monoculture of rice–wheat cropping system in major part of Uttar Pradesh is showing the syndrome of nonsustainability in terms of depleting groundwater and soil health as well as fertility. Keeping the above facts in view, the adoption of farming system would be the best approach for resource-poor small farmers (Table 8.4).

TABLE 8.4 Economics of Different Farming Systems.

Cropping/farming system	Area (0.2 ha)	Net income (Rs)	Area (0.4 ha)	Net income (Rs)
Conventional cropping system (Rice–wheat)	0.2	10,074	0.4	20,148
ICS (MRS + NBS)	0.2	21,282	0.4	42,564
Integrated farming system				
ICS+ FBS + dairy	0.1 + 0.096 + 0.004	71,383	0.2 + 0.192 + 0.008	1,42,766
ICS + poultry	0.196 + 0.004	49,530	0.392 + 0.008	99,060
ICS + mushroom	0.196 + 0.004	27,765	0.392 + 0.008	55,530
ICS + FBS + (dairy + poultry)	0.1 + 0.092 + (0.008)	1,00,057	0.2 + 0.184 + (0.016)	2,00,114

TABLE 8.4 *(Continued)*

Cropping/farming system	Area (0.2 ha)	Net income (Rs)	Area (0.4 ha)	Net income (Rs)
ICS + FBS + (dairy + mushroom)	0.1 + 0.092 + (0.008)	78,292	0.2 + 0.184 + (0.016)	1,56,584
ICS + (poultry + mushroom)	0.192 + (0.008)	56,014	0.384 + (0.016)	1,12,028
ICS + FBS + (dairy + poultry + mushroom)	0.1 + 0.088 + (0.012)	1,06,966	0.2 + 0.176 + (0.024)	2,13,932

Source: Bohra and Singh (2012).

8.9 NUTRITIONAL VALUE OF DIFFERENT ORGANIC SOURCES

On the basis of the different assumptions made, the estimated nutritional value from different sources of organic material in Uttar Pradesh is summarized in Table 8.5.

TABLE 8.5 Nutritional Value of Different Organic Sources in Uttar Pradesh.

Components	Unit	Nutrient content (%)	Major nutrients available (000' tonnes)				
			N	P	K	S	OC
FYM (414.65 lakh Bovine @1.5 tonnes/animal/year)	622 lakh tonnes	NPK (1.5:1.0:0.54)	932.95	621.96	335.86		
Biofertilizers	50 lakh packets	25–30 kg/ha	50.00				
NADEP units	59,683 units	NPK (1.2:0.83:1.83)	8.59	5.94	9.17		
Vermicompost	59,498 units	NPK (1.1:0.2:0.33)	2.62	0.48	0.79		
Green manuring	7.98 lakh ha	90 kg N/ha	71.82				
Pulses	27 lakh ha	25 kg N/ha	67.50				
Crop residues	312 lakh tonnes	NPK (0.62:0.21:1.26)	160.14	54.60	3.90		

TABLE 8.5 *(Continued)*

Components	Unit	Nutrient content (%)	Major nutrients available (000' tonnes)				
			N	P	K	S	OC
Sugar industry							
Begasse	52.23 lakh tonnes	NP and OC (0.6:0.3:25)	31.33	15.67			1302.00
Pressmud	16.37 lakh tonnes	NPS and OC (1.25:2.0:2.25:2.5)	20.57	32.92		3703.50	41.15
Total			1345.52	731.57	349.71	3703.50	1343.15

It is evident from the above data that if the potential of all available organic sources is realized, Uttar Pradesh will be able to economize the use of large amount of nitrogenous fertilizer along with surplus in potash and phosphoric fertilizers.

8.10 INITIATIVES TAKEN FOR PROMOTION OF ORGANIC FARMING

- *Paramparagat Krishi Vikas Yojana* started in 2014 to promote organic farming with an allocation of Rs 597 crore being implemented at the national level and 9186 clusters have been established against the target of 10,000.
- *Pandit Deendayal Unnat Krishi Shiksha Yojana* started in 2016 by way of 130 training programs on organic farming/natural farming and cow-based economy are playing a vital role in improving the soil fertility and soil health in 32 state agricultural universities.
- Prohibition of the burning of CRs for improving the soil health and protecting the environment.
- Promote the use of CRs, vermi and NADEP compost, GM, crop rotation involving legumes to improve physicochemical properties of soils.
- Modules of organic farming for different situations are being developed at Regional Agriculture Testing and Demonstration Stations (RATDS) with the support of National Centre for Organic Farming.
- A comprehensive program on soil health with the emphasis on GM, NADEP, vermicompost, production of biofertilizers and biopesticides are being implemented during the five-year plan.

8.11 STRATEGIC RECOMMENDATIONS

- Organic farming is a market demand-driven agriculture, aimed to cater to the foreign export and the affluent section of the Indian society. Therefore, in order to make a dent in the export market, we have to develop high-tech organic technology with strict quality control, meeting international quality parameters prescribed for organic produce.
- Increased mechanization has reduced the availability of organic material, hence, there is a need to popularize farming system approach and utilize postharvest residue to the fullest extent. However, to accomplish this objective, there is a need to develop feasible technologies for in situ rapid decomposition of green manures, agricultural wastes, and on-farm residues.
- There is a need to prepare region-specific resource inventory including animal wealth, farm residue by-products, animal wealth and their competitive uses, nonconventional nutrient sources of organic and biological origin, and others for development of crop and area specific rational technology packages of organic farming. Packages should be documented in local languages after validation of such proven technologies.
- Organic farming may be practiced in crops, commodities, and regions where the state has comparative advantage. Priority should be low-volume high-value crops, newly reclaimed/improved wastelands, and areas where consumption of chemical fertilizers and pesticide is almost negligible.
- To ensure food security, practice organic farming in phased manner, so that over a period of 20–25 years, major cropped area is brought under organic farming.
- Infrastructure facilities for commercial production of biofertilizer and biopesticides, postharvesting management and marketing strategies should be developed at local level.
- International certifying agencies should be encouraged to open certification centers in different parts of the state for offering various services. Further, to attract farmers toward organic farming, cost of certification needs to be subsidized by the state government.
- Mechanism should be developed to provide the information on export opportunities, guidance on international quality standards, training of human resources, and marketing at local level.

- Identify and develop crop varieties that can be cultivated organically under various agroecological conditions.
- How the N cycle will be affected by climate change, in turn, how that will affect C storage and availability of nitrogen and phosphorus in soils and response of plants to elevated atmospheric CO_2 under different agroecological conditions need to be studied in-depth.

KEYWORDS

- **organic farming**
- **climate change**
- **soil organic matter**
- **crop residue**
- **farming system**
- **promotion**

REFERENCES

Birthal, P. S.; Khan, M. J.; Negi, D. S.; Agarwal, S. Impact of Climate Change on Yields of Major Food Crops in India: Implications for Food Security. *Agric. Econ. Res. Rev.* **2014,** *27* (2), 145–155.

Bohra, J. S.; Singh, K. *A Study Report on "Developing Farming Model for Small and Marginal Farmers"* Submitted to UPCAR, 2012.

Boopathi, M. et al. Lessons Learnt from Permanent Manurial Experiment of Tamil Nadu Agricultural University, Coimbatore. Presented at National Seminar on "Soil Health Improvement for Enhancing Crop Production" held at the TNAU, Coimbatore, March 17–18, 2011.

Ghosh, P. K. et al. Conservation Agriculture towards Achieving Food Security in North East India. *Curr. Sci.* **2010,** *99*, 915–921.

Rama Rao, C. A.; Raju, B. M. K.; Subba Rao, A. V. M.; Rao, K. V.; Rao, V. U. M.; Ramachandran, K.; Venkateswarlu, B.; Sikka, A. K. *Atlas on Vulnerability of Indian Agriculture to Climate Chance*; Central Research Institute for Dryland Agriculture: Hyderabad, 2013; 116 p.

CHAPTER 9

Managing Soil and Water Through Organic Farming Methods for Sustainable Agriculture Production

MUNISH KUMAR

Soil Conservation & Water Management/Director, Administration & Monitoring C.S. Azad University of Agriculture and Technology, Kanpur 208002, Uttar Pradesh, India
E-mail: munish.csa@gmail.com

ABSTRACT

Depleting soil organic carbon status, decreasing soil fertility and reduced factor productivity, increase in the cost of production and deteriorating environmental quality are major issues of concern related to food production. These resources which performs the critical functions of entire life support are undergoing unabated degradation of different types of deterioration due to pollution and nutritional disorders. Vast fertile lands in the canal irrigated areas are affected by soil salinity, sodicity, and water logging thereby rendering it unsustainable. There is practically no possibility of horizontal expansion of fertile arable land. Under the above paradoxical situation, Indian agricultural sector's target of doubling farmer's income by 2022 without any adverse effect on soil health seems a formidable task.

Organic Crop Production Management: Focus on India, with Global Implications, D. P. Singh, PhD, H. G. Prakash, PhD, M. Swapna, PhD, & S. Solomon, PhD (Eds.)

9.1 INTRODUCTION

India is endowed with a vast and rich diversity of natural resources, particularly soil, water, and climate agrobiodiversity. To realize the optimum potential of the agricultural production system on a sustained basis, efficient management of natural resources is of paramount importance. Misuse and abuse of the technologies led to overexploitation of natural resources beyond their intrinsic capacity in most part of the country. Consequently, a constant decline in the factor productivity has been noticed in recent past years, which is an indicator of unsustainable agricultural production system. The growth rates for production and productivity of major crops not only declined, but also started showing negative trends. India with a geographical area of 328.72 million ha presently supports 17.6% of global population on merely 2.4% world's land area and 4.2% freshwater resources. Moreover, the population is expected to reach 1.4 billion by 2025, requiring about 310 Mt. of food grains (compared with about 200–210 Mt at present) with a projected linear decrease in per capita land availability, from 0.34 ha in 1950–1951 to 0.12 ha by the year 2025. The land-use statistics indicated a decline in net cultivated area (133 M ha) during 2002–2003 after a fairly constant figure of 140–142 M ha for nearly three decades. Agricultural production systems have been developed to meet the need of food, fiber, fuel, and feed of the increasing human population at the cost of natural ecosystem. The terms soil quality and soil health tend to be used interchangeably. Conceptually, the intrinsic health or quality of soil can be viewed simply as its capacity to function. More explicitly, soil health could be defined as "the capacity of soil to function within ecosystem boundaries to sustain biological productivity, maintain environmental quality, and promote plant and animal health" (Doran et al., 1996).

9.2 SOIL RESOURCES

The percentage of soil resources of our country as compared with the world is only 2.4%, and this small fraction fulfills the need of 17.6% of the human population and 16% of livestock population of the world. In spite of small area of soil resource, India is able to produce a large number of varieties of cereal, pulse, and oilseed crops, fruits, and vegetables owing to its diverse range of agroclimatic conditions, topography, and different types of soil. The soils of the planet have been categorized into 12 suborders, out of which 9 soil orders are present in our country, India; however, some soils have several constraints to meet the challenges and requirements of the 21st century.

Soil-related constraints are mainly severe in arid, semiarid, and hilly regions. Important constraints are low soil fertility and nutrients depletion, multinutrient deficiencies, physical degradation, and accelerated soil erosion. Apart from inherent constraints, there are several human-induced constraints, which are predominantly occurring in intensively cropped areas (Table 9.1).

TABLE 9.1 Soil Resources of India and Their Constraints.

Soil order	Land area (M ha)	% of the total area	Soil related constraints
Inceptisols	130.37	39.8	Erosion, nutrient imbalance, low soil organic matter
Entisols	92.13	28.0	Erosion, nutrient depletion, low soil organic matter
Vertisols	27.96	8.5	Massive structure, poor tilth, drought stress, water erosion
Alfisols	44.29	13.5	Weak soil structure, crusting, compaction, water erosion
Aridisols	14.07	4.3	Drought stress, nutrient depletion, wind erosion, desertification, secondary salinization
Ultisols	8.25	2.5	Erosion by water, nutrient imbalance, acidification, P fixation
Mollisols	1.32	0.4	–
Histosols	0.002	–	High organic matter
Others	9.67	2.95	–
Total	328.06	100.0	–

9.3 LAND DEGRADATION

Land degradation imposes a major threat to soil health and productivity. The imbalance between living beings and soil resources has resulted in alarming conditions for our ecosystem like different types of land degradation, environmental pollution, low rate of crop productivity, and sustainability. Other major problem arises due to such imbalance are deforestation, uses of nonagricultural land, environmental retrogression, and misplaced hydrology. The incorrect practices of hydrology have resulted in water logging, salinity, sodicity, decline in water table conditions, unguided rainwater uses, and low efficiency of other water uses. The use of agrochemicals like fertilizers, pesticide, and others cumulatively exaggerated the problem of land degradation further (Sarkar, 2005). The extent of various types of land degradation in land areas of India is shown in Table 9.2.

TABLE 9.2 Extent of Various Land Degradation Problems in India.

Kinds of degradation	Million hectares	Percent of total
Water erosion	149.037	45.3
Wind erosion	13.489	4.1
Chemical deterioration	13.818	4.2
Physical deterioration	11.515	3.5
Not fit for agriculture	18.095	5.5
Unaffected area	123.046	37.4

Source: Sehgal and Abrol (1994).

The latest reports by the National Bureau of Soil Survey and Land Use Planning, Nagpur, using Global Assessment of Soil Degradation (GLASOD) guidelines, indicate 187.8 M ha of land degraded by various degradation processes. The loss of fertile topsoil by water erosion is estimated to be about 5334 Mt/year, out of which 29% is lost permanently to sea, 10% gets deposited in irrigation reservoirs, and 59% is deposited as alluvium at different places.

The annual water erosion rate varies from <5 t/ha in arid regions of Rajasthan to more than 80 t/ha in Shiwalik hills (Singh et al., 1992). Wind erosion is a major problem of concern in arid and dry semiarid regions including the states of Rajasthan, Haryana, Gujarat, and Punjab. Here, it is important to mention that it takes from 130 to 400 years for nature to create a layer of 1 cm topsoil.

Chemical degradation by salinization or alkalinization is reported to extend to 10.1 M ha and the problem is increasing at an alarming rate in the canal irrigated areas. The industrial effluents, sewage, sludge, and agrochemical residues, either in situ or through the groundwater sources, are also degrading the soils. About 95.65 Mha of cultivated land of India suffers from physical degradation, out of which shallow soils cover 25.02 Mha, hard soils cover 20.35 Mha, highly permeable soils cover 10.77 Mha. Again, soils with high mechanical impedance at shallow depth cover 10.63 Mha, 9.43 M ha is covered by slowly permeable soils and area under other physical constraints is 9.45 M ha (Scherr and Yadav, 1996).

The impact and damage by land degradation have not been properly evaluated. Only a little bit of information about loss of productivity owing to different extents of soil degradation is available. Velayutham and Bhattacharyya (2000) have tried to relate the threatened condition of soil degradation with the loss of soil productivity (Table 9.3).

TABLE 9.3 Expected Loss of Productivity Due to Water Erosion in Different Soils.

Soil erosion class	Soil loss (t/ha)	Loss in productivity (%)		
		Alluvial soils (Inceptisols)	Black soils (Vertisols)	Red soils (Alfisols)
Nil to very slight	<5	Nil	<5	<10
Slight	5–10	<5	5–10	10–25
Moderate	10–20	5–10	10–25	25–50
Strong	20–40	10–25	25–50	>50
Severe	>40	25–30	>50	N.A.

TABLE 9.4 Highly Degraded Areas in India Due to Various Processes.

Degradation process	Severe to very severe degradation
Wind erosion	Rajasthan, Haryana, Gujarat, and Punjab
Water erosion	H.P., U.P., M.P. A.P., Rajasthan, Maharashtra
Water logging	U.P. Assam, Rajasthan, A.P., and Gujarat
Salt infestation	U.P, A.P. Tamil Nadu, Gujarat, Rajasthan, and West Bengal

9.4 SOIL ORGANIC CARBON

Soil organic matter (SOM) imparts desirable, physical environment to soils by favorably affecting soil texture expressed through soil porosity, aggregation, bulk density, and soil water storage. It also affects chemical properties of soils and nutrients availability, CEC, retention, and mobilization of metals in a significant manner. SOM can also be defined as a mixture of various biological components like microbiota in different proportions and un-decomposed plant materials. The final products of organic matter decomposition in soil accumulate as humus and disappear as CO_2. As of today, more terrestrial organic matter has been lost in the form of CO_2 than it has been sequestered in soils. This is evidenced by 28% increase (change from 280 to 365 ppm) in CO_2 load of earth's atmosphere over the years.

The global pool of SOM is estimated to contain about 1500 Pg (1 Pg = 10^{15} *g* = 1 billion tonnes) of carbon at 1 M depth (Batjes, 1997). The geological C-pool comprises 5000 with 4000 Pg C as coal, 500 Pg C as gas, and 500 Pg C as oil (Lal and Kimble, 2000). Total organic pool in soils of India (SIC) is estimated at 21 Pg to 30 cm depth and 63 Pg to 150 cm depth, which represents 2.2% of the world pool for 1 m depth.

SIC sequestration (21.8–25.6 Tg C/year) and reduction in erosion induced emission of C (4.32–7.2 Tg C/year) (Lal, 2004). The cooler and humid climate is most conductive for soils to be enriched with soil organic carbon (SOC) as in case of mollisols (Bhattacharyya and Pal, 2003). The status of different organic pool in SIC was compared with the global pool of SOM as shown in Table 9.5. Technological options that have been found to be efficient for soil C sequestration in India include green manuring, mulch farming/conservation, tillage, afforestation/agroforestry, grazing management/ley-farming, integrated nutrient management, manuring, and choice of cropping system (Lal, 2004). The quantity of residues from the principal grain-producing crops in India is estimated at about 340 Mt/year, of which wheat residue contributes to 27% and rice is of 51%. (Lal and Kimble, 2000). Approximately, 200 Mt of crop residue can be replenished to the soil annually for augmenting SOC pool and improving soil quality. The quantities of various kinds of organic wastes available in different crops cultivated in India (i.e., Nutrient Potential of Crop Residues) are given in Table 9.6 (Tandon, 1997).

TABLE 9.5 Organic Carbon Pool in Soils of India and the World (Pg)

Soil order	**India***		**World****	
	0–30 cm	**0–100 cm**	**0–25 cm**	**0–100 cm**
Alfisols	3.10	8.58	73	136
Andisols	–	–	38	69
Aridsols	0.77	1.28	57	110
Entisols	0.65	1.81	37	106
Histosols	–	–	26	390
Inceptisols	2.20	5.54	162	267
Mollisols	0.12	0.27	41	72
Oxisols	–	–	88	150
Spodosols	–	–	39	98
Ultisols	0.23	0.47	74	101
Vertisols	2.62	7.09	17	38
Total	9.77	25.04	652	1537

- The amount and quality of organic carbon are crucial factor for determining the properties of soils (physical, chemical, and biological) and its productive capacity.

- Owing to wide variations in climatic condition, topographic situation, vegetative cover, land use and management practices, the quantity, and nature of organic carbon in the diverse soils vary widely.
- Climate factors play a dominant role in deciding the concentration of SOC, as the SOC content is higher in temperate condition than the tropical environment. The cooler and humid to per humid climate is most conductive for soils to be enriched with SOC as in the case of Mollisols of subHimalayan region.
- The quantity of residue from the principal grain-producing crops in India is estimated at about 340 Mt/year, of which, wheat residue constitutes about 27% and that of rice about 51%.
- Approximately, 200 Mt of crop residue can be returned to the soil annually for augmenting soil SOC pool and improve the soil quality.

TABLE 9.6 Nutrient Potential of Crop Residues.

Sr. no.	Crop residue	Nutrient (%)				Tonne/tonne residue
		N	P_2O_5	K_2O	Total	
1	Rice	0.61	0.18	1.38	2.17	0.0217
2	Wheat	0.48	0.16	1.18	1.82	0.0182
3	Sorghum	0.52	0.23	1.34	2.09	0.0209
4	Maize	0.52	0.18	1.35	2.05	0.0205
5	Pearl millet	045	0.16	1.14	1.75	0.0175
6	Barley	0.52	0.18	1.30	2.00	0.0200
7	Finger millet	1.00	0.20	1.00	2.20	0.0200
8	Pulses	1.29	0.36	1.64	3.29	0.0329
9	Oilseeds	0.80	0.21	0.93	1.94	0.0194
10	Groundnut	1.60	0.23	1.37	3.20	0.0320
11	Sugarcane	0.40	0.18	1.28	1.86	0.0186
12	Potato tuber	0.52	0.21	1.05	1.79	0.0179

- Crop residue addition improves the soil productivity, nutrient supply, microbial and enzymatic activity, SOM, and soil physical properties.
- Crop residues are rich source of potassium. Regular return of residues in soil contributes to the proliferation of soil nutrient pool over a period of time.
- Addition of crop residues is a significant strategy to maintain SOC.
- The quality of the crop residues application varies from 2.5 to 5 t/ha.

- The total amount of nitrogen released depends upon the nature of plant residues and it degradation rate.
- The residue management under Integrated Nutrient Management System has considerable impact on soil microbial biomass carbon if residue is incorporated with use of green manure.

TABLE 9.7 The Potential of Carbon Sequestration in Soils of India.

Region	Area (M ha)	C sequestration potential (Tg/year)
Arid	52.0	0.67–1.34
Semiarid	116.4	2.33–4.66
Subhumid	86.4	3.46–5.18
Subhumid/humid	33.3	2.06–2.72
Perhumid	20.2	2.42–3.03
Subhumid/semiarid	8.5	0.34–0.51
Humid/perhumid	11.9	1.43–1.79
Total	328.7	12.71–19.23
Secondary carbonates	328.7	21.78–25.6
Erosion control	–	4.80–7.20
Total	328.7	39.29–52.03

- In India, the capacity of carbon sequestration in soils ranges from 39.3 to 52 Tg C/year, which comprises restoration of degraded soils (7.2–9.4 Tg C/year), SIC sequestration (21.8–25.6 Tg C/year), and reduction in erosion induced emission of C (4.32–7.2 Tg C/year).
- Technological options that have been found to be efficient for soil C sequestration in India include green manuring mulch farming/conservation tillage, afforestation/agroforestry, grazing management/ley farming, INP, manuring, and choice of cropping system.

9.5 SOIL HEALTH

Soil health is an assessment of ability of a soil to meet its range of ecosystem functions as appropriate to its environment. The term "soil health" is used to assess the potential of a soil to (1) maintain plant and animal productivity and diversity in sustainable manner (2) enhance the quality of water and air, and (3) support human health and habitation.

The soil health basically depends on the integration of its all physical, chemical, and biological properties and function. A healthy soil will be balanced for all physical, chemical, and biological counterparts. To interpret and measure the health of the soil, various properties, and functions should be assessed through meaningful indicators.

9.6 INDICATORS USED FOR DETERMINING THE SOIL HEALTH

The healthy soil means that each component chemical, physical, and biological are in appropriate amount so that long-term sustainable agricultural productivity is possible without or minimally affecting environmental scenario. Thus, soil health reflects impact of overall parameters that are required for soil functionality (Arias et al. 2005). A healthy soil maintains a strong and diverse microbiota, balanced nutrients, and proper pH. Assessment of soil health can be estimated by measuring values of indicators affecting all three components of soils. These indicators are listed in Table 9.8.

TABLE 9.8 Physical, Chemical, and Biological Indicators Used for Determining Soil Health.

Indicators	Measurement
Physical	Bulk density, infiltration rates, and hydraulic conductivity, water content, water-holding capacity, water release curve
Chemical	PH, EC, CEC, Organic C, macro/micronutrient analysis
Biological	
Microbial biomass	Direct microscopic counts chloroform fumigation Substrate-induced respiration (SIR) CO_2 production, microbial quotient, fungal estimation, PLFA (phospholipid fatty acids)
Microbial activity	Bacterial DNA synthesis, bacterial protein synthesis, CO_2 production
Carbon cycling	Soil respiration, metabolic quotient (q CO_2)
Nitrogen cycling	N-mineralization, nitrification, denitrification N-fixation
Biodiversity and microbial resilience	Direct counts, selective isolation plating, carbon and nitrogen utilization pattern, extracellular enzyme pattern.
Bioavailability of contaminants	Plasmid-containing bacteria, antibiotic-resistant bacteria.

9.7 ROLE OF RHIZOSPHERE IN ENHANCING THE SOIL HEALTH

Rhizosphere is the soil zone immediately surrounding the plants roots, which can be modified by increasing the number of organisms (e.g., *Rhizobia*) that reside in symbiotic or nonsymbiotic association with plant root. It is also defined as a nutrient-rich zone near roots where microbial growth is stimulated by root exudates (the rhizosphere effect). Root-induced changes in rhizosphere effect nutrient availability. These root-induced changes are related to the nutritional status of plants and thus help in improving soil health.

9.8 STRATEGIES FOR MANAGING SOIL HEALTH FOR SUSTAINABLE AGRICULTURE

Integrated Plant Nutrients Management

The basic concept underlying Integrated Plant Nutrients Management (IPNM) is the maintenance of soil fertility, sustainable agricultural production, and improving farmers' profitability through judicious and efficient use of fertilizers, organic manures, crop residues, biofertilizers, suitable agrochemical practices, and others. Soil testing is an important part of nutrient management. Integrated nutrient supply, which means the addition of chemical fertilizers along with organic matters like green manure, FYM, crop residue, and biofertilizers, help not only fulfilling the gap between nutrients displacement and replenishment but also ensure the balanced proportion of nutrients that resulted in enhanced nutrients response efficiency, which maximize the crop productivity and of desired quality.

Experiments conducted on long-term use of fertilizer in different agro-ecological zones of India clearly indicated that integrated use of chemical fertilizers and FYM, could lead to not only higher crop yield but also improved soil fertility when compared with 100% of recommended dose of NPK (Swarup, 1998).

TABLE 9.9 Effect of Integrated Nutrient Management on Grain Yield (q/ha) of Rice Grown in a Sequence (Average of Each 3 Years) at Faizabad, Uttar Pradesh.

Treatment		**Grain Yield (q/ha)**				
Rice (*Kharif*)	**Wheat (*Rabi*)**	**1984–1987**	**1987–1990**	**1990–1993**	**1993–1996**	**1996–1999**
Control ($N_0P_0K_0$)	Control ($N_0P_0K_0$)	20.0	18.6	15.9	12.9	12.0
50% NPK	50% NPK	29.8	35.0	32.7	28.1	25.3
50% NPK	100% NPK	30.2	37.9	31.7	32.1	28.0
75% NPK	75% NPK	36.7	44.6	39.7	35.7	32.4
100% NPK	100% NPK	39.4	47.6	43.7	45.7	39.6
50% NPK+50% N (FYM)	100% NPK	29.9	48.4	44.7	50.6	42.5
75% NPK+25% N (FYM)	75% NPK	37.3	51.1	44.8	43.3	43.7
50% NPK+50% N (WCS)	100% NPK	35.1	44.2	41.9	41.5	34.7
75% NPK+25% N (WCS)	75% NPK	32.6	46.9	41.3	39.8	37.1
50% NPK+50% N (GM)	100% NPK	32.3	46.7	43.4	45.2	38.9
75% NPK+25% N (GM)	75% NPK	34.8	47.8	45.2	40.5	39.4
Conventional ($N_{60}P_{13}K_{16}$)	Conventional ($N_{80}P_{13}K_{16}$)	31.6	33.0	34.7	32.4	25.6

FYM, farmyard manure; WCS, wheat cut straw, GM, green manure.

100% NPK; 120 kg N, 60 kg P_2O_5 and 40 kg K_2O/ha

- Data presented in Table 9.9 highlights that when the recommended dose of N applied in rice through combinatorial strategy of FYM or *Sesbania* green manure and fertilizers resulted in equal or more yield as compared with supplementation of 100% NPK fertilizers alone. When comparison was made between different organic sources, augmentation of green manure of *Sesbania* and FYM showed significantly superior response in term of yield as compared with wheat cut straw (WCS) in the sodic soils.
- In the beginning stage, organic sources applied to rice did not show a positive residual effect on succeeding wheat crop, but after 3 years positive residual response to FYM was observed on yield of wheat crop.
- *Sesbania* green manuring also registered its positive response on wheat yield but the magnitude was lower than that of FYM.

TABLE 9.10 Long-Term Impact of INM on Bulk Density, Infiltration Rate, Organic Carbon Content, Nitrogen, and Phosphorus Status of Soil at 2002–2003.

Tr Kharif	Rabi	Bulk density (Mg/m³)	Infiltration rate (cm/hr)	Organic carbon (g/kg)	Nitrogen (kg/ha)	Phosphorus (kg/ha)
T_1 Control	Control	1.56	0.85	4.80	156	4.89
T_2 50% NPK	50% NPK	1.53	1.04	5.51	221	18.35
T_3 50% NPK	100% NPK	1.53	1.12	5.68	230	2055
T_4 75% NPK	75% NPK	1.53	1.08	5.70	240	24.05
T_5 100% NPK (80:60:40 kg/ha)	100% NPK (100:50:30 kg/ha)	1.50	1.30	6.04	253	37.63
T_6 0% NPK+50% N (FYM)	100% NPK	1.45	1.41	6.62	261	32.20
T_7 5% NPK+25% N (FYM)	75% NPK	1.46	1.29	6.61	250	31.93
T_8 50% NPK+50% N (RS)	100% NPK	1.46	1.22	6.90	260	31.36
T_9 75% NPK+25% N (RS)	75% NPK	1.46	1.16	6.94	254	31.85
T_{10} 50% NPK+25% N (GM)	100% NPK	1.43	1.67	7.03	263	31.56
T_{11} 75% NPK+25% N (GM)	75% NPK	1.45	1.58	6.91	256	29.85
T_{12} Farmer's practice (50:30:20 NPK kg/ha)	Farmer's practice (60:40:20 NPK kg/ha)	1.51	1.05	5.51	228	22.36
CD (P = 0.05)	–	0.02	0.07	0.09	7.50	1.84
SE m±	–	0.07	0.20	0.27	22.0	5.48

Source: Bajpai et al. (2006)

Highlights of Table 9.10:

- Long-term experiment revealed that organic carbon content increased from 22.9 to 27.4% over the initial level (5.1 g/kg) and from 27.5 to 31.2% over control from continuous use of FYM/rice straw or green manure over the years.

- Findings of study clearly showed low organic carbon content of soil in control and in case of treatments, where organic manures had not supplied or supplied with half of the recommended dose of fertilizers.
- The organic carbon content of surface soil increased significantly with the addition of organic materials with chemical fertilizers as well as with 100% fertilizer treatment over control.
- The highest (7.03 g/kg) organic carbon content was observed when soil was treated with 50% N through GM and remaining 50% through chemical fertilizers.
- The highest available N status was observed with 50% N through *Sesbania aculeate* and 50% through fertilizer.
- Incorporation of 50% N through organic manure and 100% fertilizer dose (T_5) resulted in significant higher amount of available P content than control as well as over T_2, T_3 and T_4.
- Significant reduction of bulk density (1.43 Mg/m^3) was observed in case of application of 50% N through green manure + 50% N through fertilizer as compared to other treatments.
- Infiltration rate raised in a significant manner when an integrated approach for application of green manure or FYM with chemical fertilizers was applied. Integrated nutrient management during rice season had marked profound effect on the infiltration rate.

9.9 BALANCED USE OF FERTILIZERS

The average $N:P_2O_5:K_2O$ consumption ratio was 6.9:2.7:1 in 2003–2004, which is considered quite wide compared with the projected average ideal ratio 4:2:1. The loss of nutrients occurs through leaching, volatilization, run off, and fixation. Projection related to plant nutrients (NPK) addition and removal are shown in Table 9.11.

TABLE 9.11 Projected Plant Nutrient (NPK) Addition and Removal in India.

Crop	Period	Control yield (kg/ha)	Response (kg/ha)		
			N (120)	P_2O_5 (80)	K_2O (40)
Rice	1977–1978	1008	2905	500	50
	1989–1990	820	2642	925	231
	Change	−188	1263	+425	+181
Wheat	1977–1978	833	2625	617	25
	1989–1990	602	2141	1169	398
	Change	−231	−484	+552	+373

Source: Project Directorate of Cropping Systems Research, Meerut, UP, India.

9.10 ORGANIC APPROACH

Most of the Indian soils require N-fertilization for optimum crop productivity. This costly plant nutrient (N) can be supplied through the introduction of legumes in crop rotation though biological N-fixation and harvesting leftover effect on succeeding cereal crop. Results of long-term experiments indicate that the addition of 10–15 t/ha FYM annually along with 100% advocated dose of NPK, can realize additional yields in different crops when compared with application of chemical fertilization alone at the recommended rate (Swarup, 1998). Application of NPK alone or on substitution with organic sources could save fertilizers and sustain soil health. Sharma et al. (2000) observed an increase of available P, Organic C, and total N in soils of IARI as a result of incorporation of *Sesbania* or mungbean residue during summer in rice–wheat cropping system.

TABLE 9.12 Effect of FYM, Biofertilizer, Vermicompost, and Inorganic Fertilizers on Yield of Rice.

Treatment	Yield of rice (t/ha)	
	Grains	Straw
1. N through FYM 50 kg + *Azospirillum* 2 kg/ha	5.72	7.76
2. N through FYM 50 kg + *Azos.*(2 kg)+Phosphobacteria 2 kg/ha	5.95	7.99
3. N through bioslurry 50 kg + *Azospirillum* 2 kg/ha	5.98	7.89
4. N through bioslurry 50 kg + *Azos.*+Phospho 2 kg each/ha	6.18	8.17
5. N through vermicompost 50 kg +*Azos.* 2 kg/ha	6.07	8.00
6. N through vermicompost 50 kg + *Azos.*+Phospho 2 kg each/ha	6.25	8.21
7. Fertilizer N at 100 kg/ha	5.57	7.61
C.D ($P = 0.05$)	0.13	0.35

Source: Jayabal and Kuppuswamy (1999).

From Table 9.12, it can be concluded that:

- Conjoint supplementation of organic manures and biofertilizers enhanced rice crop yield significantly as compared with the supply of inorganic fertilizers alone.
- The use of *Azospirillum* along with bioslurry increased the productivity of rice in comparison to FYM (50 kg N) and *Azospirillum*.
- The highest yield of rice grain was obtained when N treatment through vermicompost (50 kg N) + *Azotobacter* + Phosphobacteria 2 kg each.

9.11 MICROBIOLOGICAL APPROACH

Biofertilizers enhance soil fertility as well as crop productivity by fixing atmospheric N, mobilizing sparingly soluble P, and by facilitating the release of nutrients through decomposition of crop residues. *Azotobacter*, free-living heterotrophic N-fixing bacteria, not only provides N but also produces a variety of growth-promoting substances. *Azospirilla* are mesophylic and have been reported to fix N in association with crops grown on soils with acidic to alkaline range. The high N-fixing capacity, low energy requirements, abundant establishment in the roots of cereals, and tolerance to high soil temperature (30°C–40°C) make *Azospirilla* suitable N fixers under tropical condition. It is claimed that blue green algae (BGA) in rice field contribute 20–30 kg N/ ha per season and may increase grain yields by 5–14%. As our soils are poor in organic matter, response of P-solubilizers has not been very promising. However, under soil conditions when P availability is decreased due to fixation and limited mobility, inoculation with filamentous P-mobilizers like VA mycorrhizae could be an advantage. Field trials conducted in high P-fixing vertisols have shown that using VA mycorrhizae may save 15 kg fertilizers P/ha. Further, there is a carryover of Vesicular arbuscular mycorrhizae inoculums through rice stubbles to the succeeding wheat crop. Tables 9.13–9.17 below describe application method of biofertilizers on various crops, profiles of different N-fixer biofertilizers, essential functions performed by the different members of the Soil Biota, various management practices used by farmers in India that may influence soil fertility through manipulation of biological processes and, state-wise production and consumption of biofertilizers in India.

TABLE 9.13 Application Method of Biofertilizers.

Biofertilizer	Crops	Method	Quantity	Time
Rhizobium	Large seeded crops	Seed treatment	1.0 kg/ha	At sowing
Azotobacter Asospirillum	Small seeded crops	Seed treatment	0.5 kg/ha	At sowing
PSB	Sugarcane, potato, vegetable	Set, tubers, buds/or soil	3.5 kg/ha	Sowing time
BGA	Rice	Broadcasting	1 kg /ha	After transplantation
Azolla	Rice	Dual crop	1 tonne	After transplantation
Vesicular arbuscular mycorrhizae	All crops	Soil	3–5 kg/ ha 20 g/tree	At sowing

Source: Tewatia et al. (2007).

TABLE 9.14 A Profile of Different Nitrogen Fixation Biofertilizers.

Biofertilizer	Function/contribution	Limitation	Use for crops
Rhizobium (symbiotic)	Fixation of 50–100 kg N/ha 10–35% increase in yield Leaves residual nitrogen	Fixation only with legumes Visible effect not reflected in traditional area Needs optimum P and Mo	Pulse legumes like chickpea, red gram, pea, lentil, black gram etc. oilseed, legumes like soybean and groundnut, forage legumes like clover and Lucerne, Tree legumes like *Leucaena*
Azotobacter (nonsymbiotic) and *Azospirillum* (associative)	Fixation of 20–25 kg N/ha 10–15% increase in yield Production of growth promoting substances	Demands high organic matter	Wheat, maize, cotton, sorghum, sugarcane, pearl millet, rice, vegetables, and several other crops
Blue Green Algae or *cyanobacteria* (phototropic)	Fixation of 20–30 kg N/ha 10–15% increase in yield Production of growth promoting substances	Effective only in submerged rice Demands bright sunlight	Flooded rice
Azolla (symbiotic)	Fixation of 30–100 kg N/ha Yield increase 10–25%	Survival difficult at high temperature, great demand for phosphorus	Only for flooded rice

TABLE 9.15 Essential Functions Performed by the Different Members of the Soil Biota.

Functions	Organisms involved
► Maintenance of soil structure	Earthworms, arthropods, soil-born fungi, mycorrhizae, plant roots, and some other microorganisms
► Regulation of soil hydrological processes	Most invertebrates like earthworms, arthropods, and plant roots
► Gas exchanges and carbon sequestration	Mostly microorganisms and plant roots, some C protected in large compact biogenic invertebrate aggregates.
► Soil detoxification	Mostly microorganisms
► Decomposition of organic matter	Various saprophytic and litter-feeding invertebrates (detrivores), fungi, bacteria, actinomycetes, and other microorganisms.
► Suppression of pests, parasites and diseases	Mycorrhizae and other fungi, nematodes, bacteria, and various other microorganisms, collembola, earthworms' various predators.
► Sources of food and medicines	Plant roots, various insects (crickets, beetle larvae, ants, termites), earthworms, vertebrates, microorganisms, and their byproducts.

TABLE 9.15 *(Continued)*

Functions	Organisms involved
▶ Symbiotic and asymbiotic relationships with plant and their roots	Rhizobia, mycorrhizae, actinomycetes, diazotrophic bacteria, and various others rhizosphere microorganism.
▶ Plant growth control (positive and negative)	Direct effects: plant roots, rhizobia, mycorrhizae, actinomycetes, pathogens, phytoparasitic nematodes, rhizosphagous insects, plant growth-promoting rhizosphere microorganisms, biocontrol agents. Indirect effects: most biota

Source: FAO-AGLL Portal: Biodiversity.

TABLE 9.16 Farmers Management Practices Influencing Soil Fertility Through Manipulation of Biological Processes.

Management practices	Biological processes influenced	Soil fertility effects
Biological Inputs		
Inoculation with nitrogen-fixing bacteria, mycorrhizae, etc. Introduce bacteria, nematodes, or insects that are predators of pest organism	Nitrogen fixation Facilitate nutrient Fauna burrowing Decomposition	Increase nutrient acquisition and H_2O uptake Increased heavy metals tolerance Improve soil structure and porosity Stimulation of nutrient release
Organic matter inputs		
Crop residues Root residues Weed residues (without seeds) Tree litters/pruning Green manure Farmyard manure Precomposting Others	Decomposition SOM synthesis Soil fauna and microflora growth	Increased nutrient availability Increased nutrient storage exchange Soil physical structure improved Soil water regimes improved Increase soil buffer capacity Toxicity diminished Macropore formation improved (macrofauna) Soil aggregation improved (microfauna)
Inorganic fertilizer inputs	Mycorrhiza function inhibited (at high P levels) N-fixation inhibited (at high N levels) Mineralization/ immobilization balanced change	Direct transfer of nutrient to plant increased Nutrient losses increased risk Acidification risk Increased nutrient availability

TABLE 9.16 *(Continued)*

Management practices	Biological processes influenced	Soil fertility effects
Tillage	Decomposition stimulated by OM incorporation SOM decay stimulated by aeration and particle size reduction Faunal and microbial population diminished	Short-term nutrient availability increased Root growth in tilled layer promoted Nutrient losses increased Long-term nutrient storage diminished
Pesticides	In general, nontarget organism populations diminished or eradicated	Destabilization of nutrient cycles Loss of soil structure

Source: FAO-AGLL Portal: Biodiversity.

TABLE 9.17 Biofertilizer Production and Consumption in Different States of India.

State	Number of units	Production	Consumption
Uttar Pradesh	17	125	125
Punjab	2	3	3
Haryana	2	10.05	10.05
Maharashtra	16+	3861	3642
Gurjarat	4	1010	880
Rajasthan	4	513	664
Madhya Pradesh	6	1236	1200
Chhattisgarh	1	NA	113.4
Jharkhand	NA	10	5.4
West Bengal	11	400	400
Odisha	NA	110	76
Andhra Pradesh	6	350	325
Karnataka	33	11839	11084
Tamil Nadu	23	1767	1767
Pondicherry	1	15	12

Source: https://www.indiastat.com

Here are some facts about the production and consumption of biofertilizers in India:

- In India, commercial production of biofertilizers started in the year 1956. The production of *Azospirillum* biofertilizer started in 1964.
- Presently, there are nearly 150 commercial biofertilizer production units with a total annual production capacity of 44,000 tonnes and capacity utilization of 50%.
- The yearly biofertilizer production and consumption are around 21.3 thousand tonnes and 20.4 thousand tonnes, respectively.

9.12 SOIL FERTILITY APPRAISAL THROUGH SOIL TESTING

In organic farming, soil fertility plays a very crucial role. Hence, the fertility of soil should be checked at regular intervals. Soil testing work should be expanded along with modern facilities available for analyzing the soil. Proper linkages should be established between soil test laboratories and research organizations. Recent Initiative of the government to facilitate farmers by establishing soil testing labs in each of *Krishi Vigyan Kendra* and also supporting the creation of soil and water analysis facilities at agriclinics would go a long way in strengthening soil testing services in India. These new soil testing labs should be, atleast, capable of soil testing for primary, secondary nutrients of soil and micronutrients besides having greater analyzing capacities. Soil testing laboratories at the division level should be upgraded as soil health clinics (Table 9.18).

TABLE 9.18 Soil Fertility Ratings Adopted in the Soil Testing Laboratories.

Soil parameter	Low	Medium	High
Organic carbon (%)	<0.5	0.5–0.75	>0.75
Available nitrogen (N kg/ha)	<280	280–560	>560
Available phosphorus (P kg/ha)	<10	10–25	>25
Available potassium (K kg/ha)	<110	110–280	>280

Source: Compendium on Soil Health, Ministry of Agriculture.

9.13 CONCLUSIONS

The fast-shrinking soil and water resource that performs the critical functions of entire life support is undergoing unabated degradation of different types of deterioration due to pollution and nutritional disorders. Vast fertile lands in the canal irrigated areas are affected by soil salinity, sodicity, and water logging thereby rendering it unsustainable. There is practically no possibility of horizontal expansion of fertile arable land. Under the above paradoxical situation, Indian agricultural sector's target of doubling farmer's income by 2022 without any adverse effect on soil health seems a formidable task. Some of the important strategies that should be embraced to manage soil health for sustainable agriculture include:

- Preventive measures for land degradation and pollution of soil resources through sound ecofriendly and cost-effective technologies.
- Restoration of biological activity of soils through amelioration of polluted and degraded soils.
- Since soil testing is the backbone of soil health indicators, it should be done on regular basis and nutrient recommendations must be implemented on the basis of soil test results.
- Proper understanding of soil–water–nutrient–plant relationships and enhancing soil productivity are the keys of good soil health.
- Efficient water management techniques should be practiced in canal irrigated areas to arrest soil salinization, soil sodification, and waterlogging.
- Care should be taken for judicious IPNM, which include the use of appropriate quantity and quality of chemical fertilizers, organic manures, green manures, crop residues, organic wastes, biofertilizers, and improvement of SOC content for maintenance of healthy soil.
- Preparation of suitable models to anticipate the likely changes in soil health/soil quality is essential for various land usage and its management implementations.
- Care should be taken to increasing use efficiency of key inputs such as nutrients/water to improve total factor productivity.
- Precautionary approach should be adopted for proper selection of crops, cropping sequences with incorporation of legume crops, and land usage according to land use capability classification.
- Criteria should be developed for assessing various degrees of soil health on the basis of agroclimatic zones of India.
- Regular monitoring of essential nutrients in different production systems under agroecological situations and timely use of corrective measures to maintain soil health for sustainable productivity.

KEYWORDS

- **soil resources**
- **land degradation**
- **soil organic carbon**
- **nutrient potential**
- **soil health**
- **balanced use of fertilizers**
- **biofertilizer**

REFERENCES

Arias, C. A.; Brix, H.; Marti, E. Recycling of Treated Effluents Enhances Removal of Total Nitrogen in Vertical Flow Constructed Wetlands. *J. Environ. Sci. Health A* **2005,** *40* (6–7), 1431–1443.

Bajpai, R. K.; Chitale, S.; Upadhyay, S. K.; Urkurkar, J. S. Long-Term Studies on Soil Physico-chemical Properties and Productivity of Rice-Wheat System as Influenced by Integrated Nutrient Management in Inceptisol of Chhattisgarh. *J. Indian Soc. Soil Sci.* **2006,** *54*, 24–29.

Batjes, N. H. A World Data Set of Derived Soil Properties by FAO-UNESCO Soil Unit for Global Modelling. *Soil Use Manag.* **1997,** *13*, 9–16.

Bhattacharyya, T.; Pal, D. K. Carbon Sequestration in Soils of the Indo-Gangetic Plains. In *RWC-CIMMYT: Addressing Resource Conservation Issues in Rice-Wheat Systems of South Asia. A Resource Book. Rice Wheat Consortium for Indo-Gangetic Plains*; International Maize and Wheat Improvement Centre: New Delhi, India, 2003.

Compendium on Soil Health, Ministry of Agriculture, Department of Agriculture & Cooperation, Govt. of India. (INM Division), 2012.

Doran, J. W.; Sarrantonio, M.; Liebig, M. Soil Health and Sustainability. In *Advances in Agronomy*; Sparks, D. L., Ed.; Academic Press: San Diego, 1996; pp 56, 1–54.

https://www.indiastat.com/industries-data/18/industrial-sector/107700/bio-fertiliser-1992-2018/452395/stats.aspx

Lal, R. Soil Carbon Sequestration Impacts on Global Climate Change and Food Security. *Science*, 2004, 304 (5677), 1623–1627.

Lal, R.; Kimble, J. M. Pedogenic Cates and the Global C-Cycle. In *Global Climate Change and Pedogenic Cates*; Lal, R., Kimble, J. M., Eswaran, H., Stewart, B. A., Eds.; CRC/Lewis Publishers: Boca Raton, FL, 2000, pp 1–14.

Sarkar, D. *Soil Erosion of Assam*, NBSS Publication No. 118, NBSS & LUP. Nagpur, 2005; p 41.

Scherr, S. J.; Yadav, S. *Land Degradation in the Developing World: Implications for Food, Agriculture, and the Environment to 2020*; Food, Agriculture, and the Environment Discussion Paper 14, 1996.

Sehgal, J.; Abrol, I. P. *Soil Degradation in India: Status and Impact*; Oxford & IBH: New Delhi, 1994.

Sharma, S. N.; Prasad, R.; Singh, R. K. Influence of Summer Legumes in Rice (*Oryza sativa*)-Wheat (*Triticum aestivum*) Cropping System on Soil Fertility. *Indian J. Agric. Sci.* **2000,** *70* (6), 357–359.

Singh, G.; Ram, Babu; Narain, P.; Bhushan, L. S.; Abrol, I. P. Soil Erosion Rates in India. *J. Soil Water Conserv.* **1992,** *47* (1), 97–99.

Swarup, A. Emerging Soil Fertility Management Issues for Sustainable Crop Production in Irrigated System. In *Long-Term Soil Fertility Management through Integrated Plant Nutrient Supply*; Swarup, A., Reddy, D. D.; Prasad, R. N., Eds.; Indian Institute of Soil Science: Bhopal, 1998; pp 54–68.

Tewatia, R. K.; Kalwe, S. P. Chaudhary, Role of Biofertilizers in Indian Agriculture. *Indian J. Fertilizer*, 2007, 5, 111–118.

Velayutham, M.; Bhattacharyya, T. *Soil Resources Management. In Natural Resource Management for Agricultural Productivity in India*; Yadav, J. S. P., Singh, G. B., Eds.; Indian Society of Soil Science: New Delhi, India, 2000.

CHAPTER 10

Improving Soil Health and Sugarcane Productivity by Managing Crop Residue and Sugar Industry By-products

S. K. SHUKLA[1], LALAN SHARMA[2], V. P. JAISWAL[3], A. GAUR[3], and S. K. AWASTHI[3]

[1]*AICRP on Sugarcane, ICAR-Indian Institute of Sugarcane Research, Lucknow 226 002, Uttar Pradesh, India*

[2]*Plant Pathology, ICAR-Indian Institute of Sugarcane Research, Lucknow 226 002, Uttar Pradesh, India*

[3]*Agronomy, ICAR-Indian Institute of Sugarcane Research, Lucknow 226 002, Uttar Pradesh, India*

**Corresponding author. E-mail: sudhirshukla151@gmail.com*

ABSTRACT

Sugarcane trash, bagasse, press mud, and molasses are important sugarcane and sugar mills waste and their efficient utilization assumes significance. Crop residue like cane trash left over the field and it take years to decompose properly. Similarly, the management of bagasse and press mud becomes cumbersome because these sugar mills by products are available in large quantity and require more space for their storage. There is an urgent need to develop appropriate methods for proper utilization for humankind. A large quantity of sugarcane waste by-products can be handled through composting, which greatly lightens the processing burden of sugar factories. Composting technology significantly reduces environmental pollution and achieves

Organic Crop Production Management: Focus on India, with Global Implications, D. P. Singh, PhD, H. G. Prakash, PhD, M. Swapna, PhD, & S. Solomon, PhD (Eds.)

resource utilization. Vermicomposting is the best way of its utilization. A huge possibility of sugarcane industry waste can be used in agriculture to cut down the fertilizer requirement and save the cost of chemical fertilizers. It can also be used in combination with inorganic chemical fertilizers and can be packed and marketed along with commercial fertilizer for a particular cropping system. Waste molasses contains many chemical elements: N, P, K, Fe, Zn, Mn, and Cu, which can improve soil structure, increase soil organic matter, promote soil permeability and enhance crop quality and yield. Press mud alone and in combination with other by-products of sugar processing industries is pre decomposed for 30 days by inoculation with combination of *Trichoderma viride, Aspergillus niger*, and *Pseudomonas striatum.* The accelerated degradation process of waste/by-products of sugar processing industry by microbial intervention produces nutrient-enriched compost product useful for sustaining high crop yield, minimizing soil depletion and value- added disposal of waste materials.

10.1 INTRODUCTION

Sugarcane is a huge biomass-generating crop, which has the potential of converting sunlight into biomass. In India, sugarcane is grown in almost 5.0 million ha of land, which contributes around 19% sugarcane area of the world. It is the second-largest sugarcane cultivating and producing country in the world after Brazil. In Brazil, it is grown in about 10.18 million ha area (FAO, 2017). The average sugarcane yield in the country is about 74 t/ha, while sugar production was around 32 million tonnes during 2018–2019. Sugarcane is being cultivated in two distinct agroecosystems in the country, namely, tropical and subtropical region. The tropical region has about 42% sugarcane area and contributes around 52% of the total sugarcane production whereas subtropical region accounts for 58% area and shares 48% of total sugar production in the country. Sugarcane is cultivated in almost all the states of the country but major sugarcane growing states are Uttar Pradesh, Maharashtra, Karnataka, Gujarat, Andhra Pradesh, Bihar, West Bengal, Haryana, and Punjab. Generally, sugarcane production and productivity of subtropical region are very much influenced by the climatic conditions throughout the year, while the tropical region is considered more congenial for its cultivation but recently identified and released varieties under AICRP on Sugarcane are playing very important role in improving cane productivity mainly in subtropical India. In India, sugar industry contributed around 6%

of the Agricultural Gross Domestic Products during 2010–2011 (Solomon, 2016). Sugarcane cultivation and sugar industry play a vital role toward socioeconomic development for the rural population by generating higher income and employment opportunities. About 6 million farmers are directly or indirectly dependent on sugarcane cultivation, harvesting, and ancillary activities in India (IISR Vision 2050). Sugarcane is grown mainly for the sugar production globally and shares around 70% of total sugar produced. Sugar beet is the second source of sugar production in the world and contributes around 30%, which is grown mainly in temperate countries of the world. Besides, the sugar production in the country, sugar juice is used for different purposes like making white sugar, brown sugar (*khandsari*), and jaggery (*gur*). These sugarcane products have high economic value but at the time of sugarcane harvesting, the crop generates huge quantity of crop residue, called cane trash, which becomes hindrance in ratoon crop initiation and other agronomic activities during field conditions. Similarly, during sugar mill crushing seasons, sugar mills generate diverse kinds of by-products like bagasse, press mud, molasses, ash, and spent wash. The sugar mills generated by-products mainly bagasse and press mud require large area for their storage, which is again very difficult task to manage it. The adoption of proper management practices/methods do not only improve the soil health and fertility but also enhance plant growth and crop yield. In this chapter, we have tried to describe methods for crop residue management and sugar industry by-products recycling so that proper sugarcane waste management can be done and could be well adopted by the sugarcane growers in the country for improving soil health and sugarcane productivity.

10.2 SUGARCANE AND SUGAR INDUSTRIAL BYPRODUCTS

Sugarcane is cultivated in around 5 million ha area of the country under diverse climatic conditions as an adoptative crop. Sugarcane cultivation produces huge quantity of cane trash, which becomes cumbersome to manage. Similarly, sugar mills by-products like bagasse, press mud, and molasses are generated in huge quantity (Pippo et al., 2011). At present, around 700 sugar factories are functional in the country (cooperative-326, private-347 and public-43). Annually, huge quantity of cane trash and sugar mills byproducts is added in the country. It is roughly estimated that around 43.8 mt of bagasse, 8 mt of press mud, 7.5 mt of molasses, and 7.4 mt bagasse ash (Dotaniya, 2016) is added.

10.2.1 SUGARCANE TRASH

Sugarcane is an efficient solar energy converter and produces quantity of biomass because of C_4 crop plant. The crop takes 10–12 months to mature. At the time of harvesting, 20–30% of generated plant biomass leftover on the field is called cane trash, which is generally burnt by the farmers but this is not a good practice. It not only reduces soil physical properties but also pollutes our environment.

10.2.2 SUGARCANE BAGASSE

Sugarcane bagasse is a fibrous content of sugarcane crop, which remains available after extraction of juice from fully matured cane. Bagasse contains 45–55% cellulose, 20–25% hemi-cellulose, 18–24% lignin, 1–4% ash, and <1% waxes. Besides this, it contains nearly 40–50% moisture, which makes it very difficult to use as fuel. Generally, 300 kg of bagasse is produced by crushing 1 tonne of matured cane (Rocha et al., 2011). A very rare percentage of bagasse is used for electricity generation, while Brazil produces 19.3% energy generation through bagasse (Hofsetz and Silva, 2012). Bagasse is multiutilized by-products of sugar mills. Due to its pulpy and fibrous nature, it is used for electricity generation, biofuel production, process papers, and building material preparation. The processed bagasse is sometimes added into human food as sugar cane fiber after proper processing and hygiene conditions. It contains soluble fiber and can help in regulating digestion.

10.2.3 SUGARCANE MOLASSES

Sugarcane molasses is a sticky and semiliquid in nature, produced during sugar production. Twenty-three liters of molasses is produced by crushing 1 tonne of sugarcane. Sugarcane molasses contains 22% water, 75% carbohydrates, and no protein or fat. Moreover, it is a rich source of minerals like manganese, magnesium, iron, potassium, and calcium (Anon., 2016). Sugarcane-produced molasses has a favorable taste and aroma in nature. It is also most economically important byproducts of sugar industries. This sugar mill by-product is used for alcohol production, preparation of animal feeds, and food stuffs. It also becomes substrate for diverse group of microbial population mainly yeast culture, *Saccharomyces cerevisae,* during fermentation. The fermented product has high economic values and can be stored for long term.

10.2.4 SUGARCANE PRESS MUD

Sugarcane press mud is organic waste product of the sugar industry, which is produced during the juice purification process. It is also known as a filter cake. It has a dark brown to brown color, which changes with time of storage (Ghulam et al., 2012). It is good source for bio-fortification. It is thumb rule that about 3 tonnes of press mud is produced by crushing hundred tonnes of sugarcane (Gupta et al., 2011). The amount of press mud production depends on the sugar filtration process. It has sufficient moisture percentage (50–70%) and used for mass multiplication of earthworms (Dominguez, 1997). As it is rejected waste material of the sugar industry that cause problem during storage and also pollutes the surroundings of sugar mills due to open storage (Bhosale et al., 2012) but it contains sugar, which helps in its decomposition speedily when applied in field condition (Dotaniya et al., 2013a, 2013b). It is an excellent source of plant nutrients, which also improves soil water holding capacity as well as soil chemical properties and conserves soil moisture for longer period (Yadav and Solomon, 2006). It also contains significant amounts of micronutrients like iron, manganese, calcium, magnesium, and silicon.

10.2.5 BAGASSE ASH

Bagasse ash is basically the dust generated from the crushed mature cane, which contains certain quantity of bagasse. Quantity of sugarcane bagasse ash produced comes to 0.3% on cane weight basis. Spreading in the field as a fertilizer has been the usual practice of its disposal as it is a rich source of silica, potash, iron oxide, and lime. Use of bagasse ash in combination with press mud and farmyard manure or with biofertilizers carriers is a common practice.

10.2.6 SPENT WASH

Spent wash is liquid in nature, which is called by various names like slops or vinasse or stillage in the different countries. Due to its high manurial value, it can be used as soil conditioner. Besides this, it is a reclaiming agent and an excellent liquid fertilizer. Diluted spent wash when applied to soil, showed an improvement in soil pH and EC values, increasing the status of total nitrogen, available P and K, total K and also improving the physical parameters of the soil.

10.3 METHODS OF CROP RESIDUE RECYCLING

Residue or by-products means the material that is left after the extraction of the main product. There are generally two types of residues in the agriculture sector: one is field crop residue and other is industrial by-product. Field residues are materials left on agricultural field, for example, leaf and top of the cane left in the field after harvesting of sugarcane stalk. Residue management in sugarcane is a big challenge and remains available for longer time in field conditions. In some cases, the crop residues can be ploughed directly into the ground or chopped and then buried. Proper management of crop residues can increase efficiency of irrigation, reduces weed population, and controls erosion. Similarly, industrial by-products are materials left after the extraction of the main product. The sugarcane industrial by-products may be like bagasse, molasses, and press mud. These industrial by-products can also be utilized for improvement of soil fertility, generation of electricity, and production of bioenergy. There are various ways by which the residues of field and industrial can be properly utilized. For crop residue and industrial byproducts like bagasse and press mud, compost preparation is considered more economical and viable method. It is of two types: the traditional method of composting is a natural method of composting. It is an approach of anaerobic decomposition and time taking processes involving several months to decompose. Second is advanced method of composting in which aerobic method of composting is applied, which brings composting period around 6–12 weeks.

10.4 TRADITIONAL METHOD OF COMPOSTING

Traditional method of composting based on passive composting approach involve simply stacking the material in piles or pits to decompose over a long period with little agitation and management. Crop residue is kept in a pit, called the pit method or makes a pile, called heap method. Some other method of composting is named as Indore method, Bangalore method, and NADEP method. Compost can be prepared from crop residue called cane trash or from sugar industry by-products like bagasse and press mud. Compost prepared from these wastes is very nutritive. Plants can take these nutrients from prepared compost very easily and quickly. In composting of sugarcane trash, sugarcane trash is very bulky and very difficult to manage. Before making the compost, the trash should be chopped in smaller size so that it can be transported to the compost yard easily. If no compost yard is available to farmer, the corner area in the sugarcane field itself can be used for making composting from cane trash. There is no necessity to make a pit for composting. Composting can be done above the soil but time varies to prepare quality compost. Due to natural decomposition process of cane trash, it takes several months to decompose properly. Similarly, sugar industry by-products like bagasse and press mud, a pit is formed and bagasse is dumped into it. Due to anaerobic decomposition process, it also takes months to proper decompose but less than the cane trash compost. In bagasse and press mud, sugar content is high, which favors diverse group of saprophytic microbes that helps in speeding up the process. Compost prepared from sugar industry by-products is more nutritive than cane trash compost and can be easily taken by the plant.

10.5 ADVANCED METHOD OF COMPOSTING

Advanced method of composting is externally added beneficial microbial inoculants and other living microorganisms to speed up the composting process. In the microbe-mediated composting method, beneficial microbes are applied to prepare quality composting. This type of composting is called phospho—composting, phospho–ulfo–nitro composting, and sometimes called microbial-enriched composting. The beneficial microbial population has specific characters like cellulose-degrading fungi such as *Trichoderma koningii, T. resii, T. viride, T. harzianum,* and *Phanerochaete chrysosporium*, which are cultured in liquid media and added in sugarcane residues such as cane trash, bagasse, and press mud. Sometimes, green leaves are also added to the compost pile since these are rich sources of nitrogen that are needed to promote accelerated growth of the microorganisms during composting process. Similarly, partially decomposed organic waste is treated with earthworms, which mineralize it very fast. After the use of worms, high quality of compost is prepared. In vermin-composting, temperature should be regulated so as to favor growth and activity of worms. Composting period is longer as compared with other rapid methods and varies between 6 and 12 weeks.

10.6 IMPROVING SOIL HEALTH AND CROP PRODUCTIVITY

Sugarcane, being a long duration and huge biomass accumulating crop, removes substantial quantities of plant nutrients from soil during its growth and development. A sugarcane crop if yielding 100 t/ha, then it exhausts 205 kg N, 55 kg P, and 275 kg K besides the 3.5 kg Fe, 1.2 kg Mn, 0.6 kg Zn, 0.2 kg Cu, and 30 kg S (Yadav and Dey, 1997). Some other micronutrients are absorbed. The sugarcane trash and sugar mill byproducts are economically important in nature and they are properly managed. For example, press mud cake has a great potential to supply plant nutrients easily (1–2% N, 2–4% P_2O_5, 0.5–1.5% K_2O, 0.83–1.98% Ca, 0.05–0.25% Mg, 0.31–0.92% Na, 0.22–0.31% S, 22.5–95 ppm Fe, 163–625 ppm Mn, 47–215 ppm Zn), besides having beneficial effects on physicochemical and biological properties of soil. These, in turn, influence the availability and uptake of plant nutrients, crop growth and development, cane yield, and even juice quality. Improvement in soil organic content can be achieved by additional application of organic manure derived from cane trash, bagasse, and press mud, which results in greater water-holding capacity and aeration. Besides this, the beneficial effects of organic manures have been attributed to suppression of some soil-borne plant diseases (Hoitink and Fahy, 1986) and to improve soil physical properties and nutrient availability (Sommerfeldt and Chang, 1985; De Luca and De Luca, 1997). The availability of natural resources for agriculture in the country is presented and compared with other sources over the period of time (Table 10.1).

TABLE 10.1 Availability of Organic Resources for Agriculture in India during 2010 and 2025.

S. no.	Resource	2010	2025
Nutrient (theoretical potential)			
1.	Human excreta (N+ P_2O_5 + K_2O in mt)	2.24	2.60
2.	Livestock dung (N+ P_2O_5 + K_2O in mt)	7.00	7.54
3.	Crop residues (N+ P_2O_5 + K_2O in mt)	7.10	10.27
Nutrients (considered tappable) *			
1.	Human excreta (N+ P_2O_5 + K_2O in mt)	1.80	2.10
2.	Livestock dung (N+ P_2O_5 + K_2O in mt)	2.10	2.26
3.	Crop residue (N+ P_2O_5 + K_2O in mt)	2.34	3.39

*Tappable; 30% of dung, 80% of excreta, 33% of crop residues.

Source: Reprinted with permission from Shukla et al. (2009). © Nova Publications, Inc.

In an experiment, the effect of cane trash recycling was studied on ratoon crop and it was recorded that recycling of sugarcane trash improves not only soil organic matter, soil structure but also enhanced cane yield (Yadav et al., 1994, 2009). In his findings, Yadav et al. emphasized that farmers generally prefer to burn cane trash before ratoon initiation which kills not the weeds but also beneficial microbes and insects. The increased ratoon cane yield and improved soil quality can be obtained with proper management of cane trash during ratoon crop cultivation. They also suggested that accelerating decomposition of cane trash using cellulolytic/lignolytic microorganisms like *Trichoderma, Trichurus*, and *Aspergillus* spp. (Saravanan and Mahendran, 2003) improved soil organic carbon content over the period of time. Similarly, Yadav et al. (2009) conducted a field experiment and observed that *Trichoderma* inoculation with trash mulch increased soil organic carbon content and phosphorus (P) content by 15.75 Mg/ha and 12.5 kg/ha over their initial contents of 15.08 Mg/ha and 11.7 kg/ha, respectively. At all the stages of crop growth, *Trichoderma* inoculation increased the soil basal respiration over untreated plot. Soil microbial biomass increased in all plots except in the plots of trash burning/removal without *Trichoderma* inoculation. The maximum increase (40 mg C/kg soil) in soil microbial biomass C, however, was observed in the plots of trash mulch with *Trichoderma* inoculation treatment which also recorded the highest uptake of nutrient and cane yield. On an average, *Trichoderma* inoculation with trash mulch increased N, P, and K uptake by 15.9, 4.68, and 23.6 kg/ha, respectively, over untreated condition. The cane yield was increased by 12.8 t/ha with trash mulch + *Trichoderma* over trash removal without *Trichoderma*. Upon degradation, trash mulch served as a source of energy for enhanced multiplication of soil beneficial bacteria and fungi and also provided suitable niche for plant–microbe interaction (Table 10.2).

TABLE 10.2 Effects of Different Treatments on SOC Content During Crop Growth and Change at Harvest (Mean Data of Two Seasons).

Treatment	Soil organic carbon[a] (Mg/ha)			Difference between SOC of soil at harvest and the SOC of soil at the start of experiment
	April	August	December (harvest)	
Trash mulching + Tv	22.80 (1.44)[b]	28.00 (2.44)	20.83 (2.00)	5.08
Trash mulching − Tv	19.81 (1.54)	24.25 (2.62)	18.88 (1.99)	3.13
Trash burning + Tv	20.83 (1.13)	25.23 (1.55)	20.46 (1.22)	4.71

TABLE 10.2 *(Continued)*

Treatment	Soil organic carbon[a] (Mg/ha)			Difference between SOC of soil at harvest and the SOC of soil at the start of experiment
	April	August	December (harvest)	
Trash burning – Tv	17.58 (1.22)	21.31 (1.64)	18.17 (1.16)	2.42
Trash removal + Tv	20.32 (1.23)	20.66 (2.14)	19.31 (1.92)	3.56
Trash removal + Tv	16.94 (1.34)	16.60 (2.43)	17.27 (1.26)	1.52
$LSD_{0.05}$				
Trash	0.56	0.69	0.58	
Trichoderma	0.45	0.54	0.45	
Trash + *Trichoderma*	0.76	0.81	0.88	

Source: Reprinted with permission from Shukla et al. (2009). © Nova Publications, Inc.

Improvement in crop productivity of various crops even in sugarcane increased due to addition of organic amendments to soil (Bevacqua and Mellano, 1994; Hallmark et al., 1995). Application of FYM and/or green manure in sugarcane established its beneficial effect in improving the production efficiency of fertilizer N and more so at its optimal level. The response of sugarcane to FYM at 25 t/ha is found to be 8.0 tonnes higher cane yield (t/ha) averaged over 258 experiments conducted at 19 AICRP centers on sugarcane (Gaur, 1992). At Indian Institute of Sugarcane Research, Lucknow, the highest cane production efficiency (418 kg cane/kg N applied) is obtained by applying 150 kg N/ha with green manuring and application of FYM. Besides this, response of various inorganic fertilizers with organics increased under different cropping systems at IISR, Lucknow, has been studied and described in Tables 10.3 and 10.4.

As we know, the crop residues are renewable and are readily available but are scattered organic resources. Intensive sugarcane-based production system besides adding huge quantities of biomass (13.32 tonnes stubble, 37.59 tons root, and 35.52 tonnes trash/year) of sugarcane, it has an enormous effect on the successive crops of cereals, pulses, and oilseeds. Since nutrients absorbed by cane plants from soil do not form the constituents of its marketable commercial product except sugar. There is a good opportunity of organic recycling in this crop. The recycling of roots/trash directly in the soil through vermin-culture, green tops/molasses through ruminants in the form of cattle dung/urine, press-mud from juice as soil amendment/sulfur source, and spent wash from distilleries as irrigation source after dilution can return multinutrients to soil from sugarcane crop itself.

TABLE 10.3 Effect of Integrated Use of Sulphitation Press Mud Cake with Fertilizer N on the Yield of Sugarcane (t/ha) Plant Crop at Different Locations.

Location	Recommended N (kg/ha)	Recommended N dose with					
		Alone		SPMC* (2 t/ha)		SPMC (4 t/ha)	
		Cane yield (t/ha)	Cane production efficiency (kg cane/kg N applied)	Cane yield (t/ha)	Cane production efficiency (kg cane/kgN applied)	Cane yield (t/ha)	Cane production efficiency (kg cane/kgN applied)
Pusa	140	44.5	318	51.7	369	59.7	426
Mandya	250	161	644	165	660	168.6	674
Thiruvella	225	98.6	438	106.9	475	112.3	499
Padegaon	250	110.1	440	113.8	455	116.9	468

Note: SPMC was applied on oven dry basis.

Source: Reprinted with permission from Shukla et al. (2009). © Nova Publications, Inc.

TABLE 10.4 Effect of Integrated Use of Mineral Fertilizers and Organics on the Sustainability of Productivity (t/ha) of Crops in a Cropping System at IISR, Lucknow.

Location	Crops in cropping system	Recommended NPK dose to each crop along with				CD at 5% level
		Alone	FYM (20 t/ha) to plant cane only	Trash + SPMC in plant cane only	GM once in crop cycle after rabi crop	
Lucknow	I ratoon	71.4	78.8	75.2	72.7	8.0
	II ratoon	64.2	68.2	70.1	64.8	5.9
	Wheat (grain)	4.26	4.49	4.86	4.25	0.73
	Plant cane (II cycle)	58.6	66.2	63.9	60.5	4.7

Note: Trash+ SPMC at 10 t/ha.

Source: Reprinted with permission from Shukla et al. (2009). © Nova Publications, Inc.

10.7 CONCLUSION

Cane trash, bagasse, press mud, and molasses are important sugarcane and sugar mills waste and their efficient utilization assumes significance. Crop residue like cane trash left over the field and it take years to decompose properly. Similarly, the management of bagasse and press mud becomes cumbersome because these sugar mills by products are available in large quantity and require more space for their storage. Even then, these cannot be stored for longer period and have no definite value. So there is an urgent need to develop appropriate methods for proper utilization in humankind. These are very good source for cultivation of microbial population. These can be used either to develop substrate for mushroom cultivation or to biofortify these by-products to organic manure like compost. Advanced methods of composting discussed in detail can be utilized. Vermicomposting is the best way of its utilization. However, fortification with beneficial microbes is also very important. This not only speeds up composting process but also provide additional nutrients to the growing crops. Sometimes bioinoculation followed by vermin-composting is applied to shorten stabilization time and improve product quality. Press mud alone and in combination with other by-products of sugar processing industries is predecomposed for 30 days by inoculation with combination of *Trichoderma viride, Aspergillus niger*, and *Pseudomonas striatum*. The accelerated degradation process of waste/ by-products of sugar processing industry by microbial intervention produces nutrient-enriched compost product useful for sustaining high crop yield, minimizing soil depletion and value-added disposal of waste materials.

KEYWORDS

- **sugarcane trash**
- **crop residue recycling**
- **composting**
- **soil health**

REFERENCES

Bevacqua, R.F.; Mellano, V.J. Cumulative Effects of Sludge Compost on Crop Yield and Soil Properties. *Commun. Soil Sci. Plant Analy.* **1994,** *24*, 395–406.

Bhosale P.R.; Chonde, S.G.; Raut, P.D. Studies on Extraction of Sugarcane Wax from Press Mud of Sugar Factories from Kolhapur District, Maharashtra. *Environ. Res. Dev.* **2012,** *6* (3A), 615–620.

De Luca, T.H.; De Luca, D.K. Composting for Feedlot Manure Management and Soil Quality. *J. Product. Agric.* **1997,** *10*, 235–241.

Dominguez, J. Testing the Impact of Vermi-Composting. *BioCycle* **1997,** *1*, 1–21.

Dotaniya, M.L.; Prasad, D.; Meena, H.M. Influence of Phytosiderophore on Iron and Zinc Uptake and Rhizospheric Microbial Activity. *Afr. J. Microb. Res.* **2013a,** *7* (51), 5781–5788.

Dotaniya, M.L.; Sharma, M.M.; Kumar, K.; Singh, P.P. Impact of Crop Residue Management on Nutrient Balance in Rice-Wheat Cropping System in an Aquic Hapludoll. *Rural Agric. Res.* **2013b,** *13* (1), 122–123.

Dotaniya, M.L.; Datta, S.C.; Biswas, D.R.; Dotaniya, C.K.; Meena, B.L.; Rajendrin S.; Regar, K.L.; Manju Lata. Use of Sugarcane Industrial By-Products for Improving Sugarcane Productivity and Soil Health. *Int. J. Recycl. Org. Waste Agric.* **2016,** *5*, 185–194.

FAO. *Statistical Yearbook. Asia and the Pacific Food and Agriculture*, Food and Agriculture Organization of the United Nations Regional Office for Asia and the Pacific Bangkok, 2017. http://www.fao.org/faostat/en/#data/QC

Gaur, A.C. Bulky Organic Manures and Crop Residues. In *Fertilizers, Organic Manures, Recyclable Wastes and Bio-fertilizers: Components of Integrated Plant Nutrition*; Tandon, H., Ed.; Fertilizer Development and Consultation Organization: New Delhi, 1992.

Ghulam, S.; Khan, M.J.; Usman, K.; Shakeebullah. Effect of Different Rates of Press Mud on Plant Growth and Yield of Lentil in Calcareous Soil. *Sarhad J. Agric.* **2012,** *28* (2), 249–252.

Gupta, N.; Tripathi, S.; Balomajumder, C. Characterization of Press Mud: A Sugar Industry Waste. *Fuel* **2011,** *90* (1), 389–394.

Hallmark, W.B.; Feagley, S.E.; Breitenbeck, G.A.; Brown, L.P.; Wan, X.; Hawkins, G.L. Use of Composted Municipal Waste in Sugarcane Production. *Louisiana Agric.* **2012,** *38,* 15–16.

Hofsetz, K.; Silva, M.A. Brazilian Sugarcane Bagasse: Energy and Non-Energy Consumption. *Biomass Bioenergy* **2012,** *4* (6), 564–573.

Hoitink, H.A.J.; Fahy, P.C. Basis for the Control of Soil Borne Plant Pathogens with Composts. *Annu. Rev. Phytopathol.* **1986,** *24*, 93–114.

http://eagri.tnau.ac.in/eagri50/AGRO301

Pippo, W.A.;, Luengo, C,A.; Alonsoamador, L.; Alberteris, M.; Garzone, P.; Cornacchia, G. Energy Recovery from Sugarcane-Trash in the Light of 2nd Generation Biofuels. Part 1: Current Situation and Environmental Aspects. *Waste Biomass Valor* **2011,** *2*, 1–16.

Rocha, G.J.M.; Martin, C.; Soares, I.B.; Souto-Maior, A.M.; Baudel, H.M.; Moraes, C.A. Dilute Mixed-Acid Pretreatment of Sugarcane Bagasse for the Ethanol Production. *Biomass Bioenergy* **2011,** *35*, 663–670.

Shukla, S.K.; Sharma, L.; Awasthi, S.K.; Pathak, A.D. *Sugarcane in India: Package of Practices for Different Agro-climatic Zones*; ICAR-*Indian* Institute of *Sugarcane* Research: Lucknow, 2017.

Shukla, S.K.; Suman, A.; Singh, A.K.; Yadav, R.L.; Lal, M. Organic Nutrition for Sustaining Soil Health, Improving Rhizospheric Environment and Sugarcane Yield in India. In *Sugar Beet Crops: Growth, Fertilization & Yield*; Hertsburg, C.T., Ed.; Nova Science Publishers: New York, 2009.

Solomon, S. Sugarcane Production and Development of Sugar Industry in India. *Sugar Tech.* **2016,** *18* (6), 588–602.

Sommerfeldt, T.G.; Chang, C. Changes in Soil Properties under Annual Applications of Feedlot Manure and Different Tillage Practices. *Soil Sci. Soc. Am. J.* **1985,** *49*, 983–987.

Yadav, R.L.; Shukla, S.K.; Suman, A.; Singh, P. *Trichoderma* Inoculation and Trash Management Effects on Soil Microbial Biomass, Soil Respiration, Nutrient Uptake and Yield of Ratoon Sugarcane under Subtropical Conditions. *Biol. Fertil. Soils* **2009,** *45*, 461–468.

Yadav R.L.; Solomon, S. Potential of Developing Sugarcane By-Product Based Industries in India. *Sugar Tech* **2006,** *8* (2&3), 104–111.

Yadav, D.V.; Dey, P. Improving Cane and Sugar Productivity of Chlorotic Sugarcane Ratoon through Foliar Application of Ferrous Sulphate. In *Souvenir with Abstracts: National Symposium on 'Sugar Recovery: Problems & Prospects'*;. Indian Institute of Sugarcane Research: Lucknow, 1997; p 67.

CHAPTER 11

Nutritional Management in System-Based Organic Farming

A. K. SINGH[1*], A. K. JHA[2], and A. SRINIVASARAGHAVAN[3]

[1]*Bihar Agricultural University, Sabour 813210, Bhagalpur, Bihar, India*

[2]*Soil Science and Agricultural Chemistry, Bihar Agricultural University, Sabour 813210, Bhagalpur, Bihar, India*

[3]*Plant Pathology, Bihar Agricultural University, Sabour 813210, Bhagalpur, Bihar, India*

[*]*Corresponding author. E-mail: technicalcellbausabour@gmail.com*

ABSTRACT

Organic farming is an efficient and promising agricultural approach for environmental sustainability as it provides yield stability, improved soil health, no environmental concerns, organic food, and a reduction in the use of synthesized fertilizers. There is a general agreement that organic farming is essential for agricultural sustainability and environmental protection. As nutrient additions on organic farms are designed to maintain soil fertility along with providing nutrients to plants, soil health is required to be improved for the adoption of organic farming. For the purpose, there is a need of system-based organic farming rather than to grow a single crop organically. Organic manure, crop rotation, vermicomposting, nitrogen-fixing microorganisms, organic residue, crop residue, bio-fertilizers, biopesticides, kitchen waste, sludge, panchgavya, biogas, and biodynamic preparation are the potential inputs for nutritional management in system-based organic farming. A variable combination of all these need to be applied for each system and location.

Organic Crop Production Management: Focus on India, with Global Implications, D. P. Singh, PhD, H. G. Prakash, PhD, M. Swapna, PhD, & S. Solomon, PhD (Eds.)

There is a need of more research to make nutrient management system more profitable in organic farming.

11.1 INTRODUCTION

An enhanced awareness and global consensus of environmental and health problems associated with the rigorous employment of chemical inputs has led to develop interest in alternative forms of agriculture in the world. Organic farming is considered as an important method of farming aiming at a sustainable farming with broad spectrum of production methodologies having coherence with the environment and human health. Owing to these facts, organic agriculture is acquiring considerable momentum across the globe. The institutions involved in agricultural research and extension are actively involved in scientific validation and promotion of organic farming practices across the world. But, sustainably feeding the growing world population is also an unprecedented challenge before the think-tanks, planners, and agricultural scientists. Both of the goals can be accomplished simultaneously by successful management of resources (Ladha, 1992). Development of nutritional modules for system-based organic farming may decrease the yield gap between organic and conventional methods even during conversion period and meeting the global food demand including nutritional security.

With enhanced health consciousness among the global population, the term "organic farming" is gaining popularity among the masses. The actual system of organic farming was initially started around 10,000 years back when prehistoric farmers started cultivation using natural resources. There is a brief mention of several organic inputs in our ancient literatures, such as *Rigveda*, *Ramayana*, *Mahabharata*, *Kautilya Arthasashthra*, and others. In fact, organic agriculture has its roots in traditional agricultural practices that evolved in countless villages and farming communities over the millennium.

Food and Agricultural Organization (FAO) of the United Nations has suggested that "organic agriculture is a unique production management system which promotes and enhances agroecosystem health including biodiversity, biological cycles, and soil biological activity, and this is accomplished by using on-farm agronomic, biological, and mechanical methods in exclusion of all synthetic off-farm inputs."

Due to emergent consciousness of health and environmental concerns, the area under organic cultivation is rapidly increasing. International Federation of Organic Agriculture Movements (IFOAM) was established at Bonn, Germany during 1972 to monitor the organic farming organization all over the world. Several other agencies and institutes including FiBL, SOEL, and others also came in existence with an objective of promoting environmentally safe and sustainable alternative, that is, organic farming across the globe. Currently, organic farming is being practiced by more than 172 countries worldwide.

11.2 GLOBAL AREA UNDER ORGANIC AGRICULTURE

Total land covered by organic agriculture adds up to approximately 33 million hectares worldwide in which Africa and Asia occupy 45 and 20% of agricultural land under organic farming, respectively. Joint survey report of FiBL and IFOAM prepared in 2012 indicates that among all countries of the world, Australia has the highest land use under organic farming. However, on the basis of percentage share of agricultural land of the country for organic farming, Falkland Islands (35.9%) tops the list. The list of the major countries that have adopted organic farming practices with actual acreage under this specialized farming is given in Figure 11.1.

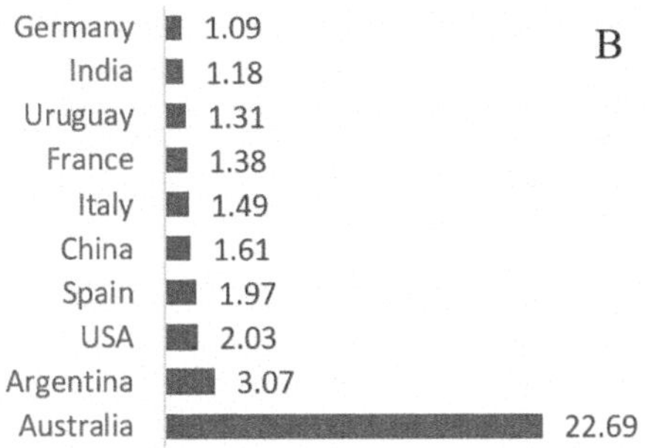

FIGURE 11.1 Top ten countries having the highest area under organic agriculture.
Source: IFOAM (2017).

11.3 STATUS OF ORGANIC FARMING IN INDIA

The Government of India has been actively promoting organic farming since 2004 by establishing National Project on Organic Farming (NPOF). Planning Commission of India has approved organic farming scheme as a pilot project for about 2 years of 10th plan period with effect from January 2004 with an outlay of Rs. 57.04 crore and continued the scheme in the forthcoming 5-year plans.

NPOF is being implemented by National Centre of Organic Farming at Ghaziabad and its six Regional Centers are located at Bangalore, Bhubaneswar, Panchkula, Imphal, Jabalpur, and Nagpur. Besides working for realization of targets under NPOF, *National Centre of Organic Farming* (NCOF) and Regional Centre for Organic Forming (RCOFs) are also performing specific roles in the promotion of organic farming.

After the establishment of NCOF, organic farming started to show a growing trend. As per the report of Agricultural and Processed Food Products Export Development Authority (*APEDA*) (2013), the total area under organic certification is 5.21 million hectares. Out of which, 10% (0.5 million hectares) is cultivable area, however, rest 90% (4.71 million hectares) is forest and wild area for collection of minor forest produces. Currently, India's organic trade is more than Rs. 2500 crore. For enhancing organic farming practice, Ministry of Commerce started National Programme on Organic Production (NPOP). This national program involves the accreditation program for certification bodies, developing norms for organic production and promotion of organic farming. The NPOP standards for production and accreditation system have been recognized by European Commission and Switzerland as equivalent to their country standards. Besides this, the Ministry of Agriculture also started various promotion schemes for small farmers. The Government of India has also approved setting up of an exclusive research institute for organic farming to extend technical support to organic growers of the country. The area under organic management in India has been depicted in Table 11.1.

TABLE 11.1 Area under Organic Certification Process (mha).

Sl. No.	Years	Area under cultivation (including in-conversion)	Wild harvest
1.	2003–2004	0.042	Data not available
2.	2004–2005	0.076	Data not available
3.	2005–2006	0.173	Data not available
4.	2006–2007	0.538	2,432,500
5.	2007–2008	0.865	32,500
6.	2008–2009	1.207	3,055,000
7.	2009–2010	1.085	3,396,000
8.	2010–2011	0.777	3,650,000

Source: Shukla et al. (2013).

Production wise, cotton (54.01%), cereals and millets (19.78%), and basmati rice (10.99%) rank first, second, and third, respectively. The production of tea and coffee are only 2.56 and 0.67%, respectively, but, the export value of these crops is very high. The commodity-wise production details of top 10 products are given in Table 11.2.

TABLE 11.2 Commodity-Wise Production Details of Top Ten Products (2011–2012).

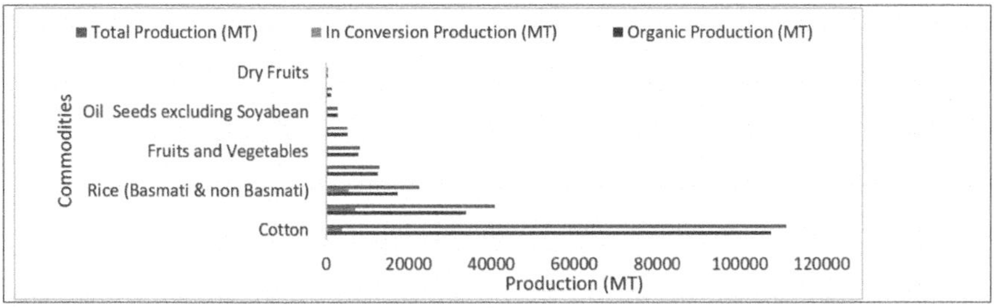

Source: APEDA (2013).

11.4 PRESCRIPTION OF NUTRIENT MANAGEMENT FOR SYSTEM-BASED ORGANIC FARMING

Organic farming may be considered as a prototype of sustainable farming (Reganold *et al.*, 2001, IAASTD, 2009). Recent meta-analyses of the literature demonstrated that organic yields of individual crops are similar to those of conventionally managed crops when good management practices are applied (Seufert *et al.*, 2012). But management is not possible for single crop. Nutrient may be managed properly in system-based organic farming, because in the system, soil is also managed for organic production.

In Bihar Agricultural College, Sabour, scented rice–potato–onion cropping pattern was selected for organic agriculture. Recommended N for these crops was supplied through vermicompost (1/3), farmyard manure (1/3) and *neem* cake (1/3), and after 11 cycles, organic farming was found significantly superior over the conventional scientific farming (Table 11.3).

At Indian Agricultural Research Institute, New Delhi, a system-based organic farming experiment (basmati rice–wheat–green gram) was conducted from 2003 to 2008 and it was found that application of *Azolla* at 1.0 t fresh biomass +BGA at 2.0 kg (dry) + FYM at 5.0 t + Vermicompost @ 5.0 t/ha was superior over all other treatment to increase the grain yield of rice (Table 11.4).

TABLE 11.3 Effect of Organic Farming Packages on Rice Equivalent Yield over the Years.

Treatment	REY (q/ha)				
	2004–2005 (Initial)	2004–2005 to 2006–2007 (3 yrs)	2007–08 to 2010–11 (4 yrs)	2011–12 to 2014–15 (4 yrs)	Pooled (11 yrs)
T_1: 50 % N_{FYM} + 50% NPK	219.7	219.5	219.7	226.9	222.3
T_2: $^1/_3 N_{FYM}$+1/3$N_{vermicompost}$ + 1/3$_{Neem\ cake}$.	182.4	189.3	212.7	230.6	212.8
T_3: T_2 + intercropping	190.8	206.3	239.9	261.5	238.6
T_4: T_2+ Weed control	188.4	192.1	215.4	230.4	214.5
T_5: 50% N_{FYM} + B.F. (N) + B. M. + P.S.B.C.	164.7	172.8	193.6	191.1	187.0
T_6: T_2+ B.F. (N & P)	186.9	192.9	216.9	235.1	217.0
T_7: 100 % NPK + S	226.0	223.0	212.3	216.2	216.6

Source: Annual Report, All India Coordinated Research Project on Cropping System, Bihar Agricultural College, Sabour, Bhagalpur.

TABLE 11.4 Effect of *Azolla* and Other Organic Treatments on Grain Yield of Rice During *Kharif* (2003–2008).

Treatment	Rice yield (t/ha)		
	2003	2006	2008
Azolla @ 1.0 t fresh biomass /ha (A)	2.87	2.36	2.06
BGA @ 2.0 kg (dry)/ha (B)	2.70	2.29	1.90
FYM @ 5.0 t/ha (F)	2.69	2.12	2.04
Vermicompost @ 5.0 t/ha (V)	2.92	2.60	2.18
A+B	3.35	3.16	2.68
A+F	3.70	3.29	2.57
A+V	4.08	3.57	2.64
B+F	3.33	3.48	2.62
B+V	3.91	3.52	2.82
F+V	3.75	3.56	3.02
A+B+F	4.05	3.68	3.34
A+F+V	4.08	3.89	3.45
B+F+V	4.10	3.93	3.64
A+B+F+V	4.19	4.16	3.68
N80P40K30	4.93	4.34	3.46
N0P0K0	2.20	1.78	1.68
C.D (at 5% P)	0.95	0.41	0.26

Source: Yadav et al. (2014).

On the basis of experimental findings of network project on organic farming, it can be concluded that for the below-mentioned cropping systems/ pattern, the following nutritional modules were found to be best suited to increase the crop yield:

Place	Cropping system	Source of nutrient
Jabalpur	Basmati rice–D wheat-GM	VC +FYM+NEOC @ 1/3 N each + *Panchgavya*
	Basmati rice–berseem	VC +FYM+NEOC @ 1/3 N each + *Panchgavya*
Coimbatore	Cotton–maize–sunhemp	FYM+NEOC @ 1/2 N each + *Panchgavya*
	Chillies–cauliflower–sunhemp	FYM+NEOC @ 1/2 N each + *Panchgavya*
Raipur	Rice–chickpea	EC + CDM + NEOC @ 1/3 N each + *Panchgavya*
	Rice–mustard–lentil	EC + CDM + NEOC @ 1/3 N each + *Panchgavya*
Dharwad	Groundnut–sorghum	VC + EC + GLM @ 1/3 N each + Biodynamic spray @ 12.5 g/ha + *Panchgavya*
	Maize–chickpea	VC + EC + GLM @ 1/3 N each + Biodynamic spray @ 12.5 g/ha + *Panchgavya*
Karjat	Rice–Red pumpkin	*Kharif*- FYM + Rice straw + glycidia leaf @ 1/3 N each *Rabi*- FYM + NC + VC @ 1/3 N each + *Panchgavya*
	Rice–cucumber	*Kharif*- FYM + Rice straw + glycidia leaf @ 1/3 N each *Rabi*- FYM + NC + VC @ 1/3 N each + *Panchgavya*
Ludhiana	Maize–wheat + gram–S. mungbean	Rice–GM+ BD + FYM Other cropsFYM + PG + BD
	Rice–wheat	Rice–GM+ BD + FYM Other crops FYM + PG + BD
Pantnagar	Basmati rice–chickpea–Sesbania	FYM + VC+ NC + EC @ 1/4 N each +*Panchgavya*
	Basmati rice–V. pea–maize + moong	FYM + VC+ NC + EC @ 1/4 N each+*Panchgavya*
Ranchi	Rice–wheat	VC + KC@1/2 N each + BD+*Panchgavya*
	Rice–potato	VC + KC@ 1/2 N each + BD+*Panchgavya*

BD: Biodynamic; CDM: Cow dung manure; EC: Enriched compost; FYM: Farmyard manure; GLM: Green leaf manure; KC: Karanj cake; NEOC: Nonedible oil cake; PG: *Panchgavya*; VC: Vermicompost.

Source: Annual Report (2011–2012), Network Project on Organic Farming, Project Directorate for Farming System Research, Modipuram, Meerut.

11.5 EPILOGUE

There is a general agreement that organic farming is essential for agricultural sustainability and environmental protection. As nutrient additions on organic farms are designed to maintain soil fertility along with providing nutrients to plants, soil health is required to be improved for the adoption of organic farming. For the purpose, there is a need of system-based organic farming rather than to grow a single crop organically. Green manure, vermicompost, farmyard manure, cakes, leaf manure, *panchgavya,* and biodynamic preparation are the potential inputs for nutritional management in system-based organic farming. A variable combination of all these need to be applied for each system and location. There is a need of more research to make nutrient management system more profitable in organic farming.

KEYWORDS

- **organic farming**
- **traditional agricultural practices**
- **global area**
- **NPOF**
- **organic certification**
- **nutrient management**
- **cropping system**

REFERENCES

APEDA. *Organic Production and Current Scenario in India.* http://www. apeda.gov.in/apedawebsite/organic/Organic_Products.htm (accessed Dec 20, 2013).

APEDA. *Organic Production and Export Scenario.* http://www.apeda.gov.in/apedawebsite/organic/Organic_products.html (accessed Dec 20, 2013).

FAO. *Good Agricultural Practices for Greenhouse Vegetable Crops: Principles for Mediterranean Climate Areas*; FAO: Rome, 2013. ISBN 978-92-5-107649-1.

FiBL-IFOAM Survey. *Organic Agriculture Worldwide: Current Statistics*; HelgaWiller, Research Institute of Organic Agriculture (FiBL): Frick, Switzerland, 2012.

Ladha, J. K. Preface. *Proceeding of 13th Congress of Soil Science on Biological Nitrogen Fixation for Sustainable Agriculture* held at Kyoto, Japan, Kluwer Academic Publishers: London, 1992.

Reganold, J. P.; Glover, J. D.; Andrews, P. K.; Hinman, H. R. Sustainability of Three Apple Production Systems. *Nature* **2001,** *410*, 926–930.
Seufert, V.; Ramankutty, N.; Foley, J. A. Comparing the Yields of Organic and Conventional Agriculture. *Nature* **2012,** *485*, 229–232.
Shukla, U. N.; Mishra, M.; Birwa, K. C. Organic Farming: Current Status in India. *Pop. Kheti* **2013,** *1* (4), 19–25.
Yadav, R. K.; Abraham, G.; Singh, Y. V.; Singh, P. K. Advances in the Utilization of *Azolla*—Anabaena System in Relation to Sustainable Agricultural Practices. *Proc. Indian Natl Sci. Acad.* **2014,** *80* (2), 301–316.

CHAPTER 12

Organic Farming Technology for Mitigating Effects of Climate Change and Ensuring Livelihood Sustainability

RAJESH KUMAR DUBEY[1] and PRIYANKA BABEL[2]

[1]*National Resource Centre – Biotechnology, MLS University, Udaipur 313 001, Rajasthan, India*

[2]*Biotechnology & Microbiology, MLS University, Udaipur 313 001, Rajasthan, India*

**Corresponding author. E-mail: directornrc8@gmail.com*

ABSTRACT

In the recent years, organic farming is gaining more and more importance as it is a sustainable method of farming and it has also resulted in an increased income source of smallholder farmers. In the past, most farmers used chemical fertilizers and pesticides for cultivation. Furthermore, it is an eco-friendly based innovative practice that is focused on diversity within the farming system and it aims to maintain a balance and healthy functioning system, improving soil organic carbon and soil structure that is helpful in reducing the impact of droughts and high intensity rainfall by improving water infiltration into the soil, and emphasizing on rainwater harvest so that the water supply to crops may improve due to better soil water retention and root growth. For sustainability of biological diversity and climate change adaptation, there is a need to adopt a systematic approach of organic farming through production management, minimizing

Organic Crop Production Management: Focus on India, with Global Implications, D. P. Singh, PhD, H. G. Prakash, PhD, M. Swapna, PhD, & S. Solomon, PhD (Eds.)

energy randomization of nonrenewable resources, and carbon sequestration. Primarily, organic farming could boost the quality of food by enhancing protein, vitamins, minerals, etc. Soil health and ecological functions such as biomass production, biodiversity maintenance, environmental protection, etc., which occur in organic farming could also be maintained or improved. In this way, it is possible for climatic aberrations could be mitigated or alleviated.

12.1 INTRODUCTION

Food is a medicine and organic products are the best nutritive food available so far in the world market. Organic farming is practiced in maximum 149 countries on more than 51 million hectares of land. India is predominantly an agricultural nation which contributes to nearly 14.6% in gross domestic product and provides livelihood to more than 55% of the country's population (GOI, 2010). More than 1 billion extremely poor people of the globe and nearly 810 million residing in the marginal areas are dependent on farming for their livelihood. The biodiversity and cultural diversity of India can be protected by promoting organic farming (Dubey, 2013a, 2013b).

Agriculture, which provides employment to majority of the population, remains the prime sector of the economy in most of the developing economics (Bage, 2005).

India ranks seventh in the largest nations of the world by geographical area. It is the second-most populous country after China with over 1.35 billion people and the largest populous democracy in the world. India is basically an agricultural economy being the 10th largest economy of the globe by nominal GDP and fourth largest economy from the point of view of purchasing power parity. India has emerged as one of the fastest growing major economies after adopting economic liberalization during 1990s and is transforming into a new industrialized country with number of new challenges due to ever-increasing population, stagnation in farm yield in intensive farming areas. This rapid and continuing increase in population implies a greater demand for food. Rapid industrialization and urbanization are increasingly causing problems of soil, affecting structure and functioning of the agroecosystems.

As we celebrate the 150th birth anniversary of father of nation, Mahatma Gandhi, we must remember he said that a free India meant the existence of thousands of self-sufficient small communities who reign themselves without creating obstacles for others. Presently, this is still quite imaginary.

The process of one-sided development can be corrected by imparting education to the rural poor and providing them access to the latest technological advancements to empower them. By rural development, we mean the economic betterment of people along with greater social transformation. People should participate more efficiently in rural development programs for providing them better prospects for the economic development. Decentralization planning, effective adoption of farm and land reforms and proving higher access to credit have been envisaged by the Government of India under its several programs for rural development. The correct dissemination of knowledge is the major constraint at every level (Dubey, 2013a, 2013b).

Organic farming is the oldest as well as the most modern eco-friendly technology most practiced worldwide. The highest number of organic farmers in India makes it the leading nation of the globe. Organic farming is growing at the rate of 18–20% per annum. Organic products are in the best demand in health industry, food industry, and medicine industry.

12.2 ORGANIC FARMING TECHNOLOGY

The principle of organic farming is supporting and strengthening biological processes without using chemical fertilizers or genetically modified organisms. Organic farming is more productive and is better than traditional agriculture. Organic agriculture is an integrated production management system which boosts the health of agroecosystem, including biological cycles, biodiversity, and soil biological activity. This integrated approach of organic farming focuses on the adoption of management practices, that include location/region/zone-specific agricultural practices. This integrated approach will include all the agronomic, biological, and mechanical methods based on natural farming. These practices will not include any chemical/ synthetic material.

12.3 ORGANIC FARMING TECHNOLOGY FOR GREEN GROWTH

Use of renewable resources and recycling means returning to the soil the nutrients found in waste products. Organic farming is contrary to modern intensive farming system wherein farmers use a range of techniques, helpful to sustain ecosystem and reduce pollution. Unscientific overuse of pesticides, especially in vegetables and fruits at certain farms or orchards resulted in the accumulation of residues above the safety levels. Similarly, the ill effects of unbalanced and unscientific overuse of chemical fertilizers were recognized in developed countries. This focused the attention of food producers to adopt organic farming. Organic farming encompasses location-specific management practices for the agroecosystem based on the soil capacity. During the last few years, organically produced agricultural products have received worldwide attention because of their multibillion trades and handsome profit to all involved in this business. The organic farming is beneficial as it provides opportunities to contribute to vibrant rural economics through sustainable development. Organic farming will provide ample opportunities for several new employment opportunities in agroprocessing and other allied sectors. Under organic farming, the peasants adopt soil carbon sequestration technique by using environmentally friendly bioenergy and biofertilizer, resulting in increasing water holding capacity of the soil in comparison to the traditional cultivation. The climate-related problems will be solved by sound application of organic farming and will be helpful in achieving higher crop production for ever-increasing population. Thus, organic farming will be helpful in increasing yield, ensuring household and national food security, and will act as a powerful tool for climate change adaptation and mitigation initiatives. The organic industry has recorded unprecedent growth during last few decades. The industry has witnessed growth to $29 billion in 2010 which was only $3.6 billion in 1997, recording the annual growth rate of 19% during the period of 1997–2008. Thus, the organic farming sector recorded the growth of 8% in 2010.The latest data clearly revealed that about 96% of nationwide organic practices will not reduce the employment opportunities during 2011 (Dubey, 2013a, 2013b). Organic farming as an ecological approach that cannot only produce safe food but is eco-friendly, farm-friendly, livelihood friendly, and sustainable.

12.4 ORGANIC FARMING TECHNOLOGY FOR MITIGATING CLIMATE CHANGE

Crop growth and development are mainly governed by environmental conditions and soil health. The success and failure of farming and qualitative improvement in agriculture depends on prevailing climatic conditions. The most imminent of the environmental change of the earth is the increase in the atmospheric temperature due to increased level of carbon dioxide and other greenhouse gases. Production of enough food for the increasing population in adverse environmental conditions, while minimizing further environmental deterioration is the challenging task of agriculture. Climate change is a tragedy of the commons. For mitigation of climate change, organic farming is the best technique which increases the production and productivity by attending cause of climate change.

For combative action against climate change, the two most important strategies are mitigation and adaptation. Mitigation is stopping the cause of warming and adaptation is addressing the effect of climate change. As organic farming is an eco-friendly technique based on minimum petroproduct-based external input supply, it involves low emission, and thus reduces the magnitude of climate change energy and biomass plantations support geoengineering and offset the global warming. Adaptation techniques such as endemic varieties and local seeds limit the vulnerability and impact of climate change. Adaptation techniques deal only with the impact on human and not on ecosystem and environment.

12.5 POTENTIAL AREAS AND NEW INITIATIVES

12.5.1 SOIL HEALTH AND SOIL FERTILITY

Inherent capacity of soil to release necessary nutrients to plants in sufficient quantity and in correct ratio for their proper growth is known as soil fertility. Fertilizer requirement is the amount of some plant nutrients required apart from the quantity released by the soil, to boost the plant growth to an optimum level. India ranks third in the globe after China and the United States of America in production and consumption of fertilizers, by accounting for 12.2% of the global nitrogen and phosphorus production and 12.6% of the global consumption of nitrogen, phosphorus and potash nutrients (Dubey, 2013a, 2013b). Use of fertilizers in India is highly distorted and shows variations at interstate and interdistrict level. Widespread deficiency of nitrogen–phosphorus–potash ratio reduces the soil fertility. Indian soils are revealing deficiency of all the macro and micronutrients. There is deficiency of NPK along with other macronutrients, such as sulfur, magnesium, and calcium and a number of micronutrients, such as like zinc, iron, boron, and copper, etc. in most of the parts of India. The extent of sulfur deficiency ranged from 20 to 55% with overall deficiency of 30% in soil. Manganese, boron, and iron deficiency emerged as a serious constraint to high productive crops. To feed increasing population of the country, Indian farmers have to adopt intensive agriculture which led to second generation challenges with respect of nutrient imbalance (Dubey, 2013a, 2013b). Some such problems include:

- Leaching of soil nutrients resulting in depletion of soil fertility.
- Emerging deficiencies in macronutrients and micronutrients.
- Depletion in water table and poor water potability.
- Declining soil humus & soil carbon–nitrogen ratio (C/N ratio).
- Improvement in physical, chemical and biological properties of soil.

12.5.2 ORGANIC FARMING FOR FOOD QUALITY

Soil fertility can be maintained through organic farming to restore the lost fertility nutrient status for optimum soil productivity. It emphasizes that the capacity of a soil to produce crops should be expressed in terms of yield. The objectives of organic farming are to produce healthy and safe foods. By adopting organic methods of production and safe methods of processing, those are in accordance with the internationally recognized principles higher growth in crops can be achieved. Continuous use of organic manures improves soil chemical, physical, and biological properties and the supply of all the essential plant nutrients—the components of soil fertility and productivity. A survey has revealed that although crop productivity is reduced by 9.2% in a certified organic farm in comparison to conventional farming, it enhances the net profit to farmers by 22.0%. This can be attributed to the availability of premium price (20–40%) for the certified organic produce and reduces the cost of cultivation by 11.7%.

The reduced productivity and yield losses along with improved soil quality parameters are recorded in organically managed farms which is an indicator of better soil health (Dubey, 2013a, 2013b). It is more economical, because its produce fetches premium price in the local and the export market, it will be adopted by the farmers. Better quality, high nutritive value, and taste of organically produced food have been scientifically established. Low yield in organic farming highlights the urgent need of priorities both national and international research activities. European countries, who are pioneer in organic agriculture research, spend nearly €60 million per annum to solve the specific challenges of organic farming. There is a need to enhance the productivity of organic crops in India also through similar research efforts. (Dubey, 2013a, 2013b). For better evaluation and to prepare research projects on organic farming, it is the need of the hour to examine its status in various cropping or farming systems of India with enough farms in each case.

12.6 NANOBIOTECH FOR INCLUSIVE DEVELOPMENT AND CLIMATE SMART TECHNOLOGY

Nanobiotechnology is a future prospective application for counting carbon credit as well as for developing the carbon-philic materials for efficient carbon sink creations. In most of the developing economies, inclusive development is given importance as one of the most important goals of socioeconomic development. Inclusive development involves benefitting poor and disadvantaged people in earning their sustenance. Modern biotechnological tools are used to increase the quantity and improving the quality of food, in the protection and management of environment and natural resource conservation. Carbon sequestration in soils is increased by C sink and decreased their decomposition. It can also be increased through production of nanoparticles for capturing carbon. Crop either removes the phosphorus or phosphorus is converted into different insoluble forms. Since the efficiency of utilization of phosphate is low and significant amount is lost from the soil through erosion. Certain biotechnological approaches such as use of wide variety of microorganism which can transform insoluble forms of nutrients in soluble forms and enhances its availability. Metagenomics-based nanotech applications can be integrated with ICT for climate smart technology in function to boost inclusive development at multilocus in both horizontal and vertical modes of technology dissemination program. Ecological improvement of rhizosphere by following rhizosphere technology also increases the metal sequestering which further enhances crop production.

12.7 STRATEGY FOR FOCUSED APPROACH

Organic farming does not only provide quality food to human beings but it also takes care of soil's health and the environment by not leaving any adverse impact on the same. Suitable crops/products on regional basis for organic production may be identified for global market demand (Yadav et al., 2013). Agricultural policies could then focus on enabling people and professionals to make use of the most of the available social and biological resources. The rising demand for organic food products in the developed nations and the support extended by the Indian government coupled with its emphasis on farm exports are the drivers for the Indian organic food industry. For promoting organic farming in India, there is need to create scientific knowledge, technical capacity, identification of constraints, and developing the strategies to overcome them (Dubey, 2013a, 2013b).

Therefore, promotion of organic farming and technical capacity building should be taken up with emphasis on following aspects:

- Scientific information and capacity building for promotion as eco-friendly and climate-resilient techniques.
- Production and promotion of organic inputs with ensuring quality control.
- Soil health assessment from organic and biological perspective.
- Technology information, development and dissemination.
- Strengthening product quality assurance system.
- Creating of mass awareness through electronic and print media.

12.8 CONCLUSION

Organic farming is an eco-friendly based innovative practice that is focused on diversity within the farming system and it aims to maintain a balance and healthy functioning system, improving soil organic carbon and soil structure that may be helpful in reducing the impact of droughts and high intensity rainfall by improving water infiltration into the soil, and emphasizing on rainwater harvest so that the water supply to crops may improve due to better soil water retention and root growth (Roychowdhury et al., 2013). The nutrients and carbon sequestration in soils can play an important role in mitigation and adaptation to climate change in various climate zones along with a wide range of location-specific conditions. For sustainability of biological diversity and climate change adaptation, there is a need to adopt a systematic approach of organic farming through production management, minimizing energy randomization of nonrenewable resources, and carbon sequestration. Therefore, the organic farming aims an effort for a slow reversal process of the effects of climate change for building resilience and sustainability by addressing the prime issues of concern (Roychowdhury et al., 2013; Mendoza and Thelen, 2008). Organic agriculture is a holistic sustainable production system which can contribute to effective climate change mitigation strategies for the farm sector.

Organic farming is employment friendly technique as it provided 10–12% more jobs on farm activities. With the integration of ICT and nanotechnology-based techniques, organic products are gaining utmost importance among consumers, researchers, administrations, policy makers, entrepreneurs, and farmers every day. Organic farming is not only considered as one of the greenest technologies because it protects the environment, restore soil health, enhance biodiversity for climatic safety but organic foods are considered as much nutritious, safer, tastier, and healthy.

There is a need to undertake research work on productivity levels and institutional environment for organic cultivation as well as on its mitigation and sequestration potential (Roychowdhury et al., 2013; Mendoza and Thelen, 2008). For popularization and developing common practice, better technology of organic farming more customized packages and practices, appropriate technology of training and capacity building modules, hands on practices, and skill development programs are required to make it inclusive technology for livelihood security and sustainability.

KEYWORDS

- **organic farming technology**
- **green growth**
- **soil health**
- **food quality**
- **nanobiotechnology**

REFERENCES

Båge, L. Statement Delivered on the Launch of the MDG Report, 2005. http://www.ifad.org/events/mdg/ifad.htmCembalo (accessed Jan 18, 2005).

Dubey, R. K. Improvising Green Growth through Agri-Organic and Biotech Innovations for Sustainable Development. Int. J. Sci. Eng. Res. **2013a,** 4 (11), 271–274.

Dubey, R. K. Innovations for Green Growth & Inclusive Development. Int. J. Eng. Res. Tech. **2013b,** 2 (10), 2698–2701.

GOI. 2010: Union Budget and Economic Survey, 2010. http://indiabudget.nic.in (accessed Sept 7, 2011).

Mendoza, U. R.; Thelen, N. Innovations to Make Markets More Inclusive for the Poor. Dev Policy Rev. 2008, 26 (4), 427–458.

Roychowdhury, R.; Banerjee, U.; Sofkova, S.; Tah, J. Organic Farming for Crop Improvement and Sustainable Agriculture in the Era of Climate Change. Online J. Biol. Sci. **2013,** 13 (2), 50–65.

Yadav, S. K.; Babu, S.; Yadav, M. K.; Singh, K.; Yadav, G. S.; Pal, S. Int. J. Agric. **2013**. http://dx.doi.org/10.1155/2013/718145.

CHAPTER 13

Organic Horticulture for Sustainable Production and Livelihood Security in Drylands

P. L. SAROJ and HARE KRISHNA*

ICAR-Central Institute for Arid Horticulture, Beechwal, Bikaner 334006, Rajasthan, India

Corresponding author. E-mail: kishun@rediffmail.com

ABSTRACT

Due to low, erratic rainfall and poor soils in India's drylands, farming systems based on synergism with nature, and thus essentially "organic," have been followed for centuries. This opens a new vista of enormous opportunities for enhancing the soil health and in turn overall environment, thereby, providing sustainable livelihoods to dwellers of these regions, by modernizing these age-old systems with new age technological interventions. The key to success in proliferating organic farming lies in enhanced collegiality and coordination among all the stakeholders, such as government agencies, NGOs, SHGs, and growers for effective implementation of the programs being run for promotion of organic farming. By working together, the goals of promoting organic practices, improving soil health and local food security and, ultimately, creating markets for organic produce both regionally and globally, can be realized for the benefit of all of India.

Organic Crop Production Management: Focus on India, with Global Implications, D. P. Singh, PhD, H. G. Prakash, PhD, M. Swapna, PhD, & S. Solomon, PhD (Eds.)

13.1 INTRODUCTION

Similar to the organic farming, agriculture system relies upon the principle of binding with nature in order to sustain humans in the preindependence era. Thereby, Indian farmers trusted on the use of practices, such as crop rotation, crop residue incorporation to soil, animal originated manures, green manures, off-farm organic wastes, and biological pest management to preserve soil productivity, retain plant nutrients, and control of insects-pests, diseases, and other weeds on their agricultural farms. In postindependence era, brisk population growth in India caused enormous pressure on land as well as on traditional farming systems. The evergreen demand for food grains has headed toward the increased use of agricultural inputs, such as fertilizers, pesticides, and water to enhance agricultural production. Being the keystone of India's agricultural accomplishments, the Green Revolution has transformed the country from the stage of food scarcity to self-reliance by exploiting high-yielding cultivars and higher quantity of inputs (Deshmukh and Babar, 2015). Green Revolution, supported by the government policies and promoted by the usage of agrochemicals, irrigation, machinery, resulted into an enhanced agricultural production and thus, generated enhanced productivity. The agriculture sector in the country has made massive head way in the last 50 years. Food grains production during the post-Green Revolution era has enhanced fivefolds, that is, from 50.82 million tonnes in 1950–1951 to 275.11 million tonnes in 2016–2017. While new technologies enormously helped to address the issues of food security of the nation on one hand (Reddy, 2010), on the other hand, undiscriminating and disproportionate usage of agrochemicals during this phase has casted aspersion on the sustainability of agriculture in the long run, seeking attention for sustainable food production. Modern farming practices, along with unmindful usage of agro-inputs over the last four decades have led not only in forfeiture of balance of natural habitat and soil health but is also responsible for many threats, for example, soil erosion, soil salinization, reduced groundwater level, water pollution due to leaching of fertilizers, fungicides, insecticides and pesticides, negative impact on biodiversity, reduced food quality, and have also fueled the cost of cultivation, rendering the farmer poorer in bygone years. Consequently, the fascination toward organic agriculture as an eco-friendly means of crop cultivation is growing leap and bounds at both national and at international levels. This change toward organic production is supported by consumers who are perceptive to associated health hazards: increasing demand for organically grown food is being recorded by 20–25%

in developed countries with higher level of awareness among the citizens. To fulfill and address the economical, ecological, and social issues together, organic farming may play a pivotal role.

Organic farming is defined as a holistic management system, which utilizes traditional as well as scientific knowledge for improving the agroecosystem health. Organic agriculture systems are based on ecosystem management instead of use of external agri-inputs (IFOAM, 2006). Similarly, FAO (2002) describes organic agriculture as an environmentally and socially sensitive food supply system. The fundamental aim of organic agriculture is to boost the health for enhancing the productivity of soil, plants, animals, and people. Therefore, organic farming is a production practice with augmented use of local resources in such a fashion that sustainability of production and wellness of the society and environment can be sustained in long run. The key features of organic farming have been highlighted here:

Key Characteristics of Organic Farming

- Preservation of soil fertility via maintaining organic matter levels, encouraging soil biological activity, and vigilant mechanical intervention.
- Endowing essential nutrients through solubilization of insoluble mineral sources by improved microbiological activity.
- Realization of nitrogen self-sufficiency through the use of leguminous crops and biological nitrogen fixation, as well as effective recycling of organic materials like crop residues and livestock manures.
- Management of weed, disease and pest primarily through crop rotations, natural predators, crop diversity, organic fertilization, resistant varieties, and minimal application of synthetic pesticides.
- The extensive livestock management as a fundamental component of organic farming.
- Crop cultivation without impairing conservation of wildlife and their natural habitats.

India's rainfed agroecosystem covers three distinct zones, namely, subhumid, semiarid, and arid zones. Arid zone represents more than 70% of the geographical land area. Since 66% of the 142 mha cultivated area is under rainfed dryland condition (Venkateswarlu et al., 2005), India has the immense potential to increase its area under organic farming (Anonymous, 2016). At present, our country is the second biggest producer of both fruits

(92.8 mt) and vegetables (175 mt) in the world, after China (Horticultural Statistics, 2017). However, Indian population is likely to grow by ~35% of the present population by 2050 and consequently, the demand for fruits and vegetables is also likely to grow by 183% and 246%, respectively by 2050 (Ghosh, 2012). This gap in demand can be met by bringing more waste lands from dryland areas into cultivation, where organic cultivation of horticultural crops is being practiced by default from centuries.

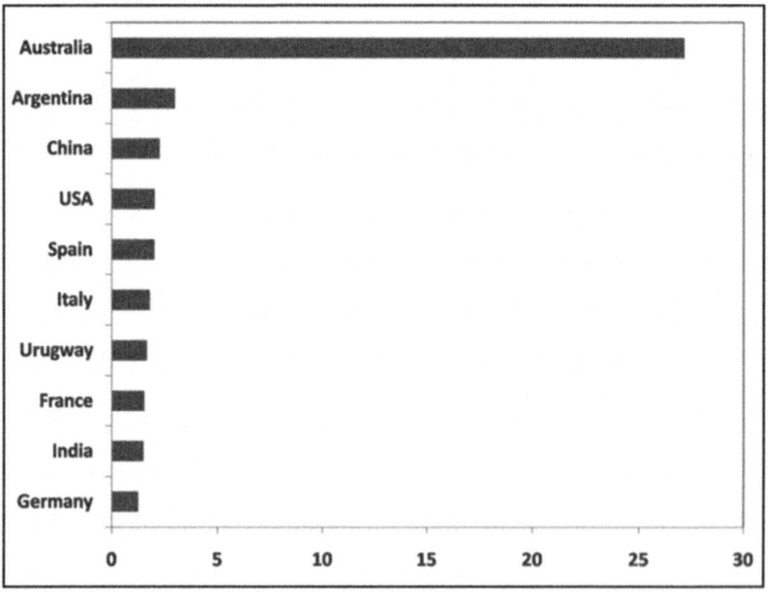

FIGURE 13.1 Top ten countries with the largest area of organic agricultural land 2016.
Source: FiBL & IFOAM, 2017

13.2 ORGANIC FARMING IN INDIA VIS-À-VIS WORLD SCENARIO

Only 1% of total global agriculture land is being used under the organic component. But it is interesting to note that nearly 11 nations in the world have more than 10% area under organic farm lands. The available data on total farm land clearly reveals that there is about 27 million ha grazing land, about 3.7 million ha area under permanent crops like orchards and others and 7.7 million ha arable crops. More than 150 countries on the globe are now exporting certified organic products. It is estimated that organic trade has been growing at the rate of 15–20% annually (Mitra and Devi, 2016). There are about two million organic farmers in the world, and Asia continent is at the top of the list with 36% of organic growers followed by Africa (29%) and Europe (17%).

India ranked ninth with respect to World's Organic Agricultural Land in 2015 (FiBL & IFOAM, 2017). The total land area under organic certification is estimated to be 5.71 mha as of 2015–2016, which comprises 26% of cultivable area with 1.49 mha and rest 74% (4.2 mha) of land includes forest and wild areas for gathering of minor forest produces.

India produced nearly 1.35 mMT of certified organic products during the year 2015–2016 which encompasses a wide gamut of food products, such as oilseeds, cereals and millets, sugar, pulses, tea, fruits, cotton, medicinal plants, spices, coffee, fresh fruits, and dry fruits, vegetables. The production has not been confined to the edible sector but has also spread to functional food items, organic cotton fiber, etc. (FiBL & IFOAM Survey, 2015).

The country is home to the largest number of organic growers in the world as per the World of Organic Agriculture Report-2018. There are 835,000 certified organic farmers in India which constitutes over 30% of the global organic growers (2.7 million).

The major organic food producing states in the country are Madhya Pradesh, Karnataka, Maharashtra, Gujarat, Rajasthan, U.P., and Odisha. In Rajasthan, the districts having potential for organic production are Jaisalmer, Barmer, Jodhpur, Jalore, Pali, Sirohi, Dungarpur, Nagaur, and Jhunjhunu. Currently, only 56,106.747 ha area (excluding wild) falls under organic certification in the State (Anonymous, 2017). At national level, Madhya Pradesh tops the list of the states which has the largest area under organic certification pursued by states of Himachal Pradesh and Rajasthan.

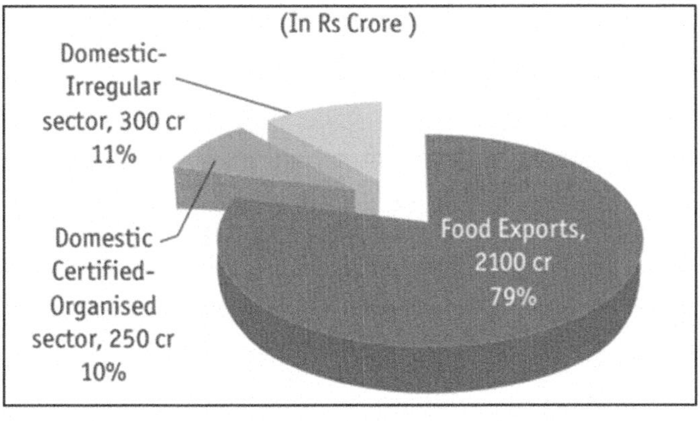

FIGURE 13.2 Indian organic food sector overview.

Countries like Uganda (210,352) and Mexico (210,000) rank as the second and third largest producers on globe, respectively (FiBL & IFOAM 2017). India's contribution to total global area (57.8 mha), under certified organic cultivation is only 2.59% (1.5 mha) of the total area. China has the lion's share which is to the tune of around 50%. India has 30% of total organic cultivable land in Asia. As per the 19th edition of the World of Organic Agriculture Report, organic agriculture area and its products value has substantially grown up. The required data was gathered from 178 nations by the Research Institute of Organic Agriculture (FiBL), the State of Sustainability Initiative (SSI), and International Trade Centre.

Organic food sector in India has been pegged at INR 2700 crore (approximately USD 415 million) that accounts for less than even 1% of the total worldwide organic food market valued at approximately USD 90 billion in the year 2015. At present, the Indian Organic food sector is exporting worth Rs. 2100 crore and an organized sector domestic certified market estimated between Rs. 250–300 crore. Global trade in organic products is recording an upward trend. However, the latest trends from India revealed something positive from the export angle as (1) During the year 2014, organic food including coffee, olive oil, soybean, bananas, honey, and wine amounting to nearly USD 1.3 billion was imported by the United States of America. India has been a principal supplier of tea, rice, ginger, and soybean to the United States of America. (2) The second largest market for organic products of the world is European Union which imports organic products mostly from 11 countries, including India, besides, Argentina, New Zealand, Australia, Tunisia, Israel, Costa Rica, the United States of America, Japan, Canada, and Switzerland. The United States of America continues to be a chief export market for Indian organic food processors and traders.

Major consumers of packaged organic products in the world are the United States of America, Germany, Italy, the United Kingdom, and France. The United States of America is the major consumer of organic products with a consumption worth around USD 15 billion, followed by Germany (USD 4.2 billion) and France (USD 3.5 billion). China, Brazil, Argentina, Turkey, Colombia, and India are emerging as other organic retail markets of the world. It is worth to mention here that in the future, China, Argentina, India, and Morocco will grow at considerable growth rates unto 20–30% in the packaged/processed food segments (Tables 13.1 and 13.2).

TABLE 13.1 Category Wise Break-Up of Organic Crops Production (2014–2015).

Category	Production (MT)	Share (%)
Sugar crops	338,193	30.9
Oilseed crops	228,414	20.8
Fiber crops	208,931	19.1
Cereals and millets	159,500	14.6
Pulses	34,717	3.2
Plantation crops	33,930	3.1
Medicinal/herbal and aromatic plants	32,663	3.0
Fruits	20,219	1.8
Spices and condiments	18,176	1.7
Vegetables	10,824	1.0
Dry fruits	7348	0.6

Source: Indian Organic Sector: Vision 2025.

Horticultural crops constitute around 11% of total organic production in the country. The maximum share of organic produce belongs to plantation crops followed by medicinal and aromatic plants, fruits, spices and condiments. The major destination of horticultural produce like fruits is the United States of America and the European Union. Similarly, the United States of America and the European Union along with Canada or Switzerland continue to be the major destinations for vegetables, spices, and condiments. The total amount of export during 2015–2016 was 263,687 MT costing around USD 298 million. European Union, the United States of America, Switzerland, Canada, Australia, New Zealand, South Korea, South Africa, Southeast Asian countries, and Middle East countries are major importing countries for organic products.

TABLE 13.2 Export Destination of Organic Produce (2014–2015).

Export category	Quantity (MT)	Major destination
Oilseed crops	160,559	USA, Canada, European Union
Cereals and millets	63,622	Canada, European Union, USA
Processed foods	23,626	European Union, USA, Canada
Sugar crops	19,450	European Union, USA
Tea	5,488	European Union, USA, Japan
Pulses	2547	USA, Canada, European Union

TABLE 13.2 *(Continued)*

Export category	Quantity (MT)	Major destination
Dry fruits	2417	USA, European Union, Canada
Spices and condiments	2403	USA, European Union, Switzerland
Medicinal, aromatic and herbal products	1223	European Union, USA, Australia
Coffee	1214	European Union
Essential and aromatic oils	866	USA, European Union, Japan
Fiber crops	397	Turkey
Tuber crops	139	USA, European Union
Edible oils	130	Switzerland, USA, European Union
Fruits	124	USA, European Union
Vegetables	109	Canada, European Union, USA
Ornamental plants and flowers	78	USA
Plantation crops (non-tea and coffee)	32	European Union
Honey	18	Australia, USA, Malaysia
Others	1165	European Union, USA, New Zealand
Total	285,608	

Source: Indian Organic Sector: Vision 2025.

An incentive of Rs 8000 ha^{-1} per farmer is being provided by the Government of Rajasthan to promote organic farming in the state. It is a cluster-based program in which assistance is given for 0.5–2 ha area. *Paramparagat Krishi Vikas Yojana* (PKVY) is also promoting organic farming through adoption of organic village by cluster approach and Participatory Guarantee Systems certification. The commercial organic production is promoted under the Scheme through certified organic farming and natural resource mobilization for traditional input production.

13.3 ORGANIC HORTICULTURE IN DRYLANDS

The projected population of the country is expected to range between 1.64 and 1.74 billion by 2050 and to feed this ever-increasing population, the demands for horticultural produce will also increase further and regions that fall under culturable waste land can be utilized for organic cultivation of arid horticultural crops in order to meet the increasing demands of quality produce. Further in the arid and semiarid regions of the Rajasthan, animal husbandry is an essential component of livelihood for large segment of its rural population. The 19th Livestock Census 2012 revealed that there were 577.32 lakh livestock, including buffalo, cattle, goat, camel, sheep, horse and ponies, mules, donkeys, pig, and 80.24 lakh poultry in India. Animal husbandry provides food, draught power, and dung as organic manure and domestic fuel and is a regular source of income for rural households. Integrating livestock sector with organic farming will have a noteworthy fruitful impact on creating employment and reducing poverty in rural belts. The majority of the arid/semiarid horticultural crops need comparatively less water than several other common field/horticultural crops, which makes them amenable for organic cultivation under drylands.

For commercial organic horticultural production, few selected vital fruit crops and their varieties are given below in Table 13.3:

TABLE 13.3 Suitable Varieties of Different Fruits Crops.

S. no.	Fruit crop	Variety
	Ber	Gola, Seb, Kaithali, Mundia, Chhuhara, Umran, Goma Kirti, Thar Bhubhraj, Thar Malti
	Bael	NB-9, NB-5, Thar Neelkanth, Goma Yashi
	Aonla	Goma Aishwarya, NA-7, Kanchan, Krishna
	Mulberry	Thar Lohit, Thar Harit
	Karonda	Pant Manohar, Pant Sudarshan, Thar Kamal, Maru Gaurav
	Lasoda	Thar Bold, Maru Samridhi
	Phalsa	Thar Pragatio

Likewise, some of the important annual crops, such as vegetables and seed spices, which can be grown for commercial organic horticultural production are presented in Table 13.4.

TABLE 13.4 Suitable arid Vegetables, Spices, and Legumes for Inclusion in Fruit-Based Cropping Systems Along with Their Cultivation Requirement.

Crop	Varieties	Seed rate	Season	Row to row × plant to plant distance	Yield
Vegetables					
Clusterbean	Thar Bhadavi	8–12 kg/ha	*Kharif*	45 × 20 cm	80–110 q/ha

TABLE 13.4 *(Continued)*

Crop	Varieties	Seed rate	Season	Row to row × plant to plant distance	Yield
Kachari	AHK-119	0.5–1.0 kg/ha	*Kharif*	200 × 50 cm	95–100 q/ha
Snap melon	AHS- 82	0.5–1.0 kg/ha	*Kharif*	200 × 50 cm	225–230 q/ha
Indian bean	Thar Maghi	4–5 kg/ha	*Kharif*	200 × 20 cm	1.73 kg/plant
Mateera	Thar Manak	4–5 kg/ha	*Kharif*	200 × 50 cm	460–500 q/ha
Bottle gourd	Thar Samridhi	4–5 kg/ha	*Kharif*	200 × 50 cm	240–300 q/ha.
Seed spices					
Fennel	PF-35, Gujarat Fennel-1, Ajmer Fennel-1, RF-101, RF-125, RF-143	10–12 kg seed/ha (direct sown crop); 4–5 kg/ha (transplanted crop)	*Rabi*	40–60 × 20–30 cm	15–25 q/ha
Coriander	Narnaul Selection, Pant Haritima, Ajmer Coriander-1, RCR-41, RCR-435, RCR-436, RCR-684, RCR-446, Gujarat Coriander	10–12 kg per hectare (spice crop) 10–15 kg per hectare (Green Leaves), 20 kg /ha (rainfed crop)	*Rabi*	20–30 × 10–20 cm	15–20 q/ha under irrigated and 6–8 q/ha in the rainfed
Fenugreek	Pusa Early Bunching, Kasuri, Ajmer Fenugreek-1, Ajmer Fenugreek-2, Ajmer Fenugreek-3, RMT-143, RMT-305, Rajendra Kranti	*Desi* – 25–30 kg/ha. *kasuri*– 10 kg/ha	*Rabi*	20–25 × 10 cm	*Desi*; 15–20 q/ha, *kasuri*: 6–8 q/ha

TABLE 13.4 *(Continued)*

Crop	Varieties	Seed rate	Season	Row to row × plant to plant distance	Yield
Cumin	RS-1, RZ-19, RZ-209, RZ-223, Gujarat Zeera-3 and Gujarat Zeera-4	10–16 kg per ha (line sowing); 15–20 kg/ha (broadcasting)	*Rabi*	20–30 × 10 cm	8–10 q/ha
Legumes					
Groundnut	T.G. 37-A, TBG-39, HNG-10, Chandra and M 13	60–80 kg/ha	*Kharif*	30 × 10–15 cm	24–32 q/ha
Guar	HG-37, RGC-936, RGC-197, RGC-986, RGC-1017, RGC-1002, RGC-1003	15–20 kg/ha	*Kharif*	30 × 10–15 cm	12–16 q/ha
Moong	SML-668, MUM-2, RMG-62, K-851, RMG-492	12–15 kg/ha	*Kharif*	30 × 10–15 cm	12–16 q/ha
Moth	RMO-257, RMO-225, RMO-435, RMO-423 and RMO-40	10–12 kg/ha	*Kharif*	30 × 10–15 cm	12–16 kg/ plant

13.4 CHALLENGES OF ORGANIC FARMING WITH SPECIAL REFERENCE TO DRYLANDS

- It is mostly targeted at the export market as growers get higher prices for their produce in the international market. Therefore, home-grown food security and employment-related problems remain unaddressed.
- It primarily makes use of rigorous agriculture practices, which are not supportable in dry areas, where land degradation is already a matter of concern.
- Unavailability of farmyard manure and other forms of organic nutrients and water in desired quantities.

- In view of low availability of organic matter and poor fertility status of Indian soils, the level of productivity decreases in the course of conversion period could be sharp in the absence of external inputs.
- Credibility of Indian certification was not universally acceptable or partly suspected in top markets of Europe and the United States of America. Although Indian certification is not an issue in minor markets like United Arab Emirates and other Asian countries.
- The processes of Indian certification processes appear to be quite cumbersome and very inefficient and identified as the primary issue impacting the growth and efficiency of the sector.
- Lack of value chains-businesses, which are primarily linked to the markets and organic consumers.
- Inadequate policy support for wide spread promotion of organic farming.
- Nonavailability of desired volume of organic supplements, biofertilizers and home-grown market for timely disposal of organic produce.
- Lack of market intelligence and professional training, possibility of reduced yield, certification and input cost clubbed with capital-driven regulation by contracting firms dissuade small farmers in following organic cultivation (Deshmukh and Babar, 2015).
- Lack of ability to meet the export requirement for organic products, especially, of the United States of America, European Union, and Japan (Gaur, 2016).
- For promoting organic crop production in arid regions, suitable strategies may involve popularization of organic farming without obligation of certification, promotion of ley farming, integrating efforts of different stakeholders (ICAR, Ministry of Rural Development, State Agriculture Departments, NGOs, etc.), adopting superior methods of composting, increasing public sentience and capacity building, promotion of high-value crops, developing organic farming clusters of villages, developing certification programs and marketing chains, conducting research pertaining to organic production systems, etc. (Sharma et al., 2015).

13.5 ORGANIC CERTIFICATION IN INDIA

Organic crop culture, a holistic system that emphasizes on amelioration of soil health, utilization of locally available inputs, and comparatively higher usage of local labor, seems to be the best proposition for drylands. India's National Project on Organic Farming (NPOF), which was launched in 2004, has also accorded utmost importance to the drylands. In the year 2001, the Government of India has also launched the National Programme for Organic Production (NPOP). The benchmarks developed by NPOP were designed following the criterion established by international organic production standards like CODEX and International Federation of Organic Agricultural Movements (IFOAM) (Deshmukh and Babar, 2015). The NPOP standards developed for production and certification system have also been accepted by the European Commission and Switzerland. These standards have been given equivalent status to their country's standards. Likewise, the United States Department of Agriculture (USDA) has given recognition to NPOP conformity appraisal procedures of certification as equivalent to those in the United States of America. With such endorsements, the Indian organic products are presently acceptable to the importing countries. The recognition of Participatory Guarantee Systems (PGS) is also being touted as a major milestone Indian organic sector. Recently, the Food Safety and Standards Authority of India (FSSAI) launched an "Indian Organic Integrity Database" to help consumers confirm the genuineness of organic food. It has also launched a common logo for organic foods with the tagline "*Jaivik Bharat.*" The database and the associated portal verify the authenticity of the organic produce—certified by either third-party systems or PGS. This development brings more numbers of small and marginal farmers in the purview of certified organic markets. Besides 1.5 million organic hectares certified through the third-party systems, there are over 200,000 ha certified under PGS. The opportunity for certification of organic farm produce is far beyond this (FiBL & IFOAM Year Book, 2017). There are a total of 24 accredited certification bodies under NPOP, namely, Indian Organic Certification Agency (INDOCERT); Export Inspection Agency (EIA), New Delhi; Rajasthan Organic Certification Agency (ROCA); ECOCERT India Pvt. Ltd., Aurangabad; Chhattisgarh Certification Society, India (CGCERT), Raipur; Odisha State Organic Certification Agency (OSOCA), FairCert Certification Services Pvt. Ltd., Gujarat; Food Cert India Pvt. Ltd., Hyderabad; Aditi Organic Certifications Pvt. Ltd. Bengaluru; Vedic Organic Certification Agency, Hyderabad; Natural Organic Certification Agro Pvt. Ltd., Bhubaneswar; Biocert India Pvt. Ltd., Indore; Madhya Pradesh State Organic Certification Agency, Bhopal; Intertek India Pvt. Ltd., Mumbai; ISCOP (Indian Society for Certification of Organic Products), Coimbatore; Control Union Certifications, Mumbai; Tamil Nadu Organic Certification Department (TNOCD),

Coimbatore; SGS India Pvt. Ltd., Mumbai; APOF Organic Certification Agency (AOCA), Pune; One Cert Asia Agri Certification (P) Ltd., Jaipur; Lacon Quality Certification Pvt. Ltd., Thiruvalla (Kerala); Uttarakhand State Organic Certification Agency (USOCA), Dehradun; Bureau Veritas Certification India (BVCI) Pvt. Ltd., Mumbai, and IMO Control Pvt. Ltd., Bengaluru, Karnataka.

13.6 ADVANTAGES OF ORGANIC AGRICUTLURE FOR THE DRYLANDS OF THE COUNTRY

Owing to low and erratic rainfall and poor soil conditions in India's arid regions, cropping systems have remained almost "Organic" by default. Soil and climatic conditions of drylands in India are favorable for organic agriculture. Such marginal soils are in fact suitable for low-input production systems, which make ample use of biodiversity (Sharma, 2005). Further in drylands, the usage of synthetic inputs, that is, fertilizers, pesticides, etc., is uneconomic and generally risky (Sharma et al., 2015). This ushers into a new era which earmarks enormous opportunities for the amelioration of the soil and environmental health and providing sustainable livelihoods in such disadvantaged areas by harmonizing these age-old systems with modern research and techniques (Chhonkar, 2003).

- Soils of semiarid and arid dryland have poor water-holding capacity as well as organic matter contents (Sharma and Goyal, 2000). However, the depth of soil in some areas is another production constraint. Agrawal and Venkateshwrlu (1989) were of view to enhanced usage of organic manure for bringing sustainability in production in such areas. This practice is helpful in obtaining some production even under prolonged dry spell (Sharma, 2005).
- For allaying the effect of climate-related insecurities, providing balanced nutrition to crops through organic source emerged to be a feasible option. Organic farming practices suitable for drylands may be adopted to mitigate the problem of moisture stress, salinization of soil, pollution, desertification, etc., and thereby, mitigating the environmental degradation (Venkateswarlu et al., 2005).
- Organic farming is labor-demanding and relies on locally available agri-inputs. It has the ability to improve local food security and also has potential to render sufficient employment opportunities for

natives besides proper utilization of precious human resource (Gupta and Sharma, 1996).

- Due to high risk associated with crop production in dryland areas, the agrochemicals are generally used in very less quantity. As a result, farmers are practicing organic farming since ages by default. Apart from it, fertilizer use has been found to be miniscule in desert areas (FAI, 1998). Pesticide usage is also quite low. Moreover, a large tract of the drylands is still hailed as "virgin," implying no usage of synthetic inputs there to date (Sharma, 2005). This enables a rapid shift to organic horticulture, with insignificant reduction in productivity, quite easier.
- Due to climatic variability in the dryland areas, diversified farming systems including annuals, perennials, and animal components are used. Though these systems have sustained production plausibly under the existing climatic uncertainties, their yield levels are quite low, which may be attributed to inefficient use of resources (Sharma et al., 2015). The basic aim of organic farming is to minimize pest incidence as well.
- Organic farming has been found to improve soil quality and water retention in contrast to modern chemical-intensive farming. Growing low water requiring but high-value crops, such as spices and legumes can further reduce the water demand (Sharma et al., 2015). Other strategies that conserve moisture and improve nutrient use efficiency in drylands include mulching or mulch cum manuring, green leaf manuring, residue management, cover cropping etc. These are also the essential components of organic production systems.
- In India, native growers have a rich volume of traditional wisdom based on keen observations (made over the generation and passed onto subsequent generations) and practices, concerning soil fertility and pest management. This can be exploited to further reinforce organic systems (Sharma and Goyal, 2000).
- Organic horticultural farming has potential to be low-cost sustainable alternative of farming in India, especially, by the small farmers in rainfed areas.
- There is a strong belief in most of the societies that organically grown produce are nutritionally superior and good for health in comparison to the farm produce grown under conventional system.
- Enhancing demand of organically produced food even at premium price may improve economic conditions of the farmers in the region.

- With regards to agri-input supply, the drylands are bestowed with home-grown resources fit for backing-up organic crop husbandry (Sharma et al., 2015).
  (i) *Neem* (*Azadirachta indica*), *aak* (*Calotropis* spp.), *karanj* (*Pongamia pinnata*) and others are some of the best sources of biopesticides, available in abundance in drylands (Rajeshwar Rao, 1999).
  (ii) Likewise, mineral resources that help improve soil conditions and supply plant nutrients, such as gypsum, rock phosphate, and lime are available abundantly in large quantities under natural conditions in Rajasthan.
  (iii) Properly treated and decomposed wastes of animal and human origin are the best organic manures available. Conventional dryland farming systems are adequately provided with both (Purohit, 1998) the wastes.

13.7 ORGANIC FARMING PRACTICES FOR DRYLANDS

13.7.1 NUTRIENT MANAGEMENT

Organic farming is not merely about replacing chemical fertilizers with organically derived fertilizers and manures. Rather, it is about nurturing the soil ecosystem and maximizing the use of the available farm resources. The overall aim is to minimize nutrient inputs and losses from within the system. Organic farmers can build up soil fertility in the orchard by planting cover crops, and by adding crop residues, animal originated manures/composts, mineral rock dusts, and special preparations, which stimulate soil biological and chemical activity. The nutrient components, such as farmyard manure, vermi-compost, green manure, vermin-wash, liquid manure, and biofertilizers (arbuscular mycorrhiza, plant growth promoting rhizobia, blue green algae, *Azotobacter*, *Azospirillum*, etc.) are vital for nutrient management in organic nutrient management.

13.7.2 WEED MANAGEMENT

Combination of cultural or husbandry techniques with direct mechanical methods are generally recommended for integrated weed management under organic agriculture. Crop husbandry practices involve adjustment of soil conditions by irrigation, various agronomic practices, diverse crop rotations, mulches, and use of smothering cultivars particularly suited for organic production. Mechanical intervention includes harrowing to control weeds in between rows. Following techniques could be used for weed management:

13.7.2.1 MULCHING

Organic mulch: Weeds during postplanting period can be suppressed by applying thick layers of organic mulch which can be helpful in creating habitats for beneficial predators like big-eyed bugs, spiders and soft-bodied flower beetles. Organic mulches are usually applied in a circle around tree trunks. However, in termite prone areas, it is recommended to avoid mulching with organic materials. Cheaper and locally available materials, such as straw, leaves, spoiled hay, yard trimmings, sawdust, and woodchips, can be used as mulch. Such materials are cheap and usually locally available.

Sheet mulch: Laying down layers of newspaper or cardboard and thin covering with organic material is known as sheet mulching which enhances the efficacy of mulch as a barrier against emerging weeds. Waxed or impregnated cardboard with fungicide, as well as color print and glossy paper should be avoided by organic farmers so that it can be compliant with the National Organic Programme standards.

Geotextile mulches: Geotextile mulches are paper or woven plastic fabrics, which play an important role in suppressing the growth of weeds. While geotextile mulches allow some air and water penetration, they may reduce water infiltration against increased infiltration in the case of organic mulches. It is being used in the western countries for weed management. However, in India also, Jute Agrotextile (JAT) has also been found quite effective in weed management. The mulching effect of Jute Agrotextile on its biodegradation helps in "greening" of the soil for a substantial period due to creation of a conducive microclimate including the retention of optional humidity of the top soil and preservation of the existing nutrients and adding plant nutrients to the soil at micro level. Usually, dense nonwoven JAT (500/1000 g) is used for this purpose.

13.7.2.2 MECHANICAL CULTIVATION

Use of mechanical tillage and weed harrowing implements are the most widely used weed management practices in fruit production which help in maintaining permanent vegetation between rows in an orchard. Under every circumstance, mechanical tillage or weed harrowing must be kept shallow to avoid any damage to tree roots. Use of a wheel hoe can also be very effective in small-scale plantings. Even birds, such as chicks and ducks can also help in weed control to some extent if allowed to forage in orchard. These birds feed upon weed seeds and their underground organs.

13.7.3 DISEASE AND INSECT PEST MANAGEMENT

Under organic farming, diseases and pests generally do not cause any remarkable trouble as healthy plants growing in healthy soil with balanced nutrition are more resistant to insect pests. Further, the dry environmental conditions of arid and semiarid regions are not much conducive for their development. Pest problems could be particularly serious in large horticultural holdings with monocrop cultivation (Mallikarjunarao et al., 2015). Pest management strategies in organic farming systems are chiefly preemptive rather than restorative. Use of natural biological agents, cultural practices, such as sanitation, insect pest resistant varieties, crop rotation, intercropping, maintenance of biological diversity or farm scaping, planting at appropriate dates, and plant spacing, good soil management practices, and use of organic pesticides of botanicals such as *neem* cake/*neem* oil and others, and microbial (*Bacillus thuringiensis* (*Bt*)), pyrethrum, and insect–parasitic fungi (*Metarhizium*, *Beauveria*, etc.) are some of the major principals of integrated pest management under organic farming (Sharma et al., 2015). Likewise, in most cases of the disease development, resistant varieties, and cultural practices are used to minimize disease incidence. Direct disease control methods are rarely necessary in most of the organic crops. Although some natural and plant extract-based fungicides are regularly applied to manage foliar diseases of horticultural crops (Narayan et al., 2016).

13.8 CONCLUSIONS

Owing to low and erratic rainfall and poor soils in dry semiarid and arid regions of the country, the prevalent crop husbandry systems are based on communion with nature, and therefore, fundamentally "organic" and have been followed from time immemorial. This opens a new vista of enormous opportunities for enhancing the soil health and in turn overall environment, thereby, providing sustainable livelihoods to dwellers of these regions, by modernizing these age-old systems with new age technological interventions. The key to success in proliferating organic farming lies in enhanced collegiality and coordination among all the stakeholders, such as government agencies, NGOs, SHGs, and growers for effective implementation of the programs being run for promotion of organic farming. The aim of promoting organic cultivation, local food security, and ultimately creation of markets for organic produce can be achieved by working in cohesion of all concerned stakeholders in synergy with each other.

KEYWORDS

- **organic farming**
- **rainfed agroecosystem**
- **global agriculture land**
- **organic food sector**
- **dryland**
- **fruit crops**
- **organic certification**
- **weed management**

REFERENCES

Anonymous. *Indian Organic Sector: Vision 2025*; APEDA, YES BANK, Ingenus Strategy and Creative Research, 2016; p 49.

Anonymous. *Rajasthan Organic Farming Policy-2017*; Govt. of Rajasthan, 2017; p 21.

Chhonkar, P. K. Organic Farming: Science and Belief. *J. Indian Soc. Soil Sci.* **2003,** *51* (4), 365–377.

Deshmukh, M. S.; Babar, N. A. Present Status and Prospects of Organic Farming in India. *Eur. Acad. Res.* **2015,** *3* (4), 4271–4287.

FAO. *Organic Agriculture, Environment and Food Security*; FAO: Rome, Italy, 2002; p 252.

Fertilizer Association of India (FAI). *Fertilizer Statistics 1997–98*; FAI: New Delhi, 1998.

FiBL & IFOAM. *The World of Organic Agriculture*; Frick & Bonn., 2017; p 348.

FiBL-IFOAM Survey. *Organic Agriculture Worldwide: Current Statistics, 2014 Helga Willer*; Research Institute of Organic Agriculture (FiBL): Frick, Switzerland, 2015.

Gaur, M. Organic Farming in India: Status, Issues and Prospects. *Sopaan-II* **2016,** *1* (1), 26–36.

Ghosh, S. P. Carrying capacity of Indian horticulture. *Curr. Sci.* **2012,** *102* (6), 889–893.

Gupta, J. P.; Sharma, A. K. *Agrihorticulture in Arid Region for Stability and Sustainability*. In *Agroforestry Systems for Degraded Lands*; Singh, P., et al., Eds.; Oxford and IBH: New Delhi, 1996; pp 164–167.

Horticultural Statistics. *Horticulture Statistics 2017*; Department of Agriculture, Cooperation & Farmers Welfare, Ministry of Agriculture & Farmers Welfare, Government of India, 2017; p 481.

Mallikarjunarao, K.; Pradhan, R.; Das, R. K. Dry Land Techniques for Vegetable Production in India-A Review. *Agric. Rev.* **2015,** *36* (3), 227–234.

Mitra, S.; Devi, H. Organic Horticulture in India. *Horticulturae* 2016, 2, 17; doi:10.3390/horticulturae2040017

Narayan, A.; Singh, P. N.; Brijwal, M.; Singh, S. K. Response of Organic Manures on Quality of Peach (*Prunus persica* Batsch) cv. Florida Prince. *Environ. Ecol.* **2016,** *34* (3), 985–987.

Purohit, M. L. Impact of Recent Spurt in Human and Livestock Population on India Arid Ecosystem. In *Desertification Control in the Arid Ecosystem of India*; Singh, S.; Kar, A., Eds.; Agrobotanical Publishers: Bikaner, India,1998; pp 198–204.

Rajeshwar Rao, B. R. Medicinal Plants for Dry Areas. In *Sustainable Alternate Land Use Systems for Drylands*; Singh, R. P.; Osman, M., Eds.; Oriental Enterprises: Dehradun, India, 1999; pp 139–156.

Reddy, B. S. Organic Farming: Status, Issues and Prospects—A Review. *Agric. Econ. Res. Rev.* **2010,** *23*, 343–358.

Sharma, A. K. The Potential for Organic Farming in the Drylands of India. *Arid Lands Newslett.* **2005,** *58*, 2–9.

Sharma, A. K.; Goyal, R. K. Addition in Tradition: Agroforestry in the Arid Zone of India. *LEISA India* **2000,** *2* (3), 19–20.

Sharma, A. K.; Patel, N.; Painuli, D. K.; Mishra, D. *Organic Farming in Low Rainfall Areas*. Central Arid Zone Research Institute: Jodhpur, 2015; p 48.

Venkateswarlu, B.; Srinivasa Rao Ch.; Ramesh, G.; Venkateswarlu, S.; Katyal, J. C. Effect of Long-Term Incorporation of Legume Biomass on Soil Quality under Rainfed Conditions, International Conference on Soil Water an Environment: Issues and Strategies, Jan 28–Feb 1, 2005; New Delhi, 2005, p 118.

CHAPTER 14

Organic Farming in Vegetable Crops

S. S. HEBBAR*, A. K. NAIR, M. SENTHIL KUMAR, and M. PRABHAKAR

Division of Vegetable Crops, ICAR-Indian Institute of Horticultural Research, Hesaraghatta Lake P.O., Bengaluru 560089, Karnataka, India

**Corresponding author. E-mail: hebbar@iihr.res.in*

ABSTRACT

Since many centuries, Indian farmers relied on the use of crop rotation, crop residues, animal manures, legumes, green manures, off-farm organic wastes, and biological pest controls to maintain soil productivity, supply plant nutrients, and control insects, pests and other weeds on their farms. Today, interest in organic agriculture as an ecofriendly system of cultivation is growing at both national and global levels. This shift towards organic production is supported by consumers who are aware of health hazards: demand for food grown organically is increasing by 20-25% in developed countries where awareness is comparatively high. In general, the organically produced vegetables fetch a higher price over conventionally produced ones. However, the industry is highly competitive and most occasions the price of the produce and income to the growers are dictated by overall supply and demand on a given day and the available organic outlets. The promotion of organic vegetable farming also requires many more scientific interventions to efficiently manage the soil health, nutrients, pests, and diseases.

14.1 INTRODUCTION

The International Federation of Organic Agriculture Movement (IFOAM) defines organic agriculture as "a production system which sustains not only

Organic Crop Production Management: Focus on India, with Global Implications, D. P. Singh, PhD, H. G. Prakash, PhD, M. Swapna, PhD, & S. Solomon, PhD (Eds.)

the health of people, but also the soils and ecosystems." It also emphasizes on the ecological processes, biodiversity, and cycles adapted to the local conditions, rather than the use of inputs with deleterious effects. In the real sense, "organic agriculture combines traditional wisdom, novelty, and science to benefit the shared environment and to promote fair relationships to lead a good quality of life for all involved..."

The organic farming system works based on the four important principles, namely:

1. Health—Organic farming should maintain the health of the soil, plants, animals, and humans and also the planet as whole without discrimination.
2. Ecology—It should be based on the existing natural systems and cycles, co-work with them, try to be like them and facilitate sustain them.
3. Fairness—It should build on the associations that make sure equal opportunity with regard to the existing environment and life opportunities.
4. Care for all including soil—It should be managed with precaution and responsibility to safeguard the health and interests of current and future generations and also the environment.

The Green Revolution which gave the food security relied mainly on the use of chemical fertilizers and pesticides, cultivation of nutrient-responsive, high-yielding varieties, greater exploitation of irrigation resources through various irrigation techniques to boost production and productivity in most cases, but the use has gradually declined in several crops. This also proved detrimental to the soil health and environment. The residues of the chemicals found in majority of the farm produce and food products cause concern over the safety of food, health of the soil, and sustainable production. Nowadays, people are interested to consume safe and quality foods and keep the crop production system sustainable one in a long run. On this count, organic farming has become an alternate way of farming ensuring quality foods, sustainability, and a profitable livelihood.

The fundamental theory of these practices go hand in hand with nature with the principle to take care of the soil health which in turn takes care of the plants. In simple terms, it is a balancing act of giving back to nature what was taken from it. With this expectation, organic farming practices are assuming importance across the globe to make sure that the harvested produce is free from pesticide residues, harmful chemicals, to reduce the level of pollution in soil, water and environment, and ultimately sustain the soil productivity.

In India, the present-day organic farming practices are the combination of science, available traditional wisdom, and refinement of a package of practices. Although the movement of organic lifestyle is almost 100 years, it gained importance only after realizing the adverse effects of modern agriculture in the last phase of 1990s. In 1905, the British botanist Sir Albert Howard, who is regarded as the father of present organic agriculture, applauded the traditional Indian farming practices as superior to the conventional agriculture science. Organic agriculture is a "holistic" food system that does not allow anyone to use synthetic fertilizers, pesticides, GM crops, and any growth hormones with a focus to minimize the pollution and optimize the overall health of plants, animals, and human beings. It stresses on the minimum use of inputs from outside and also on production practices which will facilitate to reinstate, preserve, and upkeep the ecological harmony. The term "organic" refers to the products produced by adopting the criteria laid out in the Organic Foods Production Act, 1990.

The production process of organic cultivation mainly depend on crop rotation, recycling of crop residues, use of manures of animal origin, growing green manure crops, recycling biological wastes, and mineral-bearing rocks to enrich the soil which in turn ensures the supply of plant nutrients. Weeds, pests, and diseases are managed by various mechanical, cultural practices, and bio formulations.

In India, vegetables are grown in an area of 9.6 million hectares during 2015–2016 with a production of 166.6 million tonnes with the productivity of 17.4 tonnes per ha. In fact, vegetables constitute about 59% of horticulture production. From the year 2007–2008 to 2015–2016, there was an increase of 22% in area and production increased by 29%. India ranks second in vegetable production after China. India ranks first in the production of peas and okra. It occupies the second position in the production of brinjal, cabbage, cauliflower, and onion and third place in the case of potato and tomato in the world. In India, the technological interventions have increased the per capita availability of vegetables from 264 gm/person/day in 2004–2005 to 355 gm/person/day in 2015–2016.

Several reports have shown the advantages of organic farming such as it eliminates the energy required to manufacture synthetic fertilizers and plant protection chemicals, and also encourages the use of internal farm inputs which reduce the need of fuel toward transportation. India has around 7 lakh organic producers. The cultivated area under certified organic farming has grown from 42,000 ha in 2003–2004 to 7.23 ha in 2013–2014. At this rate of growth, India can emerge as a global leader in the field of organic farming. Besides, there is a huge scope for organic vegetable cultivation as the vegetables grown organically are priced high (10–50% than the regular produce). Similarly, the market for organic products is growing at the rate of 20% whereas it is only 5% for the conventionally produced products. Organic vegetables are also preferred for export. Hence, there is a huge potential for a country like India, which has traditional wisdom and expertise for having been in the organic way of farming for ages. The research findings of the last 10 years confirm the yield advantage in many horticulture crops. However, the major impediment in the growth of organic farming in our country is the yield reduction during the early years due to the sudden shift from inorganic farming to organic cultivation and wide gap between the availability of an organic source of nutrients & requirement and the inadequate options to manage pests and diseases. The organic production system of vegetable crops is guided by a broad principle of "feed the soil to feed the plant." This basic concept is implemented through a set of standard practices which increase the soil organic matter, biological activity, and nutrient availability. In this backdrop, a minimum package of practices to be adopted for organic farming on vegetables is enlisted below for successful vegetable cultivation.

14.2 CLIMATE AND SOIL

Each vegetable has a specific climatic requirement; therefore, choosing the best season for a particular crop assumes paramount importance.

The season should be chosen in such a way that it should facilitate better crop growth, higher yield, and the best quality of produce. Some diseases and insects are prevalent at a specific period of a year and crops can be sown/planted by adjusting the dates to protect them from these high risks. The following planting schedules can be generally followed in the tropical conditions:

***Kharif*:** Brinjal, chili, onion, okra, and drumstick;

Rabi: Carrot, cauliflower, cabbage, capsicums, French beans, onion, peas, tomato, and gherkins;

Summer: Bottle gourd, cucumber, pumpkin, and ridge gourd.

However, most of the vegetables grow well during the *rabi* season in mild winter locations. It is wise to avoid tomato cultivation during the rainy period as *Alternaria*, *Phytophthora*, foliar, and fruit rot diseases are severe. Similarly during the winter season, the overall growth and development of okra (*bhindi*) are affected. High temperatures in summer favor viral diseases in okra, chillies, tomato, and beans. Besides, fruits, temperature settings also affect capsicum, chilies, tomato, and beans..

Choosing the right soil is one of the important criteria to achieve success in vegetable cultivation. The ideal soils for growing vegetables are well-drained with a pH of 5.5–6.6, fairly deep, relatively rich in organic content (1.5–2%) with a Carbon to Nitrogen ratio of 15–20:1. The choice of place should also be with a slope of around 2–3%. Raised-bed cultivation is an ideal practice to facilitate drainage and the selection of suitable crops in accordance with the soil ensures a successful crop.

Soil Preparation: Proper soil preparation is a prerequisite in organic farming for obtaining long-term benefit. Besides, unlike the conventional farming, here the practices mainly target to nourish the soil which in turn supplies the required nutrients to the plants. A perfect soil should have sand (0.02–2.0 mm), silt (0.002–0.02 mm) and clay (0–0.002 mm) in equal proportions besides 5% organic matter. However, in most of the Indian mineral soils, the organic matter content is less than 1% which needs a lot of amelioration measures to suit for the organic cultivation and make it productive.

In fact, it takes a long time to bring the soil to an ideal condition; therefore, it is not possible to get a higher yield in the initial period of cultivation. If available, clean virgin lands can be chosen. Weeds which are hard to control can be managed by soil solarization. If the farm land has compact hardpans, it ought to be broken and soils should be turned to 10–12 inches deep.

When organic matter in soil breaks down, nutrients are released and utilized by the crops. It also improves the nutrient status and water holding capacity of the soil. Similarly, the quantum of organic matter to be added depends upon the type of organic material, soil, and local weather conditions.

As much as 30–40 tonnes of organic manure/compost per acre need to be added to gain the benefit of organic matter for sandy soils of tropical and subtropical conditions. On the other hand, only 5–10 tonnes will be sufficient for an acre, if it is a region of cooler climate with receipt of less rainfall. However, frequent application of low rates of organic inputs will result in a steady improvement in soil tilth. Besides, the yearly addition of 40–100 tonnes of natural substance per hectare will be very beneficial in the upkeep of the soil tilth and establishment of plants.

Any organic matter should be tested for its nutrient content prior to its application. This will help to decide the quantity to be incorporated in a given area. At any case, fresh organic material (undecomposed) should not be applied as they are harmful to the plants and also harmful to the environment through runoff due to the presence of harmful pathogens and weeds. These composts should be composted in a scientific way and then applied.

14.3 ORGANIC NUTRIENT MANAGEMENT

Several reports revealed that the yields from organic cultivation were lower as compared with the conventional farming practices ascribing to the fact of lower nutrient availability and limited possibilities to build up the nutrient status of less fertile soils in the initial period. But over a period of time, frequent addition of green manures, composts of crop residues, and organic manures to the soils raises the soil organic matter status. Maintaining soil organic matter at an optimum level is very vital in any organic farming. Soil organic matter, in general, improves cation-exchange capacity and acts as a reservoir of nutrients for the growing crop. The incorporation of adequate organic substance, also improves soil aeration, drainage, and water-holding capacity.

It is highly economical to grow green manure crops which considerably raise the soil organic matter and provide nitrogen for the subsequent crop. They also reduce the soil erosion and extend other benefits related to the pest and disease suppression.

With regard to the availability of nutrients, much of the nitrogen (N), phosphorus (P), and sulfur (S) needed for crop nutrition can straight away be obtained through the decomposition of organic matter in soils. A fraction of N from many organic amendments is readily converted into available mineral forms. Phosphorus from organic amendments reacts quickly, gets attached with soil minerals, and moves very little from its place of application. Potassium (K), calcium (Ca), and magnesium (Mg) are comparatively soluble from plant residues or soil organic matter fractions and make available of these nutrients. Organic matter is also a reservoir of many minor elements. Nutrients are slowly released from organic matter with the pace of its decomposition and results in slow and constant availability.

The availability of nutrients in soil increases as the organic matter of soil increases. This ultimately improves the health, fertility status, and biological properties of the soil. Many soil amendments and fertilizers commonly approved for organic farming have appreciable amounts of nutrients, but only a portion of these nutrients is accessible to the current crop.

Therefore, soil fertility management has to be programmed in organic vegetable farming to maintain all the required nutrients in the soil nutrient pool at optimum level and also to supplement the pool as and when needed. Management practices should be modified to facilitate the various biological functions in the soil to ensure optimum nourishment to the crop.

14.3.1 MANAGEMENT BIOLOGICAL CHARACTERISTICS OF SOIL

A complex array of soil-dwelling microbes and animals decomposes organic matter, mineralizes organic forms of nutrients, and fixes nitrogen in the soil. These beneficial biological activities enhance the soil's ability to release nutrients needed for plant growth and to break down the plant residues. Tillage practices that aerate the soil, enhance the biological activity by providing oxygen and mixing organic matter in the tilled area. Both ammonium and nitrate are readily available sources of nitrogen to plants and soil microorganisms. A variety of soil microorganisms convert organic nitrogen to ammonium, but only a specific (nitrifying) bacterium converts ammonium to nitrate (nitrification). Besides, few soil bacteria (denitrifiers) reduce nitrates to elemental nitrogen or nitrous oxide and revert to the atmosphere due to volatilization. In nature, the majority of microorganisms have a significant role in the nitrogen cycle. Some microbes have been reported to act against soil pathogens. In support of this, it is reported that the fungus *Gliocladium virens* controls damping-off pathogens, when its population is high in the organic matter. Similarly, *Agrobacterium* reportedly restricts the growth of *Fusarium* (fungal pathogen) causing an array of diseases in vegetable crops.

In the upkeep of soil biological activity, earthworms too have a vital role. They break up the organic matter, humus, and combine them with soil particles which enhance microbial activity. Soils with high earthworm population or soils containing earthworm castings possess high water holding capacity and plant nutrients. Soils with earthworms are reportedly rich in available phosphorus, cation-exchange capacity, exchangeable calcium, magnesium, and potassium. It is also reported that earthworm activity minimizes the severity of various soil-borne diseases through physical mixing as that of mechanical soil disturbance. In most of soils, if organic matter increases, earthworm populations will increase rapidly. Continuous cropping without adding organic matter, the use of copper fungicides, soil fumigation, and frequent tillage usually reduces the earthworm populations considerably.

14.3.2 SOURCE OF NUTRITION IN ORGANIC FARMING

Cover Crops/Green Manure Crops: Cover crops are highly beneficial for an intensive organic vegetable farming in several ways. When leguminous crops are grown as cover crops, the soil nitrogen gets increased through nitrogen fixation. In general, green manures contribute 30–60 kg N/ha/crop. Sunhemp and Sesbania (*dhaincha*) could be suitable catch crop in many agro climatic situations. Similarly grasses are also useful in promoting soil structure and upkeep of its aggregate owing to fibrous root systems. Weed suppression for successive crops is another benefit. Furthermore, cover crops set an ambiance which attracts and sustains beneficial arthropods. To realize some economic returns, farmers who have irrigation facility can grow profitable vegetables like Frenchbeans or cowpea in place of green manure crops. Green manure crops can be raised in rainy summer and incorporated into soil well before of sowing/planting.

Compost: Compost is the cheapest organic source of nutrients. Crop residues, farm wastes, and other organic wastes can be decomposed to obtain composts. An ideal compost should have a carbon-to-nitrogen ratio (C:N) less than 20:1 which is the indicator of its maturity and N availability.

Manure: Farmyard manure, biogas slurry, poultry manure, and animal manures are the by-products of the integrated farming system, where output of one enterprise will be an input for the related activity. Recycling is the rule here.

Vermicompost*:* Vermicompost production is economically remunerative as it requires less cost for the production. However, the raw materials used for the production mainly determine the nutrient contents and the quality.

Other organic fertilizers: A range of approved organic fertilizers is there for the organic cultivation of vegetables. Most of them are the by-products of meat, fish, and soybean processing industries. Other simple fertilizer materials that mostly offer only one macronutrient such as sodium nitrate (mined), blood meal, rock or colloidal phosphate, potassium sulfate (mined), and green sand (for K). However, it is advisable to avoid by-products of the meat processing industry, namely, bone meal and blood meal due to food safety issues and the threat of disease transmission. Mined Chilean sodium nitrate is imported as a cheap source of N under organic fertilizer programs. However, some certifiers suggest its restricted use ascribing to its adverse effect on soil in a long run through accumulation. Minor element sources of organic fertilizer generally may have one or more elements.

Special purpose fertilizers: When a specific deficiency is observed, certain approved nutrient sources of K, Ca, and Mg will be handy to an organic vegetable farming. Gypsum, lime, and potassium magnesium sulphate are such materials used to meet the requirement as the case may be.

Biofertilizers: The beneficial role of *Rhizobium* bacteria as symbiotic nitrogen fixer is well recognized in organic farming. Besides, *Azotobacter* and *Azospirillum* are also used as nitrogen fixers. Similarly, phosphate-solubilizing bacteria (PSB) are also used for solubilizing the phosphorous along with *Arbuscular mycorrhiza* Fungi which mobilizes phosphorous. Currently, the concept of applying a consortium of biofertilizers (which has two or more biofertilizers in a single carrier) is gaining momentum to get the benefits of all these biofertilizers in a single application. ICAR-Indian Institute of Horticultural Research has also developed a consortium of biofertilizers, namely, Arka Microbial Consortium (AMC), which has given desired results, especially in vegetables.

14.3.3 SOIL NUTRIENT MANAGEMENT

1. Enrichment of farmyard manure with *Trichoderma* and biofertilizers: Well decomposed farmyard manure (FYM) is mixed thoroughly with *Trichoderma harzianum*, *Azotobacter*, or *Azospirillum* and PSB (@ 1 kg each/tonne of FYM) moistened with sprinkling water and covered with gunny bags or dried coconut fronds and kept to incubate for 15 days. This enriched FYM should be mixed with remaining FYM before applying to the field.
2. Around 35–40 tonnes of fully decomposed FYM and 1.5 tonnes of vermicompost and 250 kg of *neem* cake having 8% oil content should be added for an area of one hectare. Ridges have to be prepared for transplanting at recommended row spacing of the crop after basal application of manure.
3. For enhanced supply of nitrogen, green manuring crop can be grown before the main crop and incorporated at least 3 weeks before transplanting as N source.
4. Concentrated organic cakes (4–7% N) like castor, soybean, cotton, *etc.*, can also be used depending on the availability. At 30 days after planting, vermicompost (1.5 t/ha) and *neem* cake (250 kg/ha) applications are done followed by earthing-up operation.

5. As a source of P, rock phosphate in combination with PSB or bone meal can be used as both contain about 20–22% P_2O_5.
6. As a source of K, wood ash (2.5–3% K_2O) or sheep manure (3–4% K_2O) can be used.
7. Spray of *Panchagavya* can be followed 4–5 times at 10 days interval to supplement nutrients and also as a plant growth promoter.
8. For legume vegetables, seeds need to be treated with *Rhizobium* cultures before sowing.

14.4 CULTURAL PRACTICES

14.4.1 SELECTION OF VARIETIES

For better performance of any vegetable, varieties which are resistant or show field tolerance to important diseases and pests in that particular area are selected. Also varieties should be chosen according to the season in which their performance will be better. For example in organic cultivation of tomato during *kharif,* the major problem is *Alternaria* and leaf Spot. Therefore, a variety or hybrid having field tolerance to these diseases should be chosen. If available, a variety or hybrid having field tolerance to Tospo virus should be grown. Similarly for summer season, Tomato Leaf Curl Virus (TLCV)-resistant hybrids need to be selected and wherever there is a problem of bacterial wilt, tomato variety/hybrid resistant to this should be grown. In cabbage and cauliflower, photo- and thermo-insensitive hybrids are to be used. In India, many research institutes and private research organizations have developed many resistant varieties/hybrids for diseases, pests, biotic, and abitic stresses. In this pursuit, ICAR – Indian Institute of Horticultural Research, Bengaluru, has also released many good vegetable varieties/hybrids. Some of them are listed below with their characters for use in organic farming:

Crop	Variety/hybrid	Remarks
Tomato	Arka Abha, Arka Alok	Resistant to bacterial wilt (BW)
	Arka Rakshak (F_1), Arka Samrat (F_1)	Resistant to tomato leaf curl Virus (ToLCV), BW, and early blight
	Arka Abedh	Resistant to ToLCV, BW, and early blight and late blight
Brinjal	Arka Anand (F_1), Arka Nidhi, Arka Keshav, Arka Avinash, Arka Unnathi, Arka Harshitha	Resistant to bacterial wilt

Crop	Variety/hybrid	Remarks
Chilli	Arka Suphal, Arka Harita (F_1), Arka Khyati	Resistant to powdery mildew and tolerance to viruses
	Arka Meghana, Arka Sweta	Field tolerance to viruses
Pumpkin	Arka Suryamukhi	Resistant to fruit fly
Bottle gourd	Arka Bahar	Resistant to blossom end rot
Okra	Arka Anamika, Arka Abhay	Field tolerance to YVMV
French beans	Arka Anoop	Photo-insensitive. Resistant to both rust and bacterial blight
	Arka Arjun	Photoinsensitive and stringless and summer. Resistant to *MYMV* disease
Dolichos	Arka Jay, Arka Vijay	Photoinsensitive. Tolerant to low moisture stress
	Arka Soumya, Arka Sambhram, Arka Amogh	Photoinsensitive
Dolichos (pole type)	Arka Swagath, Arka Prasidhi, Arka Bhavani, Arka Visthar, Arka Pradhan, Arka Krishna, Arka Adarsh	Photo-insensitive
Cow peas	Arka Garima, Arka Suman, Arka Samrudhi	Photo-insensitive
Garden pea	Arka Ajit. Arka Karthik, Arka Sampoorna, Arka Priya, Arka Pramodh, Arka Apoorva	Resistant to powdery mildew and rust
Radish	Arka Nishant	Resistant to pithiness, premature bolting, root branching, and forking
Onion	Arka Kalyan	Resistant to purple blotch disease
	Arka Pitambar	Tolerant to purple blotch, basal rot diseases, and trips with long shelf life (3–4 months)
	Arka Kirthiman (F_1), Arka Lalima(F_1)	Field tolerance to diseases and pests.

14.4.2 SEEDLING PRODUCTION

For organic farming, raising quality seedlings in organic method is important. In this regard, the following methods can be adopted to obtain healthy seedlings for a profitable organic farming:

Raised-Bed Nursery Raising

Beds of 10–15 cm height with provision for drainage are prepared. During summer season, solarization of nursery beds will avoid the occurrence of nursery diseases. Farmyard manure, vermicompost and *Trichoderma* powder are mixed in a ratio of 10:1:0.1 and applied at 0.3 kg/square meter. *Neem* cake should also be added to the nursery bed at 100 g/per square meter. Line sowing of seeds with a spacing of 7.5–10.0 × 2.5 cm is advised and to be covered by vermicompost and straw after sowing. It is good to protect nursery beds with 40 or 50 mesh nylon net cover to prevent insect vectors. The seedlings will be ready in 25–45 days depending on the vegetable crop.

Raising Seedlings in Plastic Pro- or Flat-Trays

Either flat or protrays can be used for raising seedlings of tomato capsicum/chilli, brinjal, cauliflower, and cabbage. Combination of well-decomposed organic manure or coco peat or vermicompost enriched with biofertilizers and bioagents is the ideal media for filling the trays. About 1–1.25 kg coco peat is required to fill one tray. A small depression is made by finger at the center of the coco peat and one seed per cell is sown and covered again with a thin layer of coco peat, just adequate to cover up the seeds. The filled and sown trays are stacked one over the other till sprouting is observed. After sprouting, the trays are arranged in a net house and irrigated as required. A total of 20–25 days old seedlings in cabbage, cauliflower, tomato, and 30–40 days old seedlings in chilli, capsicum, and brinjal need to be transplanted for a better crop establishment.

14.4.3 TRANSPLANTING/DIRECT SOWING

BT (*Bacillus thuringiensis*) formulation (1 g or mL/L) is sprayed to the cabbage/cauliflower seedlings 1 day before transplanting. At the time of transplanting, the root portion of seedlings is dipped in 1% Burgundy solution, or seedling trays are drenched in it. Similarly, biofertilizers such as *Azotobactor* or *Azospirillum* and phosphate-solubilizing bacterial slurry can be used for root dipping of all vegetable seedlings before transplanting. When seeds are used for direct sowing, the above biofertilizers can be used for seed coating for non-leguminous vegetables while for leguminous vegetables-specific *Rhizobium* culture and PSB are recommended.

14.5 WEED MANAGEMENT

For producing seedlings from seedbeds, select a land area which was/ is free from troublesome weeds. Vegetable crops that germinate fast and grow quickly in the early weeks after transplanting perform well as they will be more competitive with weeds than crops that initially grow slow. Seedlings and transplants of large-seeded vegetables such as sweet corn and beans grow rapidly. Manual removal (i.e., pulling or hoeing) and mechanical interculture are the most widely practiced method of eliminating weeds in vegetable farming. A properly set cultivator can also be used for efficient and quick removal of small weeds. However, care has to be taken to avoid damage to the crop roots. It is better to avoid doing interculture of soil that is excessively wet (to avoid damage to the standing crop and soil compaction). Interculture/hoeing facilitates soil aeration and improves water infiltration which add further benefit to the crops. For weed management, organic mulches like dry grass, straw, bark, banana sheath, composted sawdust, and similar organic substances can be used.

14.6 ORGANIC PESTS AND DISEASE MANAGEMENT PRACTICES

Organic disease management provides satisfactory protection from many vegetable diseases that commonly occur and these practices are based on nonchemical sanitation, cultural, physical, and biological means as well as application of organically approved chemicals as an integrated approach. These practices reduce populations of fungi, bacteria, other pathogenic microorganisms, nematodes, and viruses which cause diseases. Adoption of the combination of practices is warranted, as no single practice is found effective for all the pests and diseases which jeopardize the production of any crop. Organic disease management practices also include soil solarization and removal of debris and burning them subsequently. This should be done before taking up the sowing/transplanting. To eliminate fungus spores, mycelium, bacterial cells, and any other harmful pathogens, we can disinfect stakes and cages by washing (to remove soil), followed by dipping or spraying with a copper sulfate solution before reuse. The following plant protection measures can be followed for the effective management of pests and diseases.

- Use only resistant vegetable varieties wherever available.

- Clean seeds and transplants are a must in vegetable production. To avoid problems in the subsequent season, seeds from fields or areas where diseases are prevalent should not be used. Examples of seed-borne diseases are anthracnose of lima beans, mosaic virus of southern peas, bacterial blight of snap beans, black rot of cabbage, and leaf spots of turnip and mustard greens.
- Treat the seeds with *Psuedomonas fluorescens* or *Trichoderma harzianum* or *Trichoderma viride* at 10 g/kg of seeds.
- Plant healthy and vigorous seedlings at an early date in the season and provide a healthy growing condition by reducing stress by proper irrigation and supply of nutrients in a weed-free environment.
- In crop rotation, green manure crops are raised and ploughed early so that it will get decomposed and mixed well into the soil before taking up sowing/panting of main crop.
- Use mulch to keep away the vegetables from continuous touch with soil and to avoid rots.
- Keep the field weed-free as weeds harbor most of the insect pests and disease-causing organisms.
- Crops need to be irrigated preferably at early morning hours to reduce time of plants being wet.
- Diseased plants should be removed and disposed off hygienically from time to time.
- Turn under crop residues immediately after harvesting.
- Raise barrier crop (fodder maize or sorghum) all around the vegetable field to reduce the entry of vectors which transmit dreaded virus diseases to the standing crop.
- Cover the nursery beds with 40 or 50 mesh nylon net cover to prevent insect vectors transmitting virus diseases.
- Nematodes can be managed by using cover crops, namely, southern pea and hairy indigo. They considerably reduce nematode build up. Use of mulch (30 μ) and maintaining optimum soil moisture will also reduce their number considerably. Similarly, soil solarization using large plastic covers reduces nematodes, soil-borne insects, some weeds, and some pathogens. If the type of soil is suitable for summer flooding, it will considerably reduce the problem of nematodes and other pests. Application of bioagents like *Trichoderma, Bacillus pumilus, and Paecilomyces lilacinus* mixed with farm yard manure or seed treatment with these manage the nematodes very effectively.

14.6.1 BIOAGENTS FOR FOLIAR AND FRUIT DISEASES AND SOIL-BORNE ORGANISMS

The bioagents that are used to control foliar and fruit diseases and soil-borne organisms include: *Trichoderma harzianum* (10 g/L), *Trichoderma viride*, *Pseudomonas fluorescens* (10 g/L), *Bacillus subtillis* (10 g/L), and *Paecilomyces lilacinus* (10 g/L).

14.6.2 BIOAGENTS AND BOTANICALS FOR MANAGEMENT OF INSECT PESTS

The bioagents and botanicals that are used for managing insect pests include *Beauveria bassiana* (10 g/L), *Verticillium lecanii* (10 g/L), *Metarhizium anisopliae* (10 g/L), *neem* soap (10 g/L), *Neem* oil (8 mL/L), *neem* seed powder extract (4%), Pongamia oil (8 mL/L), Pongamia soap (10 g/L), nuclear polyhedrosis virus (NPV), *Bacillus thuringiensis* on certain insects such as cabbage worms, Pyrethrins, rotenone, insecticidal oils, and soaps.

14.6.3 PROTECTED CULTIVATION OF VEGETABLES

Protected cultivation is mainly adopted to overcome many biotic and abiotic problems in the organic cultivation of vegetables. In areas that receive moderate–to-heavy rains, rain shelters will be very effective in overcoming the problems of foliar diseases of vegetable crops like tomatoes. In crops like brinjal and capsicum where controlling certain insects like shoot- and fruit-borer management is difficult under organic farming, growing these vegetables under net household a bright promise.

14.7 RESEARCH WORKS ON ORGANIC FARMING IN VEGETABLE CROPS IN INDIA

14.7.1 RESEARCH WORK ON ORGANIC SOURCES OF NUTRIENTS

Several state agricultural universities, ICAR, and various research institutes have conducted various experiments on organic farming in vegetables and come out with recommendations for the adoption of organic farming. At ICAR-Indian Institute of Horticultural Research (IIHR), Bengaluru, organic studies were carried out on many tropical vegetables and the results of a few crops are listed below:

In an organic experiment on French beans, Prabhakar et al., (2011) observed that the treatment which received 100% recommended dose of nitrogen (RDN) through organic inputs resulted with the highest pod yield (17.77 t/ha) followed by treatments which received 75% RDN through organics (17.45 t/ha) which exceeded even the conventional practices (15.9 t/ha).

In rose onion (*Allium cepa* L.), Prabhakar et al. (2012) reported that application of 100% RDN equivalent through organics resulted in the highest yield of 21.1 t/ha, followed by 75% RDN applied through organics and conventional practices (20.91 and 19.44 t/ha). Quality parameters like total soluble sugars, but the application of a higher amount of organic manure resulted in higher bulb dry matter content did not differ significantly among the treatments. The split bulbs ranged from 15 to 18.5% in the organic treatment whereas it was around 27.5% (chemical fertilizers) and 26% in the conventional practice.

The highest mean curd yield (21.23 t/ha) in cauliflower was recorded when applied with 100% recommended dose of NPK along with farmyard manure followed by 100 and 75% RDN through organics (19.36 and 18.42 t/ha). Quality parameters, namely, total antioxidants, radical scavenging ability, total flavonoids, and vitamin C were better under integrated nutrient management practices as compared to the application of only inorganic fertilizers (Prabhakar et al., 2015).

From a trial on onion with five levels of organic manure and two inorganic treatments consisting of only chemical treatment (NPK fertilizers + chemical plant protection) and conventional practice (recommended dose of farmyard manure + NPK fertilizers + chemical plant protection), Prabhakar et al., (2017) opined that the organic cultivation can yield at par with conventional cultivation methods. They also observed that conventional treatment recorded the highest plant height (50.1 cm) and bulb yield (34.8 t/ha). However, this bulb yield was on par with that of organic treatments which received farmyard manure equivalent to 50–100% RDN. According to them, there were no marked differences among the treatments for quality parameters like bulb dry matter and shelf life.

The studies conducted at ICAR-Indian Institute of Vegetable Research, Varanasi showed that the productivity of vegetables under organic farming was less in the initial years but the yields increased progressively equating the yields under the conventional inorganic farming in 4–5 years (Singh et al.,

2016; Bhattacharya and Chakroborty, 2005). In tomato and cabbage grown during winter season and okra and cowpea grown during summer season, a comparable yield under organic cultivation to the conventional system was achieved during the fourth year. However, in cowpea grown during the rainy season and pea grown during winter season, the comparable yield was recorded only in the third year of consecutive organic farming.

After practicing 5–6 years of organic farming with the soil fertility sufficiently restored, the yield realized in organic farming of vegetables is either comparable or higher than that realized in conventional farming. A long-term experiment as conducted by ICRISAT is also of the opinion that the yield of different crops in low-cost sustainable system, the annual productivity (rainy + post rainy season yields), is comparable to that of the conventional system (Rupela et al., 2004). Similarly, Rajendran et al., (1999) also reported low productivity under organic farming in the initial years, but the yields increased progressively under organic farming equating with the yields under inorganic farming by the sixth year. The association of biofertilizers with different organic manures resulted in better yield and quality of vegetables under organic farming (Kannan et al., 2006; Srivastava et al., 2007).

The yields of different vegetable crops grown organically were either on par or higher when compared to conventional farming in different parts of India as reported by Prabhakar and Hebbar (2007) for drum stick; Prabhakar and Hebbar (2008) for rose onion; Girija et al., (2006) for okra; Boomiraj and Lourduraj (2006) for okra; Ali et al., (2002) for chili; Singh (2004) for brinjal; Singh (2002) for french bean; and Thamburaj (1994) for tomato.

The results from organic trials in vegetable crops have indicated that the storage life of vegetables was improved with organic treatments in cowpea (Raghav and Kamal, 2007) and brinjal (Prasanna and Rajan, 2001). From the research experiments carried out on organic farming of vegetables in India, it is evident that irrespective of the source of organic nutrients, the yield and quality of different vegetables are found to increase by enhancing the quantity of manures supplied (Parvatha Reddy, 2008).

According to Thakur et al. (2010), higher growth and yield can be obtained in tomatoes and french beans when vermicompost and biofertilizers were combined and applied in the mid hills of Himachal.

Organic cultivation of vegetables improves the physical quality and overall health of soil by enhancing its resistance capacity against soil-borne diseases including nematodes (Chaudhary et al., 2003; Devrajan et al., 2003; Pandey et al., 2006).

14.8 MARKETING

For any organic vegetable producer, the biggest challenge is marketing. The factors such as supply and demand, pricing, perishable nature of the produce determine a grower's ability to sell his/her product. Therefore, the production and market risks both affect the profitability and economic viability of organically grown vegetables. In any situation, the risks associated with organic vegetable marketing operations should be minimized. As that of any other fresh market commodities, organic vegetables are also highly perishable. Therefore, growers must harvest, pack, and sell their products in an expeditious manner to receive satisfactory returns.

From the marketing perspective, vegetables face more risk compared to storable commodities like nuts and grains. Organic vegetable growers can manage these risks by growing vegetables like winter squash, onions, garlic, and potatoes which have more shelf life. Commodities which are grown organically fetch premium price over conventionally grown products. However, the industry is extremely competitive and the returns to growers are mainly based on the total supply, requirement, and available organic outlets. Market saturation also occurs very often. Growers will be compelled to accept lower price for their produce or market their produce at regular prices without the premium prices of organic produce.

14.9 CONCLUSION

Over the years, certain standards and certification procedures have been evolved which led to a better understanding and adoption of this eco-friendly farming practice across the world all over the world in recent years.

Organic farming mainly relies on soil health to safeguard plant health. Pest and diseases are managed through natural means and bioproducts and in a long run, it will be a farmer and eco-friendly farming practice to feed human beings and animals the safe and chemical free foods. Simultaneously, it also aims to minimize environmental pollution, promote soil health, improve fertility, and sustain the productivity.

In general, the organically produced vegetables fetch a higher price over conventionally produced ones. However, the industry is highly competitive and most occasions the price of the produce and income to the growers are dictated by overall supply and demand on a given day and the available organic outlets.

The promotion of organic vegetable farming also requires many more scientific interventions to efficiently manage the soil health, nutrients, pests, and diseases.

KEYWORDS

- **organic farming**
- **climate**
- **soil**
- **nutrient management**
- **cover crops**
- **organic fertilizers**
- **cultural practices**
- **seedling production**
- **direct sowing**
- **disease management**
- **bioagents**
- **marketing**

REFERENCES

Ali, M. A. A.; Chandrasekar, S. S.; Vardarasan,S.; Gopakumar, B.; Paramaguru, P.; Ponnusamy, V.; Muthusamy, M. Impact of Organic Cultivation of Chilli (*Capsicum Annuum* Linn) on Pests And Diseases. In *Proceedings of 15th Plantation Crops Symposium Placrosym XV*, Mysore, India, Dec 10–13, 2002; pp 451–456.

Boomiraj, K.; Lourduraj, A. C. Organic Production of Bhendi (*Abelmoschus esculentus*). *J. Ecobiol.*, **2006,** *19* (4), 389–396.

Chaudhary, R. S.; Das, A.; Patnaik, U. S. Organic Farming for Vegetable Production Using Vermicompost and FYM in Kokriguda Watershed of Orissa. *Indian J. Soil Conserv.* **2003,** *31* (2), 203–206.

Devrajan, K.; Seenivasan, N.; Selvaraj, N. Bio-Management of Root-Knot Nematode, *Meloidogyne hapla*, in Carrot (*Daucus carota* L.). *Indian J. Nematol.***2003,** *33* (1), 6–8.

Girija, I. S.; Britto, A. J., de. Comparative Study on the Effect of Organic and Inorganic Farming Methods on Clusterbean and Bhendi. *Environ. Econ.* **2006,** *24S* (Special 4), 1115–1118.

Kannan, P.; Saravanan, A.; Balaji, T. Organic Farming on Tomato Yield and Quality. *Crop Res.* **2006,** *32* (2), 196–200.

Pandey, A. K.; Gopinath, K. A.; Chattacharya, P.; Hooda, K. S.; Sushil, S. N.; Kundu, S.; Selvakumar, G.; Gupta, H. S. Effect of Source and Rate of Organic Manures on Yield Attributes, Pod Yield and Economics of Organic Gardenpea (*Pisum sativum* subsp. *Hortense*) in North West Himalaya. *Ind. J. Agril. Sci.* **2006,** *76* (4), 230–234.

Parvatha Reddy, P. *Organic Farming for Sustainable Horticulture*; Scientific Publishers: Jodhpur, Rajasthan, India, 2008.

Prabhakar, M.; Hebbar, S. S. Studies on Organic Production Technology of Rose Onion (*Allium cepa*) in Semi Arid Agro Ecosystem of India, 5th International ISHS Symposium on Edible Alliaceae, Dronten, The Netherlands, Oct 29–31, 2007.

Prabhakar, M.; Hebbar, S. S. Studies on Organic Production Technology of Annual Drumstick in a Semi-Arid Agro-Ecosystem. *Acta Hort.* **2008,** *752*, 345–346.

Prabhakar, M.; Hebbar, S. S.; Nair, A. K. Growth and Yield of Frenchbean (*Phaseolus vulgaris* L.) under Organic Farming, *J. Appl. Hort.*, **2011,** *13*, 71–73.

Prabhakar, M.; Hebbar, S.; Nair, A. K. Effect of Organic Farming Practices on Growth, Yield and Quality of Rose Onion (*Allium cepa*), *Indian J. Agric. Sci.* **2012,** *82* (6), 500–503.

Prabhakar, M.; Hebbar, S. S.; Nair, A. K.; Shivashankara, K. S.; Chinnu, J. K.; Geetha, G. A. Effect of Different Organic Nutrient Levels on Growth, Yield and Quality in Cauliflower. *Indian J. Hort.* **2015,** *72* (2), 293–296.

Prabhakar, M.; Hebbar, S. S.; Nair, A. K.; Panneer Selvam, P.; Rajeshwari, R. S.; Kumar, P. Growth, Yield and Quality of Onion (*Allium Cepa* L.) as Influenced by Organic Farming Practices. *Int. J. Curr. Microbiol. App. Sci.* **2017,** *6* (8), 144–149.

Prasanna, K. P.; Rajan, S. Effect of Organic Farming on Storage Life of Brinjal Fruits. *South Indian Hort.* **2001,** 49 (Special), 255–256.

Raghav, M.; Kamal, S. Organic Farming Technology for Higher and Eco-Friendly Cowpea Production in *Tarai* Region of Uttaranchal. *Acta Hort.* **2007,** *752*, 469–471.

Rajendran, T. R.; Venugopalan, M. V.; Tarhalkar, P. P. *Organic Cotton Farming*. Technical Bulletin No. 1/2000, CICR: Nagpur, 1999, p.37.

Srivastava, R.; Roseti, D; Sharma, A. K. The Evaluation of Microbial Diversity in a Vegetable Based Cropping System under Organic Farming Practices. *Appl. Soil Ecol.* **2007,** *36*(2/3), 116–123.

Rupela, O. P.; Gowda, C. L. L.; Wani, S. P. Lesson from No-chemical Treatments Based on Scientific and Traditional Knowledge in a Long Term Experiment, *Abstract Book, International Conference on Agricultural Heritage of Asia, Asian Agri-History Foundation*, Secunderabad, Dec, 6–8, 2004.

Singh, S. R. Effect of Organic Farming System on Yield and Quality of Brinjal (*Solanum melongena* l.) var. Pusa Purple Cluster under Mid-hill Conditions of Himachal Pradesh. *Haryana J. Hort. Sci.* **2004,** *33* (3/4), 265–266.

Singh, S. R. Effect of Organic Farming on Productivity and Quality of Frenchbean (*Phaseolus vulgaris* L.) var. Contender. *Legume-Res.* **2002,** *25* (2), 124–126.

Singh, S. K.; Yadava, R. B.; Chaurasia, S. N. S.; Prasad, R. N.; Singh, R.; Chaukhande, P.; Singh, B. Producing Organic Vegetables for Better Health. *Indian Hort.* **2016,** *61* (1), 5–8.

Thakur, K. S.; Kumar, D.; Vikram, A.; Thakur, A. K.; Mehta, D. K. Effect of Organic Manures and Biofertilizers on Growth and Yield of Tomato and Frenchbean under Mid Hills of Himachal Pradesh. *J. Hill Agric.* **2010,** *1* (2), 176–178.

Thamburaj, S. Tomato Responds to Organic Gardening. *Kissan World* **1994,** *21*, 10–49.

CHAPTER 15

Organic Cultivation of Vegetable Crops

ANITTA JUDY KURIAN[1], MINNU ANN JOSE[1], S. NIRMALA DEVI[1], P. INDIRA[1], and K. V. PETER[2]

[1]Department of Vegetable Science, College of Horticulture, Kerala Agricultural University, KAU P.O., Thrissur 680656, Kerala, India

[2]Kerala Agricultural University, KAU P.O., Thrissur 680656, Kerala, India

[*]Corresponding author. E-mail: drsnirmala@yahoo.com

ABSTRACT

Organic agriculture, a holistic system that focuses on improvement of soil health and use of local organic farm inputs. It is based on the rule of nature and operates upon the basic principles of sustainability. As a component of organic agriculture, organic vegetable farming promotes and enhances natural diversity and biological cycles on the farm rather than relying on synthetic fertilizers and pesticides, organic farming is based on making the farm self-sufficient and sustainable. Being an important natural resource, soil requires its proper management for organic production. Vegetable growers should rely on organic techniques like the use of organic or green manures, crop rotation without the use of synthetic substances, scientific management of air and water, providing proper drainage to the field, adopting integrated insect-pests and disease management including biological control.

15.1 INTRODUCTION

Organic farming can be defined as a production system, based on replenishment of ecological processes and establishment of ecological functions

Organic Crop Production Management: Focus on India, with Global Implications. D. P. Singh, PhD, H. G. Prakash, PhD, M. Swapna, PhD, & S. Solomon, PhD (Eds.)

of farm ecosystem to produce safe and healthy food. It avoids or largely eliminates the use of inorganic chemical fertilizers, fungicides, insecticides, pesticides, weedicides, growth regulators, and livestock feed additives and rely upon green manures, crop residues, animal manures, legumes, crop rotations, off-farm organic wastes, mechanical farming, mineral-containing rocks, and biological way to control pests, diseases, weeds, to maintain soil fertility, and to provide nutrients to the plants.

Food and Agriculture Organization of the United Nations has defined organic agriculture as a unique production management system that promotes and enhances agroecosystem health, including biodiversity, biological cycles, and soil biological activity, and this is accomplished by using on-farm agronomic, biological, and mechanical methods in exclusion of all synthetic off-farm inputs. The soil, plants, animals, and human beings are linked in organic farming. Organic farming is based on the concept of "farming in spirits of organic association." Therefore, its objective is to generate an integrated, eco-friendly, and economically sustainable farming system (Patle et al., 2014).

In solving malnutrition and ensuring food and nutritional security, vegetable crops have played a great role. Vegetables are a rich source of vitamins, minerals, protein, fiber as well as carbohydrates. Apart from its supply to domestic markets, vegetables have the potential for the export market also. The average yield of various vegetables in India is comparatively lower than the global average yield in spite of India being the second largest vegetables globally next to China. The per capita per day availability of vegetables of 210 g is quite lower than the recommended quantity (285 g/head/day). Thus, by exploring the possibility of vertical expansion or escalating the yield per unit area per unit time, demand for vegetables in the future with response to the rapid growth of population and reduction in cultivable land area can be solved. In this context, organic vegetable cultivation provides one of the most sustainable farming production systems with long-term soil health benefits along with better resistance against different biotic and abiotic stresses. Organic vegetables also fetch a premium price of 10–50% over traditional products. The growth rate (20%) for market demand of organic products is growing rapidly compared to conventional ones (5%). Countries like Japan, the United States of America, Australia, and the European Union have recorded the highest growth rate for market demand of organic products. India is a country that has introduced the production of crops organically way back. Thus, there is immense scope for the export of organic vegetables from India.

15.1.1 CONCEPT OF ORGANIC FARMING

The fundamental concepts of organic farming are as follows:

1. Organic farming gives much importance to the natural building up of soil fertility so that the plants absorb the nutrients from the soil as per their requirements and the formed are released with respect to the crop's requirement.
2. Promoting the application of biopesticides along with various cultural practices like mixed cropping and crop rotation for maintaining an ecological balance within the system for controlling biotic stresses like insect-pests, diseases, and weeds.
3. All the wastes and manures produced on a farm are recycled and serve as input in the production system.
4. Efforts must be made to recycle all urban and industrial wastes back to the soil for improving soil fertility.

15.1.2 ESSENTIAL CHARACTERISTICS OF ORGANIC FARMING

The following are the most important and essential characteristics of organic farming:

1. Maximum utilization of local resources in a sustainable way.
2. Excluding the use of purchased inputs.
3. Guarantee the fundamental natural functions of soil–water–nutrients–human continuum.
4. Maintaining ecological balance and economic stability through diversification of plant and animal species.
5. Risks associated with agriculture can be reduced by integrating crops with livestock systems, polycultures, agroforestry systems, and others.

15.1.3 SOIL FERTILITY MANAGEMENT

Soil fertility is the inherent ability of soil to supply vital elements for plant growth and development. It is the essence of a productive soil and is affected by the tilth of soil and quantity of nutrients it contains. Long-term optimum crop productivity can be achieved through effective cropping patterns, crop residue management, and proper crop rotation that results in a live, healthy soil, without any loss in fertility.

15.1.4 CROP ROTATION

Crop rotation is an inevitable factor of organic farming which can be defined as the cultivation of different crops in succession at the same piece of land in a year or over a long period of time. Through the use of atmospheric nitrogen-fixing leguminous crops, cover crops, and by adopting appropriate crop sequence in time and space, crop rotation in organic farming can be attained. In crop rotation, fixation of atmospheric N in the soil is a vital part. It can be achieved by growing leguminous crops after nonleguminous crops. In order to facilitate proper and uniform uptake of nutrients, crops with taproot should be followed by a crop with a fibrous root system.

15.1.5 COVER CROPPING

Cover crops include any annual, biennial, or perennial plants that are cultivated as a polyculture or monoculture which helps to prevent soil erosion, maintain soil structure, and enhance organic matter. Soil properties like soil fertility, soil quality, and so forth can be well managed by the cover crop. It can also control pests, diseases, and weeds. Cover crop can be helpful in reducing the increased level of carbon dioxide in the atmosphere resulting in enhanced soil carbon sequestration. In addition, soil carbon accumulation through crop cover has beneficial effects on soil quality and wider environmental benefits.

15.1.6 COMPOSTING

Decomposed and recycled organic matter produces compost which is used as fertilizer or organic manure to supply the essential plant nutrients in addition to improve soil structure and quality. Composting is a biological process of decomposition by microorganisms that convert organic matter to a final stable humus-like substance under controlled conditions. Soil properties like texture, permeability, and water holding capacity are improved by compost.

It feeds beneficial organisms and improves soil humus content. Balanced and slow-release of nutrients is also possible with help of compost. It enhances organic matter in the soil.

Different techniques of composting like Coimbatore method, NADEP method, Bangalore method, and Indore method have been developed. Another composting method developed by Dr. Francis Xavier of Kerala Veterinary and Animal Sciences University is the "Thumburmuzhy model of composting which is an eco-friendly livestock waste management system." It is an aerobic composting model that releases a very small amount of methane and carbon dioxide and emits zero foul smell. Since this model is a farmer-friendly and cost-effective method of waste management, it is preferred over other methods.

15.1.7 BIOFERTILIZERS

Carrier-based ready-to-use live bacterial or fungal formulations are known as biofertilizers or microbial inoculants. Application of these biofertilizers in the plants, soil, or composting pits helps in mobilization of nutrients by their biological activity. Nitrogen-fixing biofertilizers include *Rhizobium* (legumes), *Azotobacter*, and *Azospirillum* (nonlegume crops), phosphate solubilizers include *B. subtilis, B. megaterium, B. polymyxa, Pseudomonas striata*, and others. Potash solubilizers include *Fraturiaaurantia* and mycorrhiza like *Glomus, Gigaspora*are capable of absorbing more phosphorus different micronutrients and water from deep soil layers by the plants.

Choice of the appropriate combination of biofertilizers is an important task. Application of nitrogen-fixing biofertilizers, phosphate solubilizing biofertilizers, and potash mobilizing biofertilizers together in equal quantities may provide better availability of nitrogen, potash, and phosphorus. Application of biofertilizers to crops can be done through seed treatment, seedling root dip treatment, and soil treatment.

15.1.8 WEED MANAGEMENT

Weed management is one of the most difficult tasks in the organic farming that need long-term planning and adoption of different cultural practices. Weed management can be accomplished through manual weed removal, mechanical scraping, or cutting using hand or motor-operated weeders. Intercropping with legumes may reduce the weed growth to some extent. Mulching with organic mulches like crop residues, dried leaves, and others and the use of plastic mulch is also a suitable remedy for weed growth.

15.1.9 PEST AND DISEASE MANAGEMENT

The appropriate selection of crop varieties having resistance to major insect-pests and diseases and its suitability to an area are very crucial as far as insect-pests and diseases management is concerned in organic farming of vegetable crops. The use of synthetic insecticides and pesticides is generally avoided in organic farming for pest control. Thus, in organic cultivation, different eco-friendly pest management measures are used without using such synthetic products.

Through the enhancement of biodiversity of natural enemy populations by providing habitat, insect pest can be managed to some extent. Cultural practices such as crop rotation and adjusting the planting dates may reduce the build-up of harmful insects in field conditions. Therefore, physical as well as thermic methods are allowed in the organic farming. The soil used for cultivation can be thermic sterilized to control both pests and diseases. The use of all the chemical herbicides, fungicides, and pesticides should be strictly eliminated and use only traditional products of plants, animals, and microorganism's origin, which is prepared at the farm to control the pests and diseases (Wyss et al., 2005).

The use of pest- and disease-resistant varieties, control of weeds and removal of all pests and disease infested plant parts, and raising trap crops that attract insects and pathogens are encouraged. For controlling pests and diseases, adoption of biocontrol methods, and use of biopesticides and products of plant origin are suggested. *Neem*, Sabadilla, and Pyrethrum extracts are the pesticides that are recommended for use.

Pests and diseases management in vegetable crops is possible by adopting various physical, cultural, and biological methods either alone or in combination. A few of the most efficient botanical pesticides and its source and the major pests that can be controlled are listed in Table 15.1. While bioagents used in organic cultivation of vegetable crops have been listed in Table 15.2.

TABLE 15.1 Natural or Botanical Pesticides.

Botanical pesticide	Source	Nature of the product	Against which pests
1. Allicin	Garlic	Broad-spectrum pesticide	Act as antibacterial and antifungal biopesticide
2. Nemacide	*Neem* tree	Insecticide	Potato beetle, grass hopper, moth
3. Sabadilla	Sabadilla lily	Insecticide	Caterpillars, leaf hoppers, thrips, sink bug, and squash bugs
4. Pyrethrum	Chrysanthemum	Insecticide	Aphids and ectoparasites of live stocks
5. Nicotine sulfate	Tobacco	Insecticide	Aphids, thrips, spider, mites, and other sucking insects

Source: Maity and Tripathy (2004).

TABLE 15.2 Bioagents Used in Organic Cultivation of Vegetable Crops.

S. no.	Organism	Target organism
1	Predators	i. Ladybird beetles on aphids and mealy bugs ii. Chrysoperla on aphids and other soft bodied insects iii. Carabids and staphylinid beetles on vast range of insect hosts
2.	Parasitoids	i. *Trichogramma* sp. on Lepidopteran pests ii. *Apanteles* sp. on Lepidopteran larvae iii. Carabids and staphylinid beetles on vast range of insect hosts
3.	Bacteria	i. *Bacillus thuringiensis (Bt) against DBM*
4.	Fungi	ii. *Beauveria bassiana on various crop pests* iii. *Metarhizium anisopliae on Helicoverpa armigera*
5.	Viruses (nuclear Polyhedrosis virus)	i. NPV of *Helicoverpa armigera* ii. NPV of *Spodoptera litura*
6.	Nematodes	i. *Steinernema glaseri* on soil insects

Source: Maity and Tripathy (2004).

15.2 ORGANIC CULTIVATION OF SOLANACEOUS VEGETABLES

Brinjal, chilli, and tomato are the major solanaceous fruit vegetables. Care should be taken to select varieties/hybrids that possess resistance to major pests and diseases and suited for each agroclimatic region for cultivation. The season, seed rate, and spacing also vary with varieties. The following points may be kept in consideration for the organic cultivation of solanaceous vegetables.

15.2.1 NURSERY

Solanaceous vegetables are cultivated after transplanting. Therefore, seedlings for 1-ha main field are raised in a nursery area of 0.01 ha. Seeds are sown in raised seedbeds of 0.80–1.00 m width and convenient length in open space with fertile topsoil. Incorporate well-decomposed organic matter at the time of sowing. Maintenance of disease-free condition, especially damping off is necessary in the nursery. This is achieved by the addition of *Trichoderma* (1 kg) to dried farmyard manure (100 kg) and *neem* cake (10 kg). Spread the mixture under shade and water is sprinkled to maintain moisture. Keep this for 15 days with intermittent turning. Add PGPR mix (1 kg) to the nursery soil at the time of bed preparation, mulching with green leaves can be provided after seed sowing and irrigate regularly in the morning with a rose. The addition of *Pseudomonas fluorescence* at 10 g/L at the time of irrigation in frequent intervals controls fungal pathogens. Mulch should be removed after seed germination. Seedling vigor can be improved by adding 20 g/L of diluted cow dung slurry or cow urine (diluted eight times). Transplantation of seedlings to the main field is done for 1-month old seedlings. Irrigation may be withheld 1 week before transplanting. Irrigate heavily on the preceding day of transplanting.

15.2.2 LAND PREPARATION AND TRANSPLANTING

Prepare the land by ploughing or digging to form fine tilth. Based on the season, shallow trenches or ridge/leveled lands can be prepared for planting. Well rotten organic manure may be incorporated into the soil and transplant the seedlings. To prevent the transplanted seedlings from hot sun, temporary shade may be provided for the first 4–5 days.

15.2.3 MANURING

Based on the soil acidity, application of lime at500 kg/ha can be done 14 days before transplanting. Basal dose of FYM or compost (25 t/ha) mixed with *Trichoderma* (2.5 kg /ha) and PGPR-mix-1 (2.5 kg /ha) which is kept in shade for 15 days may provide sufficient nutrients to growing crops. In place of FYM, poultry or powdered goat manure (1 t/ha) may also be used. At the time of transplanting, application of *Pseudomonas* and AMF will also be beneficial for plant growth and development. Dipping of seedling roots in 2% *Pseudomonas* or PGPR-mix-1, before transplanting may positively affect the crop growth and provide resistance from diseases.

Top dressing: (soil application at 8–10 days interval).

- Fresh cow dung slurry at 1 kg/10 L (50 kg/ha).
- Biogas slurry at 1 kg/10 L (50 kg/ha).
- Groundnut cake at 1 kg/10 L (50 kg/ha).
- Powdered goat manure at 1 t/ha.
- Vermicompost/poultry.
- Vermiwash at 500 L/ha (eight times dilution).

Foliar spray of cow dung slurry/vermiwash/cow's urine also can be done.

15.2.4 AFTER CULTIVATION

- If the soil is not moist enough, provide irrigation before transplantation of seedlings to the main field.
- Irrigate the crop at 2 or 3 days interval during hot days.
- Staking of the plants, if required.
- One and two months after transplanting, weeding is done for the removal of unwanted and diseased plants followed by the use of organic manure and earthing up.
- Natural materials like plant residues, green leaves, coconut husk, decomposed coir pith, straw, and others can be used as mulch throughout the crop growth period in the field.

15.2.5 PLANT PROTECTION

TABLE 15.3 Pests and Disease Control.

Crop	Pest	Control measures
Chilli	Aphids	Spray tobacco decoction or *neem* oil–garlic emulsion (2%) or Nattapoochedi (*Hyptis suaveolens*) emulsion (10%). Spray *Verticillium lecanae* or *Fusarium pallidoroseum* (1010 conidia/L). Release green lacewing bugs at 50,000 eggs/ha
	Jassids	Spray neem oil–garlic emulsion (2%) or lemongrass/ginger extract (10%)
	Thrips	Spray Kiriyath (*Andrographis paniculata*) extract (10%)
	Mite	Apply *neem* oil 5% or *neem* oil + garlic emulsion 2%. Spray diluted rice water once in 10 days against mite
Chilli and tomato	Whitefly	Spray *Verticillium lecanae* (1010 conidia/L) tomato or garlic emulsion (2%). Place Sticky yellow traps
Brinjal	Shoot and fruit borer	Protect the seedling in the nursery with net. Mechanical hand fruit borer picking and destruction of the affected part along with the larvae. Place pheromone traps at 100 nos./ha. Spray *neem* garlic emulsion (2%). Spray *Bacillus thuringiensis* available as Dipel, Delphin, Halt, Bioasp, Biolep (0.7 mL/L). Use S-NPV (250 LE/ha). Spray leaf extract of Ailanthus and Cashew (10%)
	Hopper	Spray *neem*-garlic emulsion (2%) or products like Nimbicidin/Econeem/Uneem (2 mL/L). Spraying of lemongrass/ginger extract (10%) is also effective
	Epilachna beetle	Spray soap–garlic–castor oil emulsion (2%). Collect and kill beetle all stages of the pests. Spray *Clerodendron* plant extract 4–8% or custard apple seed extract 2–5%
Tomato	Fruit borer	Spray *Neem* seed kernel extract 5%. Use *Ha*NPV (250 LE/ha). Spray *Bt*. Spray Pongamia oil (2%). Apply Pongamia or *neem* cake 250 kg/ha at planting and repeat two or three times at 30–45 days interval
	Red spider mite	Spray water using sprayer. Spray rice gruel water on under mite surface of leaves. Spray castor oil–soap emulsion or neem oil–garlic emulsion (2%)
	Serpentine leaf miner	Spray neem oil–garlic emulsion (2%) before 8'O clock in the leaf miner morning. Apply *neem* cake to soil (250 kg/ha). Spray *neem* oil, marotti oil, or illupai oil 2.5% or spray *neem* seed kernel extract 4%
Chilli, brinjal, and tomato	Nematode	Apply Eupatorium and *neem* leaves, *neem* cake, rice husk, brinjal and wood shavings, castor cake at 100 g/m^2. Apply VAM, tomato Plant Growth Promoting Rhizobacteria, Paceilomyces to soil at 2 kg/ha. Seed treatment with *Bacillus macerans* at 3% w/w (2.5 kg/ha) and drenching with *B. macerans* at 3% solution 30 days after sowing

TABLE 15.3 *(Continued)*

Crop	Pest	Control measures
Chilli and tomato	Damping off	Sow the seeds in raised beds prepared in brinjal open area and off during summer months. Preinoculation of AMF in tomato furrows at 200 g/m^2. Apply lime in nursery bed. Use Trichoderma, *Pseudomonas fluorescens*, and PGPR mix II. *Neem* cake can be applied at 250 kg/ha to reduce soil innoculant
	Leaf spot	Spray *Pseudomonas fluorescens* (2%) Spray Bordeaux mixture (1%)
	Bacterial	Cultivate resistant varieties. Use lime in the field. Cultivate wilt marigold in field prior to tomato cultivation. Soil application of *Pseudomonas fluorescens* or PGPR mix II at 20 g/L at 15 days interval. Seedling root dip and foliar spray of *Pseudomonas fluorescens* 1–2%
Chilli	Leaf curl virus	Spray *neem*-based insecticides (2 mL/L) to control the vectors. Grow resistant varieties like *Punjab Lal* and *Pusa Sadabahar*
Tomato		Spray *neem*-based insecticides (2 mL/L) to control the vectors. Grow five to six rows of maize around the crop at least 50 days before transplanting tomato. Keep the plot weed free.

Source: The ad hoc package of practices recommendations for organic farming (KAU, 2009).

15.3 ORGANIC CULTIVATION OF CUCURBITACEOUS VEGETABLES

Pumpkin, ash gourd, snake gourd, bitter gourd, little gourd, bottle gourd, ridge gourd, cucumber, watermelon muskmelon, round melon, and long melon are the important cucurbitaceous vegetables cultivated in India. The requirements for organic cultivation of cucurbitaceous vegetables include the following.

15.3.1 SOWING/PLANTING

- Seeds can be sown in pits/furrows/ridges/beds depending on season and methods of irrigation. Pits having dimensions of 60 cm diameter and 30–45 cm depth are dug at a distance of 2 × 2 m. The furrows/ridges can be taken at the distance of 1.5 m and convenient length. The seeds are sown at a distance of 0.75 m. Well rotten farmyard manure or other organic manure at12 t/ha is properly mixed with topsoil in the pit and four to five seeds are sown in a pit. In furrows/ ridges, the seeds are sown at a distance of 0.75 m. Stem cuttings with three to four nodes from female plants of the little gourd are used at two to three cuttings/pit. Remove unhealthy plants after 2 weeks and only three plants may be retained in a pit. Two splits of manures may be applied at the stage of vining and flowering. At fortnightly intervals from flowering, fresh cow dung slurry prepared at1 kg/L of water should be applied.

15.3.2 CROP MANAGEMENT

The crop should be irrigated at 2–3 days interval during the initial stages of growth and on alternate days during flowering and fruiting. Irrigating the crop at 15 mm CPE (approximately at 3 days interval for sandy loam soils) has been found more economical than one irrigation in 2 days, particularly during summer months.

For obtaining better growth and yield, bitter gourd, snake gourd, little gourd, ridge gourd, and sponge gourd are trailed on trellis/pandals. Cucumber, muskmelon, watermelon, bottle gourd, pumpkin and ash gourd, long melon, and round melon can be trailed on dried twigs spread on the ground. Regular weeding should be done before the application of manures. Materials such as plant residues, green leaves, decomposed coir pith, straw, coconut husk, etc., may be used for mulching throughout the crop growth period.

15.3.3 CROP PROTECTION

Fruit fly (*Bactrocer acucurbitae*), aphids, green jassid, whitefly, mites, leaf and flower feeder (*Diaphania* sp.), American serpentine leaf miner, and Epilachna beetle are the major pests on cucurbits.

15.3.3.1 FRUIT FLY

1. Remove and destroy infested fruits and cover the healthy fruits.
2. At planting and after its 1one month, use *neem* cake 250 kg/ha or 100 g/pit.

3. Any of the following fruit fly traps/baits may be used:
   Adult fruit flies can be trapped by hanging plywood blocks comprising a mixture of ethyl alcohol, cuelure, and malathion in the ratio of 6:4:1 at10 number/ha, and it should be reset at 4 months interval. We can also make fruit fly trap by mixing 20 g banana pulp with beer 3 mL and palm oil three drops in a coconut shell. Coconut shell traps (painted with yellow color) containing carbofuran smeared ripe banana pieces can be placed at 2 m spacing in the field from flowering till final harvest to trap the flies. There is a need for replenishment of traps once in 7 days. Fish meal traps can also be used to control fruit fly infestation.
4. Spraying *Beauveria bassiana* 10% WP, *Paecilomyces lilacinus* 5% WP, and a combination of leaf extracts of Ailanthus 10% and cashew 10% is also recommended.

15.3.3.2 APHIDS, WHITEFLIES, GREEN JASSIDS, AND MITES

1. Spraying 2% *neem* oil + garlic emulsion spray or1.5% fish oil soap.
2. Add soap solution (prepared by dissolving 60 g soap in 150 mL warm water) slowly to *neem* oil and castor oil and mix well. Dilute it with 6 L of water. Add 120 g garlic paste. Spray this extract for managing aphids, whiteflies, green jassids, and mites.
3. Apply 10% magnesium sulfate on leaves for providing strength for plants and preventing mites, planthoppers, and jassids.

15.3.3.3 LEAF AND FLOWER FEEDERS (DIAPHANIA SP.)

Collection and destruction of larvae and application of *Beauveria bassiana* 10% WP and *Paecilomyces lilacinus* 5% WP.

15.3.3.4 AMERICAN SERPENTINE LEAF MINERS

Spray 4% *neem* seed kernel emulsion (NSKE) in the morning before 8 O'clock.

15.3.3.5 EPILACHNA BEETLES

1. Egg masses, grubs, and adults present on the leaves may be removed and destroyed.
2. Predator (*Chrysocaries johnsoni*) of larvae and pupae may be used.
3. Use *Paecilomyces lilacinus* 5% WP *and Beauveria bassiana* 10% WP.
4. Spray leaf extract of ailanthus and cashew (10%).
5. Spray *Neem* oil mixed with garlic emulsion (2%).

Diseases

Mosaic

1. Mosaic-affected plants may be removed and collateral hosts should be done.
2. Spray *Neem* based insecticides for controlling the vector.

15.4 ORGANIC FARMING OF AMARANTH (*AMARANTHUS* SPP.)

The most popular leafy vegetable, amaranth can be cultivated around the year, although summer has been found the best season. There are red-leaved varieties like *Arun*, *Krishna Sree*; green-leaved Co-1, Co-2, Co-3, *Mohini*, and *RenuSree* having mixed leaf color. Amaranth can be cultivated by direct sowing or transplanting. The requirements for organic cultivation of amaranth vegetables include the following.

15.4.1 NURSERY

For disease management, nursery bed should be solarized before sowing. Seed should be treated with *Pseudomonas* (10 g/kg seed) for controlling nursery diseases. Application of 10 kg FYM enriched with 50 g each *Trichoderma* and *Neem* cake, PGPR-mix-1–100 g, and AMF 200 g/mL is recommended to get healthy seedlings

15.4.2 MAIN FIELD

Shallow trenches of 30–35 cm width are taken at 30 cm apart. In the shallow trenches, transplanting of seedlings of 20–30 days old is done at a distance of 20 cm in two rows. Planting shall be done on raised beds during the rainy season. Dip the roots of the seedlings for 20 min in a solution containing *Pseudomonas* 20 g/L before planting.

15.4.3 MANURING

FYM or compost at 25 t/ha may be applied as a basal dose. In FYM, *Tricho-derma* PGPR-mix-1 each at 2.5 kg/ha are mixed and kept for 10–15 days in a cool atmosphere. It is applied to the soil as a basal dose. Topdressing of fresh cow dung slurry at 1 kg/10 L water or diluted (eight times) cow's urine at 50 kg ha or diluted (eight times) vermiwash at 500 L/ha or application of vermicompost at 1 t/ha or groundnut cake at 1 kg/10 L (50 kg/ha) can be done. After each harvest, cow dung slurry/vermiwash/cow's urine may be used through foliar application.

Irrigate at regular intervals, in case, there is not sufficient moisture. Various materials like green leaves, plant residues, decomposed coir pith, straw, and others should be used as a mulch throughout the crop season. Regular weeding and earthing up also should be done in rows during the rainy season.

15.4.4 PESTS

Collect the leaf webber and leaf roller and destroy them mechanically. Spray Dipel or Halt (0.7 mL/L) for the management of leaf webber. In total, 4% of leaf extract of *neem*, thevetia, or clerodendron mixed with soap water may be applied.

15.4.5 DISEASES

During the rainy season, leaf spot is the major disease that can be managed to a great extent by using an integrated approach that includes growing of leaf spot resistant varieties like Co-1; seed treatment with *Pseudomonas* at8 g/kg seed; soil application of *Trichoderma* as enriched cow dung–neem cake manure and soil application of green manures, such as *neem* cake (100 kg/ha) + sun hemp/glyricedia + *Trichoderma* (1–2 kg/ha).

15.5 ORGANIC CULTIVATION OF OKRA (*ABELMOSCHUS ESCULENTUS*)

High-yielding disease varieties/hybrids are selected for cultivation. The seed rate varies from 7 to 9 kg/ha. The following major recommendations should be followed for the organic cultivation of okra.

15.5.1 SOWING

The seed rate of 8.5 kg/ha has been recommended for the January–February sown crop, which harvests during summer, while seed rate of 7 kg/ha is sufficient for *kharif* crop. Seeds should be soaked in water in double volume for 2 h for improving the germination and vigor of the seedlings. Seed treatment with *Pseudomonas* at 8 g/kg of seed has also been found effective in improving germination and vigor of the seedling.

15.5.2 MANURING

Trichoderma and PGPR-mix-1 each at 2.5 kg/ha are mixed with the FYM at 25 t/ha and kept at a cool temperature for 15 days. After that, it can be applied in the soil as basal dose along with *Pseudomonas* at 2 kg/ha. Foliar spray of a supernatant solution of cow dung slurry/vermiwash/cow's urine can be given up to flowering.

15.5.3 PLANT PROTECTION

Jassids, fruit, and shoot borer along with root-knot nematode are the major insect-pests.

15.5.3.1 JASSIDS

Nimbicidine (2 mL/L)/*neem* oil-garlic mixture (2%)/econeem (2 mL/L)/ uneem (2 mL/L), and lemongrass suspension (10%) can be used for controlling jassids.

15.5.3.2 *FRUIT AND SHOOT BORER*

The affected shoots and fruits should be removed and destroyed. Spraying *neem* kernel suspension (5%)/ginger suspension (10%)/neem leaf extract (4%) has been found effective in controlling jassids. The use of *Trichogramma chilonis* and *Trichogramma japonicum* cards at 1each/200 m^2 followed by a spray of *Bacillus thuringiensis* (Delphin/Bioasp/Halt-0.7 mL/L) is also an effective method.

15.5.3.3 *BHINDI LEAF ROLLER*

Collection and destruction of the *bhindi* leaf rolls and application of *Beauveria bassiana* 10% WP will control the pest.

15.5.3.4 *ROOT-KNOT NEMATODE*

In the case of heavily infestation of root-knot nematode, seed should be treated with *B. macerans* at 3% w/w and drenched with *B. macerans* at 3% solution 30 days after sowing for reducing nematode infestation. *Neem* leaves or eupatorium leaves should be applied at 250 g/plant a week before planting and should be irrigated daily. During the summer season, the nematicidal effect of the spray will persist upto 75 days after sowing. Application of *neem* cake/castor cake at 1 t/ha or cultivating marigold as a trap crop in between okra plants is also an effective method.

15.5.3.5 *DISEASES*

Yellow Vein Mosaic

Vein chlorosis or vein clearing of leaves are the characteristic features of yellow vein mosaic disease of cowpea. The yellow network of vein is very conspicuous and veins and veinlets are thickened. Fruits of the diseased plants become small and yellowish-green in color. Whitefly (*Bemisia tabaci*) is the vector of this virus. For prevention of yellow vein mosaic, disease-resistant varieties/hybrids should be selected and host weeds such as *Ageratum* sp. and *Croton sparsiflora* should be destroyed. Spraying of *neem* oil–garlic mixture (2%) or nimbicidine/econeem/uneem at 2 mL/L has been found very effective.

15.6 ORGANIC CULTIVATION OF LEGUMINOUS VEGETABLES

15.6.1 COWPEA

There are two types of vegetable cowpea, namely, yard long bean (*Vigna unguiculata subsp. sesquipedalis*) and bush type (*Vigna unguiculata subsp. unguiculata*).

15.6.1.1 VARIETIES

There are bush, semitrailing, and trailing varieties of cowpea.

Bhagyalakshmy, *Pusa Komal*, *Kashi Kanchan*, and others are bushy-type varieties, while *Kanakamony* and *Anaswara* are semitrailing varieties. *Malika*, *Vaijayanthi*, *Sharika Lola*, *Vellayani Jyothika*, and *Velayati Geetika* are trailed on pandals or trellis to get higher yields.

Bushy varieties are sown at a row spacing of 30 cm and plant to plant spacing of 15 cm. While semitrailing varieties should be sown at a spacing of 45 × 30 cm. Varieties trailing on pandals or trellis should be sown in pits at three plants/pit at 2 × 2 m spacing or in channels at 1.5 m × 45 cm spacing.

15.6.1.2 SEASON

Cowpea can be cultivated round the year in different parts of the country. Cowpea is sown in the month of June as a rain-fed crop. The second to fourth week of June has been found to be the most suitable time for sowing the crop. Sowing can also be done during the second crop season of September–October. Cowpea can be sown during January–February for taking harvest during summer.

15.6.1.3 SEED TREATMENT

Before sowing, seeds of cowpea seeds should be inoculated with *Rhizobium* and pelleted with lime.

15.6.1.4 RHIZOBIUM INOCULATION

The *Rhizobium* inoculum is mixed on the seeds for a thin uniform covering by using jaggery or minimum quantity of 2.5% starch solution for ensuring adhesiveness of the inoculants with the seed. The seed coat should not be damaged. Inoculated seed should be dried under shade over a clean cloth/ paper or gunny bag. Sowing of these seeds should be done at the earliest. The seeds inoculated with *Rhizobium* culture should not be mixed with any of the chemical fertilizers. The seeds inoculated with *Rhizobium* culture or *Rhizobium* culture as such should not be exposed to direct sunlight or heat, otherwise, the inoculation will not be useful as beneficial bacteria may be killed under heat or direct sunlight. The seeds can also be coated by vermicompost.

15.6.1.5 LIME PELLETING

To moist fresh *Rhizobium* treated seeds, these are mixed with finely powdered (300 mesh) calcium carbonate for 1–3 min and mixed until each seed is uniformly pelleted. The requirement of lime varies with the seed size. Small seeds require lime at 1.0 kg/10 kg of seed. Medium-sized seeds are mixed with the lime at 0.6 kg/10 kg of seed, while the large-sized seeds require lime at 0.5 kg/10 kg of seed.

The pelleted seeds are spread on a clean cloth or paper for its hardening and sowing is done at the earliest. Under unavoidable circumstances, lime pelleted seeds can be stored in a cool place for a period of 1 week before sowing. Although fertilizer can be mixed with the lime pelleted seeds, but the duration of contact between the fertilizer and the pelleted seeds should be the minimum. Pelleted seeds should be sown in moist field only.

15.6.1.6 MANURING

Basal application of FYM at 20 t/ha, lime at 250 kg/ha, or dolomite 400 kg/ ha is recommended. The combination of FYM/Cow dung at 2 t/ha + rock phosphate at 100 kg/ha or compost at 4 t/ha + rock phosphate at 70 kg/ ha or vermicompost at 2 t/ha + rock phosphate at 110 kg/ha or greenleaf at 3.5 t/ha + rock phosphate at 100 kg/ha or poultry manure at 1.5 t/ha + rock phosphate at 50 kg/ha can also be used as supplement. Depending upon the further need, the additional organic manures can be used in splits at 15-days interval.

Growth promoters like *panchagavyam* or vermiwash should be used as foliar spray at 15 days interval has been found helpful in enhancing marketable yield. Application of AMF/phosphorus solubilizing microorganisms at 1 g per plant at the time of sowing enhances the availability of phosphorus.

15.6.1.7 AFTER CULTIVATION

As excessive vegetative growth hampers the prospects of good yield, therefore, pruning excessive vegetative growth has been found beneficial for enhancing flowering and fruiting. Yard long bean should be provided pandal or trellis for trailing immediately at the start of vining. Heavy irrigation should be avoided at the vegetative phase. Better flowering and fruit sets are induced by irrigation at the stage of flowering.

15.6.1.8 PLANT PROTECTION

Pests

Aphids (*Aphis craccivora*) are major sucking pests. Spraying of Neemazal T/S 1% at 2 mL/L at 15 days interval is quite effective for managing aphid in cowpea. *Fusarium pallidoroseum* fungus may also be used for aphid control. Bran-based fungus at 3 kg/400 m^2 may also be used once just after infestation. One liter *Hyptis suaveolens* extract mixed with 60 g soap in half liter of water and diluted 10 times can also be sprayed to control aphids.

Spray of 5% neem seed kernel extract can effectively control jassids and whiteflies. American serpentine leaf miner (*Liriomyza trifolii*) is also a common pest in cowpea. Weed host plants, namely, *Achyranthus aspera, Amaranthus viridis, Cleome viscosa, Heliotropium indicum*, and *Physalis minima* should be destroyed and *neem* oil, marotti oil, or illupai oil at 2.5 % may be used, if required. Pod borer can be controlled by spraying the dilute solution of cow's urine + asafoetida + bird chilli extract, or *neem* seed kernel extract 5%. Pod bugs can be controlled by collecting and destroying different stages of the bug. Alternate host plants should be removed. Spray *amruthneem* 5 mL/L or nimbicidin 2 mL/L or *neemazal* 2 mL/L or *neem* seed kernel extract 5%. Spraying *neem* oil 5%/ *neem* oil garlic emulsion 2%/garlic emulsion 2%/fish oil soap 2.5% will control red spider mite. For controlling root-knot nematode and reniform nematodes, *neem* or eupatorium leaves at15 t/ha should be used 15 days before sowing.

Diseases

1. Soil-borne diseases and nematodes.

Soil Solarization

Soil solarization method is used for managing soil-borne diseases and nematodes during the summer months of May and June. The 150-gauge clear polythene sheets are used to cover the soil after slightly moistening the soil. The soil temperature of the soil surface increased up to 52°C reduces the soil inoculum load by destroying the soil-borne fungi, bacteria, nematodes and weeds, and thereby. For protecting the crop from fungal diseases, soil drenching with 1% Bordeaux mixture or 2% *Pseudomonas* is recommended.

Web blight (*Rhizoctonia solani*) and collar rot are the major diseases of cowpea. These can be controlled by neem cake at 250 kg/ha or *Pseudomonas* (2%) application. Application of *Trichoderma* enriched organic manure in the main field has also been effective for controlling collar rot.

Fusarium wilt (*Fusarium oxysporum*) can be controlled by uprooting the whole affected plants and burning. Seed treatment with *Trichoderma viride* at 2 g/kg seed + soil application of 2.5 kg/ha at 1 month after sowing along with soil application of *neem* cake at 150 kg/ha at the time of land preparation can be done to reduce the extent of wilt.

Dry root rot incidence can be controlled by soil drenching of *Pseudomonas* at 2% (20 g/L). Soil application of *neem* cake at 250 kg/ha or seed treatment with *Trichoderma virid*e at 4 g/kg or *Pseudomonas fluorescens* at 10 g/kg has also been found quite effective.

The above-mentioned organic farming production methods of cowpea are also applicable to other leguminous vegetables like cluster bean, winged bean, and *Dolichos* bean.

15.7 CONCLUSION

Organic farming is based on the rule of nature and operates upon the basic principles of sustainability. Being an important natural resource, soil requires its proper management for organic production. One should rely on organic techniques like the use of organic or green manures, crop rotation without the use of synthetic substances, scientific management of air and water, providing proper drainage to the field, adopting integrated insect-pests and disease management including biological control.

KEYWORDS

- **organic farming**
- **soil fertility**
- **crop rotation**
- **cover cropping**
- **composting**
- **natural pesticides**
- **lime pelleting**

REFERENCES

Maity, T.; Tripathy, P. Organic Farming of Vegetables in India: Problems and Prospects, 2004. https://www.researchgate.net/publication/237641960

Patle, G. T.; Badyopadhyay, K. K.; Kumar, M. An Overview of Organic Agriculture: A Potential Strategy for Climate Change Mitigation. *J. Appl. Nat. Sci.* **2014,** *6* (2), 872–879.

Wyss, E.; Luka, H.; Pfiffner, L.; Schlatter, C.; Uehlinger, G.; Daniel, C. Approaches to Pest Management in Organic Agriculture: A Case Study in European Apple Orchards. *Org. Res.* **2005,** (May), 33–36.

CHAPTER 16

Organic Farming in Vegetables: An Opportunity for Sustainable Production and Livelihood Enhancement

B. SINGH and S. K. SINGH*

ICAR-Indian Institute of Vegetable Research, Post Box No. 01, P.O. Jakhani (Shahanshahpur), Varanasi, 221 305, Uttar Pradesh, India

**Corresponding author. E-mail: skscprs@gamil.com*

ABSTRACT

Organic agriculture can contribute to meaningful socio-economic and ecologically sustainable development, especially in the developing countries. In the last five decades, environmental problems arising due to misuse of synthetic fertilizers, pesticides, and machinery has caused land degradation and irreversible ecosystem damage. Organic farming of vegetables has potential for food and nutritional security and repairing and strengthening the soil fertility with organic residues. Certified organic vegetable products provide the opportunity to farmers for high-income options and, therefore, can be instrumental in promoting environmental-friendly farming practice.

16.1 INTRODUCTION

Indian economy is agrarian by nature where agriculture plays a vital role. In India, the contribution of the agriculture sector to the country's GDP is about 17%. Nearly two-thirds of the population of the country depends on

Organic Crop Production Management: Focus on India, with Global Implications, D. P. Singh, PhD, H. G. Prakash, PhD, M. Swapna, PhD, & S. Solomon, PhD (Eds.)

agriculture for employment. Agriculture is the principal means of livelihood for 65–70% of the rural households in the country. In spite of rapid growth in various sectors, agriculture still remains the mainstay of the Indian economy. A large proportion of farmers in India lives in a less favored, marginal or submarginal, and more complex environment. In these areas, vegetable cultivation provides economic and nutritional security, especially to the resource-poor farmers.

India with its varied agroclimatic zones is amenable to grow a different types of horticultural crops including vegetables, tuber and root crops, spices, fruits and condiments, flowers, ornamental plants, medicinal and aromatic plants, plantation crops, and mushrooms. These crops constitute a considerable part of the total agricultural production in the country. More than 60 popular and about 30 less known vegetable crops are being grown in different parts of the country due to varied climates. A number of technologies have been developed and demonstrated, which has made it possible to double vegetable production in the country in the last one and half decades. India is the second-largest vegetable producer in the world after China. India having vegetables production to the tune of 169.478 million metric tonnes contributes about 14% of the total world vegetable production. Important vegetables grown in India are potato, onion, tomato, cauliflower, cabbage, bean, eggplants, cucumber and garkin, peas, garlic, and okra. The total export of fresh vegetables other than onion, during 2015–2016 was 699,600.34 MT worth Rs. 2119.50 crores. The key destinations for Indian vegetables are Bangladesh, Malaysia, Nepal, The Netherlands, Pakistan and Qatar, Sri Lanka, Saudi Arabia UAE, and UK.

According to a report of the Food and Agricultural Organization of the United Nations, about 17% of Indians are too malnourished to lead a productive life. About 51% of men and 48% of women take diets that are high in fat. Nearly three out of five men and women take a very low quantity of fruits and vegetables. Vegetable consumption in India is 252 g per capita per day which is far less than recommended (300 g). This highlights the need for balanced nutrition in the diet. Vegetables being nutritionally rich, have a major role in eradicating hunger and malnutrition in the country. Many of the vegetables are short-duration crops and can fit very well in the different multiple and intercropping systems prevalent in the country. Vegetables have the potential of giving very high production and economic returns to farmers in a short period of time, besides generating on-farm and off-farm employment. Therefore, in recent times, the main emphasis is given for the commercial exploitation of vegetable crops. However, continuous use of chemicals fertilizers, and nonjudicious use of pesticides has resulted in a decline in factor productivity of various vegetable crops as worsening of soil health and environments.

The nonjudicious use of chemical inputs in agriculture particularly in vegetable crops has resulted in the occurrence of widespread contamination of vegetables with agrochemicals that may lead to different kinds of health problems. Hence, there is an urgent need to produce vegetables free from contaminants. Organic agriculture is the production method that encourages and supports the safety of environment and restricts the use of synthetic inputs. The organic farming has emerged as an alternative system of farming due to its capacity to deliver a long-term sustainable production system with a good quality environment. The increasing awareness about the safe quality of foods produced from organic farming has added a profitable livelihood option to the organic growers. Organic farming of vegetables is gaining momentum all over the world and now organic farming is being practiced in almost all the countries of the world. India with its diverse climate and different types of soils has an immense potential of producing organic vegetables and generate revenue through export. However, the technologies and research findings of organic farming in vegetable production have not reached to farmers, due to deficiencies of delivery systems or lack of economic incentives. The main problem faced by vegetable growers to sell organic produce is the variation in price due to sudden increase and decrease in supply and demand. The vegetable production under protected cultivation or during off-season production and facility of the cool chain may provide a better opportunity to get better price through organic vegetable cultivation.

The numerous research findings have shown that organically produced vegetables have low levels of pesticides and hormonal residues and in most of cases, lower contents of nitrate. The influence of organic farming on natural resources allows interactions within the agroecosystem that are important for both agricultural production and environmental conservation. Continued use of manures improves the content of organic matter in soil, is supportive to the soil micro-, meso-, and macro-fauna of soil, and makes it a living body. Organic matter also augments soil structure and improves water-holding capacity, which is vital under dry farming situations. Therefore, under the scenario of climate change, organic cultivation of vegetables can improve the livelihood of poor resource constraint farmers.

Organic farming systems have lower consumption of fossil fuel energy, lower carbon dioxide and nitrous dioxide emissions, less soil erosion, and increased carbon stocks. The consumption of energy in organic farming has been found to be 10–70% less in European countries and 28–32% less in the United States of America in comparison to high-input inorganic systems, except for crops such as potato and apple where the energy use is equal or higher. Greenhouse gas emission under organic systems is 29–37% lower because of nonuse of chemical fertilizers and pesticides as well as lower use of high energy feed. Carbon sequestration efficiency under organic farming in temperate climates is almost twice (575–700 kg carbon per ha per year) as compared to conventional farming. Thus, organic cultivation gives an opportunity to achieve the twin goals of nutritional safety and environmental sustainability.

16.2 ORIGIN OF ORGANIC FARMING

Green revolution or intensive cultivation practices that included higher use of synthetic agrochemicals like fertilizers and pesticides, increased use of nutrient-responsive, high-yielding varieties of crops, higher use of irrigation water, and others during the 1970s accelerated the production output in most cases. However, continuously indiscriminate use of these high energy inputs has resulted in a reduction in production and productivity of different crops as well as the decline of soil health and environments. A large quantity of pesticide is being used to control insect-pests and diseases in a number of vegetable crops like brinjal, tomato, cauliflower, chilies, sweet pepper, cucurbits, and so on. Vegetable growers are spraying pesticides at higher quantity and frequent intervals than recommended to control pests. Such contaminated vegetables carrying pesticide residues are causing several problems in humans like cancer, infertility, Parkinsons, baldness, and so forth. Besides, these pesticides are also polluting soil and groundwater. Herbicides and chemical fertilizers are the other harmful inputs that are frequently used for conventional vegetable production. The need for an alternative agriculture system is realized now, which can function in eco-friendly system while sustaining the soil and crop productivity. Organic farming is considered as the best-known substitute to the conventional agriculture. Organic production offers a better possibility to address soil, human, and environment health and to accomplish the larger goals of sustainable production with the least use of chemicals and increased use of biological agents, recycling of nutrients, and harnessing solar energy in an eco-friendly manner.

16.3 SCOPE OF ORGANIC FARMING IN INDIA

There are three schools of thought about the suitability of organic agriculture in India. The first one simply dismisses it as a fad or craze. The second group comprises many scientists and growers who advocate that there are merits in the organic farming but one need to proceed carefully considering the national requirement and conditions in which Indian agriculture functions. They know the environmental problems arising due to conventional farming. However, many of them are of the opinion that during the initial period yields are lower in organic cultivation, and also the cost of labor is higher. The third party advocates adoption of organic farming wholeheartedly. In their opinion, future ecology and environment are more important than present farm benefits of conventional farming There are strong views, both in favor and against of organic farming.

Organically produced vegetables fetch a higher price of 10–50% over-produce obtained through conventional farming. Presently, the organic vegetable market is growing at the rapid rate of 20% in comparison to conventionally produced vegetables which is around 5%. This rate of growth is the highest in Japan followed by the United States of America, Australia, and the European Union.

Indian farmers have inculcated the skill of growing vegetables organically since time immemorial which gives an edge to the Indian farmers in the production of organic vegetables. Therefore, India has great scope and potential in export of organic vegetables. The contribution of organic vegetables in total organic export of the country was roughly 1%. There is ample scope to improve the organic vegetable export. Considerable opportunity for organic farming exists in the north eastern region of India due to the least utilization of chemical inputs in these areas. According to an estimate, there are 18 million hectares of land now available in the north east, which can be put under organic farming. India has the ample scope to become an important organic vegetable producer country because of the existence of different agroclimatic regions for the cultivation of a number of vegetables, the ever-increasing size of the domestic market, and also having a long tradition of living in environment-friendly agriculture.

16.4 BENEFITS OF ORGANIC FARMING

The agricultural practices under organic farming are based on optimum harmonious association with nature.

16.4.1 HIGHER NUTRITIVE VALUE

Organic produce contains more vitamins, minerals, enzymes, trace elements, and even cancer-fighting antioxidants than conventionally grown food (Bhattacharya and Chakraborty, 2005). The quality, taste, and flavor improve in organically produced vegetables mainly through increased dry matter, Vitamin C, protein content and quality, decreased free nitrates in vegetables, decreased storage losses and disease (IIVR-Vision 2050). In a study conducted at IIVR, Varanasi, it was found that the Vitamin C content in spinach, fenugreek, cabbage, and pea grown organically increased by 51.12, 25.76, 41.31, and 18.68%, respectively, and the protein content in cowpea improved by 30% while lycopene content in tomato improved by 39% (Singh et al., 2017). Similarly, the total phenolic compounds and peroxidase activity also improved by 44 and 38%, respectively, in organically produced cabbage (Singh et al., 2017). Organic farming also improved the physical attributes of vegetables. The organically produced cowpea, okra, cabbage, and tomato had better color, luster, and texture (Singh et al., 2017) (Table 16.1).

Smith (1993) stated that organic food is more nutritious. He found that apples, potatoes, pears, wheat, and sweet corn grown organically averaged 63% higher in calcium, 73% higher in iron, 118% higher in magnesium, 178% higher in molybdenum, 91% higher in phosphorus, 125% higher in potassium, and 60% higher in zinc. The organic food averaged 29% lower in mercury than the food produced from the conventional system. Research work also showed that organically grown vegetables have higher Vitamin C, total carotenoids, higher mineral levels, and higher phytonutrients, which can be effective against cancer (Worthington, 2001; Bahadur et al., 2003, 2006a, 2006b, 2009). A detailed scientific analysis of organic fruits and vegetables (Baker et al., 2002) showed that organic foods have significantly less pesticide residues than conventionally grown foods. It is well known that the nitrate content of organically grown crops is significantly lower than in conventionally grown products (Worthington, 2001).

16.4.2 IMPROVEMENT IN SOIL QUALITY

The foundation of organic farming is soil quality. Soil fertility is built and maintained through the organic farming practices. The different methods adopted for this include multiple cropping, crop rotations, use of organic manures and organic pesticides, minimum tillage, and so forth. To

TABLE 16.1 Quality of Organic Produced Vegetables.

Nutrients	Spinach		Fenugreek		Cabbage		Peas	
	Organic[a]	Inorganic[b]	Organic[a]	Inorganic[b]	Organic[a]	Inorganic[b]	Organic[a]	Inorganic[b]
Ascorbic acid (g/100 g)	0.65	0.43	1.66	1.32	234.31	165.81	264.95	223.25
Total phenol (g/100 g)	10.48	6.52	12.97	9.35	48.85	34.33	179.84	136.87
Antioxidant (%)	42.73	30.80	93.02	65.63	14.35	12.54	17.15	11.78

[a]Organic = (Vermicompost at 10 t/ha); [b]inorganic = recommended dose of inorganic fertilizer.

supplement the nutritional requirement of crops green manures, farmyard manures, composts, and plant residues are added. Organic carbon builds up was noticed in vegetable fields fertilized organically. The addition of organic manure on a regular basis improves soil fertility and quality. On an average in the organic field, there was 39% increase in organic carbon and 22.3% increase in soil carbon stock of the soil as compared to the conventional system over a period of only 3 years (Singh et al., 2017). The carbon sequestration was 301.1 kg/ha/year under organic farming while it was only 42.6 kg/ha/year under conventional system in cabbage (Singh et al., 2017). Organic carbon improves the physical and biological properties of soil and also acts as reservoir for nutrients.

The nutrients loss from organic manure is less due to its slow release. The phosphorous use efficiency was higher in organic soils due to the slow rate of release as well as fixation of phosphate ions in organic soils. In the organic field, the bulk density of the soil was lower as compared to the soil of conventional system. The lower bulk density of soil is a sign of better soil texture and soil physical condition.

Besides, the increased organic matter content also supports micro-, meso-, and macrofaunas of the soil and makes it a living body. In a study at IIVR Varanasi, the microbial activity measured, in terms of dehydrogenase activity, alkaline phosphatase, and microbial biomass carbon recorded higher values in organic soils by 32, 26.8, and 22.4%, respectively, as compared with the conventional system in cabbage (Singh et al., 2017). The higher activity of microbes in the soil of organic fields helps in transformation of nutrient and availability of nutrients to the plants.

Improvement in the organic matter content of the soil also increases the water-holding capacity of the soil. The water-holding capacity and hydraulic conductivity of soil were significantly improved in organic fields as compared to conventional system. In a study at IIVR, Varanasi, it was found that the soil moisture content was 25–27% higher and relatively uniform in 0–45 cm of soil profile in organically fertilized field as compared to conventional system in tomato and cabbage crop (Singh et al., 2017). The soil moisture content varied from 11.56 to 12.26 cm in 0–45 cm depth of soil in organic field, while it ranged from 9.1 to 9.78 cm in conventional system (Singh et al., 2017).

Regular addition of organic manure assures a continuous supply of micronutrients. A slowly increase in Zn, Cu, Fe, and Mn was recorded in organic plots as compared to conventional system. It is a well-established fact that there is a significant positive correlation exists between organic matter and micronutrient availability. Similarly, improvement in soil health by the addition of organic manures in organic farming has been reported by Ramesh et al. (2008). Thus, organic farming improves soil fertility, restore soil health, and effective soil microorganisms.

16.4.3 INCREASED CROP PRODUCTIVITY AND INCOME

It is reported that during the conversion period, the crop yield is low compared with the conventional management. However, the yields under the organic farming system start increasing from the second year onward and equating to that of the conventional system by the fifth to sixth year depending on the crop. Due to 20–25% higher premium prices under the organic cultivation, the net income starts increasing progressively from the fourth year onward. The cost of organic vegetable production is usually lower by 50–60% than that of conventional farming. Results obtained from 1050 organic field demonstration trials in different parts of India showed an increase of 4% yield in plantation crops, 7% in fruit crops, 9% in wheat and sugarcane, 10% in millet and vegetable, 11% in fiber, condiments and spice crops, 14% in oilseeds and flowers, and 15% in tobacco (Bisoyi et al., 2003).

16.4.4 INCREASE IN POPULATION OF BENEFICIAL INSECTS

As the pests management tactics under organic farming rely on cultural, physical, mechanical, and biological methods as well as botanical-based chemicals, there is a huge build-up of population of beneficial insects (predators, parasites, antagonists, etc.) in the field, which keep the harmful pests under control. The study on organic cotton cultivation revealed that the mean monthly counts of eggs, larva, and adults of American bollworms were far lesser under organic farming than under the conventional method (Sharma, 2003).

16.4.5 EMPLOYMENT OPPORTUNITIES

A number of studies have reported that organic farming requires more labor input than the conventional farming system. In India, there is the problem of unemployment and underemployment. In such situation, organic farming can solve these problems to some extent. Besides, organic farming encourages diversification of cropping system and growing of more number of crops simultaneously in the same field to promote biodiversity. The diversification of crops with their different planting and harvesting schedules require relatively high labor input, thereby mitigating the problem of periodical unemployment to some extent.

16.5 PROBLEMS IN ADOPTION OF ORGANIC FARMING IN INDIA

- Low yields.
- Lack of awareness.
- Shortage of biomass.
- Problems in marketing of outputs.
- High input costs.
- Lack of financial support.

16.6 COMPONENT FOR ORGANIC VEGETABLE PRODUCTION

16.6.1 SOIL NUTRIENT MANAGEMENT

The basic idea of soil nutrient management in organic farming is to fulfill the requirement of nutrients through the use of permitted inputs. A number of organic sources of nutrients are available for the purpose and among them, green manures, compost/FYM, vermicompost, oil cakes, biofertilizers, and biodynamic preparations are some of the important sources that can be used in vegetable crops.

16.6.1.1 COMPOSTS, VERMICOMPOSTS, AND FARMYARD MANURES

These bulky organic manures, although supply a low quantity of major nutrients, have the capacity to supply all essential nutrients for longer periods. They improve physical, chemical, and biological properties of the soil, such as soil structure, soil water holding capacity, and soil microbial population. Application of well-rotten farmyard manure and compost at 20–25 tonnes/ha/year and/or vermicompost at 5–7 tonnes/ha/year is recommended for most of the vegetable cultivation. These sources are usually incorporated in soil about 15–30 days before sowing/planting. The use of undercomposed or raw manures is usually not allowed in organic certification. FYM alone or in combination with biofertilizers is found to enhance the vitamins content in several vegetable crops (Worthington, 2001; Bahadur et al., 2003, 2004, 2006a, 2006b, 2009) with the yield at par to conventional farming system.

16.6.1.2 GREEN MANURING

The fast-growing and nitrogen-fixing crops like *Dhaincha* (*Sesbania* sp.), sunhemp and mungbean, cowpea, Lucerne, and others are used as a green manure which fix atmospheric nitrogen to the extent of 60–100 kg/ha. The green manure crop of *Dhaincha* (*Sesbania esculenta, S. rostrata*) and sunhemp (*Crotolaria juncia*) are generally ploughed into the soil about 45–50 days after sowing, when sufficient vegetative growth is attained. In addition to supplying nutrients, green manures also improve the physical and microbial properties of the soil. The nutrient content of some of the major green manure crops has been presented in Table 16.2.

TABLE 16.2 Nutrient Content (Dry wt. Basis) in Some Green Manure Crops.

Green manure crops	N (%)	P_2O_5 (%)	K_2O (%)
Dhaincha	2.1	0.5	2.4
Sunhemp	2.16	0.46	2.2
Mungbean	2.0	0.44	2.5
Urdbean	2.0	0.40	2.0

16.6.1.3 BIOFERTILIZERS

Biofertilizers are the carrier-based preparation containing beneficial microorganism in a viable state intended for seed, seedling, or soil inoculation. Some biofertilizer have the capacity to mobilize through biological processes the nutritionally essential elements from nonusable to usable form. Nowadays, biofertilizers have become an important component of the IPNM (integrated plant nutrient supply system). Biofertilizers that hold great promise in vegetable production include N-fixer (*Azotobacter*, *Azospirillum*, and *Rhizobium*), P-solubilizer (*Bacillus*, *Pseudomonas*, *Aspergillus*, etc.), K-solubilizer (*Bacillus* spp.), and arbuscular mycorrhizal fungi (*Glomus mosseae* and *Glomus intraradices*). Proper inoculation of beneficial microorganisms can enhance the atmospheric nitrogen fixation, decompose organic wastes and residues, detoxify pesticides, suppress plant diseases and soil-borne pathogens, enhance nutrient cycling, and produce some biologically active compounds, such as vitamins, hormones, and enzymes that stimulate plant growth. Several workers have reported that application of biofertilizers especially in combination with organic manures improves the productivity and quality in several vegetable crops (Bahadur et al., 2003, 2004, 2006a, 2006b, 2009).

16.6.1.4 CROP RESIDUES

It is essential to use crop residues in the organic cultivation of vegetable crops as it increases the content of organic matter in soil, maintains soil fertility status, and thus, increases the crop yield. Besides nutrients such as N, P, K, S, Zn, Mn, Fe, and Cu from plant wastes can be recycled through scientific composting. There are technologies available for recycling waste into compost but it requires refinement as per the location and situation. Vegetable crops generate a large amount of crop residues after harvesting of the economic part. The unused vegetable waste can be successfully converted in to valuable vermicompost and NADEP compost. In a study conducted at IIVR, Varanasi, on production of NADEP compost and vermicompost by utilizing crop residues, it was found that vegetable residues are a good source for the production of organic manures. The efficiency of the composts prepared from vegetable wastes was evaluated in field experiments on organic vegetable production. The study revealed that vermicompost produced through mixture of nonleguminous and leguminous vegetable waste in 1:1 ratio along with cow dung (40–50%) provided the major nutrients in a more balanced proportion (Singh et al., 2017) (Tables 16.3 and 16.4).

TABLE 16.3 Quality and Recovery of Vermicompost Produced from Vegetable Wastes.

Vegetable residues	C:N ratio	Nutrient content (%)			Dry matter content (%)	Recovery of VC(%)
		N	P	K		
Waste of solanaceae (brinjal, tomato) + *Leguminosae* (garden pea, French bean, Indian bean) in 1:1 ratio	25.17	1.72	0.74	1.32	46.45	46.24
Waste of *cruciferae* (cabbage, cauliflower)+ *Leguminosae* in 1:1 ratio	26.20	1.73	0.75	1.34	45.62	42.54

TABLE 16.3 *(Continued)*

Vegetable residues	C:N ratio	Nutrient content (%)			Dry matter content (%)	Recovery of VC(%)
		N	P	K		
Waste of *cucurbitaceae* (bottle gourd, pumpkin, spong gourd, bitter gourd) + *Leguminosae* in 1:1 ratio	27.32	1.62	0.69	1.31	45.12	45.86
Leguminosae + cow dung only	22.14	1.74	0.81	1.36	38.27	48.56
Cow dung only	26.84	1.54	0.76	1.20	40.23	51.20

TABLE 16.4 Quality and Recovery of NADEP Compost Produced from Vegetable Wastes.

Vegetable residues	C:N ratio	Nutrient content (%)			Dry matter content (%)	Recovery of C (%)
		N	P	K		
Brinjal	36.14	0.69	0.32	0.67	72.51	65.32
Cucurbits	32.52	0.61	0.24	0.54	74.31	64.27
Cowpea	25.42	1.11	0.46	0.72	76.12	66.4
Crucifers	34.21	0.71	0.31	0.74	68.23	61.12

16.6.1.5 BIODYNAMIC FARMING

The late anthroposophist, Rudolf Steiner, founded *Biodynamic farming* which has grown and developed in popularity since 1922. Biodynamic farming is a component of organic agriculture, capable of affording long-term sustainability to agriculture, and particularly to the ecosystem. A very small dose of a few biodynamic preparations has shown significant effects on growth, plant metabolism, crop yield, and product quality. The basic principle in biodynamic system is to harness the synergy between Cosmos, Mother Earth, Cow, and Plant. Most of the practices in biodynamic system can be done at farm level by simple training.

16.6.1.6 CORRECTING DEFICIENCIES ORGANICALLY

Due to unfavorable conditions, such as a prolonged dry spell or excessive moisture, or miscalculation of crop nutrient requirements, a nutrient deficiency in the crop may result. Nutrient deficiency during a critical crop growth period can decline plant growth, predisposing the crop to pest and disease attack, and a heavy yield loss could result. Therefore, it is essential to correct any deficiency quickly. Leaf analysis is the most commonly used method of detecting deficiencies. Organic farmers use foliar sprays (such as fish and seaweed extracts), molasses, compost teas, and trace elements to correct temporary deficiencies. Sometimes foliar application with fish emulsion, seaweed, biostimulants, and compost or weed teas is done to correct such anomalies. Foliar application of apple cider vinegar in ratio of 1:100 may stimulate flowering if delayed by weather or soil conditions (Table 16.5).

TABLE 16.5 Concentrated Manures/Fertilizers Used in Organic Farming.

S. no.	Input	N%	P_2O_5%	K_2O%
1.	Chilean nitrate	16	0	0
2.	Blood meal	12	0	0
3.	Guano	9–12	3–8	1–2
4.	Pelleted chicken manure	2–4	1.5	1.5
5.	Bone meal	2	15	0
6.	Soft rock phosphate	0	15–30	0
7.	Potassium magnesium sulfate	0	0	22

16.6.2 WEED MANAGEMENT

In organic vegetable cultivation, management of weed is the most difficult task, even with the most suitable crop rotations. Weed intensity could be particularly high, when partial composted cow dung manures are used to supply nutrients. Failure to manage weeds successfully leads to reduction in crop yield. The critical period of weed control for the most of the vegetables is about 4–6 weeks after transplanting. The weed competition during this period must be minimized to avoid reduction in yield. The knowledgeable organic farmers depend on different cultivation practices and tools to deal with various weed and crop combinations. The cultivation practices that lower weed menace, such as proper composting of manures, use of plastic mulch, and cover cropping with smother crops should also be employed in addition to mechanical removal of weeds.

The most effective means for managing weeds under organic vegetable production are through suitable crop rotation. Suitable crop rotations prevent the weed species to grow. The dissimilar type of crops and variations in their management practices in a rotation gives less opportunity to a weed species to become dominant over several years.

The use of organic mulches not only suppresses weed but also adds nutrients and improves soil organism's activity by providing feed during decomposition, besides increasing the number of predatory beetles, and spiders in the field. The use of organic mulches containing seeds, rhizomes, and other propagules of weed and grasses should be avoided to check the weed infestation. The zero tillage practice with mulches of cover crop or previous crop residues can suppress both inter- and intra-row weeds. It was observed that tomato crop transplanted on the crop residue mulch benefitted due to weed suppression, soil moisture retention, and the slow-release of nitrogen during the residue mulch decomposition.

16.6.3 INSECT-PESTS MANAGEMENT

In organic farming, the use of chemical plant protection measures is not permitted, hence, appropriate integrated pest management strategies need to be followed for effective control of pests and diseases.

16.6.3.1 CULTURAL PRACTICES

Crop rotation and crop diversification are effective tool to minimize pest problems. To minimize pest problems in *solanaceous vegetables such as* potatoes, brinjal, chilies and peppers, crop rotation with nonsolanaceous crops for 3 years is usually recommended.

Trap crops attract and hold insect pests where they can be managed more efficiently. It also prevents or reduces the movement of the pest into crops. Thus, it diverts pests away from the main crops. African marigold (*Tagete-serecta*) planted in tomato field attracts tomato fruit borer and similarly mustard or Chinese cabbage planted in cabbage field attracts diamondback moth.

Pheromone traps are available to control several pests of vegetables. Pheromone traps are used to attract males and prevent them from mating, thus reducing the next-generation pest populations. Pheromone trap is being commercially utilized to control shoot and fruit borer of brinjal and fruit fly in cucurbits.

The physical methods such as handpicking of pests, use of sticky boards or tapes for flying insects and various trapping techniques are used for control of insect pests. The use of reflective plastic mulches has been effective for the prevention of early aphid infestation in vegetables, such as tomato, brinjal, pumpkin, and squash.

The cultural practices such as the timing of plantings to avoid peak insect pressure, soil solarization, or heating soils by covering with clear thick plastic sheets prior to sowing are nonchemical methods employed for the management of pest and disease problems. The regular monitoring of soil and scouting for insect pests are critical to know when cultural practices alone are not enough to provide sufficient insect control and application of an organic pesticide is necessary treatment. As a practical matter, however, its use is limited to small-scale operations.

16.6.3.2 BIOLOGICAL CONTROL

The management of pest populations by natural enemies (predators, parasitoids, and pathogens) is termed as biological control. The organic farming practices offer many opportunities for enhancing biological control factors. Since in organic cultivation, farmers do not apply chemical pesticides, it is possible to build-up large number of beneficial parasites and predators that may help in control of pests in many crops. Ladybird beetles and lacewing insects are free-living predator species that devour a large number of preys during their lifetime. Immature stage of some parasitoids (certain wasps and flies) develops on or within a single insect host, ultimately killing the host.

In organic agriculture, the natural enemies are the most important and easily available biological control practice. Natural enemies occur in all vegetable production systems. Natural enemies can be fairly selective in their prey or they can utilize many different species of insects as food. Since predators and parasitoids are very mobile in search of their prey and spend a considerable amount of time in moving across the plants, they are more likely to contact the insecticide as a result, natural enemy is more adversely affected by chemical insecticides than the target pest. Ladybird beetles are considered generalists because they feed on many types of preys, whereas *Trichogramma* are microscopic wasps specialized to parasitize eggs of moths. Depending upon the severity of the pest, it may require multiple weekly releases of biocontrol agents (Table 16.6).

TABLE 16.6 Botanical Extracts Used for Control of Pests.

Sl. No.	Botanical extracts	Effective against
	Vitex nugundo (2%) + *Neem* Kernel Cake (1–2%)	Thrips
	Vitex nugundo (2%) + *Calatropips gigantia* (2%)	Aphids
	Neem Kernel Cake (3–4%)	Leaf eating caterpillar
	Nerium thevitifolia (2%) + *Vitex nugundo* (2%)	Fruit borers
	Andrographis panaculata (1%) + *Neem* cake (1%)	Fruit borers and stem borers
	Parthenium sp., (3%) + *Vincarosea* (1%)	Thrips
	Lantana camera (2.5%) + *Nerium thevitifolia* (1%)	Aphids
	Calatropis gigantia (2%) + *Lucas aspera* (1%)	Leaf eating caterpillar
	Nicotiana tobaccum (1%) + *Vitex nugundo* (1%)	Leaf roller
	Calatropis gigantia (1%) + *Nerium thevitifolia* (1%)	Leaf roller
	Gingiber officianale (3%)	Thrips and aphids
	Occimum sp. (Tulasi) leaf extracts (3%)	Caterpillars and spotted leaf beetles
	Soak turmeric 1 kg in 10 L of cow urine for 2 days and then dilute it to 100 L for an acre of crop	Caterpillars and aphids

The following are some of the examples of the pest control through organically approved materials:

- The population of aphids can be checked through the use of oils, soaps, and pyrethrum/rotenone combinations. In case, where aphids are protected from predators and parasites by ants, the most effective control could be obtained by controlling the ants and permitting biological control to take care in the crop.
- The populations of whitefly can be reduced by multiple applications of soaps and oils. However, pupa stage of whitefly is resistant to the above-mentioned material, therefore, to prevent more pupal stages from occurring, two applications of the spray are required.
- The population of leaf miner may be checked by frequent sprays at short intervals of solution containing azadirachtin, pyrethrin, and rotenone; however, this spray may also check the build of native wasp, which is a natural parasite of leaf miner.

- The mites population can be managed with sprays or dusting of sulfur. Drenching of the underside leaves with *neem* seed oil is also effective to check mite population.
- The damage infected by the borers can be reduced through use of *Bacillus thuringiensis* (*Bt*) formulation. The *Bt* formulations should be applied at an early stage to achieve effective control because borers must be controlled before they enter inside the fruits or deep in the foliage.

16.6.4 DISEASES MANAGEMENT

In organic cultivation of vegetables, the management of diseases is achieved through combinations of different preventive techniques. It consists of a combination of various practices such as soil management practices, cultural practices, IPM practices, natural remedies, and limited use of permitted chemicals. The cultural practices such as wide row spacing or drip irrigation are helpful, as it promotes quick drying of foliage, thereby, minimizing the chance of disease infection. In heavy soils, raised beds and subsoiling are preferred as they facilitate drainage of excess moisture. The most effective disease control measure is the use of resistant varieties. There are a number of biological fungicides available containing species of *Trichoderma*, *Bacillus*, *Pseudomonas*, *Gliocladium*, *Streptomyces*, and other beneficial microbes. Many of them are useful and affordable, like the T-22 strain of *Trichoderma*, which are frequently used even on conventional farms. The use of *Trichoderma viride*—10 g/kg seeds or *T. harzianum* at 10 g/L for spray or 10–12 kg/ha for basal dressing is found effective against wilt and rot diseases in vegetables. Similarly, bioagent *Pseduomonas fluorescens* at 5 g/kg seeds, 3 g/L for spray, or 5 kg/ha for basal dressing is also effective for many vegetable diseases.

16.6.4.1 PREVENTION AND SANITATION

In disease management, the preventive and sanitation measures are very effective and important tools. Removing and destroying the stalks, vines, and other plant parts after the crop maturity are quite effective in control of many diseases. Crop residues may be incorporated in soil, destroyed through burning, or properly digesting in composts. Removal of diseased plants parts, sterilizing the plant stakes before reuse, and regular cleaning of tools and implements should be done to prevent spread of disease between fields.

The fundamental golden rule of controlling plant disease is to grow healthy crops through disease-free pure seed and healthy seedlings/transplants. A crop grown through diseased or infected seed and planting material results in entire field getting infested and contaminated with disease and pests. The diseased planting material serves as primary source of inoculums in the field and increases the chance of early-season epidemics, causing reduction in yields and deterioration in quality of produce. Therefore, the use of healthy transplants and seeds is of immense importance for organic vegetable production. Soil solarization of nursery bed using transparent polythene mulch reduces the chance of occurrence of soil-borne diseases.

Rotating different crops in the field, avoid the build-up of many plant pathogens in the soil. The longer duration crop rotations reduce the chance of an early-season disease outbreak. It has been observed that the disease-causing organisms attack quite easily the crops belonging to the same family. Therefore, growing successive crops of the same family should be avoided. It is best to grow unrelated crops in the rotation that may include cereal crops like rice or wheat or sweet corn to beans and legume vegetables, leafy vegetables to cucurbits, brassicas to okra, and others. Rotation of legume vegetables and beans with cereal crops like wheat, rice, or maize or with a fodder crop is effective for control of root rot disease. Growing grain crops for 1 or 2 years good enough to prevent severe root rot disease in the field. Some of the soil-borne pathogens are not easily controlled by crop rotations. Such pathogens produce certain structures that remain in the soil over longer period even without specific hosts, for example, clubroot of crucifers, *Phytophthora* blight, and *Fusarium* wilt of solanaceous vegetable crops. There are some other pathogens that have a wide range of hosts where they can survive for a quite long time because a number of crops and weed species serve as hosts to them, for example, *Sclerotinia*, *Rhizoctonia*, *Verticillium*, and root-knot nematodes.

16.6.4.2 CHEMICAL CONTROL

In spite of adopting very good cultural and healthy management practices, diseases occur quite frequently, which is the most serious challenge in the organic cultivation of vegetables. The severity of disease incidence depends mostly on local/regional environmental conditions. Chemical options are very narrow in organic cultivation, copper and sulfur-based products are the only labeled fungicides allowed in certified organic production. Bordeaux mixture (1% spray) can be used to control some vegetable diseases. Coppers are used for anthracnose, bacterial speck, bacterial spot, early and late blight, gray leaf mold, and septoria leaf spot. Sulfur is used for control of powdery mildew. The use of sulfur reduces the predator/parasite complex in check as it has mild insecticidal and minor fungicidal properties. Care should be taken in the use of sulfur as the plant may show sign of burn due to a rise in air temperatures. In organic vegetable cultivation, application of copper is quite common practice for disease control. Copper can act as both fungicide and bactericide. The formulations that are allowed in organic cultivation include Bordeaux, basic sulfates, hydroxides, oxychlorides, and oxides.

16.6.4.3 USE OF DISEASE-RESISTANT VARIETIES

The use of disease and pest-resistant varieties is a very effective tool to reduce the yield losses due to the incidence of insect pests and diseases. Many varieties of major vegetable crops have been developed and identified for cultivation (Table 16.7).

TABLE 16.7 List of Resistant Varieties of Vegetables Identified Through AICRP-VC.

Crop	Variety	Resistance to
Tomato	BT-10, BWR-5 (*Arka Alok*), BRH-2, LE-415	Bacterial wilt
	H-24, H-86, LE-79-5	Tomato leaf curl virus
Brinjal	BB-7, SM-6-7, BB-44, BB-64, CHES-309, BWR-12, SM-6-6, VNR-218	Bacterial wilt
Okra	NDO-10, HRB-55, HRB-9-2 , HRB-107-4, Sel-10 (A. *Anamika*), Sel-4 (A. *Abhay*), VRO-3, VRO-4, VRO-6, VRO-5, IIVR-11, VRO-22, JNDOL-03-1, JOL-2K-19, P-7, PB-57	Okra yellow vein mosaic virus
Pea	NDVP-4, DVP-250, KTP-8, FC-1, (*Ajeet*), VRPMR-11, JP-4, JP- 83, PRS-4, KS-245, DPP-68, DPP-9411, VP-233, VP-434	Powdery mildew
Muskmelon	DMDR-1	Cucumber green mottle virus (CGMV)
	DMDR-2	Downy mildew + CGMV

16.7 FUTURE PRIORITY FOR IMPROVING LIVELIHOOD OF ORGANIC VEGETABLE GROWERS

In rural areas, farmers are gradually moving toward organic cultivation of vegetable due to higher profit realized in organic farming in comparison to conventional farming. The farmers will be encouraged to do so till they get more profit from it. The vegetables being highly perishable in nature, require, government initiative/intervention to provide/ensure the minimum support price. This will force the middlemen to reduce the holding of stocks and farmers will be benefitted to get remunerative price in open market and it will also reduce the price volatility. Vegetables crops are more vulnerable to adverse weather, leading to higher risk of crop failure (Singh et al., 2017). Therefore, crop insurance scheme to the farmers may be of great help. There are nearly 6000 cold storages in the country; however, they are not very well suited for the storage of fresh vegetables except for potatoes. Farmers have to sell their produce just after harvesting on current market price, thus, are deprived of the profit that can be achieved by storing the produce for some time. The increase in the cool chain facility will help the produce to reach in distant markets of country where demand is high.

16.8 CONCLUSION

Environmental problems arising due to misuse of synthetic fertilizers, pesticides, and machinery can cause land degradation and irreversible ecosystem damage. Organic farming of vegetables has potential for food and nutritional security and repairing and strengthening the soil fertility with organic residues. Certified organic vegetable products provide the opportunity to farmers for high-income options and, therefore, can be instrumental in promoting environmental-friendly farming practice.

KEYWORDS

- **organic farming**
- **nutritive value**
- **soil quality**
- **beneficial insects**
- **biofertilizers**
- **biodynamic farming**
- **weed management**
- **chemical control**

REFERENCES

Bahadur, A.; Singh, J.; Singh, K. P. Response of Cabbage to Organic Manures and Biofertilizers. *Indian J. Hort.* 2004, *61* (3), 178–179.

Bahadur, A.; Singh, J.; Singh, K. P.; Rai, M. Plant Growth, Yield and Quality Attributes in Garden Pea as Influenced by Use of Organic Amendments and Biofertilizers. *Indian J. Hort.* **2006a,** *63* (4), 464–466.

Bahadur, A.; Singh, J.; Singh, K. P.; Upadhyay, A. K.; Rai, M. Effect of Organic Amendments and Biofertilizers on Growth, Yield and Quality Attributes of Chinese Cabbage (*Brassicapekinensis* Olsson). *Indian J. Agric. Sci.* **2006b,** *76* (10), 596–598.

Bahadur, A.; Singh, J.; Singh, K. P.; Upadhyay, A. K.; Rai, M. Morpho-physiological, Yield and Quality Traits in Lettuce (*Lactuca sativa*) as Influenced by Use of Organic Manures and Biofertilizers. *Indian J. Agric. Sci.* **2009,** *79* (4), 282–285.

Bahadur, A.; Singh, J.; Singh, K. P.; Upadhyay, A. K.; Singh, K. P. Effect of Organic Manures and Biofertilizers on Growth, Yield and Quality Attributes of Broccoli (*Brassica oleracea* L var. italic Plenck). *Veg. Sci.* **2003,** *30* (2), 192–194.

Baker, B.; Benbrook, C. M.; Groth III, E.; Lutz Benbrook, K. Pesticide Residues in Conventional, IPM-grown and Organic Foods: Insights from Three U.S. Data Sets. *Food Additives Contaminants* **2002,** *19* (5), 427–446.

Bhattacharya, P.; Chakraborty, G. Current Status of Organic Farming in India and Other Countries. *Indian J. Fertilizer* **2005,** *1* (9), 111–123.

Bisoyi, B. N.; Majumdar, P.; Srivastava, R. S. M.; Creep, S. Prospects of Organic Input Production in India. Paper Presented at National Seminar on Organic Farming with Special Reference to Organic Inputs; Aug 21–22, 2003; Bangalore.

Ramesh, P.; Panwar, N. R.; Singh, A. B.; Ramana, S. Effects of Organic Manure on Productivity Soil Fertility and Economics of Soybean- Durum Wheat Cropping System under Organic Farming in Vertisols. *Indian J. Agric. Sci.* **2008,** *78*, 1033–1037.

Sharma, P. D. Prospects of Organic Farming in India. In *Proceedings of National Seminar on Organic Products and Their Future Prospects*, Sher-e-Kashmir University of Agricultural Sciences and Technology: Srinagar, 2003; pp 21–29.

Singh, S. K.; Yadav, R. B.; Singh, J.; Singh, B. *Organic Farming in Vegetables*. Technical Bulletin No. 77, ICAR-IIVR, Varanasi, 2017; 33 p.

Smith, B. Organic Foods vs. Supermarket Foods: Element Levels. *J. Appl. Nutr.* **1993,** *45*, 35–39.

Vision-2050. Indian Institute of Vegetable Research, Varanasi, UP, Indian Council of Agricultural Research, New Delhi. 2015, 33 p.

Worthington, V. Nutritional Quality of Organic *versus* Conventional Fruits, Vegetables and Grains. *J. Altern. Compl. Med.* **2001,** *7* (2), 161–173.

CHAPTER 17

Organic Pest Management: Emerging Trends and Future Thrusts

S. SITHANANTHAM*

Sun Agro Biotech Research Centre, Chennai 600116, Tamil Nadu, India

**Corresponding author. E-mail: sabrcchennai@yahoo.co.in*

ABSTRACT

The principal methods of organic farming include crop rotation, green manure and compost, mechanical cultivation and biological pest control. Farmers are faced with myriad of production challenges where the most common problems are pests, which include insects, diseases, and weeds. They integrate cultural, biological, mechanical, physical and chemical practices to manage pests. There is need to adopt multiple and varied tactics in the cropping system design to prevent damaging levels of pests, thereby minimizing curative solutions.

17.1 INTRODUCTION

Pest management in organic farming is focused on augmenting natural processes and on tactics that produce long-term effects. There are often far fewer materials available from commercial sources necessitating advance planning and decisions on inputs made carefully, so as to fully comply with regulations for organic certification (Weintraub et al., 2017). There is need to adopt multiple and varied tactics in the cropping system design to prevent damaging levels of pests, thereby minimizing curative solutions. Wyss et al.

Organic Crop Production Management: Focus on India, with Global Implications, D. P. Singh, PhD, H. G. Prakash, PhD, M. Swapna, PhD, & S. Solomon, PhD (Eds.)

(2005) proposed a conceptual model for the development of an insect pest management program for organic crop production. In this model, indirect, preventative measures are of higher priority to be considered early in the adoption process, followed by more direct and curative measures when needed. Arthropod pest management strategies for organic crops, with priority given to preventative strategies, which are considered first, followed by more direct measures (Table 17.1).

TABLE 17.1 Arthropod Pest Management Strategies for Organic Crops.

First phase	Cultural practices compatible with natural processes, such as crop rotation, soil management, nontransgenic host plant resistance, farm/field location
Second phase:	Vegetation management to enhance natural enemy impact and exert direct effects on pest populations
Third phase	Inundate and inoculate releases of biological control agents
Fourth phase	Approved insecticides of biological and mineral origin and mating disruption

Source: Weintraub et al., 2017.

17.2 INSECT BIOCONTROL AGENTS (IBAS)

Many commercial and community-based mass production units have now been established for multiplying several parasitoids and predators, which are required for release at farm level, as a component of organic pest management (OPM). While there is emerging initiatives by private sectors, there have been quite a few inventions for the automation in the mass production of biocontrol agents (BCAs), which have addressed some of the bottlenecks for their economic and efficient production and have facilitated the mass production of BCAs. Now, the BCAs can be mass produced at small scale/ cottage industries just on the lines of sericulture or apiculture. Simple, low-cost and down-to-earth technology can be used for their mass production and can be marketed in the same region where they are produced, besides simpler transportation options developed, which is also likely to increase their acceptability in the rural areas (Kumar et al., 2017).

The two major thrusts of recent R&D that have facilitated the availability and impact potential of IBAs for a wider adoption of OPM are the availability of warm temperature tolerant strains of the BCA and improving the shelf life to promote their mass production. Improving the shelf life is useful for sustaining commercial mass production while there have been successes in inducing diapause for some other *Trichogramma* species elsewhere in recent years. Recently, Pandiyan (2016) has succeeded in extending the shelf life of *T. pretiosum* by exposing the late larval instar to photoperiod of 8:16 of L:D under three acclimation temperature of 10°C for 20 days for satisfactory extent of adult emergence (62%) up to 90 days of cold storage at 3°C. Similarly, the first successful induction of diapause in *T. chilonis* and in *T. pretiosum* has been recently accomplished in a DBT funded project hosted at Sun Agro Biotech Research Centre, Chennai (Sithanantham et al., 2017).

17.3 BOTANICAL BIOPESTICIDES (TA)

Nelson (2006) had reviewed the status and scope of botanical biopesticide in India and shortlisted several plants, which had earlier been identified (Table 17.2).

TABLE 17.2 Examples of Plants with Insect Pest Control Potential.

S. no	Common name	Scientific name
1	African marigold	*Tagetes erecta*
2	American False Hellebore	*Veratrum viride*
3	Angel's Trumpet	*Datura metel*
4	Black pepper	*Piper nigrum*
5	Castor bean	*Ricinus communis*
6	Chinaberry, Persian Lilac	*Melia azedarach*
7	Chrysanthemum	*Chrysanthemum cinerariifolium*
8	Cockroach plant	*Haplophyton cimicidium*
9	Custard apple	*Annona reticulate*
10	Derris	*Derris elliptica*
11	Devil's shoestring	*Tephrosia virginiana*
12	European White Hellebore	*Veratrum album*
13	French marigold	*Tagetes patula*
14	Ginger	*Zingiber officinale*
15	Goatweed	*Ageratum conyzoides*
16	Indian Aconite	*Aconitum ferox*

TABLE 17.2 *(Continued)*

S. no	Common name	Scientific name
17	Jimsonweed	*Datura stramonium*
18	Mammey apple tree	*Mammaea Americana*
19	*Neem* tree	*Azadirachta indica*
20	Peanut	*Arachis hypogaea*
21	Purging cotton	*Croton tiglium*
22	Sabadilla	*Schoenocaulon officinale*
23	Southern Prickly Ash	*Zanthoxylum clava-herculis*
24	Sugar apple	*Annona squamosa*
25	Sweetcane	*Mundulea suberosa*
26	Sweet flag	*Acorus calamus*
27	Tobacco	*Nicotiana tabacum*
28	Tung tree	*Aleurites fordii*
29	Vogel Tephrosia	*Tephrosia vogelii*
30	Wild Tobacco	*Nicotiana rustica*

Among these, *neem* (*Azadirachta indica*), has been used as a potential biopesticide and the commercial formulations of *neem* have gone up to the field level adoption by the farmers. Several other botanical products developed by farmers and private companies are also available for access in local outlets. The new botanical insecticide from *Acorus calamus* is found useful in managing rice ear head bug.

17.4 MICROBIAL BIOPESTICIDES

The commonly used microbial biopesticides include fungi (like *Metarhizium, Beauveria*, and Lecanicillium), viruses (NPV of *Helicoverpa* and *Spodptera*), besides *Bacillus thuringiensis (Bt).*

17.4.1 BACULOVIRUSES

Rabindra (2004) has reviewed the potential for the use of insect viruses in pest control. Baculoviruses containing rod-shaped DNA capsids inside polyhedral (NPV) or granular (GV) occlusion bodies are considered as potential microbial insecticides for the management of many lepidopteran pests like *Helicoverpa armigera*. Through intensive research in the past couple of decades, several of these viruses have been developed as microbial insecticides for the management of important crop pests in India and other countries.

17.4.2 BACILLUS THURINGIENSIS

Vimala Devi et al. (2005) have studied the multiplication of *Bacillus thuringiensis* (*Bt*) using barley *Hordeum vulgare* as the carbon source, which led to the development of a protocol for the cost-effective mass production of *Bt*. Vimala Devi et al. (2014) have developed a suspension concentrate formulation of *Bt* var. *kurstaki* for effective management of *Helicoverpa armigera* on sunflower. The protocol is suitable for promoting localized production of *Bt* at the village/district level.

17.4.3 FUNGAL ENTOMOPATHOGENS

There is a resurgence of interest in the use of fungal biopesticides for biological control of insect pests. Balakrishnan et al. (1994) reported the natural infection by the entomopathogenic fungus, *Beauveria bassiana* on coffee berry borer, *Hypothenemus hampei,* and coffee shoot borer, *Xylosandrus compactus*. Sreedhar and Devaprasad (1995) reported epizootics of *Nomuraea rileyi* on *Spodoptera litura* and *Helicoverpa armigera* in *Kharif* groundnut.

17.5 ATTRACTING INSECTS IN TRAP SYSTEMS

Insect attractants may include chemical substances or attractants such as allomones, kairomones, pheromone, and repellents or mixtures produced by an organism that conveys information to another insect. Pheromones are used for monitoring, mass trapping, or mating disruption (Harari et al., 2016). Traps are just kept in fields to attract specific target pests and so compatible with organic management practices.

Parapheromones are used as insect attractants although they are occurring naturally in the two common attractants being methyl eugenol (ME) and cue lure (CL). Standard white plastic jar traps with optimum size and access vents have been found promising for mango fruit flies in India (Sithanantham, 2011). A mixture of ME and CL attractant traps have been found promising for monitoring and mass trapping of *B. dorsalis* and *B. cucurbit* males in mixed crop systems involving cucurbit vegetables and bitter gourd. Amsa et al. (2015) have found that the catches of *B. cucurbitae* could be enhanced significantly by blending CL and ME in particular ratios. Amsa and Sithanantham (2017) have recently shown that blending of the recommended parapheromone-CL at 3:1 or 1:1 ratios with ME can significantly enhance the *B. cucurbitae* catches in fruit fly traps kept in cucurbit crop. This can confer the twin benefit of enhanced catch plus reduced lure cost, since ME is cheaper than CL.

17.6 INSECT EXCLUSION SYSTEMS

Physical barriers such as nets can be used around vents in solid structure greenhouses and act as exclusion barriers between the plant and the pest. These, in consequence, are an effective tool to limit the entrance and dispersal of aphids on protected vegetable crops. There is a range of the screen hole sizes that can exclude insects physically, especially in greenhouse cultivation, which are available for aphids (0.34–0.341[1]; 0.266 × 0.818[2,3]), whiteflies (0.46–0.462[1]; 0.266 × 0.818[2] ; 0.230 × 0.900[3]), thrips (0.19–0.192[1] ; 0.150 × 0.150[2]), and dipteran leaf miners (0.61–0.64[1] ; 0.266 × 0.8) (Weintraub et al., 2017). Screens with higher ultraviolet-absorbing properties are more effective barriers for a variety of pest insects (aphids, leafhoppers, thrips, and whiteflies) that reduce virus spread in the greenhouse (Weintraub, 2009).

17.7 INDUCED RESISTANCE BY NUTRIENT APPLICATION

While utilization of genetically expressed insect resistance among crop genotypes is reckoned as more desirable (Painter, 1968), there is also scope for induced resistance by applied nutrients to manifest as adversity to the development or survival of pest insects. The scope for inducing resistance to sugarcane stalk borer, *Eldana saccharina*, with applied silica, manifest as physical resistance to stalk penetration by young larvae, causing increased mortality and slower larval growth (Keeping et al. 2009) while, studies by Bhavani et al. (2011) concluded that varietal resistance to early shoot borer (ESB), *Chilo infuscatellus* is positively correlated with leaf silica content in sugarcane varieties. The recent pilot study by Prabakaran et al. (2015) has also shown that the application of potassium silicate could significantly reduce the ESB infestation due to induced resistance. The silica dose-dependent reduction in shoot borer infestation is a new finding and the related SEM-EDAX results are also in conformity with earlier studies on applied silicon as a promoter of pest resistance in sugarcane (Kvedaras et al., 2005), besides studies on potassium silicate as 10–20 ppm sprays to optimize the dose regimes and extending to locally promising sugarcane varieties may culminate in more holistic IPM tactics (Prabakaran et al., 2017).

17.8 ECOLOGICAL ENGINEERING PUSH–PULL SYSTEMS

The strategy for ecological engineering is selective planting of crops with such beneficial plants, which complement in terms of biological pest suppression. The push–pull strategy is an extension of this approach by suitably planting those plants that are repulsive. A "push–pull" or stimulo-deterrent diversionary strategy is used for minimizing damage due to stem borers developed in maize-based farming systems for small-scale farmers, which could plant both a trap crop to attract stem borer colonization away from the cereal plants and another crop as intercrops to repel the pests. The more successful trap crop plants Napier grass, *Pennisetum purpureum* and Sudan grass, *Sorghum vulgare* sudanesis attracted greater oviposition by stem borers than cultivated maize. The intercrops giving maximum repellent effect were molasses grass, *Melinis minutiflora* and two legumes, silverleaf, *Desmodium uncinatum* and Greenleaf *Desmodium introtrum*. "Push–pull" trials using the trap crops and repellent plants significantly reduced stems borer attack and increase in maize yield (Khan et al., 2003).

17.9 MODULES OF OPM

Experiments were conducted by Patel et al. (2011) to evaluate the effects of organic manures on *melongena* L. Many modules are now available for several crops in terms of OPM. For pests of several spice crops have provided illustrations of different components for OPM. As listed by Lingappa et al. (2001) in Table 17.3, OPM module is also available for an intensively protected crop such as cotton.

TABLE 17.3 Successful Examples of Organic Pest Management in OPM of Cotton Bollworms*

OPM–module of organic methods	Results
NPV (0.5 × 10^4 POB/ml) + Microsporidian suspension (1 × 10^5 spores/mL)	Efficacy of microsporidian could be doubled by adding NPV
Trichogramma sp., *C.carnea*, NPV	Effective for the control of *H. armigera*
Combined use of *T.chilonis, C.carnea,* NPV, *Bt*	Reduced *H. armigera* below EIL
Cotton + intercrop, *T. chilonis* @ 1.5 lakh/ha, *C.carnea* @ 50,000/ha	Higher pest suppression and yield
NPV +10% extract of plants (*Tagetes patula, Prosopis juliflora, Calotropis gigantean, Vitex negundo*) + NSKE 1%	Increased the bioefficacy of NPV
T. chilonis @ 1.5 lakh/ha/week + NPV(250 le/ha) at fortnightly interval from 50 to 100 DAS	Proved very effective
Bt @ 1.0 kg/ha–three sprays; *T. chilonis* @ 2.0 lakh/ha-3 releases *Bt* @ 1.0 kg/ha–three sprays; *T.chilonis* @ 2.0 lakh/ha-twice.	378 and 385 kg/ha seed cotton yield, respectively, 370.5 kg/ha seed cotton yield
T. chilonis (1.5 lakh/ha) 40 DAS +NPV (450 le/ha) at 65 DAS + *C.carnea* (1.0 lakh/ha) at 100 DAS + *neem* oil spray at 120 DAS	Proved effective for the control of *H. armigera*
C. carnea @ 50,000/ha, eight releases of *T. chilonis* @ 1.5 lakh/ha, *HaNPV* @ 250 le /ha	15.76 q/ha seed cotton yield
Cotton + groundnut *T. chilonis* @ 50,000/ha twice, *C. carnea* @ 50,000/ha twice, NSKE 1.5%, *HaNPV* @ 450 le /ha, *Btk* (*Bacillus thuringiensis* var. *Kurstaki* de Barjac and Lemille) II @ 2.0 kg/ha	17.60 q/ha seed cotton yield + 5.3 q/ha groundnut and cost-benefit ratio (CBR) of 1:6.27
T. chilonis at 1.0 lakh/ha *C. carnea* @ 50,000/ha SlNPV at 250 le/ha	14.50 q/ha seed cotton yield and 1:4.17 CBR
Eight release of *T. chilonis* at 1.5 lakh/ha (1 & 2 bit/100 m^2, i.e., 100 and 200 strips/ha) and adult release	At GAU, TNAU, and PAU, adult release performed better while at APAU, 100 strips/ha was superior.

*Lingappa et al. (2001).

17.10 CRITICAL THRUSTS FOR FUTURE R&D

Future R&D in OPM may focus on the following areas:

(i) Beneficial crop combinations and practices like mulching, besides supportive on-farm plantings for in situ conservation of IBAs.
(ii) Enhancing the performance of IBAs by developing/ identifying strains adapted/tolerant to climatic and chemical stresses.
(iii) Search for more efficacious predators and parasitoids for pests that currently have no commercial BCAs and identify methods to improve their shelf life.
(iv) Elucidation of new and selective biopesticides that can be registered for organic production, with improved and innovative delivery systems.
(v) Promote crop nutrition-based induced resistance to major pests especially by silica application/availability in crops like rice and sugarcane.

KEYWORDS

- **organic crops**
- **insect biocontrol agents**
- **botanical biopesticides**
- **microbial biopesticides**
- **trap systems**

REFERENCES

Amsa, T.; Nalina Sundari, M. S.; Suganthy, M.; Venkatachalam, A.; Sithanantham, S. Potential for Improvement of Fruit Fly Trapping System in Cucurbit Crops Ecosystem. In *Proceedings of International Conference on Innovative Insect Management Approaches for Sustainable Agro Eco System* (IIMASAE), Tamil Nadu Agricultural University, Madurai, January 27–30, 2015, 211–213.

Amsa, T.; Sithanantham, S. Scope to Enhance the Catches of Major Fruit Flies in Cucurbit Vegetable Ecosystems by Blending of Two Parapheromones. In *Proceedings of National Conference on New Vistas in Vegetable Research towards Nutritional Security under Changing Climate Scenario*, Tamil Nadu Agricultural University, Coimbatore, Tamil Nadu, Dec 6–9, 2017.

Balakrishnan, M. M.; Sreedharan, K.; Bhat, P. K. Occurrence of Entomopathogenic Fungus, *Beauveria bassiana* on Certain Coffee Pests. *Indian J. Coffee Res.* **1994,** *24*, 33–35.

Bhavani, B.; Reddy, K.; Dharma, R. N.; Venugopala, L. M.; Bharatha. Biochemical Basis of Resistance in Sugarcane to Early Shoot Borer, *Chilo infuscatellus. Indian Journal of Plant Protection*, 2011, 39(4), 264–270.

Harari, A. R.; Weintraub, P.; Sharon, R. Disruption of Insect Reproductive Systems as a Tool in Pest Control. In *Advances in Insect Control and Resistance Management*; Horowitz, A. R., Ishaaya, I., Eds.; Springer: The Netherlands, 2016; pp 93–119.

Keeping, M. G.; Reynolds, O. L. Silicon in Agriculture: New Insights, New Significance and Growing Application. *Ann. Appl. Biol.* **2009,** *155*, 153–154.

Khan, Z. R.; Hassanali, A.; Pickett, J. A. Exploting Chemical Ecology in a 'Push-Pull' Strategy for Management of Cereal Stemborers in Africa. In *Proceedings of the National Symposium on Frontier Areas of Entomological Research*, IARI, New Delhi, India, Nov 5–7, 2003.

Kumar, P.; Kaur, J.; Sekhar, J. C.; Lakshmi, S. P.; Suby, S. B. Mass Production of Bio-control Agents of Insect Pests. In *Industrial Entomology*; Omkar, Eds.; Springer: Singapore, 2017; pp 451–465. https://doi.org/10.1007/978-981-10-3304-9_17;Print ISBN978-981-10-3303-2

Kvedaras, O. L.; Keeping, M. G.; Goebel, R.; Byrne, M. Effects of Silicon on the African Stalk Borer, *Eldana saccharina* (Lepidoptera: Pyralidae) in Sugarcane. *Proc S. Afr. Sug. Technol. Assn.* **2005,** *79*, 359–362.

Lingappa, S.; Brar, K. S.; Yadav, D. N. In Augmentative Biological Control within Cotton IPM Indian Scenario. In *Proceedings of the ICAR-CABI Workshop*, Project Directorate of Biological Control (ICAR), Bangalore, June 29–July 1, 2000; Singh, S. P., Murphy, S. T.; Ballal, C. R., Eds.; CABI Bioscience, UK & ICAR, 2001; pp. 57–91.

Nelson, J. Research Status and Scope of Botanical Biopesticides in India. In *Proceeding of National Seminar on Organic Crop Protection Technologies for Promoting Agri-horticulture*; Sithanantham et al., Eds.; Sun Agro Biotech Research Centre and Gill Research Institute: Tamil Nadu, India, 2006; pp 107–129.

Painter, R. H. *Insect Resistance in Crop Plants*; University Press of Kansas: USA, 1968; p 520.

Pandiyan, R. Studies on Extending the Shelf-life of Trichogrammatid Egg Parasitoids for Promoting their Commercial Mass Production in India. PhD Thesis, Tamil Nadu Agricultural University, Coimbatore, Tamil Nadu, India, 2016.

Patel, J. J.; Patel, H. C.; Kathiria, K. B. Effect of Organic and Inorganic Fertilizers on the Incidence of Pest Complex of Brinjal *Solanum melongena* L. In *Insect Pest Management: A Current Scenario*; Ambrose, D. P., Ed., Entomology Research Unit, St. Xavier's College: Palayamkottai, India, 2011; pp 485–487.

Prabakaran, M.; Sithanantham, S.; Kandasamy, R.; Subramanian, K. S.; Gunasekaran, K.; Kannan, M. Foliar Application Effect of Silicon in Sugarcane Early Shoot Borer Infestation. *Hexapoda* **2015,** *22*, 59–62.

Prabakaran, M.; Sithanantham, S.; Subramanian, K. S.; Vijayaprasad, P.; Punna Rao, B. V.; Babu, B. Study of Silica Application Effect on Sugarcane Early Shoot Borer in South India. In *Proceedings of Sucro Symposium*; ICAR-Sugarcane Breeding Institute: Coimbatore, India, 2017; pp 393–396.

Rabindra. Genetic Improvement of Baculoviruses for Microbial Control of Lepidopteran Insect Pests. In *Proceedings of National Symposium on Frontier Area of Entomological Research*; Subrahmanyam, B., Ramamurthy, V. V., Singh, V. S., Eds.; IARI: New Delhi, 2004; pp 1–12.

Sithanantham, S. 2011. Recent Progress in Improving Parapheromone Based Trap-Lure Systems for Fruit Flies (Tephritidae:Diptera) in India: Overview. In *Insect Pest Management A Current Scenario*; Ambrose, D. P., Ed.; Entomology Research Unit, St. Xavier's college: Palayamkottai, India, 2011; pp. 372–388.

Sithanantham, S.; Ramaraju, K.; Kuttalam, S.; Nataraja, N. Ecofriendly Insect Pest Management, Highlights of Phase I (2012–15), Department of Agricultural Entomology, Centre for Plant Protection studies, Tamil Nadu Agricultural University, Coimbatore, 2017.

Sreedhar, V.; Devaprasad, V. Mycosis of *Nomuraea rileyi* in Field Populations of *Spodoptera litura* in Relation to Four Host Plants. *Indian J. Entomol.* **1995,** *58*, 192–195.

Vimala Devi, P. S.; Vineela, V. Suspension Concentrate Formulation of *Bacillus thuringiensis* var. *kurstaki* for effective management of *Helicoverpa armigera* on Sunflower (*Helianthus annuus*). *J. Biocontrol Sci. Technol.* **2014,** *25* (3), 329–336.

Vimala Devi, P. S.; Ravinder, T.; Jaidev, C. Barley-based Medium for the Cost-effective Production of *Bacillus thuringiensis*. *World J. Microbiol. Biotechnol.* **2005,** *21* (2), 173–178.

Weintraub, P. G.; Kleitman, S.; Mori, R.; Gan-Mor, S.; Ganot, L.; Palevsky, E. Novel Application of Pollen to Augment the Predator Amblyseius Swirskii Athias-Henriot on Greenhouse Sweet Pepper. *IOBC/WPRS Bull.* **2009,** *50*, 119–124.

Weintraub, P. G.; Recht, E.; Mondaca, L. L.; Harari, A. R.; Diaz, B. M.; Bennison, J. Arthropod Pest Management in Organic Vegetable Greenhouses. *J. Integr. Pest Manage.*, **2017,** *8* (1). https://doi.org/10.1093/jipm/pmx021

Wyss, E.; Luka, H.; Pfiffner, L.; Schlatter, C.; Uehlinger, G.; Daniel, C. Approaches to Pest Management in Organic Agriculture: A Case Study in European Apple Orchards. *Org. Res.* **2005** (May), 33–36.

CHAPTER 18

Ecoorganic Agriculture Toward Climate Resilience and Livelihood Security: True Agrarian Development Perspective

THOMAS ABRAHAM[1*] and SURYENDRA SINGH[2]

[1]*Department of Agronomy, Sam Higginbottom University of Agriculture, Technology and Sciences, Prayagraj (Allahabad) 211007, Uttar Pradesh, India*

[2]*Guru Angad Dev Veterinary and Animal Science University, Ludhiana 141001, Punjab, India*

[*]*Corresponding author. E-mail: thomas.abraham@shuats.edu.in*

ABSTRACT

Innovations toward food, nutritional, and livelihood security for the rapidly growing population in the perspective of myriads of susceptibilities and adaptation competence in the context of climate change are needed. Soil health management is important for sustainable farming and the concepts of organic farming implicate productivity and economic stability by promoting agrobiodiversity as potential for compensation among components of farming system.

18.1 INTRODUCTION

In October, 2017, the "World Food Day" coincided with a catastrophic fact that the population hunger increased in a decade in the world for the first

Organic Crop Production Management: Focus on India, with Global Implications, D. P. Singh, PhD, H. G. Prakash, PhD, M. Swapna, PhD, & S. Solomon, PhD (Eds.)

time, affecting around 11% of the population (FAO, 2017). The increase is largely due to increased violent conflicts and natural calamities which are mainly accountable for distress migration. Climate change is one of the reasons for migration challenge.

UNAPCAEM/ESCAP (2012) has rung the alarm bell by pronouncing, "our present farming practices based on chemical inputs, damaging our planet's ecosystems. The planet functions as our cornucopia, but this horn of plenty now serves as our refuse basket as well. If we continue with our current agricultural practices, this will be producing contaminated waste rather than food and we should not experiment with our planet Earth.

First time in the history, people's safety on the Earth is under threat due to climate change resulting from our existing farming practices. In the year 2012, the global emission of greenhouse gases was around 52 Gt CO_2e. There is need to drop the level of these emissions to a level of 41 Gt CO_2e if we want to limit this global warming up to 1.5°C, above, which we cannot dare to pass it ". Rodale Institute's White Paper (2014)

The Hindustan Times' editorial of October 16, 2017 had a highly intriguing article entitled, "India must detoxify its poisoned farmlands," which is again another warning for the global citizens, regardless of fairly insurmountable challenge it holds for India. Crane et al. (2010) strongly emphasized that research on agrarian adaptation to climate change and climate variability needs a greater emphasis on farmers' creative adaptive capacities and socio-cultural institutions. Climate change creates new dynamics and uncertainties into agricultural production.

Lal and Abraham (2012) underscored that in the postdiamond anniversary era of independence, agriculture will depend upon the resilience to acquiesce and implement what the first Prime Minister of India, Shri Jawaharlal Nehru, stated in the letter and spirit, that "Everything else can wait but not agriculture." A multipronged approach with its focal point fundamentally on adoption of scientific agriculture, holds a positive promise for the sustainability of agrarian livelihoods of this continent, in general and of India, in particular.

UNAPCAEM/ESCAP (2012) stated that if the first green revolution was an innovation, then we can also now have a truly sustainable "emerald, jade, olive, and lime colored revolution." These inventions in zero budget farming are existing, however, other must be found, communicated, and analyzed. Innovations toward food, nutritional, and livelihood security for the rapidly growing population in the perspective of myriads of susceptibilities and adaptation competence in the context of climate change are needed (Abraham et al., 2018). Soil health management is important for sustainable farming and the concepts of organic farming implicate productivity and economic stability by promoting agrobiodiversity as potential for compensation among components of farming system.

18.2 MITIGATION, RESILIENCE, AND REGENERATIVE POTENTIAL OF ECOORGANIC AGRICULTURE

Organic farming has the potential to mitigate climate change by improving farming systems resilience to changing weather conditions, improving farm agrobiodiversity, improving soil health, declining eutrophication, and water pollution IFOAM-EU and FiBL (2016) (Fliesbach et al., 2007), Brodin (2016). Regenerative approach is indispensable in this current era, as the persistence of ill-health will lead to ecosystem demise. Organic farming management practices can sequester higher amount of carbon to soil and can reverse the climate change (Rodale Institute, 2014).

18.3 ECOLOGICAL SERVICES AND MORE

As per FAO (2010), organic farming provides a range of ecological properties and facilities, which may be summarized as follows:

i. **Soil**
   (a) Stability of natural soils, which decrease the problem of soil degradation.
   (b) Ecoorganic agriculture soils have higher number of beneficial microorganisms and good structure as well as soil fertility.

ii. **Water**
   (a) No risk of water pollution with agrochemicals and pesticides.
   (b) Less nitrate leaching per unit area in ecoorganic agriculture compared with traditional agriculture.

iii. **Air**
   (a) Ecoorganic agriculture helps in improving air quality by reducing greenhouse gas emissions and sequestering carbon to the soils.

iv. **Energy**
   (a) Energy use efficiency is higher under ecoorganic agriculture system because of less use of synthetic inputs, that is, fuel, agrochemicals, and others at farm.

v. **Biodiversity**
   (a) Ecoorganic agriculture systems improve and sustain the biodiversity by improving soil health and fauna, agrobiodiversity, providing shelter to other species at farm, and reducing water pollution.

vi. **Ecological Services**
   (a) Organic farming provides safe foods and shelter for wild species, thereby, less insect-pest problems at farm.

vii. **Landscape**
   (a) Ecoorganic farming makes varied landscapes, which gives natural habitats for species within and around these production systems that helps in conserving nature.

18.4 STRATEGIES WORTH FOCUSING

According to Crane et al. (2008, 2010), the top six management decisions that could potentially be influenced by seasonal climate forecasts mentioned by farmers, in order, of priority were (1) changing crop selection, (2) changing planting dates, (3) adjusting input management, (4) changing land management practices, (5) changing varietal selection, and (6) adjusting marketing practices.

Nevertheless, it is inevitable that focus must be definitely given to some of the vital strategies. Abdalla (2013) in his research review observed that conservation tillage practices along with improvement in soil structure, also diminishes CO_2 emissions as well as increases SOC. Further, conservation tillage practices a boon to lessen GHGs emissions when it is followed in ecoorganic agriculture unlike N_2O productions (Stöckle et al., 2013; Skinner et al., 2014), which are rampantly practiced under conventional systems. In fact, Rodale Institute (2014) quoted the findings of Khan et al. (2007) and Jasper et al. (1979) inorganic nitrogenous fertilizers application enhances carbon dioxide microbial respiration though phosphorus application prevent the growth of fungi root symbioses, which is vital for soil carbon storage in long term.

The fact that organic farmers replace synthetic fertilizers with biomass management results not only in enhanced soil fertility, which also helps in sequestering carbon to soil (FAO, 2011). For addressing climate change consequences, much prominence require to be imparted toward cover crops as around 50% of the soil carbon is secure in aerial parts of the crop (Montagnini and Nair, 2004), Growing cover crops and using crop residues is the need for carbon accumulation in the soil. Cover crops grown between crops rotations, or can be planted at the same time. Cover crops decrease soil erosion, improve soil health, enhance water conservation, smother weeds, help in controlling insect pests and diseases, and increase biodiversity (Hartwig and Ammon, 2002).

However, since vast expanses of crop lands and even uncultivable areas fall under a warm to hot ecosystems, and even with severe moisture unavailability, identifying indigenous, under exploited species, preferably with legume like features, will enable more efficient systems to promote cover cropping and ley framing systems for Indian subcontinent. Leaving soils bare and uncovered, has to be tackled on a war footing. The accrued benefit of revitalizing soils was emphasized by Pandey and Begum (2010), who stated that both use of cover crops and intensive crop sequences ensure continuous covering, which also enhances soil biological activities. Furthermore, the significance of residue retention has to be popularized and efforts to foster this under the extremely popular RWCS (rice–wheat cropping system), which have become unsustainable (Lal and Abraham, 2012).

18.5 INDIA'S BIG CONFRONT AND COMPASS

Though as per the available statistics of 2016, India ranked ninth in terms of World's Organic Agricultural land (Willer et al., 2018), regrettably, in India the shift toward organic farming even in the most recent past, that is, 2015–2016 is very dismal. According to the APEDA (2018), the 5.71 million hectares are organically certified comprised of arable and nonarable land. The total organic production was 1.35 million MT. During 2015–2016 around 2.63 lakh MT produce was exported. Globally requirement for these foods is growing at 20–25% per annum (Pandit, 2017). India's market is growing itself at 40–50%. A huge scope for expansion lies in the millennia, and the current era, which is otherwise known as "*anthropoceneera*," is highly relevant to the situation of adhering to policies and practices in the background of responsibility toward fulfillment of the SDGs (sustainable development goals) by a dependable nation.

18.6 EVIDENCE AND FUTURE PERSPECTIVES IN THE CONTEXT OF CLIMATE CHANGE

The proposition that the soils under the organic system are easily adapted to weather extremes is unquestionably not a myth. Smith et al. (2005) declared that the only system of farming, which enhances carbon stocks on croplands is ecoorganic agriculture. Evidence from farming system trials worldwide revealed that ecoorganic agriculture system enhances carbon stocks, though productivity is lower in comparison to agrochemical-based agricultural production systems.

As a rule, organic agriculture uses less energy, in terms of both per unit area and production (Reganold and Wachter, 2016; Meier et al., 2015). Nemecek et al. (2005) found energy use in Swiss organic systems on unit area basis to be 46–49% less compared to mineral fertilizer-based system and 31–35% less in manure-based system. On a crop unit basis, variances were 36–43% and 10–20%, respectively. Greenhouse warming potential in ecoorganic agriculture system per unit area basis was 29–32% less than mineral fertilizer system and 35–37% less in manure-based system. The globally applied organic agriculture is, therefore, highly significant especially in coming 20–40 years which is critical in policy terms for delivering major greenhouse gas reductions, transitioning to a low-carbon economy as well as overall avoiding catastrophic climate change (Anon., 2009).

Based on results and observations of long-term comparative field trials, Niggli et al. (2009) estimated that the additional annual soil carbon sequestration for various agricultural production systems (arable, pasture, and permanent crops) if these manage organically. A soil sequestration rate of 200 kg C/ha/yearfor cultivable and enduringly covered soils and 100 kg C/ha/year for pastures use organic inputs such as compost and manure was suggested. They further suggested that in cultivable soil figures could be increased to 500 kg C/ha/year, if appropriate tillage with organic inputs integrated cropping systems. In fact, earlier, soil C addition of 1000 kg and 2000 kgC/ha/year (Hepperly et al., 2006) above conventional farming consistently demonstrated in long-term comparative field trials in temperate areas of the united states of America. These rates are achieved when cover crops and compost are utilized in combination with suitable tillage operation. Production systems based on organic inputs preserve soil health and maintain or even increase organic matter in soils under drought, irregular rainfall events with floods, and rising temperatures like situations. Soils under organic management retain significantly more rainwater due to the "sponge properties" of organic matter.

18.7 FEASIBLE INTERVENTIONS FOR ORGANICS

Certain important indicators of resilience are necessary, to have a tangible proof in order for agroecology concepts to be evident.

An efficient approach of *microsite improvement* is the necessity in the current era of paucity and degradation (Abraham, 2015), particularly for addressing expanded belts with ameliorative measures, which are rather very high burden of cost, labor, and time. Further, ecointensification in different dimensions, namely, rejuvenation of soils through rhizosphere engineering and SRI for rice, System of Wheat Intensification (SWI) for wheat, scaling up to an array of crops, and building it up to system of crop intensification (SCI) holds an important key toward livelihood security for food-based production systems.

The viewpoint of becoming entrepreneurs—in short span of a season, becomes a reality, once the farming systems are able to cross threshold of optimal production with ample diversity (Abraham et al., 2016). Thus, quality addition to livelihoods through the marketing of surplus produce from a homestead or farm-stead system lies in the diversity (Abraham et al., 2013). Some of the enterprises at the SMOF, which have been found successful, have also been emulated to a limited extent by the farming community, particularly the cooperative members of the Allahabad Organic Agriculture Cooperative.

Choice of enterprises for any farming community is highly desirable, which enables flexibility and adaptable mechanism to work. At the SMOF, persistent endeavors to explore feasible enterprises are attempted (Abraham et al., 2016). Within the limited area, the diverse land-use system by raising the woody perennials with shrubs and herbs through alley cropping, boundary plantings, and live hedges as a buffer zone, enables harvest of fuelwood and green leaf manure, besides aiding as windbreaks and enriching soil. The woody portion of prunings from leucaena, guava, gooseberry, mango, and others is chipped at the wood chipper-cum-fermenter unit for utilizing as oyster mushroom substrate, besides the usual rice and wheat straw.

Rice–wheat cropping system (RWCS) requires vital technological interventions, innovations, and support services, which are indispensable to make the RWCS sustainable and profitable, and the work which has focused on raising productivity, reducing labor costs, and developing ecofriendly pest and diseases management practices, now needs to be replicated at a faster rate, with appropriate adaptations in respective ecosystems.

The SRI has proved beyond doubt that the increased tillering, greater root growth, higher grain quality and greater grain weight, less lodging, less pest and disease infestation, seed saving, and lower production cost are innately motivating. Most of all, the overall reduction of the risk factors to which majority of the farmers are apprehensive in the current climate and economic disparity has been challenged with a victorious triumph (Abraham et al., 2018). Organic production system (OPS) may also produce equal yield attributes as conventional system, which was evident at certified organic plots in the faculty of agriculture at SHUATS. Once the soil fertility is built-up sufficiently, organic system may also produce equal yields as conventional system.

Reddy (2017) reported from field trials that the second-highest value of 3.99 t/ha of rice was registered in treatment with Rice Duck System with OPS under manual transplanted conditions, which may be accepted as fairly clear evidence that OPS has an efficiency potential at par with conventional production system.

Under certified OPS, Mithilesh and Abraham (2017) observed that the SWI (system of wheat intensification) plots showed up to 50% tolerance to lodging and associated damages (shrinking of grain). Crops with larger, more effective root systems in association with more abundant and diverse life in the soil are more resilient when subjected to drought, storm damage, and other climatic hazards. Buffering of such effects has been seen frequently with SRI management for rice (Uphoff, 2012). These findings also corroborate with Dhar et al. (2015) who reported about wheat crop in the NWPZ of India.

In another trial at Allahabad, Wahab (2016) observed N uptake in baby corn increased by 19 and 10.95% in association with cluster bean as well as greengram (Fig. 18.1). Further the K uptake in baby corn increased by 15.40 and 9.46% in association with cluster bean and green gram, respectively. Substantial improvement in dry matter accumulation accompanied with high cob yield and fodder production could be possible reason for greater nutrient uptake. It might also be due to high nitrogen fixation by intercropped legumes leading to an increase in N uptake by the associated baby corn (Haymes and Lee, 1999).

18.8 SOIL ORGANIC MATTER CONTENT UNDER ORGANIC FARMING

The soil OM status is a core issue for the vitality of soil, which is foundational and has consequences rolling out in multiple directions. Assessment at farms and in long-term trials showed that under OPS s, soils OM improves considerably. It is estimated that under Northern European conditions, if ecoorganic agriculture adopted soil OM at the rate of 100–400 kg/ha annually during the first 50 years would be increased. After 100 years, a stable level of soil OM may be reached (Foereid and Høgh-Jensen, 2004).

In the organically manured plots of SMOF, significant and highest organic carbon (0.55%) recorded in treatment with duck component. Even in the control plot (without duck component under OPS) also, there was a noticeable change in organic carbon.

Further, in the farmers' field, which were converted recently toward OPS, treatments with duck showed the highest soil organic carbon (0.51%), which was 6.25% higher when compared to initial value of organic carbon (0.48%). Surekha et al. (2013) reported that "once the soil fertility is built up sufficiently, organic system also may produce equal yields as conventional system." Mithilesh (2017) reported the highest organic carbon (0.62%) in treatment with SWI planting + FYM (16.00 t/ha).

The application of biological and organic manures not only supply balanced amount of micronutrients but also improves the soil health, thus, creating a conducive environment for crop production (Abraham and Lal, 2002).

C (Cropping system): C_1: Sole babycorn; C_2: Babycorn + Greengram; C_3: Babycorn + Clusterbean; M (Manures): M_1: Goat manure; M_2: Poultry manure; M_3: FYM; M_4: Goat manure + Poultry manure; M_5: Goat manure + FYM; M_6: Poultry manure + FYM.

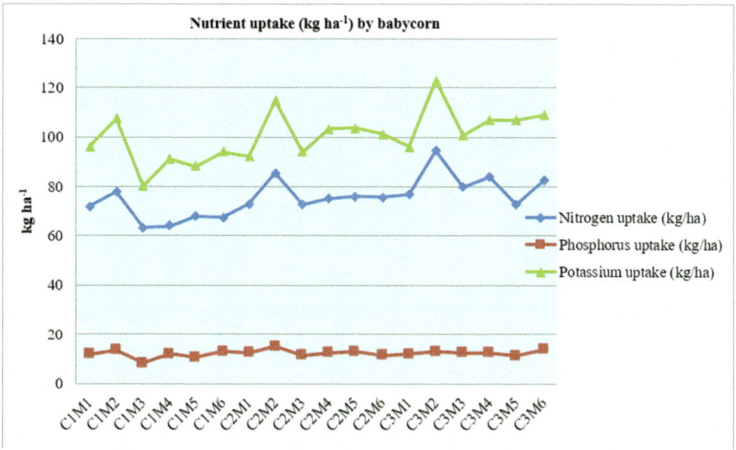

C (Cropping system): C_1: Sole babycorn; C_2: Babycorn + Greengram; C_3: Babycorn + Clusterbean; M (Manures): M_1: Goat manure; M_2: Poultry manure; M_3: FYM; M_4: Goat manure + Poultry manure; M_5: Goat manure + FYM; M_6: Poultry manure + FYM

FIGURE 18.1 Effect of intercropping and nutrients management on nutrient uptake by organic baby corn.

The positive results of the build-up of soil OM, besides the general firmness of the soil in multifarious dimensions, can particularly become evident by the intensive microbial activity. From a study, Sagar et al. (2017) demonstrated that OPS enriches soil bacterial variety and soil health, rendering the agroecosystem less reliant on outdoor inputs with huge potentials for stability of these systems. They observed good bacterial range at the SMOF soil.

The Indian subcontinent too must contemplate on implementing similar suggestions that were specified for the EU, which is becoming relevant for its true development. The International Federation of Organic Agriculture Movements-EU as well as FiBL (Research Institute of Organic Agriculture, 2007) seems to approve the suggestion that the EU should engage in a farming system transition, equal to the energy transition, and move agriculture toward agroecological approaches, for example, ecoorganic agriculture and agroforestry (Hilbeck and Oehen, 2015) as food transition toward agroecology can go a long way to the European Union meeting its commitments to implement the 2030 Agenda for Sustainable Development and the SDGs. It can be euphemistically or realistically concluded that India can become the change maker by fulfilling the SDGs through a transformational revolution, which arrests "anthropocene" (Abraham, 2017), to attain livelihood security by vehemently adopting organic approaches, especially in its vastness of agrarian sector. This, ultimately, will be "true development."

KEYWORDS

- **regenerative potential**
- **ecological services**
- **climate change**
- **soil organic matter**

REFERENCES

Abdalla, M. Conservation Tillage Systems: A Review of its Consequences for Greenhouse Gas Emissions. *Soil Use Manag.* **2013,** *29*, 199–209.

Abraham, T.; Lal, R. B. Sustainable Enhancement of Yield Potential of Mustard through Integrated Nutrient Management in a Legume Based Cropping System for the Inceptisols. *Cruciferae. Nr.* **2002,** *24*, 99–100.

Abraham, T.; Miura, T.; Kumar, S. In Potential for Small Holders' Entrepreneurship for Sustainability through Organic Farming in the Indo-Gangetic Plains of Asia-Pacific—Models Worth Emulation. In *Abstracts of Asia-Pacific Regional Symposium on 'Entrepreneurship and Innovation in Organic Farming*; Bangkok, Thailand, December 2–4, 2013; p 63.

Abraham, T.; Miura, T.; Kumar, S.; Debbarma, V. Prospects for Small-holders' Entrepreneurship through Eco-Organic Agriculture. In *Extended Summaries, 4th International Agronomy Congress on Agronomy for Sustainable Management of Natural Resources, Environment, Energy and Livelihood Security to Achieve Zero Hunger Challenge*; ISA & ICAR: New Delhi, Nov 22–26, 2016, *1*, 158–159.

Abraham, T.; Sagar, A.; Singh, S.; Ramteke, P. W. Potentials for Scaling-up Efforts through Farmers' Adaptation of System of Rice Intensification (SRI)—Approach to Tackle Unsustainable RWCS (Rice–Wheat Cropping System) of North Eastern Plain Zone. In *System of Rice Intensification*; Bhatt, K. N., Bhargava, P., Eds.; Studium Press: New Delhi, 2018; pp 180–203.

Abraham, T. 2014. Indispensability of Pre-eminence for Smallholder Farmers' Issues for Re-moulding India's Future. In *Brainstorming Workshop on Family Farming: Mainstreaming Smallholder Farmers' Agenda in Agricultural Education, Research and Extension (Souvenir)*; Abraham, T., Dawson, J., Wesley, J., Eds.; SHIATS-DU: Allahabad, 2014; pp 41–46.

Abraham, T. Organic Agriculture towards Climate Resilience and Livelihood Security, *In: Souvenir of the National Conference on Organic Farming for Sustainable Agriculture and Livelihood Security under Changing Climatic Conditions*, CSAUA&T, Kanpur, Dec 12–13, 2017; pp 58–63.

Abraham, T. Eco-Organic Agriculture and SRI. *LEISA India* **2013,** *15* (1), 37.

Abraham, T. Eco-Organic Agriculture: Concepts and Practices. In: *Souvenir on Training on Organic Farming for Agriculture Sustainability*; GI, IPRs and Organic Facilitation Cell, UPCAR: Lucknow, Aug 27–28, 2015; pp 34–43.

Abraham, T. *Final Report of National Project on Organic Farming* (NPOF). National Centre of Organic Farming, Ghaziabad and Ministry of Agriculture, Government of India: New Delhi, 2011; pp 451+.

Abraham, T. On-Farm Adaptive Research for Food and Nutritional Security in the context of Climate Change—A non-conventional approach. Paper presented at the 3rd Uttar Pradesh Agricultural Science Congress on Strategic Governance & Technological Advancement for Sustainable Agriculture, June 14–16, 2015; SHIATS, UPCAR & UPAAS: Allahabad, 2015a; pp 580–585.

Abraham, T. Smallholder Farmers' issues: Potential Interventions, Paper presented in the South Asia Conference on Smallholder Farmers [Resilient Future for Small Farmers] Organized by the South Asia Coordination Unit of the SAF-BIN Project under *Caritas India* at India International Centre (IIC), New Delhi, March 10–12, 2015b.

Anonymous. *Organic Agriculture: A Guide to Climate Change & Food Security*; IFOAM, FiBL: Bonn, Germany, 2009; p 23.

APEDA. http://apeda.gov.in/apedawebsite/organic/Organic_Products.htm. (accessed Feb 23 2018.

Crane, T.; Roncoli, C.; Paz, J.; Breuer, N. E.; Broad, K.; Ingram, K. T.; Hoogenboom, G. Seasonal Climate Forecasts and Risk Management Among Georgia Farmers. In *Southeast Climate Consortium Technical Report Series*; Ingram, K. T., Ed., South-East Climate Consortium: Tallahassee, FL, 2008; p 64.

Crane, T. A., Roncoli, C.; Paz, J.; Breuer, N. E.; Broad, K.; Ingram, K. T.; Hoogenboom, G. Forecast Skill and Farmers' Skills: Seasonal Climate Forecasts and Risk Management among Georgia (U.S.) Farmers. *Weather Clim. Soc.* **2010,** *2*, 44–59.

Dhar, Shiv; Barah, B. C.; Abhay, K.; Vyas; Uphoff, N. T. Comparing System of Wheat Intensification with Standard Recommended Practices in the North Western Plain Zone of India. *Arch. Agron. Soil Sci.* **2015**.

FAO. *"Climate-Smart" Agriculture*; FAO: Rome, 2010.

FAO. *Organic Agriculture and Climate Change Mitigation* (A Report of the Round Table on Organic Agriculture and Climate Change); Natural Resources Management and Environment Department: Rome, Italy, 2011; pp 68 + iv.

FAO. Organic Agriculture and Climate Change. http://www.fao.org/DOCREP/005/Y4137E/y4137e02b.htm#89 (accessed Oct 12, 2017).

FAO. *Organic Agriculture: African Experiences in Resilience and Sustainability*; FAO: Rome, 2013.

FAO. World Food Day Ceremony. http://www.fao.org/world-food-day/2017/wfd-ceremony/en/ (accessed Oct 12, 2017).

FiBL (Research Institute of Organic Agriculture). *Organic Farming and Climate Change*: Technical Paper; Monograph. Doc. No. MDS-08-152.E; FIBL, Germany, 2007; pp 30+vi.

Fliesbach, A.; Oberholzer, H. R.; Gunst, L; and Mader, P. Soil Organic Matter and Biological Soil Quality Indicators after 21 Years of Organic and Conventional Farming. *Agric. Ecosyst. Environ.* **2007,** *118*, 273–284.

Foereid, B.; Høgh-Jensen, H. Carbon Sequestration Potential of Organic Agriculture in Northern Europe—A Modelling Approach. *Nutr. Cycl. Agroecosyst.* **2004,** *68* (1), 13–24.

Hartwig, N. L.; Ammon, H. U. Cover Crops and Living Mulches. *Weed Sci.* **2002,** *50*, 688–699.

Haymes, R.; Lee, H. C. Competition between Autumn and Spring Planted Grain Intercrops of Wheat (*Triticium aestivum*) and Fieldbean (*Viciafaba*). *Field Crop Res.* **1999,** *62*, 167–176.

Hepperly, P.; Douds, D. Jr.; Seidel, R. The Rodale Farming Systems Trial, 1981 to 2005: Long-Term Analysis of Organic and Conventional Maize and Soybean Cropping Systems. In *Long-Term Field Experiments in Organic Farming*, Raupp, J., Pekrun, C., Oltmanns, M., Köpke, U., Eds.; International Society of Organic Agriculture Research (ISOFAR): Bonn, Germany, 2006; pp 15–32.

Hilbeck, A.; Oehen, B., Eds. *Feeding the People: Agroecology for Nourishing the World and Transforming the Agri-Food System*; IFOAM EU: Brussels, 2015.

IFOAM EU and FiBL. *Organic Farming, Climate Change Mitigation and beyond: Reducing the Environmental Impacts of EU Agriculture*. http://www.ifoam-eu.org/sites/default/files/ifoameu_advocacy_climate_change_report_2016.pdf (accessed Dec 5, 2016).

Jasper, D. A.; Robson, A. D.; Abbott, L. K. Phosphorus and the Formation of Vesicular Arbuscular Mycorrhizas. *Soil Biol. Biochem.* **1979,** *11* (5), 501–505.

Khan, S. A.; Mulvaney, R. L.; Ellsworth, T. R.; Boast, C. W. The Myth of Nitrogen Fertilization for Soil Carbon Sequestration. *J. Environ. Qual.* **2007,** *36*, 18–21.

Lal, R. B.; Abraham, T. Scientific Agriculture for Sustainability of Agrarian Livelihoods: An Approach Paper. *Allahabad Farmer* **2012,** *58* (1), 1–6.

Meier, M. S.; Stoessel, F.; Jungbluth, N.; Juraske, R.; Schader, C.; Stolze, M. Environmental Impacts of Organic and Conventional Agricultural Products—Are the Differences Captured by Life Cycle Assessment? *J. Environ. Manage.* **2015,** *149*, 193–208.

Mithilesh. *Effect of Planting System and Organic Manures on Growth and Yield of Wheat* (*Triticum aestivum* L.). M.Sc. (Ag) Agronomy Thesis. Sam Higginbottom University of Agriculture, Technology and Sciences: Allahabad, 2017; pp 65 + xxvi.

Mithilesh; Abraham, T. Agronomic Evaluation of Certified Organic Wheat (*Triticum aestivum* L.). *Int. J. Curr. Microbiol. Appl. Sci.* **2017,** *6* (7), 1248–1253. https://doi.org/10.20546/ijcmas.2017.607.151.

Montagnini, F.; Nair, P. K. R. Carbon Sequestration: An Underexploited Environmental Benefit of Agroforestry System. *Agroforestry Syst.* **2004,** *61–62* (1), 281–295.

Nemecek, T.; Huguenin-Elie, O.; Dubois, D.; Gaillard, G. *Ökobilanzierung von Anbausystemenim Schweizerischen Acker- und Futterbau*; Zürich, **2005,** p 156.

Niggli, U.; Fließbach, A.; Hepperly, P.; Scialabba, N. *Low Greenhouse Gas Agriculture: Mitigation and Adaptation Potential of Sustainable Farming Systems*; FAO, April 2009. ftp://ftp.fao.org/docrep/fao/010/ai781e/ai781e00.pdf (accessed Dec 2, 2017).

Pandey, C. B.; Begum, M. The Effect of a Perennial Cover Crop on Net Soil N Mineralization and Microbial Biomass Carbon in Coconut Plantations in the Humid Tropics. *Soil Use Manage* **2010,** *26*, 158–166.

Pandit, V. By 2020, Area under Organic Farming May Treble to 20 Lakh Hectares https://www.thehindubusinessline.com/economy/agri-business/by-2020-area-under-organic-farming-may-treble-to-20-lakh-hectares/article9616068.ece (accessed Dec 2, 2017).

Reddy, Y. C. *Agronomic Evaluation of Rice [Oryza sativa (l.) Sub species japonica] as Influenced by Rice Duck Farming*. M.Sc. (Ag) Agronomy Thesis. Sam Higginbottom University of Agriculture, Technology and Sciences, Allahabad, 2017; pp 75 + xxv.

Reganold, J. P.; Wachter, J. M. Organic Agriculture in the Twenty-first Century. *Nat. Plants* **2016,** *2*, 1–8.

Rodale Institute. *Regenerative Organic Agriculture and Climate Change: A Down-to-Earth Solution to Global Warming.* White Paper, **2014,** p 25. https://rodaleinstitute.org/assets/WhitePaper.pdf (accessed Dec 2, 2017).

Sagar, A.; Debbarma, V.; Abraham, T.; Shukla, P. K.; Ramteke, W. P. Functional Diversity of Soil Bacteria from Organic Agro Ecosystem. *Int. J. Curr. Microbiol. Appl. Sci.* **2017,** *6* (12), 3500–3518.

Skinner, C. et al. Greenhouse Gas Fluxes from Agricultural Soils under Organic and Non-Organic Management—A Global Meta-Analysis. *Sci. Total Environ.* **2014,** *468–469*, 553–563.

Smith, P.; Andren, O.; Karlsson, T.; Perala, P.; Regina, K.; Rounsevell, M.; van Wesemael, B. Carbon Sequestration Potential in European Croplands has been Overestimated. *Global Change Biol.* **2005,** *11*, 2153–2163.

Surekha, K.; Rao, K. V.; Rani, N. S.; Latha, P. C.; Kumar, R. M. Evaluation of Organic and Conventional Rice Production Systems for their Productivity, Profitability, Grain Quality and Soil Health. *Agrotechnology* **2013,** *11*, 1–6.

UNAPCAEM/ESCAP. *Organic Agriculture Gains Ground on Mitigating Climate Change and Improving Food Security: Healthy Food from Healthy Soil.* UNAPCAEM (United Nations Asian and Pacific Centre for Agricultural Engineering and Machinery). ESCAP (United Nations Economic and Social Commission for Asia and the Pacific) Policy Brief Issue No. *2*, 2012(May–August), p 14.

Uphoff, N. Supporting Food Security in the 21st Century through Resource-Conserving Increases in Agricultural Production. *Agric. Food Sec.* **2012,** *1*, 18.

Wahab, A. H. *Study on Babycorn (Zea mays L.) Based Legume Intercropping under Certified Organic Production System*. Ph.D. Agronomy Thesis. Sam Higginbottom University of Agriculture, Technology and Sciences, Allahabad, 2016; pp 304 + xiv.

Willer, H.; Lernoud, J.; Kemper, L. The World of Organic Agriculture: Statistics Summary. In *The World of Organic Agriculture: Statistics and Emerging Trends*; Willer, H.; Lernoud, J., Eds.; Research Institute of Organic Agriculture (FIBL); International Federation of Organic Agriculture Movements- Organics International (IFOAM-OI): Bonn, Germany, 2018. https://shop.fibl.org/CHen/mwdownloads/download/link/id/1093/?ref=1

CHAPTER 19

Organic Jaggery Production

PRIYANKA SINGH[1], S. I. ANWAR[2], M. M. SINGH[1], and B. L. SHARMA[1]

[1]*Uttar Pradesh Council of Sugarcane Research, Shahjahanpur 242001, Uttar Pradesh, India*

[2]*ICAR-Indian Institute of Sugarcane Research, Lucknow 226002, Uttar Pradesh, India*

**Corresponding author. E-mail: priyanka.vishen75@gmail.com*

ABSTRACT

Jaggery or gur is a solid/semi-solid form of sugar obtained by concentration of sugarcane juice in an open pan. Jaggery is important in Indian diet, which is consumed either directly or used in preparation of various sweet based foods and is generally called as "medicinal sugar" because of its use in Ayurveda as well as its comparison with honey . Jaggery when produced from sugarcane grown in soils rich in organic content, by adopting organic farming and by the use of vegetative clarificants, is found superior in quality and have better settling property and storability. Jaggery, in contrast with white sugar, contains a robust quantity of iron and copper percentage. The liberal vitamin and mineral content makes jaggery a superior class of natural sweeteners. The color of jaggery may vary from light golden brown to dark brown but it will have great market potential due to the liking of health-conscious people for organic and chemical-free jaggery.

19.1 INTRODUCTION: ORGANIC JAGGERY

The value of jaggery is considered to be the best when prepared organically. It fetches more price like other organically made products. Organic jaggery not

Organic Crop Production Management: Focus on India, with Global Implications, D. P. Singh, PhD, H. G. Prakash, PhD, M. Swapna, PhD, & S. Solomon, PhD (Eds.)

only retains all the carbohydrates but also all the other natural nutrients, such as iron, calcium, phosphorus, magnesium, and others and trace amount of vitamins. Due to tremendous nutritional and medicinal properties of jaggery, globally its market is growing (Fig. 19.1). The commercially available jaggery is made by adding certain chemicals to regulate the hardness and color of jaggery, but it is considered hazardous from the point of view of health. The organic jaggery is supposed to be made from sugarcane that is grown without the use of inorganic fertilizers or pesticides. The juice clarification during jaggery making also should not use any chemical of harmful nature. Organic jaggery is therefore free from harmful chemical contamination. Jaggery is pure, unrefined sugar, and increasing awareness for its health and medicinal advantages makes it a precious traditional product. It contains all the natural minerals and vitamins present in cane juice, such as calcium, iron, potassium, copper, magnesium, zinc, and phosphorous. Traditional *ayurvedic* medicine of India suggests that consumption of jaggery in daily diet purifies the blood, improves digestion, and strengthens the lungs bones and nervous system. Jaggery is one of the good sources of energy and nutrients for healthy person, however, it is more beneficial for health when produced organically. Organic jaggery is produced without the use of synthetic chemicals and is found to contain mineral content 50-folds higher than the white refined sugar. A spoonful of organically produced jaggery contains calcium (3–5 mg), phosphorous (3–5 mg), magnesium (6 mg), and potassium (45 mg) (Singh et al., 2014). Like conventional jaggery, organic jaggery could be directly added into cereals or coffee. It can be substituted for granulated white sugar in many recipes. Health benefits of organic jaggery is tremendous; thus, it helps in the purification of blood, maintains good health, and also helps as a curing agent in rheumatic afflictions and bile disorders.

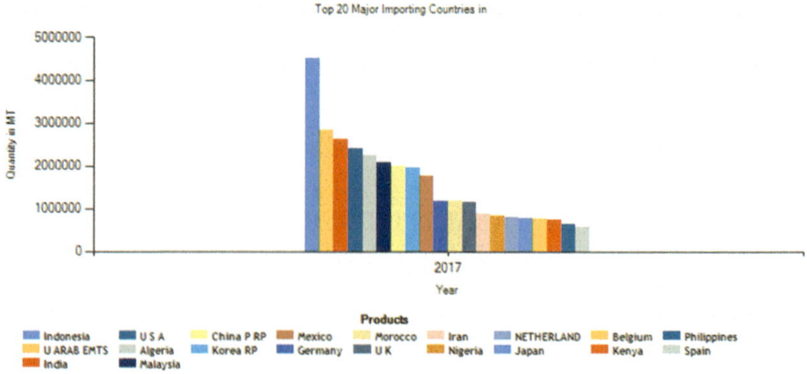

FIGURE 19.1 Top 20 major jaggery importing countries.
Source: COMTRADE, United Nations, (2017).

19.2 ORGANIC SUGARCANE CULTIVATION FOR JAGGERY PRODUCTION

The chemical contamination in jaggery comes not only in its preparation stage, but also right from the sugarcane cultivation. Being a profitable cash crop, sugarcane farming is often carried out with lots of inorganic fertilizers, weedicides, pesticides, and other plant protection chemicals throughout its life cycle to ensure higher productivity and profitability. Good quality organic jaggery essentially requires not only chemical-free manufacturing but also organic sugarcane cultivation (Fig. 19.2).

For organic sugarcane cultivation, supplementation of nutrients through farmyard manure (FYM), vermicompost (VC), and enriched press mud (EPM) at 33% of each equivalent to 100% recommended dose of nitrogen along with soil application of biofertilizers (*Azospirillum* and PSB @ 10 kg/ha each), sett treatment with *Trichoderma* @ 5 g/L + *Beejamruth* (cow dung 100–200 kg + cow urine 100 L + jaggery 500 g + water 300 L is mixed and fermented for 10 days), coupled with liquid manures like *jeevamruth* (cow dung 10 kg + cow urine 10 L + jaggery 2 kg + gram flour 2 kg + Water 100 L mixed and fermented for 5–7 days), and foliar application of *panchagavya* (fresh cow dung 1 kg + cow urine 3 L + cow milk 2 L + curd 2 L + cow *desi ghee* 1 kg + sugarcane juice 3 L mixed and ferment for 7–10 days), at bimonthly interval was found to be good for obtaining excellent grade and better quality jaggery with overall acceptable organoleptic characteristics than nutrient management practices (Kuri and Chandrashekara, 2014). The sugarcane crop, which is utilized for organic jaggery production, is protected from insect and pests by spraying cow urine diluted to 5%. Intercropping of peas with autumn cane and cowpea with spring planted cane and harvesting of crop in the month of January, February, and March, respectively, proved to produce jaggery of improved and superior quality.

FIGURE 19.2 Good quality organic jaggery requires not only chemical-free manufacturing but also organic sugarcane cultivation.

19.3 IMPORTANCE OF SOIL STRUCTURE FOR QUALITY ORGANIC JAGGERY

Although higher sugarcane yields are realized through application of higher doses of nitrogenous fertilizers, jaggery produced has lower export potential due to poor quality. The studies on performance of recently released sugarcane genotypes and organic nutrient management practices on jaggery quality and recovery are very limited. There is a need to popularize scientific and eco-friendly methods of sugarcane cultivation and jaggery manufacturing.

The vital ingredients for marketing of jaggery is its external appearance which includes the color of the jaggery, its texture, and shelf life, and these characters of jaggery, in turn, depends on the quality of sugarcane. The quality jaggery is produced from the variety with high sucrose content, high purity, and low colloid content. The quality of the cane majorly depends on the soil properties. Soil is a natural medium for plant growth and it provides the root system nutrients availability. In India, sugarcane is grown on varied soil types, ranging from sandy to clayey soils. Mineral content of the soil and microclimate of rhizosphere plays a vital part in determining the chemical composition and quality of sugarcane juice affecting the quality of jaggery manufactured from cane juice. Nature and quantity of chemical fertilizer also affects the quality of juice and jaggery. Integrated nutrient management is a potential tool to sustain soil fertility and maintain juice quality for the production of quality jaggery. Generally, low nitrogen doses are preferred for getting better juice quality as well as jaggery. High nitrogen doses adversely affect the juice quality as it increases glucose content. Jaggery prepared from the juice of such canes has more invert sugars which is hygroscopic in nature and as a result, such jaggery when stored in rainy season, deteriorates rapidly (Singh et al., 2014).

19.4 VARIETAL SCREENING FOR ORGANIC JAGGERY PRODUCTION

Sugarcane varieties play a major role in quality jaggery production (Fig. 19.3). The quality of the jaggery depends on the sugarcane variety used for its manufacturing. However, a good quality jaggery is comprised of less amount of reducing sugars with high sucrose and purity. It has been found that there is a wide variation in quality of jaggery depending on varieties used for its preparation. On the basis of the results of varietal studies made at UPCSR, Shahjahnpur, the under mentioned varieties were found suitable for the purpose of good quantity as well as quality organic jaggery production. These varieties could be used for jaggery production on commercial scale yielding maximum jaggery recovery (% cane as well as % juice basis), minimum invert sugars, and the color (Singh et al., 2019).

FIGURE 19.3 Sugarcane varieties play a major role in quality production of jaggery.

TABLE 19.1 Comparative Performance of Different Varieties of Sugarcane for Yield and Quality of Organic Jaggery Production.

S. no.	Varieties	Cane yield (t/ha)	Jaggery yield (t/ha)	Jaggery (% cane)	Jaggery (% juice)	Pol % jaggery	Invert sugar (%)	Color reading
1	CoS 08276	97.06	11.80	12.16	21.43	85.4	3.27	104
2	Co 0238	96.91	12.00	12.39	21.33	92.2	4.68	127
3	UP 05125	89.96	10.36	11.52	21.53	71.8	3.46	121
4	CoSe 01434	89.04	9.33	10.48	19.63	82.2	3.74	127
5	CoS 08279	95.83	10.53	10.99	20.51	84.2	3.34	116
6	Co 0118	91.20	11.23	12.31	21.87	85.2	3.79	110
7	CoS 07250	92.59	11.01	11.90	21.57	77.4	3.66	107
8	CoS 08272	90.27	11.03	12.22	21.68	85.6	3.77	112
9	CoSe 03234	75.77	8.67	11.45	21.37	83.6	3.70	113
10	CoS 767	84.10	9.63	11.45	21.15	83.2	3.88	123

Source: Singh et al. (July–Aug 2019).

19.5 MINERAL CONTENT IN JAGGERY

Quality traditional sweetener jiggery, which is prepared by evaporating raw sugarcane juice, is composed of high amount of sucrose and purity with less reducing sugars. Quality of the jaggery is directly influenced by the sugarcane varieties used for its manufacturing (Rakkiyappan and Janki, 1996; Singh et al., 2016). Organic jaggery extracted from canes grown without chemical fertilizers or pesticides is free from lead and has considerable amount of medicinal as well as nutritional values because of the vitamins and minerals present in it (Singh et al., 2016). However, study carried out by Mishra (1992) revealed that the quality and external appearance of the jaggery directly depends on the type of sugarcane varieties used for its production. The factors responsible for enhancing the quality of the juice also influence the quality of jaggery produced from it. Study by Singh et al. (2019) reveals that regardless of the boiling and clarification method, the mineral composition in jaggery is mainly influenced by the varieties used and the chemical composition of its juice (Table 19.2).

TABLE 19.2 Mineral Composition of Jaggery Produced from Different Sugarcane Varieties.

S. No	Varieties	Mineral content (mg/100 g of jaggery)			
		Zn	Fe	Mn	Cu
1.	Co 0238	0.41	8.52	0.38	0.56
2.	Co 0118	0.26	12.74	0.26	0.78
3.	UP 05125	0.27	6.18	0.38	0.22
4.	CoS 08272	0.69	6.66	0.38	0.56
5.	CoSe 03234	0.43	10.87	1.22	0.78
6.	CoS 08276	0.29	8.52	0.50	0.34
7.	CoSe 01434	0.22	11.33	0.62	0.45
8.	CoS 07250	0.29	4.78	0.26	0.34
9.	CoS 08279	0.31	5.71	0.74	0.67
10.	CoS 767	0.45	8.99	0.38	0.56

Source: Singh et al. (July–Aug 2019).

19.6 MECHANICAL CRUSHING OF SUGARCANE

In jaggery making, sugarcane juice is extracted by dry crushing process. The crushers used for juice extraction vary in size and roller orientation. In early days, the cane was crushed mostly by bullock-operated wooden pestle–mortar (Fig. 19.4a) assembly which was later replaced by stone. Later, two-roller iron crusher came into existence and now a conventional crusher has three rollers (Fig. 19.4b). Number and orientation of rollers, power source, etc., are main factors that are considered while defining a crusher (Anwar, 2009). Many studies have been carried out by different centers, namely, Kharagpur, Kanpur, Bilari, Bhopal, and Lucknow for improving extraction efficiency of crushers during the past few decades (Anonymous, 1954). The horizontal crusher was observed to be 2–4% more efficient in juice extraction than vertical ones. Re-absorption of juice is less in horizontal crushers as compared with vertical crushers (Anwar, 2017a). Higher production of jaggery is directly linked with the higher juice extraction. On an average, a good crusher gives 65–70% juice extraction on cane basis.

(a) (b)

FIGURE 19.4 (a) Bullock-operated wooden pestle–mortar. (b) The horizontal crusher, 2–4% more efficient in juice extraction than vertical ones.

19.7 HEATING, BOILING, AND CONCENTRATION OF EXTRACTED JUICE

In jaggery industry, sugarcane juice is processed over open pan furnaces. Earlier, the juice was concentrated in open pan on simple circular pits made in ground with support of bricks on boundaries for giving draft. Furnaces underwent a series of development during the past few decades at different places in the country due to which various improved furnaces came into existence and were adopted (Singh and Singh, 1996; Anwar et al., 2014). Some of the improved-designs furnaces have been established in Meerut, Bijnor, Lucknow, Poona, Godawari, Puglur, and South Bihar (Baboo, 1993). The jaggery manufacturers/farmers require a furnace that should be self-sufficient in fuel with capacity matching with the juice output of crusher to achieve optimum productivity. Figure 19.5 illustrates the steps involved in the production of organic jaggery. For improvement in jaggery productivity, some innovative ideas have also been carried out for improving furnace efficiency to save fuel and processing time (Anwar, 2010, 2015). The time to reach the striking point depends on the capacity of pan and design of furnaces working mostly on natural draught utilizing the air-dry bagasse (Singh, 1997, 1998).

Another important factor that influences the quality of jaggery is the stage at which the concentrated sugarcane juice is transferred from boiling pan to cooling pan. This striking point is difficult to judge but scientifically it depends on temperature, viscosity, and the type of jaggery being prepared. It is 105°C–108°C, 114°C–117°C, and 118°–120°C for liquid, solid, and powder/granular jaggery, respectively. However, artisans judge this striking point based on their experience and traditional methods.

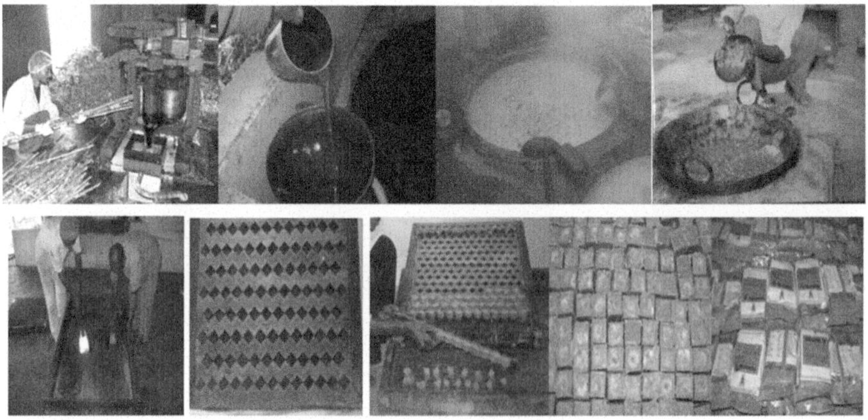

FIGURE 19.5 Different steps involved in jaggery preparation.

19.8 CLARIFICATION OF JUICE FOR ORGANIC JAGGERY

The extracted juice of sugarcane needs clarification during boiling for preparing hard, crystalline light colored, and hygienic jaggery. Alum, which is used for sedimenting the impurities present is juice, has also shown to impart improvement in color of the jaggery. Due to unawareness of health hazardous effects of chemicals, market competition and ignorance or limited knowledge of available good vegetative/herbal clarificants, most of the farmers are using several chemical clarificants, such as *hydros*, sodium formaldehyde sulfoxylate (*chakke*), sodium bicarbonate (baking soda), sodium carbonate (washing soda), super phosphate, phosphoric acid, alum, salicylic acid, and lime in higher concentration and dose. The sodium hydrosulfite and sodium carbonate are liberally used to get attractive color of the jaggery without knowing the deleterious effects on human health. The jaggery which is prepared using higher quantity of sodium hydrosulfite and sodium formaldehyde sulfoxylate contains more than 500 ppm of sulfur dioxide in the jaggery which is well above the prescribed norms of 50 ppm as per Indian standards (IS-12923):1990. This amount of sulfur dioxide is detrimental to the beneficial intestinal microflora leading to digestive disorders and gastrointestinal problems (Chandrashekar*a* et al., 2014).

However, there are a wide range of vegetative/herbal/organic clarificants available which could be very well utilized to get a better quality healthy organic jaggery. These clarificants are cheap, effective, and do not leave any harmful ingredient behind as an end product. *Sukhlai* (*Kydia calycina*) extract and groundnut (*Arachis hypogaea* L.) are good clarificants. Castor (*Ricinus communis*) seed extract could be used for the jaggery of better luster, attractive color, and relatively higher sucrose. Colocasia (*Colocasia esculanta*) proved to be beneficial in obtaining good-colored jaggery (Anonymous, 1957–58). Use of groundnut, castor seed extract, and soybean tablets yielded better quality jaggery than liming from cane juice of pH 6.4. A study also showed that the liming of cane juice to pH 6.4 and addition of *bhindi* (*Hibiscus esculentus*) mucilage (Fig 19.6a and b) was superior for jaggery of good color and consistency (Anonymous, 1957–58). Processed soybean flour, alum, and castor oil, in the proportion of 10:1:2, proved to be relatively superior to soybean flour in combination with monocalcium phosphate. Clarifying efficiency of *deola* (*Hibiscus ficulneus*) was found to be better as compared with groundnut, *sukhlai*, *semal* bark, and castor seed (Joshi and Pandit, 1959). It has also been found that mucilage of *phalsa* (*Grewia asiatica*) and *ambadi* (*Hibiscus canabinus*) gave good-colored jaggery than *Chikani* (*Sida carpinifolia*), *kateshevari* (*Bombax malabaricum*), soybean, and tapioca (*Manihot esculentus*) (Vaidya et al. 1984). A study carried out with *bhindi* mucilage tend to remove the maximum amount of scum from juice, however, it produced medium quality jaggery (Anonymous, 1995). The results of the studies conducted at different states revealed that out of various vegetative and chemical clarificants, *deola* (Fig. 19.6c and d) is found to be more effective at juice pH 6.0. It produces superior quality jaggery with high sucrose content, and low reducing sugars absorbing minimum moisture and having relatively better shelf life. The studies show that there is a marked favorable effect of the clarificant on the quality of jaggery. Studies on different doses of the *deola* showed significant improvement in the quality of jaggery, pertaining to higher pol purity and lesser percentage of invert sugar and ash content in jaggery. The best dose of *deola* appeared to be 150–160 g per four quintal of juice (Lal and Sharma, 1983). Banerji (2014) listed some of the vegetative clarificants along with part of plant and quantity to be used for good quality jaggery. Efforts are still continuing to evolve ready to use vegetable clarificants for making good quality organic jaggery.

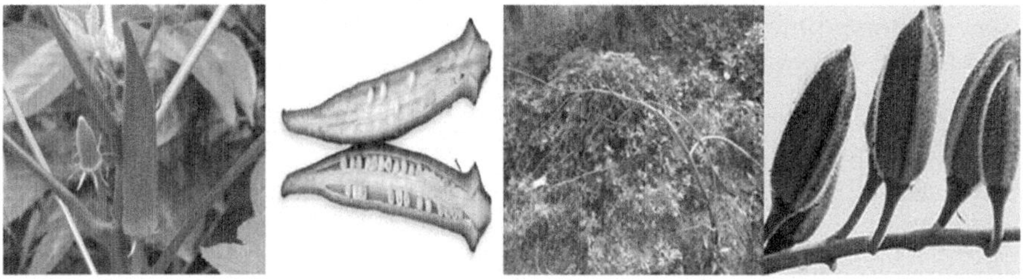

(a) (b) (c) (d)

FIGURE 19.6 Addition of *bhindi* (*Hibiscus esculentus*) (a and b) and *deola (Hibiscus ficulneus)* (c and d) mucilage is considered superior for jaggery of good color and consistency.

19.9 DIFFERENT SHAPES OF JAGGERY

Since time immemorial, jaggery is made in different shapes, sizes, and weights. The ratio of surface area-to-weight should be minimum because it was reported that these aspects are used to affect the drying and storability quality as well as their market value (Baboo and Ghosh, 1985; Lal and Sharma, 1983). After removing concentrated juice from the boiling pan, it is puddled and cooled in the cooling pan and is molded in desired shape and size using different molds (Fig. 19.7a). Some of the common shapes include *laddoo, bheli, dhayya, pari, chaukunta/chausera, pansera, dhansera, balti,* basket, *chaku, and khurpapad* (Anwar et al., 2014). Shape and size of jaggery lump should be easy to mold and dry, convenient in handling and packaging, and should cause minimum losses during storage, transport, and distribution. In order to standardize the shape and size, ICAR-Indian Institute of Sugarcane Research (IISR), Lucknow has developed jaggery molding frame for molding jaggery in 500 g bricks (Baboo et al., 1988). Further molding frame for the production of jaggery in 1-inch cubes was developed

(Baboo and Anwar, 1995) and was further modified (Anwar, 2017b). The hot concentrated juice transferred from concentrating pan to the cooling pan is puddled, cooled, and poured in molding frames and leveled up. The frame is dismantled after 40–50 min and the cubes are removed.

Studies have also standardized the process for making quality jaggery powder. For this, the juice is heated up to striking point of 120°C–122°C. Concentrated juice is then removed and allowed to cool by mixing, after this for the formation of crystals of jaggery, it is transferred from pan to platform and left without stirring. Immediately after solidification of the concentrated mass, the powder of the jaggery is prepared manually with the help of wooden scrapers. The powder is then sieved through 1–3 mm sieves and dried until 1% moisture content before storing. This form has better storability due to lower moisture content (Fig. 19.7b).

For preparation of liquid jaggery, the striking temperature of 105°C–106°C is most suited and liquid jaggery thus produced has good quality with minimum microbial growth and crystallization (Fig. 19.7c). However, the use of certain chemicals, such as 0.1% potassium metabisulfite or 0.5% benzoic acid for increase in keeping quality and 0.04% citric acid for minimizing crystallization with increase in glucose/fructose and improvement in color is also required (Singh, 1998).

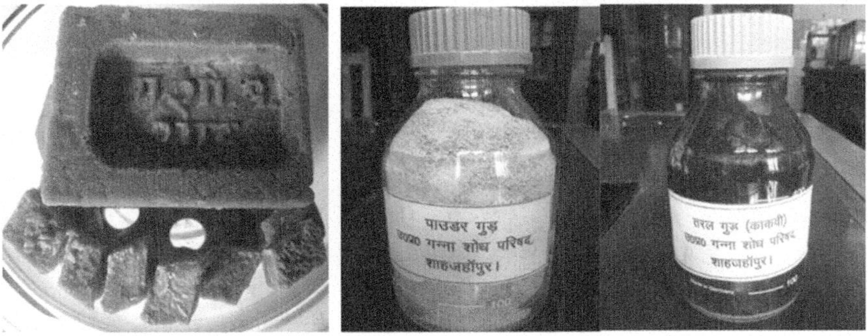

(a) (b) (c)

FIGURE 19.7 Solid, powder and liquid forms of jaggery

19.10 STORAGE AND PACKAGING OF JAGGERY

Early studies conducted at Shahjahanpur revealed that jaggery obtained from ratoon crop could be kept for longer period than that obtained from plant crop. Jaggery from top portion of cane deteriorated quickly as compared with that of the lower portion. Lodged cane yielded jaggery of poor keeping quality. For storage purposes, jaggery prepared during January to March was superior in quality. The common practice for storing jaggery earlier was in wheat straw, ordinary earthen pots, hessian bags, and also in open room where jaggery is used to liquefy at higher RH in monsoon season. Improved methods of jaggery storage, viz., painted earthen pots, drying cum storage bin, masonry bin-cum-bed, and drying cum-storage godown improved the storability of jaggery (Fig. 19.8).

FIGURE 19.8 Drying cum storage godown improved the storability of jaggery.
Source: ICAR- IISR, Lucknow, India.

Uniform brick-shaped jaggery of 500, 250, and 125 g and cubes of 25 mm made in molding frames and packed in double-layered butter and glazed papers giving cellophane wrapping in hygienic way which in turn helps in fetching good price from consumers (Baboo & Anwar, 1995). The keeping quality of jaggery by storing in cold storage, tin or at room temperature after wrapping in alkathene was found to be better, combined packing with alkathene and lesion cloth had a more advantageous effect (Fig.

19.9). Jaggery packed in tin foil and covered with polyethylene or hessian cloth remained in good condition with fewer changes in physicochemical characteristics for longer period of time. Withdrawal after 14 days from cold storage recorded lower deterioration in pol %, color, invert sugar %, and moisture % of stored jaggery wrapped with alkathene than those stored in ordinary room temperature. Polyester film packing proved superior for storage of well-dried jaggery in ordinary storage. Mixing of ginger powder at 2% with jaggery was found the most suitable for storage as it improves the keeping quality of jaggery.

FIGURE 19.9 Packaging of jaggery for better storability.

19.11 CONCLUSION

It could be concluded that jaggery, being an important sweetener and food material, has good nutritive and medicinal value and has more scope to be promoted as health food. Jaggery when produced from sugarcane grown in soils rich in organic content, by adopting organic farming and by the use of vegetative clarificants, is found superior in quality and have better settling property and storability. The color of jaggery may vary from light golden brown to dark brown but it will have great market potential due to the liking of health-conscious people for organic and chemical-free jaggery.

KEYWORDS

- **organic sugarcane cultivation**
- **soil structure**
- **mineral content**
- **mechanical crushing**
- **storage and packaging**

REFERENCES

Anonymous. *Annual Report*; SBI: Coimbatore, India, 1954; p 45.

Anonymous. *Annual Progress Report*; Sugar Research Scheme: Andhra State, India, 1957–58; pp 117–120.

Anwar, S. I. *Gur Evam Khandsari Udyog Mein Ganna Kolhu Ka Rakhrakhao*; Kheti, 2009 April; pp 29–32.

Anwar, S. I. Fuel and Energy Saving in Open Pan Furnace Used in Jaggery Making through Modified Juice Boiling/Concentrating Pans. *Energy Conv. Manage.* **2010,** *51*, 360–364.

Anwar, S. I. Improving Thermal Efficiency of Open Pan Jaggery Furnaces—A Novel Concept. *Indian J. Sugarcane Technol.* **2015,** *29* (01), 32–34.

Anwar, S. I. A Simple Yet Effective Method of Improving Juice Extraction in 3-Roller Power Driven Horizontal Sugarcane Crushers. In *Technical Compendium of 51st Annual Convention of ISAE and National Symposium*; Feb 16–18, 2017; ISAE: Hisar, 2017a; p 114.

Anwar, S. I. Development of Modified Jaggery Moulding Frame for Cubical Shape Jaggery. *Agric. Eng. Today* **2017b,** *41* (1), 44–47.

Anwar, S. I.; Kumar, D.; Singh, J. *Gur Ka Moolyavardhan Evam Dhalai. In:* Uttam *Gur Utpadan—Prashikshan Pustika*; IISR: Lucknow, 2014; pp 36–42.

Anwar, S. I.; Kumar, D.; Verma, P. *Jaggery: Evolution to Revolution—An Overview of Jaggery Research at IISR, Lucknow, Souvenir-cum Proceedings of National Meet on Modernization of Jaggery Industry in India*; ICAR-IISR, Lucknow, Nov 1–2, 2014; pp 36–44.

Baboo, B.; Anwar, S. I. *AICRP on Processing, Handling and Storage of Jaggery and Khandsari*; Tech. Bull (IISR/JRS/94/9); IISR: Lucknow, 1995.

Baboo, B. *Manufacture of Good Quality Gur (Jaggery) and its Safe Storage*; IISR Tech Bull. No. 33; IISR: Lucknow, 1993.

Baboo, B.; Ghosh, A. K. 1985. Effect of Shapes on Artificial Drying of Jaggery. *Maharashtra Sugar* **1985,** *10* (5), 47–50.

Baboo, B.; Zahoor, M. A.; Garg, S. K. Design and Development of *Gur* Moulding Frame. *Indian J. Sugarcane Technol.* **1988,** *5* (2), 89–92.

Banerji, R. *Uttam Gur Utpadan Ke Liye Ganne Ke Ras Kee Safai Evam Sandrikaran. In: Uttam Gur Utpadan -Prashikshan Pustika*, IISR: Lucknow, 2014; pp 30–35.

Chandrashekara, C. P.; Patil, S. B.; Nadagouda, B. T.; Desai, S. R.; Biradar, D. P.; Khadi, B. M.; Thippannavar, P. S.; Kambar, N. S. In Establishment of modern hygienic SS 304 jaggery processing unit at Jaggery Park, Mudhol and Sankeshwar. In *Jaggery: Evolution to Revolution, Souvenir-cum-Proceedings of National Meet on Modernization of Jaggery Industry in India*; ICAR-IISR: Lucknow, Nov 1–2, 2014; pp 63–65.

Joshi, K. C.; Pandit, S. N. Improved Technique of *Gur* Manufacture and the Role of Various Clarificants. *Indian J Sugarcane Res. Dev.* **1959,** *4* (1), 42–50.

Kuri, S.; Chandrashekara, C. P. Effect of Organic Nutrient Management Practices on Sugarcane Genotypes with Special Reference to Jaggery Production and Quality. In: *Jaggery: Evolution to Revolution, Souvenir-cum Proceedings of National Meet on Modernization of Jaggery Industry in India*; ICAR-IISR: Lucknow, Nov 1–2, 2014; pp 152–155.

Lal, U.; Sharma, R. K. Role of *Gur* Shapes in its Deterioration during Storage. *Co-operative Sugar* **1983,** *15* (1), 7–9.

Mishra, A. Parameters for Selection of Sugarcane Varieties for Jaggery Quality. *Indian J. Agric.* **1992,** *37* (2), 391–392.

Rakkiyappan, T.; Janki, P. Jaggery Quality of Some Commercial and Promising Sugarcane Varieties. *Co-operative Sugar* **1996,** *27* (12), 909–913.

Singh, J. *Proc. National Symposium on Sugar Recovery Problems and Prospects*; IISR: Lucknow, Sept 13–15, 1997; pp 53–55.

Singh, J. *Jaggery and Khandsari Research Digest,* IISR (JK Cell); *Tech. Bull.* 98–99; IISR: Lucknow, 1998; p 140.

Singh, J.; Singh, R. D. *Report of CCM of AICRP on Renewable Energy Sources*; CIAE: Bhopal, Sept 30–Oct 01, 1996.

Singh, P.; Singh, M.; Sharma, B. L. Screening of Sugarcane Varieties for Quality Jaggery in North India, Strategic Governance and Technological Advancement for Sustainable Agriculture. In *Proceeding of 4th Uttar Pradesh Agricultural Science Congress*, Kanpur, Mar 2–4, 2016; p 45.

Singh, P.; Bhatnagar, A.; Singh, M. M.; Singh, A. Validation of Elite Sugarcane Varieties for Quality Jaggery Production in Sub-tropical India. *Sugar Technol.* **2019,** *21* (4), 682–685. https://doi.org/10.1007/s12355-018-0647-6.

Singh, R.; Rai, R. K.; Singh, P.; Shrivastava, A. K.; Solomon, S.; Arya, N. Organic Jaggery: A Healthy Alternative. In *Jaggery: Evolution to Revolution, Souvenir-Cum Proceedings of National Meet on Modernization of Jaggery Industry in India*, Nov 1–2, 2014; ICAR-IISR: Lucknow, 2014; pp 120–121.

Vaidya, B. R.; Kadlage, A. D.; Hapse, D. G. The Efficiency of Different Vegetable Clarificants for Quality Jaggery Preparation. *Maharashtra Sugar* **1984,** *9* (4), 69–71.

CHAPTER 20

Organic Vegetable Production: Needs, Challenges, and Strategies

B. S. TOMAR*, GOGRAJ SINGH JAT, and JOGENDRA SINGH

Division of Vegetable Science, ICAR-Indian Agricultural Research Institute, New Delhi 110012, India

Corresponding author. E-mail: bst_spu_iari@rediffmail.com

ABSTRACT

Organic agriculture is a production system that sustains the health of soils, ecosystems and people. It relies on ecological processes, biodiversity and cycles adapted to local conditions, rather than the use of inputs with adverse effects. Organic agriculture combines tradition, innovation and science to benefit the shared environment and promote fair relationships and a good quality of life for all involved. Organic vegetable cultivation helps in maintaining sustainability of the cropping systems with recurring advantages of long-term soil health, on the other, it provides stability in production due to better resistance against various biotic and abiotic stresses. The basic purpose of the farmers involved in organic cultivation of vegetables should supply those crop and animal nutrients that take care of the soil health, increase soil life, and conserve nutrients. These practices involve development of both long-term and short-term strategies to advance soil health and supply crop nutrition. However, the primary goal of organic vegetable production is to optimize the health and productivity of interdependent communities of soil, plant, animal and people.

Organic Crop Production Management: Focus on India, with Global Implications, D. P. Singh, PhD, H. G. Prakash, PhD, M. Swapna, PhD, & S. Solomon, PhD (Eds.)

20.1 INTRODUCTION

India is largely an agricultural country with total net sown area of 141.4 million ha and gross cropped area of 200.9 million ha with a cropping intensity of 142% (Annual Report, DAC&FW, 2016–2017). India is the second largest populous country in the world with an estimated population of 1.31 billion after China. It is estimated that India will be recording annual growth rate of 1.2% per annum by 2050. India has the highest population of 1.7 billion in the world. International Food Policy Research Institute (IFPRI), Washington, 2016, revealed that 15.2% of the Indian citizens are malnourished and deprived of enough quantity of food (quantity and quality). In India, per capita land resource is 0.12 ha which is shrinking further due to the tremendous pressure of the developmental activities, therefore the only option is to attain higher productivity in a sustainable manner. To grapple with the consistently increasing demand of vegetables for a colossal population, there is an increasing demand of high-yielding varieties, extension of irrigated areas, use of high analysis NPK fertilizers and increase in cropping intensity, and excessive application of agrochemicals. This ultimately leads to loss of genetic diversity, enhances vulnerability of crops to insect pests and diseases, soil erosion, low soil fertility, contaminated soil, low organic matter of soil, and water shortages that further lead to the reduction in the availability of nutritious food crops to consumer. New pests have evolved with the increased resistance to insect pests to insecticides. The increased intensity of pests has also been realized. Health hazards such as pesticide residues in food products and groundwater contamination with intensive modern agriculture are matters of concern. Extensive use of fertilizers leads to the occurrence of multinutrient deficiencies and overall decline in the soil fertility. Thus, there is an urgent need to develop a sustainable strategy for increasing vegetable production with meagre resources with minimum use of fungicide, insecticide, pesticide, and chemical fertilizers without causing any harmful effect on the soil and environment. Organic vegetable cultivation is one of the most sustainable farming systems with recurring benefits to long-term soil health and also enhanced stability in production by incorporating higher resistance to various biotic and abiotic stresses (Palaniappan, and Annadurai,1999).

Organic farming has been defined by the United States Department of Agriculture as a system which avoids or largely excludes the use of synthetic inputs (such as fertilizers, pesticides, hormones, feed additives, etc.) and to the maximum extent feasible, rely upon crop rotations, crop residues, animal manures, off-farm organic waste, mineral grade rock additives, and biological system of nutrients mobilization and plant protection. Food and Agriculture Organization, a United Nations Agency has described the organic agriculture as a unique production management system which promotes and enhances agroecosystem health, including biodiversity, biological cycle, and soil biological activity, and this is accomplished by using on-farm agronomic, biological, and mechanical methods in exclusion of all synthetic off-farm inputs." In simple words, organic agriculture is the farming system without the addition of synthetic chemicals (Shiva et al., 2004; Yadav, 2005).

20.2 NEEDS OF ORGANIC VEGETABLE PRODUCTION

1. Vegetables are fast-growing and eaten raw. Therefore, any contamination and chemical residue may result in various health problems.
2. Vegetables are suited to high cropping intensity, hence, require more organic content for sustainable production.
3. Exhaustive use of fertilizers and pesticides increases cost of production.
4. High environmental pollution due to excessive input of agrochemicals in vegetable production.
5. Ensuring supply of heavy metals and chemicals-free vegetables.
6. Generating additional income through marketing of organically grown vegetables on cheaper rates in the high-income societies.
7. Intensive use of chemical fertilizers and pesticides leads to the deterioration of the environment quality, ecological stability, and sustainability of vegetables production.
8. Organic production system can enhance the crop productivity and restore natural base.

20.3 CHALLENGES IN ORGANIC VEGETABLE PRODUCTION

1. Lack of availability of organic seeds.
2. Low yield during conversion period.
3. Limited availability of organic manure, namely, FYM, vermicompost, cakes, and others.
4. Higher price of organic manure.
5. Limited availability of organic-based pesticides.
6. Quality assurance of organic manure and pesticides

7. Need and greed of quick higher returns of famers.
8. Limited awareness for organic certification of produce.
9. Lack of consumer awareness for organic vegetable and their products.
10. Higher price of organic produce affects the consumer's choice in market.
11. Confidence about the organic produce in the consumer's mind (Is organic really synonymous with pure?).
12. Farming organically on an industrial scale is difficult.

20.4 CONVERSION PERIOD

Transition period is the minimum period of 3 years which is taken for the conversion of nonorganic/conventional farm to organic farm which starts with the signing date of the contract with the Certification agency. Produce can be sold as "in conversion to organic agriculture" after 1 year and after 2 years, annual crops can be sold as "organic."

20.5 STRATEGIES FOR ORGANIC VEGETABLE PRODUCTION

20.5.1 SELECTION OF FARM

For successful organic production of vegetables, site selection is the most important criterion which includes suitable soil and climate, regular supply of good quality irrigation water, availability of plenty of labor, transportation, and marketing facility. While selecting the site for organic vegetable cultivation, care should be taken so that there are no pests and diseases infestations of crops on the site. For example, in the areas where there is higher rainfall and a single crop is grown on a very large area, there may be greater risks for cultivating organic vegetables as the crop may face severe infestation of pests and diseases.

20.5.2 SOIL

Majority of the vegetable crops are cultivated in a proper drained loam or clay–loam soil with the pH ranging between 6.0 and 7.5. Before planting the vegetable crops, there is a need to make some adjustments for balanced nutritional requirements. There should be testing of chemical fertilizers and pesticides residues and heavy metals in the soil. In the case of presence of the above materials in the unacceptable levels in the soil, the produce may be discarded from organic certification. Such soil should not be used for the cultivation of root vegetables.

20.5.3 NUTRIENT MANAGEMENT

For the successful production of organic vegetables, the soil fertility of the field should be high. Thus, the basic purpose of organic farming is to build up reserves of soil nutrients and to maintain a proper system of nutrient cycling. The farmers involved in organic cultivation of vegetables should ensure supply of crop and animal nutrients to the crop by adopting practices that take care of the soil, enhance soil life, and also conserve nutrients. These practices involve the development of both long-term and short-term strategies to advance soil health and supply crop nutrition. A cafeteria of practices, such as organic manure, green manure, lime, rock phosphate, and other rock materials and supplemental organic fertilizers are known as organic fertility system.

20.5.3.1 ORGANIC MANURES (COMPOSTS, VERMICOMPOST, AND FYM)

Compost which may include pesticide and heavy metals-free crop residues and animal manures is an essential nutritional requirement for organic vegetable production. In compost, the C:N ratio should be about 20:1 so that it improves the soil structure and stimulates the beneficial microorganisms. For the production of compost, a particular area of the farm should be used. The organic manures, such as FYM, fish manures, poultry manures, sheep compost, and so on should be given as a basal dose at 25–38 t/ha. Use of organic cake from *neem*, *Pongamia*, castor, mustard, and groundnut hasalso been found beneficial for organic vegetable production.

20.5.3.2 USE OF BIOGAS SLURRY

Bioslurry is a by-product obtained after the digestion of dung or other biomass for the generation of methane-rich gas from the biogas plant.

Digested bioslurry contained organic nitrogen (mainly amino acids), abundant mineral elements (macro and micronutrients) and low molecular mass bioactive substances (hormones, humic acid, and vitamins).

20.5.3.3 GREEN MANURING

Green manure crops are raised and then ploughed or turned into the soil at the time of flowering, that is, 40–45 days after sowing for decomposition to improve the soil physical structure and fertility, enhances organic matter or organic nutrients to the soil and also increases microbial activities. Various leguminous crops, such as sun hemp (*Crotalaria juncea*), *dhaincha* (*Sesbania aculeata*), pillipesara (*Phaseolus trilobus*), cowpea (*Vigna unguiculata*), and cluster bean (*Cyamopsistetragonoloba*). A luxuriant, vigorously growing legume sward (*Dhaincha*, sunhemp, and fababeans) contains large amount of 80–120 kg nitrogen. It should be incorporated 20–30 cm deep into the field using a tractor or a mold board plough. Some farmers cultivating organic vegetables also apply foliar sprays of sugar, molasses, or compost teas prior to turning of the green manure crop. It is believed that this practice provides added energy for soil microorganisms, which favors rapid breakdown of green matter prior to planting the subsequently vegetable crop.

20.5.3.4 USE OF BIOFERTILIZERS

Biofertilizers are the carrier-based preparation containing beneficial microorganism in a viable state intended for seed, seedling, or soil inoculation. Biofertilizers play an important role in vegetable production by enhancing atmospheric nitrogen fixation, decomposing organic wastes, increasing soil health, reducing environmental pollution, as well as cost of production. Inoculation by improved strains of nonsymbiotic nitrogen-fixing bacteria, that is, *Azotobacter* significantly enhanced the productivity of vegetables, namely, potato, onion, brinjal, tomato, chilli, cabbage, cauliflower, and okra. *Azospirillum* increased nitrogen uptake and also helped in reduction of nitrogenous fertilizers. Application of PSB results in enhanced availability of phosphorous to the plants.

20.6 APPLICATION OF BIOFERTILIZERS IN VEGETABLE CROPS

There are many methods through which biofertilizers are applied in vegetable crops at various stages. These biofertilizers are discussed in the following sections.

20.6.1 SEED TREATMENT

Seeds of leguminous vegetable crops are treated with *Rhizobium* culture (250 g/acre) by mixing the culture well with 5% of the *gur* solution. This mixture is first poured on seeds which is spread on a cement floor or a polythene sheet and then properly mixed with hands so that there is a thin and uniform coating on each seed. These seeds are then spread in shade for drying for at least 10–15 min. After this, they are immediately sown in the soil.

20.6.2 SEEDLINGS TREATMENT

This method is mainly used for the seed treatment in the case of tomato, onion, and chilly, and others. The culture suspension is prepared in an approximate ratio of 1:10 by mixing 1 kg of culture in 10 L of water. This culture suspension is sufficient for 1-acre area. Seedlings are first arranged into bundles and dipped in suspension for 15–20 min and transplanted in the soil immediately.

20.6.3 DIRECT APPLICATION

Approximately, 2–3 kg of biofertilizer mixed in 40–60 kg of soil, compost, or FYM is sufficient for broadcasting in 1-acre area either at the time of sowing or at least 24 h before sowing. This is the most common method for using phosphate solubilizers.

20.6.4 ACCOMPLISHMENT OF NUTRIENT REQUIREMENTS

For successful production of organic vegetables, the organic vegetable crop may require additional dose of nutrients for the successful production which can be supplied during the crop's growth period. The details are given in the next section.

20.6.4.1 NITROGEN

The short-duration vegetables may be able to complete their entire requirement of nitrogen from green manure, compost, or organic fertilizers that have been applied before planting. Crops which remain in the field for the period beyond 6–8 weeks may need additional nitrogen, applied as side dressing or foliar spray. Commonly used source with readily available nitrogen are fish emulsion, worm juice, and compost teas (concentrated organic liquid fertilizer is made from steeping biologically active compost in aerated water).

20.6.4.2 PHOSPHORUS

Organic sources of phosphorus are rock phosphate, guano (accumulated excrement of seabirds, seals, or cave-dwelling bats), fish meal, and bone meal.

20.6.4.3 POTASSIUM

Compost, seawood, basic slag, wood ash, and sulfate of potash are the organic sources of potassium.

20.6.4.4 OTHER

Lime is a source of calcium, dolomite a source of calcium and magnesium and gypsum is the source of calcium and sulfur.

The details of some of the concentrated organic manure and the availability of nutrients are depicted in Table 20.1.

TABLE 20.1 Concentrated Manures/Fertilizers Used in Organic Farming.

S. no.	Manure/fertilizer	N (%)	P_2O_5 (%)	K_2O (%)
1.	Blood meal	12	0	0
2.	Bone meal	2	15	0
3.	Chilean nitrate	16	0	0
4.	Guano	9–12	3–8	1–2
5.	Rock phosphate (Soft)	0	15–30	0
6.	Potassium magnesium sulfate	0	0	22
7.	Pelleted chicken manure	2–4	1.5	1.5

20.7 SELECTION OF CROP AND VARIETY

Availability of seed and seedlings, resistance to insect pests and diseases, availability of nearby market for supplying the fresh produce supply, physiological characteristics, and environmental suitability are the major criteria which should be taken care of during selection of the crop and the variety of the vegetables. The seeds of the selected variety of the crop can be obtained from organically certified seed or seedlings.

The high-yielding, pest- and disease-resistant, and superior seedling vigor varieties, which are the most popular among the vegetable growers, should be selected. Varieties performing well under limited resources availability and having resistance to major biotic and abiotic stresses including pest and diseases must be preferred for cultivation. To reduce the cost of cultivation of organic vegetables, it is recommended to use these varieties. These varieties can meet the standards of organic vegetable cultivation, as they do not require any chemical insecticide for the pest management (Table 20.2). It has been observed that vegetable crop varieties, which are early in vigor commonly, hamper the growth of weed plants.

TABLE 20.2 Varieties and F_1 Hybrids of Vegetable Crops Having Resistance/Tolerance Against Major Diseases and Pests.

S. no.	Crop	Disease	Varieties/F_1 hybrids
	Tomato	Bacterial wilt	ArkaAbha, Arka Abhijit, Utkal Pallvai (BT1), Utkal Kumari (BT10), Arka Alok, Arka Vardhan
			F_1 hybrids-ArkaRakshak, Arka Ananya, Arka Samrat, Arkashreshta
		Late blight	TRB1 and TRB 2
		Leaf curl virus	Hisar Anmol (H-24), Hisar Arun (H-36), H-88, F_1 hybrids-Arka Rakshak, Arka Ananya, Arka Samrat
		Early blight	F_1 hybrids-Arka, Samrat, Arka, Rakshak
		Root knot nematode	Pusa-120, Pusa Hybrid-2, Pusa Hybrid-4, Hisar Lalit (NRT8), Punjab-NR-7
			F_1 hybrids—ArkaVardan
	Brinjal	Phomopsis blight	PusaBhairav, Pusa Purple Cluster
		Bacterial wilt	Arka Keshav, Arka Neelkanth, Arka Nidhi, Pusa Purple Cluster, Utkal Tarini, Utkal Madhuri, Annamalai
		Little leaf	Pusa Purple Cluster, Hisar Shyamal, Pant Rituraj
		Shoot and fruit borer	SM 17-4, Punjab Barsati, ARV 2-C, Pusa PurpleRound, Punjab Neelam

TABLE 20.2 *(Continued)*

S. no.	Crop	Disease	Varieties/F_1 hybrids
	Chilli	Bacterial wilt	Ujjwala, Anugraha, Pant C-1, Punab Lal
		CMV, TMV, TLCV	PusaJwala, Pusa Sadabahar, Pant C-1
		Thrips	PusaJwala, Pusa Sadabahar, Pant C-1
		Mites	Punjab Lal, Pant C-1
		Aphids	Punjab Lal, Pant C-1, Pusa Jwala
	Sweet pepper	Bacterial wilt	Arka Gaurav
	Okra	Yellow vein mosaic virus	Pusa Bhindi-5, Varsha Uphar, Arka Anamika, Arka Abhay, Punajb Padmini, Punjab-7, Punjab-8, Hisar Unnat, Azad Kranti, Utkal Gourav
	Cabbage	Black rot	Pusa Mukta
		Black Leg	Pusa Drum Head
	Cauliflower	Black rot	Pusa Shubra
	Onion	Purple blotch	Arka Kalyan, Nasik Red
		Thrips	Arka Niketan, Pusa Ratnar
	Pea	Powdery mildew	JP-83, JP-4, Arka Ajit, JP179, Arka Karthik, Arka Sampoorna (snap pea), JP9
		Fusarium wilt	Kalanagini, JP179, Pusa Vipasa
		Rust	JP. Batri Brown 3, JP. Batri Brown 4, JP179, Arka Karthik, Arka Sampoorna (snap pea)
	Cowpea	Bacterial blight	Pusa Komal
	French bean	Powdery mildew	Contender, Pusa Parvati
		Wilt	Jampa
		Rust	Pant Anupama, Arka Bold, Pant Bean-2
		Angular leaf spot	Lakshmi, Pant Anupama
		Common bean mosaic	Pant Anupama, Pant Bean 2
	Musk melon	Downy mildew	Punjab Rasila
		Powdery mildew	Arka Rajhans
		Cucumber green mottle mosaic	DVRM-1, DVRM-2
	Watermelon	Anthracnose, downy mildew, powdery mildew	Arka Manik

20.8 CROPS TO BE TAKEN IN ROTATION

It is recommended to include any leguminous crop like beans, cowpea, peas, etc. in the crop rotation for enhancing soil fertility due to their naturally in-built capacity of fixing atmospheric nitrogen into the soil and enhancing the productivity of the crop by 30–35%. It will be better to further improve the nitrogen-fixing ability of these crops by inoculating the seeds of leguminous crops with crop-specific rhizobial strains available in SAUs/ICAR Institutes/state government's Department of Agriculture. The quantity of N fixed by different crops has been depicted in the Table 20.3.

TABLE 20.3 Amount of Biological Nitrogen Fixed by Leguminous Crops.

S. no.	Crop	Nitrogen (%)	Biomass productivity (t/ha)	Estimated N (kg/ha)
1.	Sunhemp	0.43	12–13	52–56
2.	*Dhaincha*	0.43	20–22	86–95
3.	Cowpea	0.49	15–16	74–78
4.	Cluster bean	0.34	20–22	68–75
5.	*Berseem*	0.43	15–16	65–69
6.	Mung bean	0.53	08–09	42–48

20.9 MULCHING

In mulching, the outer surface of the soil is covered with a layer of an organic or an inorganic material. Mulching facilitates to transform environment of the soil, checks soil erosion, prevents weed growth and increase the soil microorganism activity. Crop or animal residues or by-products, such as straw, sawdust, animal manure, dry leaves, sugarcane bagasse, and others are the example of organic mulch. The application of organic mulch improves physical and chemical properties of the soil by enhancing the organic carbon content of the soil. Mulching with black plastic film is helpful in reducing weed intensity by inhibiting their germination and growth due to reduced solar radiation. Reflective plastic mulches (yellow, white, silver, aluminum-coated, etc.) have been reported to reduce aphids and other pest population in vegetable field.

20.10 WEED MANAGEMENT

To restore soil fertility, locally available organic manure is used which causes the problem of weeds. Weed competition must be suppressed during the critical period of organic vegetable cultivation. In addition to the various mechanical methods, namely, deep ploughing during the summer, manual hand weeding, various cultural practices, such as composting of manures, use of black plastic mulch and cover cropping with other crops are the most common method to reduce weed pressure in vegetables. Crop rotation also an effective means for suppressing weeds in organic vegetable production.

20.11 INSECT PEST MANAGEMENT

The insect pests management depends upon different integrated management practices to check the population of the pests to the economic threshold level pest. For organic vegetable cultivation, a site which does not have the history of pest or disease incidence in the past should be selected. The plants of weeds act a carrier for several diseases carrying pathogens and pests should be removed or destroyed.

20.12 TRAP CROPS

Trap crops are also known as sacrificial crops. The crops which are planted to attract insect and pests and can be efficiently managed are known as sacrificial crops or trap crops. Planting of trap crops minimizes pest damage in the main crop. These crops can be planted in the periphery or in the middle of the main crop. For example, Chinese cabbage or collard planted in the field of cabbage attracts diamond back moth, onion, and garlic planted in carrot decreases the outbreak of carrot root fly and African marigold planted in tomato field attracts tomato fruit borer.

20.13 PHEROMONE TRAPS

In pheromone traps, pheromones are used in the trap to lure insects. The sex pheromones are most commonly used pheromones. Pheromone traps are used to monitor mass-trapping of male insects as male insects are traps due to female pheromone, and thus they disrupt the mating process of insect pests.

Pheromone traps are effectively used for the management of *Helicoverpa* in leguminous crops, brinjal shoot and fruit borer, tomato fruit borer, and fruit fly in cucurbits.

20.14 BIOLOGICAL CONTROL OF PESTS

When insect and pests are controlled by natural enemies, such as predators, parasitoids, and pathogens, this method of pest management is known as biological control. A number of natural biological control agents have been used (Maity and Tripathy, 2004) for effective control of insect pests in organic vegetable production (Table 20.4).

TABLE 20.4 Biological Agents Used for Controlling Different Vegetable Pests.

S. no.	Crop	Pest	Biocontrol agents	Field application
	Cole crops	Diamond back moth	Larval parasitoids (a) *Cotesia plutellae* (b) *Diadegma semiclausum*	Release adult parasitoids at 15,000 / ha at weekly interval during the initiation of larval damage
		Tobacco caterpillar	Entomopathogenic fungus (a) *Nomura erileyi*	The fungus is diluted in water and mixed in Tween 80 (0.04%) and sprayed on the crop during evening hours
			Nuclear Polyhedrosis virus of *S. Litura* (SI NPV)	Spraying of SI NPV at 250–300 Larval Equivalent (LE) mixed in 250 L of water, 1% jaggery and 0.1% teepol during evening hours
		Aphids	Lady bird beetles (*Coccinella septempunctata*)	Adult beetle at 30/sq. m
2.	Tomato	Tomato fruit borer	Egg parasitoid: *Trichogramma chilonis*	Parasitoids are released as adult or in parasitized egg form at 50,000/ha in six releases starting from 45 days after transplanting

TABLE 20.4 *(Continued)*

S. no.	Crop	Pest	Biocontrol agents	Field application
			Larval parasitoid: *Campoletis chlorideae*	at 15,000 adults/ha
			Nuclear polyhedrosis virus of *H. armigera* (*HaNPV*)	Spraying of *HaNPV* at 250–300 Larval Equivalent (LE) mixed in 250 L of water, 1% jaggery and 0.1% teepol during evening hours
3.	Okra and bean	Mites	Predatory mites (*Amblyseius tetranychivorus*)	at 10–60 mites/plant or 100 mites/m^2

20.15 DISEASE MANAGEMENT

Disease management in organic vegetable production is based on the combination of organic soil management practices, cultural practices, IPM practices, natural remedies, and limited use of permitted chemicals. Fungicides that may be allowed organically includes many copper and sulfur compounds and biological fungicides containing species of *Trichoderma, Bacillus, Pseudomonas, Gliocladium, Streptomyces*, and other beneficial microbes. Application of *Trichoderma viride* at 10 g/kg seed or *Trichoderma harzianum* at 10 g/L for spray or 10–12 kg/ha for basal dressing is effective against wilt and rot diseases in vegetables. Copper and sulfur-based products are the only labeled fungicides allowed in organic certification. Copper is labeled for anthracnose, bacterial speck, bacterial spot, early and late blight, gray leaf mold, and septoria leaf spot. Similarly, sulfur is labeled for the management of powdery mildew.

20.16 APPLICATION OF *NEEM*-BASED PRODUCTS IN ORGANIC VEGETABLE PRODUCTION

Rather than damaging digestive or nervous system of the insect and pests, the *neem*-based products work on the hormonal system of the insects. Thus, resistance to major insect pests is not developed in future generations. Belonging to a general class of natural products, these compounds are known as liminoids. These liminoids available in *neem* make it an eco-friendly and effective insecticides, pesticide, fungicide, nematicide, and others. Azadirachtin, salanin, meliantriol, and nimbin are the most important liminoids present in *neem* with well-proven ability to block growth of insects. For managing insects, azadirachtin has emerged as a main agent of *neem,* resulting in about 90% damage to most of the pests. Azadirachtin does not kill insects immediately but it repels as well as disrupts the growth and reproduction of insects. It repels or decreases the feeding of number of species of insect pests along with few nematodes. In organic vegetable production, *Neem*-based products may be used in various ways, the details of which are given in the following section.

20.16.1 MANURE AND FERTILIZER

Neem manure is biodegradable, eco-friendly, excellent soil conditioner which also enhances the content of nitrogen and phosphorus in the soil. Being a rich source of sulfur, nitrogen, potassium, and calcium, and so on *neem* manure works as a biofertilizer and facilitates in supplying the needed nutrients to plants.

20.16.2 SOIL CONDITIONER

Granules or powdered form of *neem* seeds are used to prepare the soil conditioner which can be used at the time of sowing of plants or can be sprinkled and raked into the soil. Sprinkling of soil conditioner should be followed by proper irrigation to enable the product reaching the root zone.

20.16.3 FUMIGANT

Available in gaseous state, *neem*-based pest fumigant is used as a disinfectant and pesticide by killing pests and adversely affects feeding and oviposition deterrence, inhibition of growth, and mating disruption, and so on. Not developing resistance by the pests is the major advantage of using n*eem* pest fumigant. The antibacterial and germicidal properties possessed by *neem* oil and seed extracts have been found very beneficial for protecting the plants from incidence of various insect and pests without leaving any residue on the plants. Being nontoxic and eco-friendly, the *neem* fumigants do not cause any harm to other microorganisms and do not contaminate the environment. Resistance is not developed among the pests and there is no side effect of it. Apart from being cheaper pest repellent, pest reproduction controller and no side effect, pests do not develop resistance to it.

20.16.4 PESTICIDES

Neem-based pesticides play a major role in insect pests management, and therefore are being used on a large-scale in organic vegetable production. The germicidal and antibacterial properties of *neem* oil and seed extracts are beneficial for the plants to protect them from a number of insect pests (Subbalakshmi et al. 2012). Not leaving any residue on the plant is the major benefit of using *neem*-based pesticides and insecticides.

20.17 HOW TO GET CERTIFICATION OF YOUR PRODUCE

20.17.1 ORGANIC CERTIFICATION

A well standard certification process involves a set of production standards for crop cultivation, their storage, conversion into processed products, packaging and shipping that exclude the use of synthetic chemical inputs (e.g., pesticides, fertilizer, antibiotics, food additives, etc.). Genetically modified organisms have been structured for organic food producers and related organic agricultural products. Any organization involved in business related to organic food products, including seed suppliers, farmers, food processors, retailers and restaurants can be certified. It allows the use of only that field which remains chemical-free for the past few years (generally three or more), keeping records in detail for production and sales, strictly maintaining the organic products from noncertified products and periodic inspection of the field.

20.17.2 OBJECTIVE OF CERTIFICATION

Certification of organic produce is necessary for increasing worldwide demand of organic food. It is required to assure product quality and also for the prevention from fraud products. Certification gives an identity to organic producers for supplying the approved products for use in certified operations. The tag "certified organic" serves the purpose of product assurance to the consumers. Similarly, "low fat," "100% whole wheat," or "no artificial preservatives" tags also assure the customers for the quality assurance. The organic standards set by the National Government's minimum requirements are maintained by most of the certification agencies.

20.18 CONCLUSION

With available natural resources for organic vegetable production, a natural balance needs to be maintained with a sustainable strategy to produce more vegetables from limited resources with lesser application of chemicals, fertilizer, and pesticides without any detrimental effects to the soil and environment. On one side, organic vegetable cultivation helps in maintaining sustainability of the cropping systems with recurring advantages of long-term soil health, on the other, it provides stability in production due to better resistance against various biotic and abiotic stresses. The basic purpose of the farmers involved in organic cultivation of vegetables should supply those crop and animal nutrients that take care of the soil health, increase soil life, and conserve nutrients. These practices involve the development of both long-term and short-term strategies to advance soil health and supply crop nutrition. Varieties of the vegetables performing well with scarce resources and having resistance to major biotic and abiotic stresses including pest and diseases should be selected for the cultivation. These varieties can be grown to reduce the cost of cultivation in organic vegetable production. These varieties can meet the standards of organic vegetable cultivation.

KEYWORDS

- **organic vegetable production**
- **conversion period**
- **organic manures**
- **biogas slurry**
- **weed management**

REFERENCES

Annonymous Indian Horticulture Database 2014. National Horticulture Board. Ministry of Agriculture, Govt. of India, 2015.

Dahama, A. K. *Organic Farming for Sustainable Agriculture*; Agribios: Jodhpur, India, 2002.

Maity, T. K.; Tripathy, P. Organic Farming of Vegetables in India: Problems and Prospects. Department of Vegetable Crops Faculty of Horticulture Bidhan Chandra Krishi Viswavidyalaya, 2004; pp 1–23.

Palaniappan, S.P.; Annadurai, K. *Organic Farming Theory and Practice*; Scientific Publications: Jodhpur, India, 1999.

Shiva, V.; Pande, P.; Singh, J. *Principles of Organic Farming (Renewing the Earth's Harvest)*; Navdanya Publisher: New Delhi, 2004.

Lokanadhan, S.; Muthukrishnan, P.; Jeyaraman, S. Neem Products and Their Agricultural Applications. *J. Biopest* 2012, *5*, 72–76.

Yadav, A. K. Organic Farming (Concept, Scenario, Principals and Practices): Relevance, Problems and Constraints. National Project on Organic farming; Department of Agriculture and Cooperation, Govt of India, 2005.

CHAPTER 21

Traditional Kalanamak Rice-Based Organic Production System in Northeastern Uttar Pradesh

B. N. SINGH

Centre for Research and Development (CRD), Gorakhpur, Uttar Pradesh, India; E-mail: baijnathsingh08@gmail.com

ABSTRACT

Kalanamak is one of the finest quality aromatic rice grown in India. This scented variety has been cultivated since 600 BC i.e., since and probably before the Buddhist era. There is need to promote small grain aromatic traditional Kalanamak notified variety KN-3 rice under organic farming cultivation for domestic consumption and export like Basmati. Kalanamak being a traditional tall variety, is more suitable for organic farming as its nitrogen requirement is low. APEDA, NABARD, and other government agencies should support for its cultivation, processing, certification, bagging, labeling, and marketing. The Government of India has started Participatory Guarantee System (PGS) in which groups of farmers certify each other's production and product. For the first 3 years, a "Green certificate" is given and later, regular "Blue Logo" is given with organic certification. Organically grown seed should be used for organic cultivation of Kalanamak rice. The state government's initiative should be linked with purchase, millers, and marketing, so that its cultivation is profitable and no product mixing is done.

Organic Crop Production Management: Focus on India, with Global Implications, D. P. Singh, PhD, H. G. Prakash, PhD, M. Swapna, PhD, & S. Solomon, PhD (Eds.)

21.1 INTRODUCTION

Kalanamak is a black husk (*Kala*) colored traditional rice variety, which is tall, and has lodging type plant. It is a photosensitive variety, with medium slender aromatic grains. It can yield 3.0 t/ha of paddy. It is grown in the Northeastern Uttar Pradesh, and its area has reduced from 50,000 ha in pre-HYV era to mere 2000 ha at present (Chaudhary et al., 2017). Kalanamak flowers in third and fourth week of October, irrespective of sowing/transplanting. Kalanamak fossilized/carbonized rice grains have been reported from Aligarhwa, Dist. Siddharthnagar, Uttar Pradesh located at India–Nepal border (Singh et al., 2003, 2005). The Kalanamak in this region has been grown since Lord Buddha period, that is, sixth century B.C. During the British *Raj,* efforts were made to produce Kalanamak *variety* for export to England from Alidpur, Birdpur, and Mohana areas of Siddharthnagar district. From Uska Bazar Mandi, the Kalanamak rice has been exported through Dhaka via sea route.

It has been given GI (Geographical Indication) tag in Zone 6 of the Northeastern Uttar Pradesh, spreading between Ghaghra and Rapti rivers from Bahraich in the west to Deoria in the east. GI was registered for Kalanamak in August 2014 by the Government of India. It consists of 11 districts, namely, Bahraich, Barabanki, Balrampur, Basti, Gonda, Sant Kabir Nagar, Siddharthnagar, Maharajganj, Kushinagar, Gorakhpur, and Deoria. This variety now has legal protection and cannot be grown other than the notified areas under GI for export purposes. The white-milled rice during cooking gives pleasant aroma in kitchen and surroundings, and it also has good taste, quality, and palatability. Aroma development is influenced by both genetic factors and the environmental factors. A single recessive gene has been reported to control aroma, but there are also reports of multiple factors/polygenes, controlling aroma. Aroma is due to certain chemicals present in the endosperm and such chemicals are also found in the vegetative parts, which emit aroma in the standing crop in some cases even at early stages of growth. Aroma gene is $badh_2$ and the main aromatic compound is 2-acetyl-1-pyrroline (2-AP) for fragrance. Aroma is quantified through peaks by gas chromatography. In addition to 2-acetyl-1-pyroline, there are 100 other volatile compounds, including hydrocarbons, acids, alcohols, aldehydes, ketones, esters, phenols, and some other compounds, which are associated with the aroma development in rice (Singh et al., 2005).

Due to its lower yield than the semidwarf modern cultivars, its area has reduced, and farmers only grow for their own consumption. In market, its grains fetch three times higher price, Rs. 70–80 per kg than coarse and medium grain is sold at Rs. 20–25 per kg. Kalanamak has also shown tolerance to saline and saline–alkaline soils. The plant height also reduces under saline soils, so that it does not lodge. It is also rich in iron and zinc, which is good for pregnant and lactating mothers, and young children. Being a late duration photosensitive variety, Kalanamak is also suitable for double transplanting (Kalam), after damage of first transplanted crop by flood, the seedlings from first crop can be re-transplanted once the flood recedes. Selected from traditional Kalanamak types, KN-3 (Kalanamak-3), a new variety notified in August 2010. Kalajoha is a similar variety from Assam, which is being promoted organically for export through APEDA.

On-station and on-farm trials are being carried out in the five different districts of Eastern Uttar Pradesh, Bihar, and Jharkhand. The site of experiments is located at 26°42′45.50″N latitude, 83°36′36.6″E longitude and 83 m above the mean sea level. CRD is also working on developing Integrated Crop Management (ICM) package for organic cultivation of traditional Kalanamak variety KN-3 at its farm at Gaunar Uaraha, in Gorakhpur district. *Dhaincha* (*Sesbania aculeata*)–Rice–Wheat is the cropping system being followed in 2000 m^2 plot. Nucleus seed of KN-3 is being produced, and breeder seed is also being multiplied by using mustard cake, Azolla, blue-green algae, vermicompost, liquid biofertilizer (*Azotobactor*, Zinc, and Potash) as plant nutrients. For organic farming of Kalanamak rice, *neem* (*Azadirachta indica*) is planted as biofence, while *neem* leaves and seeds, mustard cake, and *neem* cake are used as biopesticide. For the cultivation of Kalanamak rice, 50 landraces from 29 villages of Siddharth Nagar district were selected in 2014. Since Kalanamak variety is affected by panicle and neck blasts (*Magnaporthe grisea*) and stem borers, proper screening is done for their resistance/tolerance. Dwarf Kalanamak developed at Genetics Division, IARI, New Delhi, is also being evaluated at CRD farm for increasing its yield and profitability. Due to low yield, there is a plan to promote organic Kalanamak rice to be sold at Rs.100 per kg.

21.2 MATERIAL AND METHODS

For improving yield of KN-3 variety, research on four themes is being carried out at CRD, Gorakhpur. These are collection and evaluation of Kalanamak traditional genetic resources (KTGR), nucleus, and breeder seed production of KN-3 variety, development of ICM practices for traditional organic Kalanamak, and evaluation of Dwarf Kalanamak for its yield potential and grain quality traits. Wheat varietal trials are also conducted for selecting Karnal bunt resistant cultivars for organic cultivation. As Kalanamak is harvested late in November end or early December, following efforts were made to select suitable variety for surface seeding, so that wheat sowing may not be affected due to late seeding.

- **Collection and Evaluation of Kalanamak Traditional Genetic Resources**: In March–April 2014, 50 collections of seeds were made from 29 villages of Birdpur block of Siddharthnagar district. These collections were mostly from farmers' house and a few samples from each village. Names and mobile numbers of farmers and villagers whose samples were collected were also recorded, so that if they need in future, seeds can be supplied back to them. These collections were purified during *kharif* 2014 by harvesting true to type 20 panicles as bulk. Collections were evaluated for their grain yield, agronomic traits, and quality traits in augmented design with KN-3, Pusa-1176, Pusa Basmati 1, and Badsahbhog. Due to variation in flowering dates and grain type, five collections were selected. During 2016 *kharif*, these collections were further evaluated in RCBD, and further selections were made based on yield and quality traits. During 2017 *kharif*, 11 KTGR collections are being evaluated to select better lines in quality than KN-3. Three collections, namely, KTGR2, 9, and 26 that are shorter in height and nonlodging type were found better in grain yield, cooking, and quality traits than KN-3. In early 2018, these collections will also be preserved in PPV and FRA, New Delhi, with a tag of names of their farmers for further evaluation and promotion.
- **Nucleus and breeder seed production**: To produce uniform quality rice product, KN-3 is the only released and notified variety. For organic farming, the nucleus seeds are produced by taking 100 panicles from different plants and grown as head rows in 5 m plots. Rouging is done at vegetative, flowering, and grain maturation stages, and bulk is grown for breeder seed production.
- **Development of ICM for organic Kalanamak production**: Use of *Dhaincha* for soil incorporation, mustard and *neem* cake, Azolla, BGA, liquid fertilizers are being used to produce organic traditional Kalanamak (Table 21.1).

- **On-farm organic Kalanamak cultivation**: Farmers are being encouraged to grow organic Kalanamak variety KN-3 for the verification of ICM package of practices, grain quality, aroma, and yield. One such trial was conducted by Col. (Retd.) Viswa Nath Singh at a village in Block Siswa Bazar of District Maharajganj, Uttar Pradesh, in nearly 0.90 acre plot. He had grown *Dhaincha*, followed by KN-3, and wheat cropping system using *neem* and mustard cake as biopesticides.
- **Evaluation of Dwarf Kalanamak**: Yield trials are being conducted since 2015 *kharif* season to select dwarf Kalanamak variety in collaboration with IARI, New Delhi. SL-03 is found to be a promising dwarf Kalanamak variety with yield potential of 4.0 t/ha. It has similar aroma as traditional Kalanamak, and hold promise for cultivation under organic farming.
- **Evaluation of wheat varieties**: Varietal trials were conducted since 2014, to select foliar blight, and Karnal bunt resistant high-yielding varieties. Foliar diseases are major problems and tolerant varieties with higher yields are selected.

21.3 RESULTS AND DISCUSSION

Although Kalanamak is a traditional aromatic black husk variety, genetic variation for plant height, days to 50% flowering, grain yield, effective panicles per m^2, grain length, and grain type, and resistance/tolerance to various biotic and abiotic stresses has been reported (Chaudhary et al., 2014). In the present study, variation for grain type was also observed, and efforts are on-way to select better grain quality types than KN-3, the only notified variety selected from traditional Kalanamak type (DAC & MOA, 2010). Cooking and eating quality tests will be carried out for this purpose through collections from the different villages of Birdpur area of Siddharthnagar in the Eastern Uttar Pradesh.

For uniform grain quality, it is important to produce good quality seed, so that cooked rice looks uniform, and consumers prefer it for their household consumption. As KN-3 is a notified variety, its nucleus seed and breeder seeds are produced organically by CRD, so that it can be promoted on large-scale promotion in national and international market.

TABLE 21.1 Integrated Crop Management for Production of Organic Traditional Kalanamak.

S. no.	Month	Field operations	Detail activities
1	May second fortnight	Sowing of *Dhaincha* in main field	i. Deep ploughing of main field by soil turning plough after pre-monsoon rains.
			ii. Procure organic KN-3 seed, *Dhaincha*, *neem* cake, mustard cake, and liquid biofertilizer
			iii. Seed rate *Dhaincha* (*Sesbania aculeata*): 25 kg/ha.
			iv. Plough the field by rotavator. Apply FYM at 10 t/ha or *neem* and or mustard cake at 200 kg/ha for *Dhaincha* crop. Mix the two in equal proportion. Liquid biofertilizer can also be mixed with cakes.
2	June end	Nursery sowing	i. Nursery seeding of KN-3 in June end.
			ii. Seed rate: KN-3: 15 kg/ha.
3	July Second fortnight	Transplanting	i. Incorporate 45–60 days *Dhaincha* crop 1 week before transplanting by rotavator.
			ii. Transplant 20–30 days old KN-3 seedlings.
4	August	First weeding	By hand
5	August	First top dressing	Apply Vermicompost at 2 t/ha or *Neem* and or Mustard cake at 100 kg/ha.
6	September	Second weeding	By hand
7	September	Second top dressing	Apply *neem* and or Mustard cake at 100 kg/ha.
8	October	Irrigation	i. In lack of rain, mild irrigation should be provided to avoid moisture stress.
			ii. Flowering of Kalanamak at third week of October.

TABLE 21.1 *(Continued)*

S. no.	Month	Field operations	Detail activities
9	November	Paddy crop	Grain filling and dough stage
10	December first week	Harvesting	2.5 t/ha(1.6 t/ha milled rice yield)
11	December First fortnight	Wheat seeding	Sowing of wheat seed at 125 kg/ha by Zero till or surface seeding or 100 kg/ha after conventional tillage. Use of mustard mixed with *neem* cake as basal dose at 100 kg/ha. Cake is mixed with potash-solubilizing liquid biofertilizer. Presowing irrigation should be given for germination.
			ii. In surface seeding, basal dose of cake is applied 15 days after sowing.
12	January	First irrigation in wheat	Irrigation after use of vermicompost as first top dressing. Zinc solubilizing liquid biofertilizer is mixed with vermicompost or with *neem* and mustard Cake at 100 kg/ha.
13	February	Second irrigation in wheat	Irrigation after use of mustard mixed with neem cake at 100 kg/ha as topdressing. *Azotobacter* liquid biofertilizer is mixed with cake.
14	March	Third irrigation in wheat	Irrigation and use of vermicompost as top dressing.
15	April second fortnight	Harvesting	Wheat crop is harvested by mid-April.
			Wheat yield : 3.0 t/ha

Package of practicing organic cultivation needs to be developed and continuously improved, so that its maximum yield potential can be obtained in the farmers' field (Chaudhary et al., 2008a, 2008b, 2008c, 2014, 2017).

A target of yielding 2.5 t/ha and beyond has been fixed to realize the KN-3 yield so that 1.6 t/ha milled rice can be obtained. Since organic cultivation will always be in cropping system mode, *Dhaincha*–rice–wheat cropping system has been developed. Azolla, *Azotobacter* liquid biofertilizer, potash-solubilizing liquid biofertilizer, and zinc-solubilizing liquid biofertilizer from IARI, New Delhi are also being used along with mustard and *neem* cake. Vermicompost and blue-green algae are also being added in field (Table 21.1). On-farm trial by Col. (Retd.) Viswa Nath Singh at Siswa Bazar, Maharajganj district, during 2017 *kharif* season produced a yield of 1.86 t/ha. After 13 days of transplanting, and 15 days water lodging, the crop was completely submerged for 8 days.

As the grain yield of traditional Kalanamak variety KN-3 is low, efforts are underway to reduce the height, and improve the nonlodging trait with higher yield. IARI, New Delhi, has developed some guidelines through hybridization and backcrossing to give rise to dwarf Kalanamak. These guidelines are being evaluated at CRD farm, Gorakhpur. Efforts are under way to harvest a paddy yield of 4.0 t/ha and beyond with good aroma.

In wheat varietal trial, foliar blight tolerant varieties with higher yield are being selected. During 2016–2017 *rabi* season, analyses were made for Carnal bunt infection, but no Karnal bunt was found in any samples. For Karnal bunt control, Propiconazole (Tilt as trade name) is the only effective fungicide. So, further efforts are needed to select resistant varieties. PBW-677 for normal sowing, and BHU Genhu-6, a high-zinc content variety were found suitable for late seeding, resistant to Karnal bunt, and tolerant to foliar diseases and higher yield.

21.4 OUTLOOK

There is need to promote small grain aromatic traditional Kalanamak notified variety KN-3 rice under organic farming cultivation for domestic consumption and export like *Basmati*. Kalanamak being a traditional tall variety, is more suitable for organic farming as its nitrogen requirement is low. APEDA, NABARD, and other government agencies should support for its cultivation, processing, certification, bagging, labeling, and marketing. Rice mills should be identified where farmers can send their harvest. Minimum support price for Kalanamak paddy has to be announced, so that farmers are not cheated after harvest. A minimum support price (MSP) of INR 3000/- per quintal should be fixed for organic traditional Kalanamak, so that farmers are encouraged to grow it. The MSP will reduce the adulteration, and farmer's profitability can be restored. The Government of India has started Participatory Guarantee System (PGS) in which groups of farmers certify each other's production and product. For the first 3 years, a "Green certificate" is given and later, regular "Blue Logo" is given with organic certification. Organically grown seed should be used for organic cultivation of Kalanamak rice. The state government's initiative should be linked with purchase, millers, and marketing, so that its cultivation is profitable and no product mixing is done.

KEYWORDS

- **breeder seed production**
- **dwarf kalanamak**
- **organic traditional kalanamak**

REFERENCES

Chaudhary, R. C.; Mishra, S. B.; Dubey, D. N. Scented Rice Variety Kalanamak and Its Cultivation for Better Quality and High Yield. *Rice India* **2008a,** *18* (8), 23–25.

Chaudhary, R. C.; Mishra, S. B.; Dubey, D. N. Cultivation of New Variety of Rice "Kalanamak". *Indian Farming* **2008b,** *58* (6), 21–24.

Chaudhary, R. C.; Mishra, S. B.; Dubey, D. N. Scented Rice Variety of New Kalanamak and Its Cultivation Package (in Hindi). *Kheti* **2008c,** *6* (6), 10–12.

Chaudhary, R. C.; Gandhe, A.; Mishra, S. B. *Revised Manual on Organic Production of Kalanamak Rice* (English and Hindi). PRDF: Gorakhpur, India, 2014; p 64.

Chaudhary, R. C.; Kumar, S.; Mishra, S. B. Dwarf Kalanamak: A New Scented Rice Variety (in *Hindi*). *Kahaar Magazine* **2017,** *4* (1–2), 9–13.

Department of Agriculture and Cooperation & Ministry of Agriculture (DAC & MOA). Notification of Kalanamak 3 (KN 3) (IET 21268) for State of Uttar Pradesh. The Gazette of India, New Delhi. http://www.seednet.gov.in (accessed Aug 31, 2010).

Singh, U. S.; Singh, N.; Singh, H. N.; Singh, O. P.; Singh, R. K. Rediscovering Scented Rice Cultivar Kalanamak. *Asian Agri-History* **2005,** *9* (3), 211–219.

CHAPTER 22

On-Farm Production of Quality Inputs for Organic Production of Horticultural Crops

R. A. RAM

ICAR-Central Institute for Subtropical Horticulture, Rehmankhera, Lucknow 226101, Uttar Pradesh, India
E-mail: raram_cish@yahoo.co.in

ABSTRACT

It is now an established fact that the on-farm produced inputs or organic matter could be a potent source of nutrients and beneficial microbes which could help in improving the soil fertility, crop productivity, and produce quality. These bioenhancers could be a potential alternative for fertigation which is becoming common in most of the crops. But care should be taken that bio enhancers that are used in limited quantities should meet the entire nutrient requirement of the crops. Microorganisms and organics are generally helpful in catalyzing quick decomposition of organic wastes into humus, hence, incorporation of enough biomass preferably the combination of monocot and leguminous crops along with supplementation of animal wastes would be advantageous for quality humus production which is the very essential component for improving soil fertility and crop yield. Humus along with organic manures and regular application of bioenhancers and biopesticides can address many challenges of horticulture production and will be helpful to show a way for sustainable production through organic resources.

Organic Crop Production Management: Focus on India, with Global Implications, D. P. Singh, PhD, H. G. Prakash, PhD, M. Swapna, PhD, & S. Solomon, PhD (Eds.)

22.1 INTRODUCTION

Use of agrochemicals was initiated in commercial cultivation of horticultural crops after the advent of green revolution. Significant increase in production, productivity, and area expansion under horticulture crops has occurred during the past two decades under the influence of chemical intensive systems. Imbalanced use of chemical fertilizers, especially nitrogenous fertilizers, has resulted in some regions manifesting adverse effects on the whole environment through polluting groundwater and soil resources. Soil quality, especially that of organic matter and micronutrients deficiencies are becoming ubiquitous, threatening sustainability, and quality of produce impacting nutritional security. Further, indiscriminate use of pesticides has led to the development of resistance in pests to pesticides, while destroying irretrievably the beneficial ones, namely, honeybees, pollinators, parasitoids and predators, besides causing harmful pesticide residues in the end product adversely impacting productivity and food safety. Increasing awareness about conservation of ecosystem as well as of health hazards caused by the excess use of agrochemicals has brought a major shift in consumer preference toward food quality, particularly in the developed countries. Organic foods consumers are increasingly globally considering it safe and hazard-free. The organic production of horticultural crops leads to excellent taste, wide varieties with nutritive, and therapeutic values, which has immense scope in domestic and export markets. Preliminary attempts for organic production in different crops in various regions have shown spectacular response. It is imperative to note that with proper follow-up, all the organic production systems, that is, *Rishi Krishi*, biodynamic farming, *Panchagavya Krishi,* Natueco *Krishi*, Homa organic farming and natural farming have indicated a high-quality production without the use of agrochemicals. There is need to integrate a few compatible techniques from these established systems to develop package of practices which can easily be adopted by the common growers to ensure sustainable organic production. Availability of quality organic inputs in the markets are not assured and they do not meet the standard set by the BIS. Emphasis should be given to on-farm production of quality organic inputs for cost-effective and sustainable production. There are few organic farming practices in which all the inputs, namely, composts, bioenhancers, and biopesticides are produced with locally available materials. Composts are rich source of nutrients, bioenhancers are rich source of beneficial microbes and biopesticides are effective source for the management of various insect pests.

22.2 ON-FARM PRODUCTION OF ORGANIC INPUTS: IMPORTANCE OF ORGANIC MATTER

The term "organic matter" is generally used for the dead and decomposing remains of living things like plant and animal waste, and manures. These crucial components of the soil provide food for soil fauna specifically for plants which is a primary source of nitrogen. Soil without organic matter may be sterile dust only. Organic matter of soil is gradually decomposed by soil creatures and by natural oxidation which is replenished in the natural cycles of life and death. Humus which is the end product of decomposed organic matter works as a valuable reservoir of water and plant nutrients and as a constituent of the soil structure. Various organic matters/inputs that are produced on-farm have been described in the next sections

22.2.1 COMPOST

Crop waste and organic residue burning is the most common practices in today's farming system. These organic wastes can easily be converted into good quality composts, such as biodynamic, vermi, NADEP, and vermicompost, and others for soil health management.

22.2.1.1 BIODYNAMIC COMPOST

Biodynamic compost, a good manure, is an important source of nutrients for various crop production. Green (nitrogenous material) and dry leaves (carbonaceous material) in the ratio of 40:60 are used for making biodynamic compost. The process of decomposition is enhanced by the integration of cow dung slurry and BD-502-507 in the compost. Depending upon the prevailing temperature and moisture in the atmosphere, organic wastes are decomposed by the composition of air, moisture, and warmth. Depending upon the prevailing temperature and moisture in the atmosphere, the compost is ready for use after 100–120 days (Fig. 22.1).

FIGURE 22.1 Biodynamic compost.

22.2.1.2 VERMICOMPOST

Earthworms are natural creatures of the soil for quick recycling of organic wastes which help in effectively harnessing the advantageous soil microflora, reduce soil pathogens, and convert organic wastes into valuable products, such as biofertilizers, biopesticides, enzymes, vitamins, growth hormones, antibiotics, and proteinous biomass. Vermicompost can easily be produced by using livestock's dung and locally available organic waste (Fig. 22.2). If vermibeds are slanted, say 5–10%, then vermiwash can also be obtained without any expenditure. In an experiment, application of biodynamic compost (30 kg/ tree) +bioenhancers (CPP 100 g, BD–500 and BD-501 as soil and foliar spray) in 30 years old tree of mango cv. Mallika yielded 160 kg fruit per tree with improvement in fruit quality TSS (26.36 ^{0}Brix), total carotenoids (6.40 mg/100 g) and fluorescence recovery after photobleaching (FRAP) (74.478 µmol/L) with BC ratio of 5.10 (Ram et al., 2017).

FIGURE 22.2 Vermicompost bed and vermicompost ready for use.

22.2.1.3 NADEP COMPOST

Sri Narain Deorao Pandari Pande, a farmer of Maharashtra, has initiated this aerobic method of composting. A 2 × 3.3 × 1.25 m brick aerobic structure is built at elevated portion of the farm and layering of organic wastes, cow dung, and fertile soil are done in the structure. After completion of filling of organic wastes, 45 cm above the structure, the heap is covered by cow dung and mud. Due to incorporation of fertile soil, proper ratio of green and dry biomass in aerobic composting process, nutritional status of the end product are better than the FYM. For composting, organic wastes, such as cow dung, wheat/paddy straw, dry/green grasses, and weeds) along with virgin farm soil should be used (Fig. 22.3).

FIGURE 22.3 NADEP compost.

22.2.1.4 MICROBE-MEDIATED COMPOST

In Japan, Prof. Teruo Higa developed an effective microorganism consortium during the early 1980s. Effective microorganism is a group of beneficial microbes that acts as microbial inoculants in the soil and helps in developing congenial environment for the plants. There is lactic acid and photosynthetic bacteria, filamentous fungi, yeast, and actinomycetes, and others, in the microbe-mediated compost. These are anaerobic and aerobic in nature and can survive in saline as well as acidic conditions. These effective microbes are used for quick conversion of dung and farm wastes in to good quality compost.

In different parts of India, several methods of composting have been standardized. Any of this can be practiced as per the facilities available at the farm. These composts contain more nutrients than common farmyard manure (Table 22.1).

TABLE 22.1 Nutrient Analysis of Various Organic Preparations (% on Dry Weight Basis).

S. no	Compost	N (%)	P (%)	K(%)	Ca(%)	Zn (ppm)	Cu (ppm)	Fe (ppm)	Mn (ppm)
1	Vermicompost	2.15	1.29	0.53	1.72	168	61	3545	252
2	Biodynamic compost	1.68	0.17	1.23	1.20	96	45	357	3352
3	NADEP compost	0.98	0.35	1.00	1.25	162	56	430	230
4	Microbe-mediated compost	1.54	0.51	1.06	1.35	140	45	433	275
5	Farmyard manure	0.70	0.19	0.37	0.24	75	34	222	235

Source: Ram and Pathak (2016).

22.2.2 BIOENHANCERS

Organic preparations prepared by quick decomposition of animal and crop residues over specific period are known as bioenhancers which are concentrated medium in solid or in liquid form and are valuable source of microbial consortia, micronutrients, macronutrients, and plant growth promoting substances. Bioenhancers are used for treating seed/seedlings, resulting in increased decomposition of organic materials and enriched soil with improved plant vigor. Bioenhancer could also be a potent tool for fertigation in various crops (Pathak and Ram, 2012).

Characteristics of Bioenhancer

- It is a major source of macronutrients as well as micronutrients.
- It contains plant growth promoting factors.
- It also contains plant immunity promoters.
- It has pesticidal property.
- Results may vary with the use of inputs and preparation methods.

- It is used for seed/seedling treatment, improving decomposition, soil fertility, and productivity.
- Very potential input for use in fertigation.

Vrikshayurveda is the science of plant life which deals with nourishing plants with liquid bioformulation (prepared with crop and cattle biproducts). There are several versions (e.g., *Brahatasamhita* of *Varahamira*, Surpala's *Vrikshayurveda*, and *Upavanavinoda* of *Sharangadhara*), which provide information on irrigation methods using water mixed with herbal products obtained from different plant species and animal products to increase crop production (Sadhale, 1996). According to Sanskrit dictionaries, *Kunapajala* is a combination of two words "Kunapa" which means dead or decaying matter and "Jala" means water derived from this dead matter. Application of *Kunapajala* promotes growth, flowering, and crop yield and is also used for plant protection measures (Majumdar, 1935). For using as a common booster of plant vigor, *Kunapajala* is prepared by boiling of animal flesh, fat and marrow, such as deer, goat, sheep, pig, fish in water, placing it in earthen pot, and adding milk, ghee, hot water, powders of sesame oil cake, black gram boiled in honey and decoction of pulses (Nene, 2007). *Kunapajala* is used in the plants for improving their growth and production. Due to unavailability of supporting data/evidences on implications of *Kunapajala,* its use in present scenario is a cumbersome process. In a study, Sarkar et al. (2014) have reported that *Panchagavya* and *Kunapajala* individually as well as in combination proved their efficacy in promoting the growth and yield of vegetables crops. Although there was variations in the level of efficacy of individual treatments but *Panchagavya* with *Kunapajala* has emerged as the best for efficient utilization of leaf nitrogen, efficient photosynthetic activity, and improving the yield. The preparation of *Kunapajala* is a complex procedure, the other bioenhancers which are easy to prepare are being used by majority of the Indian farmers, and have been discussed in the next section.

22.2.2.1 COW DUNG

Since the period of Kautilya (300 BC), cow dung has been used for seeds dressing, cut ends plastering of vegetatively propagated sugar cane, plastering of wounds, and sprinkling of diluted solution on crops. ICAR-Central Institute for Subtropical Horticulture, Lucknow, India, has identified the presence of four potent strains of *Bacillus subtilis*, which have shown strong antipathogenicity against several diseases in mango, guava, and papaya rots (Pathak et al., 2009). Actinomycetes as *Streptosporangium pseudovulgare* isolated form cow dung has shown antipathogenic potential against *Colletotrichum gloeosporioides* (anthracnose pathogen) and *L. theobromae* (gummosis, stem end rot and die back pathogens) (Garg et al., 2003, 2012). Microbial analysis of fresh cow dung showed that it contains nitrogen-fixing, phosphorus-solubilizing bacteria, and other beneficial microbes in plenty (Table 22.2) (Ram et al., 2017).

TABLE 22.2 Microbial Population in Fresh Cow Dung.

S. no.	Microbial type	Population (cfu/g)
	Actinomycetes	3.96×10^6
	Azotobacter	3.48×10^5
	Azospirillum	1.43×10^5
	Rhizobium	7.5×10^8
	Pseudomonas	1.44×10^8
	Phosphate-solubilizing microbes	1.9×10^8

22.2.2.2 COW URINE

In India, cow urine (*gau-mutra)* has been used since time immemorial. It is a liquid with innumerable therapeutic values. Cow urine is an efficient source of curing several incurable diseases in human beings as well as plants. Number of bioenhancers and biopesticides are being prepared from cow urine which are effectively improving soil fertility, quick decomposition of organic wastes, and management of several insect pests in various groups of crops.

22.2.2.3 COW HORN MANURE (BD-500)

Cow horn manure is basically decomposed cow dung which has been found very useful for improving soil fertility and renewal of degraded soils. Fresh cow dung filled in cow horn is buried in rich soil (Fig. 22.4). Cow horns are buried in September–October and taken out in March–April. During the invention of biodynamic farming, this manure has been generally the first preparation used. BD-500 is a basic biodynamic preparation which can be sprayed in the field. During dung decomposition in winter, the cow horn has the potential to absorb life energies.

The spraying of these specially prepared preparation helps in the development of the plants through vitalizing the soil, enhancing seed germination and root formation. It has been suggested to use it four times in a year. Autumn (October) is the best period for spraying. Spray should be repeated during the spring (February and March). About 25 g of BD-500 is mixed in 13.5 L of water in plastic bucket by making vortex in clockwise and anticlockwise movement for 1 hour in the evening for spraying. During the process of stirring clockwise and anticlockwise, cosmic forces are absorbed by making the water active and oxygenation of the preparation. "Dynamization" is a process of stirring little material in large quantity of water which helps in transferring the forces of energy from the preparation to the water itself. By natural brush or tree twigs, we can spray the solution. It needs to fall on the soil in droplets BD-500 should be sprayed in the evening at the field preparation time during descending period of the Moon. Microbial analysis of BD-500 (Ram et al., 2018) showed that it contains beneficial microbes responsible for improving the soil fertility (Table 22.3).

FIGURE 22.4 Preparation of BD-500.

TABLE 22.3 Isolation of Microbes from BD-500.

S. no.	Type of microbes	Microbial population (cfu/g)
	Actinomycetes	22.6×10^6
	Pseudomonas	0.7×10^6
	Phosphate- solubilizing microbes	38.6×10^5
	Azotobacter	33.2×10^5
	Azospirillum	53.7×10^5
	Rhizobium	6.0×10^7

22.2.2.4 BD-501

BD-501 is prepared by incubating silica powder in cow horn during April–October in the ascending period of Moon. Use of BD-501 improves the photosynthetic process in the leaf by enhancing chlorophyll content. It strengthens the plant and improves the quality of produce and enhances the fruit and seed development. For accruing the maximum benefits, BD-501 is to be applied once at the time of germination at the stage of four-leaves. It should be applied second time at flowering and fruit growth stage. It may be sprayed on the foliar parts of plants in the form of fine "mist" at the time of sunrise and it works best when the Moon is opposite to Saturn. The quality of fruit in mango is generally improved significantly by the use of BD-501 (Pandey et al., 2003). Few types of the microbes isolated from cow horn silica (BD-501) are presented in Table 22.4 (Ram et al., 2018).

FIGURE 22.5 Preparation of BD-501.

TABLE 22.4 Microbial Analysis of BD-501.

S. no.	Type of microbes	Microbial population (cfu/g)
	Actinomycetes	3.3×10^6
	Pseudomonas	4.9×10^6
	Phosphate-solubilizingmicrobes	24.2×10^5
	Azotobacter	49.0×10^5
	Azospirillum	0.7×10^5
	Rhizobium	6.0×10^7

22.2.2.5 COW PAT PIT

Cow pat pit (CPP), which is also known as "soil conditioner," is a special field preparation, prepared from fresh cow dung from lactating and grazing cows. Cow dung is decomposed along with egg shells powder (calcium) and bentonite (clay) dust mixed and placed in pit size of 3 × 2 × 1.5' (Fig. 22.6). Two sets of BD-502-507 preparations are added for enhancing the composting process. In a study, CPP, a concentrated source of beneficial organisms, revealed the highest bacterial load (4.8×10^{6}) per g, *Rhizobium* (1.9×10^{6}), *Azospirillum* (0.2×10^{6}), *Azotobacter* (8.0×10^{5}) and fungi (2.5×10^{6}) (Ram et al., 2010). The highest amount of *B. subtilis* (1.9×10^{6}) responsible for disease tolerance is also present in CPP (Proctor, 2008). Microbes isolated from the CPP are presented in Table 22.5 (Ram et al., 2018).

FIGURE 22.6 Preparation of cow pat pit.

TABLE 22.5 Microbial Analysis of Cow Pat Pit.

S. no.	Microorganisms	Microbial population (cfu/g)
	Actinomycetes	13.1×10^{6}
	Pseudomonas	6.8×10^{6}
	Phosphate-solubilizing microbes	8.4×10^{5}
	Azotobacter	28.6×10^{5}
	Azospirillum	224.0×10^{5}
	Rhizobium	310.0×10^{7}

22.2.3 BIODYNAMIC LIQUID MANURE/PESTICIDES

Biodynamic liquid pesticides are prepared by locally available materials, such as urine, cow dung, and leaves of leguminous tree, *neem* leaves, fish waste, caster leaves, and other medicinal plant parts. Besides cow dung and cow urine, one set of biodynamic preparations (502–507) are also added.

The crops are sprayed with the liquid manures for improving the vigor and quality produce. Liquid manure/pesticide generally requires 8–10 days for maturation, on an average. In an experiment, mango hopper management with biodynamic liquid pesticide was effectively done. Before spray, the hopper population was 3.07 hoppers/panicle, which helped in reducing the insect population up to 15th SMW (Standard Meteorological Week) with 0.95 hoppers/panicle. Second spray was done at 14th SMW, resulting in reducing the hopper population by 0.4 hopper/panicle up to 19th SMW (Ram et al., 2017).

FIGURE 22.7 Preparation of biodynamic liquid pesticide.

22.2.3.1 PANCHAGAVYA

Panchagavya is a rich source of nutrients, gibberellins, auxins, and microbial fauna. For enriching the soil to induce plant vigor with quality production, *Panchagavya* acts as a strong tonic which is prepared using five different products obtained from cow (cow dung, cow urine, cow milk, cow *ghee*, cow milk curd) along with banana, coconut water, and palm wine (Fig. 22.8). It gets ready in 18 . *Panchagavya* is highly effective for all types of plants, milch animals, goat, poultry, fish, and pet animals. In a number of fruits, vegetables, and floriculture crops, such as mango, guava, acid lime, banana, spice turmeric, jasmine cucumber, spinach, etc., its application has produced excellent results. The application of *panchagavya* on chilies produces dark green colored leaves within 10 days (Sreenivasa et al., 2009). The mixed culture of naturally occurring beneficial microbes are present in *Panchagavya*. Most of the lactic acid bacteria (*Lactobacillus*), photosynthetic bacteria (*Rhodopsuedomonas*), yeast (*Saccharomyces*), actinomycetes (*Streptomyces*), and certain fungi (*Aspergillus*) are the major microorganism present in *Panchagavya*. There are several naturally occurring beneficial microorganisms in *Panchagavya,* some of which are nitrogen fixers and P-solubilizers (Swaminathan, 2005, Sreenivas et al., 2011), this can be considered as an ideal organic growth promoter. But this is suggested to apply it within 1 month of its preparation to achieve better results (Patnaik et al., 2012). Application of *Panchagavya* has been found more profitable than recommended dose of fertilizer and agrochemicals spray. Kumar et al. (2015) reported that among the different treatments tested against the management of insect pest in teak (*Tectona grandis*), 7% and 5% diluted *Panchagavya* application were found efficient for the management of insect pests. Seven percent diluted *Panchagavya* has been found a very cheaper and affordable organic substance for the tree growers in a cost-effective analysis.

Systematic microbial analysis of *Panchagavya* from 0 to 25 days suggested that *Panchagavya* contains maximum microbial load on 18th day of preparation. Therefore, use of *Panchagavya* on the 18th day will be more effective than other days (Ram et al., 2017) (Table 22.6).

FIGURE 22.8 Preparation of *Panchagavya.*

TABLE 22.6 Different Microbial Populations in *Panchagavya.*

S. no.	Type of microbe	Multiplication factor	Microbial population (cfu/mL) after days of preparation							
			0	3	6	9	14	18	20	25
	Actinomycetes	10^6	0.19	0.30	1.70	1.80	1.40	2.20	8.00	7
	Pseudomonas	10^6	1.89	1.20	1.42	2.40	6.00	47.00	57	3.1
	Rhizobium	10^6	1.48	6.74	1.55	2.05	1.92	2.43	4.14	2.42
	Phosphate-solubilizing microbes	10^6	0.29	0.15	0.15	0.16	1.40	3.20	2.42	2.13
	Azotobacter	10^6	4.50	3.93	0.01	0.092	0.07	0.14	0.15	0.15
	Azospirillum	10^5	1.12	0.29	0.06	0.49	0.72	1.03	1.60	4

22.2.3.2 JEEVAMRITA

Jeevamrita is prepared by decomposing 10 kg cow dung, 5 L cow urine, 2 kg jaggery, 2 kg pulse flour, 100 g virgin soil in 200 L of water by simple facilities created at the farm with minimum expenditure (Fig. 22.9). The credit of recipes for *Jeevamrita* and its extension can be attributed to Palekar (2006). It can be used at 15–30 days interval through the irrigation water coupled with mulching (green/dry {monocot + dicot}) and proper soil aeration. *Jeevamrita* is a rich bioenhancer containing consortia of beneficial microbes (Ram et al., 2018). Systematic microbial analysis of *Jeevamrita* from the day zero to 20th day of preparation suggests that formulation should be used within 6–9 days for maximum benefits (Table 22.7) (Ram et al., 2017). Drenching can be done by mulch or drip irrigation or through spraying which is very effective in early decomposition of organic wastes if applied with irrigation water during preparation of the field. More area to the tune of 3–4 times can be covered with 200 L of *Jeevamrita* by microirrigation.

FIGURE 22.9 Preparation of *Jeevamrita.*

TABLE 22.7 Different Microbial Populations in *Jeevamrita.*

S. no.	Type of microbe	Multiplication factor	Microbial population (cfu/mL) after days of preparation					
			0	3	6	9	14	20
	Actinomycetes	10^6	2.70	8.00	3.10	3.10	0.50	0.30
	Pseudomonas	10^7	0.19	2.60	2.89	5.09	0.005	0.30
	Rhizobium	10^6	1.66	10.80	12.41	75.10	35.0	0.71
	Phosphate-solubilizing microbes	10^6	1.20	3.80	3.94	5.04	1.40	0.03
	Azotobacter	10^5	5.00	0.10	0.30	1.12	0.30	0.10
	Azospirillum	10^5	9.00	0.50	0.60	0.01	Nil	Nil

22.2.3.3 BEEJAMRITA

Beejamrita is a potent bioenhancer prepared with locally available materials for seed and seedlings treatment. Being very cost-effective, this can easily be produced and used by small and marginal farmers. It is prepared using 5 kg cow dung, 5 L cow urine, 50 g virgin soil, 50 g slacked lime and 20 L of water (Fig. 22.10). It is advisable to prepare fresh *Beejamrita* and use it for treatment of seeds/seedlings and other plant parts before sowing/planting/

transplanting in earthen/mud pot of suitable size. Microbial analysis of *Beejamrita* showed that it should be used on the seventh day of preparation. In the case the crop is transplanted, it is advisable that seeds are treated with *Beejamrita* before sowing and the roots of the seedlings are dipped in *Beejamrita* before transplanting. Microbial analysis of *Beejamrita* is presented in Table 22.8.

FIGURE 22.10 Preparation of *Beejamrita.*

TABLE 22.8 Different Microbial Population in *Beejamrita.*

S. no.	Microbes	Population at seventh day of preparation (cfu/mL)
1	Actinomycetes	15.25×10^6
2	*Pseudomonas*	28.32×10^6
3	*Azotobacter*	78.98×10^6
4	*Azospirillum*	44.93×10^5
5	Phosphate-solubilizing bacteria	1.36×10^6
6	*Rhizobium*	17.72×10^6

Source: Ram et al. (2017).

22.2.3.4 AMRITPANI

This bioformulation is rich source of nutrients and beneficial microbes (Fig. 22.11). Deshpande (2003) advocated for all the ingredients to be used in preparation of *Amritpani* and its intensive use which may be used to improve seed germination, soil fertility, and plant vigor. Systematic microbial analysis of *Amritpani* from 0 to 20 days suggests that *Amritpani* should be utilized from sixth to ninth day of preparation for maximum effectiveness. Its application in soil, improves humus content, earthworm's activity which improves soil fertility and crop productivity. Microbial analysis of *Amritpani* is presented in Table 22.9.

TABLE 22.9 Materials Required for Preparation of *Amritpani*.

S. no	Ingredients	Quantity
1	Cow dung	10 kg
2	Cow *ghee*	250 g
3	Honey	500 g
4	Water	200 L

FIGURE 22.11 Preparation of *Amritpani*.

TABLE 22.10 Different Microbial Populations in *Amritpani*.

S. no.	Type of microbe	Multiplication factor	Microbial population (cfu/mL) after days of preparation					
			0	3	6	9	14	20
	Actinomycetes	10^7	0.66	0.73	0.10	1.31	0.37	2.00
	Pseudomonas	10^7	0.71	1.47	1.29	1.53	4.80	3.10
	Rhizobium	10^6	0.50	0.40	1.20	3.03	1.12	1.64
	Phosphate solubilizing microbes	10^6	0.72	2.20	3.20	4.80	2.93	2.00
	Azotobacter	10^7	2.97	0.26	0.01	0.001	0.28	0.27
	Azospirillum	10^6	2.01	0.80	0.02	Nil	0.20	0.01

Source: Ram et al. (2017).

22.2.3.5 VERMIWASH

Vermiwash, a collection of excretory products and mucus of earthworms along with nutrients, is a liquid leachate collected using excess water for saturating the vermicomposting substrate (Fig. 22.12). Vermiwash was found to contain enzyme cocktail of proteases, amylases, urease, and phosphatase. Vermiwash contains several nitrogen-fixing bacteria, such as *Rhizobium* sp., *Agrobacterium* sp. and. *Azotobacter* sp., and few phosphate-solubilizing bacteria. Laboratory-scale trial showed the effectiveness of vermiwash on cowpea plant growth (Zambare et al., 2008). Total heterotrophs, that is, *Nitrosomonas* 10.1×10^3, total fungi 1.46×10^3 along with *Nitrobacter* 1.12×10^3 are the major constituents of vermiwash.

Protease present in the soil helps in early and good seed germination, while amylase plays a vital role in the availability of simple carbon source for enhancing the plant vigor as well as crop yield. Soil borne microflora are very much required for plant growth as they help in decomposition and mineralization of organic nitrogenous compounds and phosphorus by fixing and phosphate-solubilizing bacteria (Table 22.11). The presence of a large number of beneficial microorganisms helps in plant growth by protecting from a number of pathogens in the field. Repeated spray of vermiwash has been found very effective even in the management of thrips and mites in chilies (George et al., 2007). If needed, vermiwash may be mixed with cow urine (i.e., mix vermiwash, cow urine, and water in the ratio 1:1:8) and used as foliar spray for nutrients and pesticidal properties.

FIGURE 22.12 Preparation of Vermiwash.

TABLE 22.11 Microbial Analysis of Vermiwash.

S. no.	Microbial types	Population (cfu/mL)
	Actionomycetes	1.04×10^6
	Pseudomonas	0.01×10^7
	Rhizobium	0.07×10^5
	Phosphate-solubilizing microbes	0.06×10^6
	Azotobacter	0.14×10^6
	Azospirillum	0.007×10^6

Source: Ram et al. (2017).

22.3 CONCLUSION

It can be concluded in brief that on-farm produced inputs or organic matter could be a potent source of nutrients and beneficial microbes which help in improving the soil fertility, crop productivity, and produce quality. Bioenhancers could be a potential alternative for fertigation which is becoming common in most of the crops. But care should be taken that bio enhancers that are used in limited quantities should meet the entire nutrient requirement of the crops. Organic wastes are generally helpful in catalyzing quick decomposition of organic wastes into humus, hence, incorporation of enough biomass preferably the combination of monocot and leguminous crops along with supplementation of animal wastes would be advantageous for quality humus production which is the very essential component for improving soil fertility and crop yield. Humus along with organic manures and regular application of bioenhancers and biopesticides can address many challenges of horticulture production and will be helpful to show a way for sustainable production through organic resources.

KEYWORDS

- **biodynamic farming**
- **biodynamic compost**
- **NADEP compost**
- **bioenhancers**
- **cow dung**

- **cow urine**
- **BD-501**
- **cow**
- **pat pit**
- **panchgaya**
- **jeevamrita**
- **beejamrita**
- **amritpani**

REFERENCES

Deshpande, M. D. *Organic Farming wrt. Cosmic Energy, Non Violence Rishi—Krishi*; Khede-Ajra: Kolhapur, Maharashtra, 2003; p 65.

Pandey, G.; Singh, B. P.; Ram, R. A.; Pathak, R. K. Studies on Pre-Harvest Spray of Bd 501 on Fruit Quality and Shelf Life of Mango (*Mangifera indica* L.); Paper presented at National Symposium on Organic Farming in Horticulture for Sustainable Production, CISH, Lucknow, Aug 29–30, 2003; p 74.

Garg, N.; Prakash, Om; Pathak, R. K. Cow Dung: A Source of Potent Bio Control Agents; Paper presented at the National Seminar on Cow in Agriculture and Human Health. Agri History Society, Udaipur, Rajasthan, 2003; p 82.

Garg, N. Cow Dung as Source of Bio Agents. Paper presented at the National Conference on Managing Threatening Diseases of Horticultural, Medicinal, Aromatic and Field Crops in Relation to Changing Climatic Situations & Zonal Meeting, Indian Phytopathological Society, Nov 3–5, 2012; pp 85–87.

George, S.; Giraddi, R. S.; Patil. Utility of Vermi Wash for the Management of Thrips and Mites in Chili (*Capsicum Annum*) Amended by Soil Organics, *Karnataka J. Agric. Sci.* **2007,** *20*, 657–659.

Majumdar, G. *Upavana Vinodha*; The Indian Research Institute: Calcutta, 1935; 58p.

Nene, Y. L. Utilizing Traditional Knowledge in Agriculture, Paper presented at the National Seminar on "Organic Agriculture: Hope of Posterity", UPCAR and NCOF, July 13–14, 2007; pp. 6–10.

Palekar, S. *The Philosophy of Spiritual Farming*; Zero budgets Natural Farming, Amrit Subash Palekar, Amravati, Maharashtra, 2006.

Pathak, R. K.; Ram, R. A. *Nutrients Levels of Different Compost: Manual on Jaivik Krishi.* Bulletin No. 37, 2009; p 31.

Pathak, R. K.; Ram, R. A. Bio enhancer: A Potential tool to Enhance Soil Fertility and Crop Productivity. In *Souvenir & Abstracts in International Conference on Organic Farming for Sustainable Horti-Agriculture and Trade Fair*; Jharkhand State Horticulture Mission, Nov 8–9, 2012.

Pathak, R. K.; Ram, R. A. Bio-enhancers: A Potential Tool to Improve Soil Fertility and Plant Health in Organic Production of Horticultural Crops. *Progress. Hort.* **2013,** *45* (2), 237–254.

Pathak, R. K.; Ram, R. A.; Garg, N.; Kishun, R.; Bhriguvanshi, S. R.; Sharma S. Critical Review of Indigenous Technologies for Organic Farming in Horticultural Crops. *Org. Farm. News Lett.* **2010,** *6* (2), 3–16.

Patnaik, H. P.; Dash, S. K.; Shailaja, B. Microbial composition of Panchagavya. *J. Eco-friendly Agric.* **2012,** *7* (2), 101–103.

Proctor, P. *Biodynamic Farming and Gardening*. Other India Press: Goa, India, 2008.

Ram, R. A.; Pathak, R. K. *Bio-enhancers*; LAP Lambert Academy Publishing: Germany, 2017.

Ram, R. A.; Pathak, R. K. Soil Health Management in Organic Production of Horticultural Crops. In *National Seminar on Integrated Development of Horticulture in Sub-tropical and Hill Region at HRS*; AAU, Kahikuchi, Feb 17–19, 2016; pp 128–150.

Ram, R. A.; Verma, A.; Gundappa; Vaishe, S. Studies on Yield, Fruit Quality and Economics of Organic Production of Mango cv. Mallika. In *Innovative Research on organic Agriculture*; WOC: New Delhi, India, 2017.

Ram, R. A.; Singha, A.; Verma, A. K. *Annual Report on Development of Organic Package of Practice in Mango cv. Mallika*; ICAR: New Delhi, India, 2017; pp 2–8.

Ram, R. A.; Singha, A.; Vaish, S. A. Microbial Characterization of On-farm Produced Bio-enhancers Used in Organic Farming. *Ind. J. Agric. Sci.* **2018,** *88* (1).

Ram, R. A.; Singha, A.; Kumar, A. Microbial Characterization of Cow Pat Pit and Other Biodynamic Preparations Used in Biodynamic Agriculture. *Indian J. Agric. Sci.***2018,** *89* (2), 42–46.

Sadhale, N. *Surpalas* Vrikshayurved; Asian Agri-History Foundation: Secunderabad, AP, 1996; pp 35–39.

Sarkar, S.; Kundu, S. S.; Ghorai, D. Validation of Ancient Liquid Organics-*Panchagavya* and *Kunapajala* as Plant Growth Promoters. *Indian J. Traditional Knowledge* **2014,** *13* (2), 398–403.

Sreenivasa, M. N.; Nagaraj, N.; Bhat, S. N. Beneficial Traits of Microbial Isolates of Organic Liquid Manures; Paper presented at the First Asian PGPR Congress for Sustainable Agriculture; ANGARU, Hyderabad, June 21–24, 2009.

Swaminathan, C. Food Production through *Vrikshayurveda* Way. In *Technology for Natural Farming*; Swaminathan, C., Ed.; Agricultural College & Research Institute: Madurai, Tamil Nadu, India, 2005; pp 18–22.

Zambare, V. P.; Padul, M. V.; Yadav, A. A.; Shete, T. B. Vermiwash: Biochemical and Microbial Approach as Ecofriendly Soil Conditioner. *ARPN J. Agric Biol. Sci.* **2008,** *3* (4), 1–4.

CHAPTER 23

Organic Farming for Sustainable Production of Sugarcane

R. B. KHANDAGAVE

S. Nijalingappa Sugar Institute, Belgaum 591108, Karnataka, India
E-mail: dr.khadagave_r@yahoo.com

ABSTRACT

Organic farming of sugarcane emphasizes management practices involving substantial use of organic manures and bio-fertilizers, residue recycling and management of insect pests and diseases through the use of non-synthetic pesticides and practices. Thus, it prohibits the use of harmful chemicals and promotes the use of renewable organic resources. To maintain the higher sugarcane yields on a long-term basis, it is essential to evolve a system whereby adequate supplies of bulky manures coupled with biofertilizers are ensured. The addition of bulky manures and biofertilizers is inevitable in the nutrient management program for achieving sustainable sugarcane production. Application of FYM, cane trash, press mud, biocompost, green manuring, and vermicompost are essential for organic farming in sugarcane. Besides increasing the soil properties, it improves the profit of cane production. Intercropping and incorporation of green manures such as dhaincha and sunhemp are helpful in increasing the efficiency of nitrogen, enhancing the cane yields, and improving the different soil properties. It is necessary to ensure a better price for sugarcane grown on organics to compensate for the lower cane yields during initial periods.

Organic Crop Production Management: Focus on India, with Global Implications, D. P. Singh, PhD, H. G. Prakash, PhD, M. Swapna, PhD, & S. Solomon, PhD (Eds.)

23.1 INTRODUCTION

Sugarcane crop (*Saccharum spp.* hybrids) is one of the commercial crops in Indian agriculture and has a major share in the national economy by sustaining the second largest organized agroindustry next to textile. Sugarcane gives raw materials for sugar, alcohol, cogeneration units besides biomanures. Noble cane was grown till the 20th century by vegetative propagation. In the process of nobilization, *Saccharum officinarum* L. is crossed with a wild relative *Saccharum spontaneum* to evolve aneuploid hybrid sugarcane (Mumtaz et al., 2011). Although India is the largest consumer of sugar in the world and is trailing behind Brazil to occupy first place in sugar production as it is being ranked the second position. India, producing annually 25–28 million tons in recent past, shares about 15% of world sugar production. From 2006–2007 to 2015–2016, sugarcane area, its production, and productivity on all-India basis are depicted in Figure 23.1. Cyclicality nature was observed in cane production and productivity. The increasing trend in productivity was observed from 2012–2013 to 2014–2015 and declined thereafter from 72 to 71 tonnes/ha in 2014–2015 and 2015–2016, respectively. This fluctuating trend is also noticed in terms of cane area. Major sugarcane growing areas have witnessed two back-to-back droughts are likely to reduce India's sugar production by about 25 million tonnes (a fall of about 3 million tonnes compared to last year) in 2016–2017 sugar season.

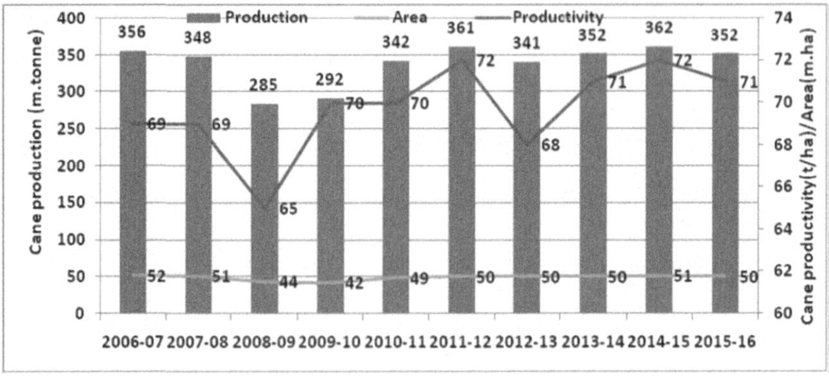

FIGURE 23.1 Cane area, cane production, and cane productivity in India from 2006–2007 to 2015–2016.

The fluctuating cane productivity trend is an indication to adopt sustainable practices. However, intensive agriculture practices involving crops like rice, wheat, sugarcane, cotton, and others have enhanced the usage of chemical manures, pesticides, irrigation, and many other technologies. Of late, these technologies have failed to improve or stagnate per ha yields of food crops besides creating problems like contamination of groundwater, food contamination, and problematic soils. Under these conditions, the use of organics has gained the importance. The variations in the cultural, ecological, and geophysical parameters have forced to adopt regional specific technologies over the past few decades in organic farming.

For optimum productivity of biomass, sugarcane requires higher nutrients, moisture, and more sunlight as it is a C_4 plant with a longer crop duration. The crop removes 0.56–1.20 kg of nitrogen, 0.38–0.82 kg of phosphorus, 1.00–2.50 kg of potassium, 0.25–0.60 kg of calcium, 0.20–0.35 kg of magnesium, 0.02–0.20 kg sodium, and 2.0–2.7 kg of sulfur besides micronutrients for every ton of cane produced. An average crop of sugarcane removes 208, 53, 280, 30, 3.4, 1.2, 0.6, and 0.2 kg nitrogen, phosphorus, potassium, sulfur, iron, manganese, and copper, respectively, to yield about 100 t of cane/ha. The major components of organic sugarcane (OS) are as depicted in Figure 23.2.

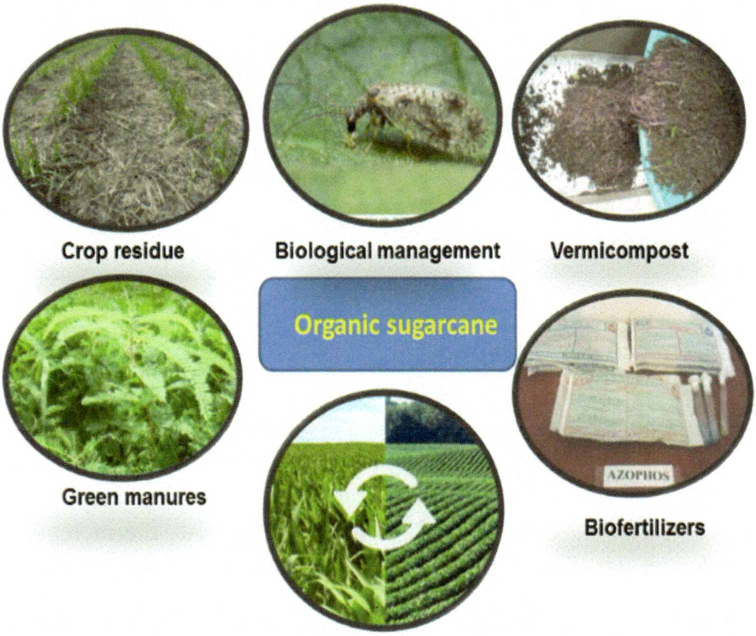

FIGURE 23.2 Major components of organic sugarcane.

23.2 USE OF ORGANIC MANURES

Organic manures such as farmyard manure, composts, and crop residues are more commonly used. They are bulky, low in nutrient status, and response of the crops to them is low mainly because of their short duration. These organic manures have to undergo several reactions to become an available form of nutrients, which require more time (Table 23.1). Sugarcane which is 10–16 months crop has a better opportunity to absorb nutrients from the organic manures during prolonged availability.

TABLE 23.1 Nutrient Content of Different Organic Manures.

Category	Source	Nutrient content (%)		
		N	P_2O_5	K_2O
Animal wastes	Cattle dung	0.3–0.4	1.10–0.15	0.15–0.20
	Cattle urine	0.30	0.01–0.02	0.5–0.70
	Sheep and goat dung (mixed)	0.65	0.5	0.03
	Night soil	1.2–1.5	0.8	0.5
FYM composts	Farmyard manure	0.5–1.0	0.15–0.20	0.5–0.6
	Poultry manure	2.87	2.90	2.35
	Town compost	1.5–2.0	1.0	1.5
	Rural compost	0.5–1.0	0.2	0.5
Oil cakes	Groundnut	4.5	1.7	1.5
	Karanj (*Pongamia pinnata*)	3.9–4.0	0.9–1.0	1.3
	Neem (*Azadirachta indica*)	5.2	1.0	1.4
Factory by-product	Press mud	0.56	0.34	0.34

23.2.1 FARMYARD MANURE

Farm manure is decomposed compost derived from the mixture of dung, urine, farm litter, and leftover or roughages unused materials or fodder. Nearly 258 experiments carried out country-wide in 19 locations, which revealed that 25 t/ha of FYM application given to cane, the increased cane yield by 8 t/ha with the minimum increase being at 3.7 and the maximum at11.7 t/ha. Combined application of farmyard manure at 10 t/ha FYM and at 50 kg urea/ha proved better for a plant crop and four ratoons performance (Radder, 1998). Sharma et al. (2005a) reported that FYM application on an equal N-basis helped in increasing the shoot population, number of millable canes, and cane yield over others treatments without affecting the sugar content of cane. Gana and Busari (2006) showed that cane yields were generally lower under the legume-treated soils (38.7–53.4 t/ha) when compared with yields recorded under FYM (58.6–70.1 t/ha) or even inorganic fertilizers (55.7–66.9 t/ha). The highest mean germination (43.89%), number of tillers (167,087), millable canes (125,055) and cane yield per hectare (78.85 t), sucrose percentage in juice (19.05%), and purity coefficient (86.58%) were on par with that obtained with FYM + groundnut cake + urea treatment (Vedprakash et al., 2009).

23.2.2 CROP RESIDUES/TRASH

Sugarcane crop produces 10–12 tons of dry leaves/ ha/year that contains an appreciable amount of NPK, secondary and other micronutrients. In India, about 282 million tonnes of sugarcane is produced, of which trash contributes 10% of total biomass. The soil incorporation of sugarcane trash releases nutrients after decomposition. This may build up the fertility status and changes in the physical characters of the soil. With the decomposition of trash, the organic acids are released that would help in the solubilization of the different nutrients and thereby enhance their availability. Incorporation of sugarcane trash at 3 t/ha, compared without trash, enhances the availability of K from 329 to 338 kg/ha in the soil (Kumar and Sagwal, 1998). This trash can be used in sugarcane farming by incorporating it along with cellulolytic fungi before planting. Usage of trash in sugarcane serves many purposes: as a nutrient upon decomposition, acts as mulch, conserves water by reducing evaporation, and hinders the weed growth. Trash used on the surface and as vertical mulch gave higher cane yield compared with no trash plots (Table 23.2).

TABLE 23.2 Usage of Trash as Surface and Vertical Mulch on Cane Yield Substantially.

Treatment	Sugarcane yield (t/ha)		
	Plant crop	Ratoon crop	
		I	II
Surface mulch	64.8	55.6	42.2
Vertical mulch	63.6	57.2	45.9
Surface + vertical mulch	67.7	60.0	60.0
Control	54.6	49.5	39.1

23.2.3 GREEN MANURE

Green manuring helps in increasing crop productivity and improving the soil fertility. The addition of organic matter by green manure after decomposition might supply nitrogen and other macro and micronutrients during the course of the growing period of the crop. Initially, green manure residues have wider carbon to nitrogen ratio which favors the immobilization of applied nitrogen; however, the same will be released slowly over a period. Organic matter postdecomposition contributes by increasing the available nitrogen as well as phosphorus. Sugarcane is a wide-spaced crop and takes more time to cover the canopy due to its slower growth during the early stages. Incorporation of green manure helps in producing organic acids which will solubilize the fixed phosphorus along with continuous availability of nitrogen. These conditions favor obtaining higher cane yields. The spatial and temporal opportunities of sugarcane can be effectively utilized for cultivating green manure crops. Green manuring crops are leguminous crops and these crops fix larger amounts of nitrogen (at 30 kg N/ha). Different leguminous crops can be used as green manure crops. In sugarcane, sunhemp is the ideal crop and in 45–60 days, it gives at 22 kg ammonium sulfate per ha which is equivalent to 30 tonnes of FYM/ha. The cane yield can be increased by 30–45% due to the incorporation of green manure (Table 23.3). Green manuring also helped in retaining soil fertility even after 12 years. Other green manuring crops are soybean, green gram, and black gram.

TABLE 23.3 Effect of Green Manure on Cane Yield and Cane Response to Nitrogen.

Treatment	N addition by green manure crops	Cane yield (t/ha)		
		Mean cane yield (t/ha)	Increase over control	Average response of cane/kg N
Control (No GM crops)	–	47.20	–	–
Dhaincha	55	61.50	30.4	2.6
Guar	41	60.20	27.5	3.2
Cowpea	71	67.50	43.0	2.8
Sunhemp	70	67.40	42.6	2.9

23.2.4 CROP ROTATION AND INTERCROPPING

Adoption of wide row spacing and slow growth rate of sugarcane at the initial stages offer ample scope for accommodating short durational intercrops. Pulses like black gram, soybean, green gram, and French beans provide extra monitory returns. The benefits of increasing nitrogen use efficiency are observed by the incorporation of intercropped legumes, such as French bean, soybean, and sunhemp. Pest incidence can be minimized by discouraging the growing of cereal crops, namely, corn and sorghum as intercrops. Hence, intercropping with leguminous crops, namely, peanut, soybean, *mung* bean, and others or green manure crops, namely, sunhemp, *Dhaincha*, or vegetables, namely, tomato, brinjal, pepper, cabbage, cowpea, cauliflower, and others can be grown. Intercropping can increase the farmers' income and also improve the microclimate, creating an ecological environmental balance suitable to the survival of predators/parasites, and minimizing the damage by pests and diseases.

23.2.5 VERMICOMPOST

Vermicompost is an organic manure produced by using earthworms and waste organic material. The earthworms that live in the soil absorb the organic waste and excrete it as digested material in the form of manure. It is estimated that 1800 worms, which occupies 1 m^2 area, are able to feed on 80 tonnes of humus per year. The compost is rich in macro- and micronutrients, vitamins, growth hormones, and immobilized microflora. The average nutrient content of vermicompost is substantially higher than that of FYM. In comparison with FYM, the vermicompost contains two times more of N and four times more of P_2O_5 and K_2O. It has 1.60% N, 5.04% P_2O_5, and 0.80% K_2O along with micronutrients. Vermicompost facilitates easy availability of essential plant nutrients to crop. Vermicompost preparation technique is very easy for sugarcane farmers as sugarcane trash is available in plenty. Yadahalli (2008) reported that the addition of *neem* cake helps in realizing a significantly higher germination (68.5%) than other amendments.

The studies carried out in Vasantdada Sugar Institute, Pune, revealed that application of vermicompost prepared from sugarcane trash and press mud cake, which are mixed in the ratio of 1:2 (at 5 t/ha) to sugarcane gave a maximum cane yield of 96.5 t/ha as compared to 83.39 t/ha obtained from control plot. The cost-benefit ratio recorded was 1:1:63. Similarly, vermicompost at 5 t/ha gave cane yields on par with the treatments having no vermicompost (Table 23.4).

TABLE 23.4 Effect of Vermicompost on Cane and Sugar Yield.

Sl. No	Treatment	Cane yield (t/ha)	Sugar yield (t/ha)
1	Vermicompost @5 t/ha and 100% RDF	97.85	12.73
2	Vermicompost @5 t/ha and 75% RDF	97.32	12.89
3	Vermicompost @5 t/ha and 50% RDF	96.87	12.62
4	Vermicompost @5 t/ha and 25% RDF	82.70	10.80
5	Only vermicompost @5 t/ha	78.95	10.15
6	Control (only chemical fertilizers)	84.95	10.89
C.D at 5%		5.30	0.10

23.2.6 USE OF PRESS MUD/BIOCOMPOST

Press mud also provides the essential elements besides 22–25% organic carbon and its application also reclaims problematic soils. Press mud is produced to the extent of 3% on cane weight. During 2003–2004, in India, sugarcane production was 287.38 M tonnes out of which 194.4 m tonnes of total cane was crushed which has produced 5.2 M tonnes of press mud. This sizeable quantity can be utilized for producing organic manure. The fresh press mud, when incorporated directly, may cause burning in the crop. Therefore, the addition of decomposed press mud is more beneficial. The biocompost prepared by using press mud is helpful in maintaining soil fertility and enhances the crop production despite improving the soil's physical, chemical, and biological properties because it is a rich source of essential plant nutrients, namely, organic carbon, nitrogen, phosphorus, potassium, calcium, and magnesium along with traces of micronutrients, namely, zinc, iron, copper, and manganese. The manufacturing of biocompost is achieved by drying the press mud in a phased manner and mixing the compost with beneficial microbes and then eventually packing the same or heaping it for delivery. In some of the cases, the biocompost is also enriched with biomethanated spent wash, which is also a rich source of nitrogen, phosphorus, potassium, and sulfur.

On an oven-dry weight basis, the press mud cake contains 1–3.1% N, 0.6–3.6% P, and 0.3–1.8% K, in addition, secondary and trace elements are also present in it. The production of press mud cake amounts to about 3–7% of the amount of cane crushed in a sugar factory. Press mud at 25 t/ha produced significantly better cane yield over lower doses of press mud and control. Biocompost prepared by using press mud and distillery spent at 5 t/ha, produced equivalent cane yield (101.9 t/ha) to that with press mud at 25 t/ha (91.0 t/ha) (Table 23.5).

Similarly, the compost made out of sugarcane trash and press mud can be compared with the FYM. In sugarcane, press mud compost gave results at par, when compared with recommended inorganic fertilizers (Table 23.6). Press mud cake being a carrier of more of phosphatic fertilizer can also act as an ameliorant for acid and saline-sodic soils.

TABLE 23.5 Influence of Different Levels of Press Mud and Biocompost on Cane Yield

Treatment	Organic manures (t/ha)					Mean (t/ha)
	PM_1 (12.5 t)	PM_2 (25 t)	BC_1 (2.5 t)	BC_2 (5 t)	Control	
Chemical fertilizers						
RDF_1 (100%)	107.8	109.5	109.2	129.4	104.7	112.1
RDF_2 (75%)	93.4	100.7	95.1	111.8	90.4	98.3
RDF_3 (50%)	77.8	84.9	78.4	95.7	73.5	82.1
RDF_4 (Zero%)	54.6	68.9	54.5	67.3	52.9	59.6
Mean	83.4	91.0	84.3	101.9	80.3	88.2
Source of variation			Cane yield (t/ha)			
SEm + for chemical fertilizer			1.77			
CD at 5% for chemical fertilizer			4.91			
SEm+ for organic manures			1.98			
CD at 5% for organic manures			5.93			
SEm+ for interaction			3.96			
CD at 5% for interaction			10.97			

TABLE 23.6 Effect of Compost Press Mud, Green Manure on Cane and Sugar Yield.

Treatment	Cane yield(t/ha)	Sugar yield(t/ha)
Control	104.0	13.2
Fertilizer (NPK)	118.0	14.6
(250: 115: 115 kg/ha)	128.0	16.3
Press mud compost	120.0	15.2
Green manure	123.0	15.6
Fertilizer + PMC	130.0	16.3
Fertilizer + GM	122.0	15.7
PMC + GM	133.0	16.4
Fertilizer + PMC + GM	13.8	1.40
C.D. at 5%		

Of the organics evaluated, biocompost at 10 t/ha provided the greatest enhancement of cane yield (97.1 t/ha). The higher yield response due to biocompost is attributed to better nutrient content and its positive effect on soil microorganisms. Further, farmyard manure at 25 t/ha and green manuring as an intercrop produced equivalent yields indicating green manuring can serve as an alternative to the farmyard manure, which is usually in short supply. Trash incorporation at 8 t/ha and vermicomposting at 5.0 t/ha also produced more cane yield (77.9 and 77.1 t/ha, respectively) over the control plot (Table 23.7).

TABLE 23.7 Response of Sugarcane Yield (t/ha) to Organic and Chemical Fertilizers.

Treatment	Organic manures						Mean (t/ha)
	TR (8 t/ha)	FYM (25 t/ha)	VC (5 t/ha)	BC (10 t/ha)	GM	Control	
Chemical fertilizers							
RDF_1 (100%)	102.4	106.0	98.7	117.2	108.0	95.1	104.6
RDF_2 (50%)	75.0	85.4	71.4	100.6	81.1	68.5	80.3
RDF_3 (Zero%)	58.4	61.7	61.4	73.4	63.6	50.9	61.6
Mean	78.6	84.4	77.2	97.1	84.3	71.5	–
Source of variation	SEM ±		C.D at 5 %				
Organic manures	1.94		5.33				
Chemical fertilizers	1.24		3.44				
Interaction	3.03		8.41				

23.3 USE OF NONTRADITIONAL ADDITIVES

23.3.1 BIOLOGICAL NITROGEN FIXING ORGANISM AND PHOSPHATE-SOLUBILIZING MICROORGANISM

Biofertilizers like *Azotobacter* at 5 kg/ha, PSB (*Bacillus*, Achromobacter, Acrobacter) and fungus (*Aspergillus penicillium*), or press mud at 4 t/ha are also recommended. The use of biofertilizers in agriculture has found an economically viable and ecologically sustainable source for reducing external inputs while improving the quality and quantity of internal sources. Biofertilizers are less expensive, eco-friendly, and sustainable. These are the microbes that are capable of fixing atmospheric nitrogen when inoculated with suitable crops. Some of them are also capable of mobilizing nutritive elements from nonusable form and convert them to usable form through biological process.

In the rhizosphere of sugarcane, there is considerable nitrogen fixation taken by non-nodulating bacteria. This will help in adding 30–75 kg N/ha/year. The promising N-fixers in sugarcane are *Azospirillum* and *Azotobacter*. In clay soils, *Azospirillum* is more effective and in loams and sandy types of soils, the *Azotobacter* is more effective in improving cane yield and quality. The group of microorganisms known to have the ability to convert insoluble P into soluble and their use has shown significant increases in yields of crops like wheat, rice, maize, chickpea, pigeon pea, soybean, groundnut, and *berseem*. These organisms are also useful in sugarcane cultivation.

23.3.2 CONCENTRATED ORGANIC MANURE

Organic manures that are organic and bulky in nature and have higher concentration of nutrients, namely, as nitrogen, phosphorous, and potash compared to other bulky organic manures are called as concentrated organic manures. These concentrated manures are derived from animal or plant origin. To prepare the concentrated organic manures, oil cakes, blood meal, fishmeal, meat meal, horn, and hoof meal are used. After extraction of oil from the oilseeds, the remaining solid portion is dried as the cake that can be used as manure (Table 23.8).

TABLE 23.8 Nutrient Content of Oil Cakes and Animal Base Organic Concentrates.

Oil cakes	Nutrient content (%)		
	N	P_2O_5	K_2O
Nonedible oil cakes			
Castor cake	4.3	1.8	1.3
Cotton seed cake (undecorticated)	3.9	1.8	1.6
Karanj cake	3.9	0.9	1.2
Mahua cake	2.5	0.8	1.2
Safflower cake (undecorticated)	4.9	1.4	1.2
Edible oil cakes			
Cotton seed cake (undecorticated)	3.9	1.8	1.6
Karanj cake	3.9	0.9	1.2
Mahua cake	2.5	0.8	1.2

TABLE 23.8 *(Continued)*

Oil cakes	Nutrient content (%)		
	N	P_2O_5	K_2O
Safflower cake (undecorticated)	4.9	1.4	1.2
Cotton seed cake (undecorticated)	3.9	1.8	1.6
Karanj cake	3.9	0.9	1.2
Mahua cake	2.5	0.8	1.2
Safflower cake (undecorticated)	4.9	1.4	1.2
Animal-based concentrated organic manures			
Blood meal	10–12	1–2	1.0
Meat meal	10.5	2.5	0.5
Fish meal	4–10	3–9	0.3–1.5
Horn and hoof meal	13	–	–
Raw bone meal	3–4	20–25	–
Steamed bone meal	1–2	25–30	–

23.4 WEED AND PEST MANAGEMENT

The requirement of chemicals in sugarcane crops is lesser compared to other commercial crops like cotton. During the initial stages of sugarcane, the weed flora is higher and the same can be overcome easily by going in for hand weeding or by cultivating intercrops. The practice of earthen-up takes care of the weeds. After 90–120 days of growth, the sugarcane growth is vigorous and weeds are suppressed easily. Pest and disease problem is also very less in sugarcane. Many of the diseases can be prevented by using good seed material while pests are prevented by early planting, resistant varieties, proper earthen-up, heavy irrigation, growing intercrops like garlic and onion.

Among the biotic stresses, pests and diseases pose a major threat to sugarcane cultivation. About 210 insects are reported to infest the crop, of which borers and sucking pests are of economic importance (Salin and Srikanth, 2011). Following measures are suggested for weed management in sugarcane.

23.4.1 STRICT PLANT QUARANTINE PREVENT PESTS FROM ARTIFICIALLY SPREADING

Plant quarantine is to prevent the agricultural product, seeds, and seedlings during their transportation through applicable laws. The purpose is to prevent dangerous pests, diseases, and weeds to spread by imports. The transport of sugarcane seed canes must be examined, and the transport must be stopped or the seed canes must be dealt with strict quarantine regulations when pathogenic organisms are noticed. Soaking the setts at 24–48 h with 2% lime water can help in cleaning the pathogens transmitted through sugarcane seeds (Wei, 1996; Feng, 2003).

23.4.2 CROP ROTATION AND INTERCROPPING

Sugarcane is intercropped with legumes or green manure crops or vegetables, such as peanut, soybean, *mung* bean, legume green manure crops, tomato, hot pepper, Chinese cabbage, and others. Intercropping increases income, and also can improve the microclimate by creating an ecological environmental balance suitable to the survival of parasites and other bioagents acting on different insects and pathogens (Wei, 1996; Feng, 2003). If sugarcane is cultivated continuously for successive years, borer development will be asynchronous as borers will have numerous overwintering populations, which may cause severe damage. So crop rotation with legumes and noncereal crops such as soybean is to be encouraged as an intercrop. In the on-farm trial with soybean as intercrop, the range of incidence of shoot borer in intercrop and monocrop varied with the age of the sugarcane. In a 25-day old crop, the incidence was significantly ($p < 0.001$) higher in intercrop (range: 3.9–12.6%) than in monocrop (range: 0.2–1.3%). In most of the intercrop studies, pest abundance was less in intercrop which is, generally, short in stature and duration.

Trap crops or catch crops are species of plants that are planted to attract and retain pest species or to provide a more favorable habitat than the main crop to increase natural enemies. Biocontrol agents *Trichogramma chilonis* is used against borers in sugarcane, *Dipha aphidivora* and *Micromus* sp. are used against sugarcane woolly aphid. Similarly, the microbe *Metarhizium anisopliae* is used to control white grub in sugarcane and viral insecticides, namely, GV of *Chilo infuscatellus controls* shoot borer of sugarcane.

23.5 ORGANIC SUGAR PRODUCTION—A REAL OPPORTUNITY

A common definition of organic sugar is the sugar produced by processing juice obtained by crushing organic sugar cane without the application of inorganic synthetic chemicals. Organic sugar cane is produced according to the certain organic standards, that is, they are grown without the use of conventional pesticides and artificial fertilizers and free from contamination by human or industrial waste. Consumer demand for organically grown foods in several developed countries is opening new opportunities for agricultural producers around the world. A small but steadily growing number of sugar producers around the world have embraced the organic movement over the past few years having successfully implemented certified organic production and milling techniques.

23.6 IMPACT OF ORGANIC MANURES ON SOIL PROPERTIES

The addition of organics helps to improve the soil characters. Among the organics evaluated, application of biocompost, trash incorporation, farmyard manure, green manuring, and vermicompost increased organic carbon by 61.4, 36.11, 36.11, 25.00, and 22.22%, respectively, over the no-organic control. Additionally, soil bulk density was significantly lower due to the application of all organics compared to the control (1.44 g/cc).

Significantly higher microbial numbers of P-solubilizer, *Actinomycete*, fungi, *Azotobacter*, and others were observed due to biocompost over the other organics and the control. Soil organic carbon due to the addition of the organics enhances soil microbial activity and microbial biomass. These microbes are responsible for degradation of complex organic matter and also act as powerful antifungal, antibacterial agents by competition, antagonism, and hyper parasitism (Table 23.9). Substantial improvement in the uptake of nutrient (nitrogen, phosphorus, and potassium) was recorded due to organic nutrition modules compared with that of control, both for autumn and spring planted sugarcane crops. During both the seasons, significantly the highest quantities of nitrogen (227.7 and 185.4 kg/ha for autumn and spring cane) were removed under SPM 10 t/ha + FYM 10 t/ha application compared with the uptake of 152.6 and 141.7 kg/ha nitrogen under control condition. Phosphorus uptake was significantly higher (17.8 and 13.4 kg/ha) under this treatment for autumn as well as spring cane (Srivastava et al., 2008). The addition of 100% N through organics brought about substantial increase in the organic carbon content of the soil. The highest enhancement in organic carbon content (0.65 over initial 0.40%) at ratoon harvest was observed in the treatment having 100% N through organic + biofertilizers + intercropping of legume with *Rhizobium* + pests/diseases control by either synthetic pesticides or biopesticides (Singh and Srivastava, 2011). Soil-organic carbon at sugarcane harvest over the initial content was the highest (70.73 %) because of FYM at 20 t/ha + *Trichoderma viride* + lentil intercropping (1:2) or SPM 10 t/ha + lentil intercropping (1:2) in autumn sugarcane, and the corresponding highest increase in spring sugarcane (65.9%) due to the application of SPM at 10 t/ha + *Azotobacter* inoculation. The high content of cellulosic compounds in SPM and crop residues may be due to an increase in the organic carbon content under these treatments.

TABLE 23.9 Effect of Organics on Physical and Biological Properties of the Soil.

Treatment	Bulk density (g/cc)	Organic carbon (%)	P-solubilizer (10^4/g soil)	Actinomycete (10^4/g soil)	Fungi (10^4/g soil)	*Azotobacter* (10^4/g soil)
Trash (8 t/ha)	1.36	0.49	16.75	22.75	25.75	16.00
FYM (25 t/ha)	137	0.49	13.75	19.75	21.50	13.50
Vermicompost (5 t/ha)	1.42	0.44	14.25	21.25	24.50	15.50
Biocompost (10 t/ha)	1.34	0.58	19.25	25.25	28.50	20.00
Green manuring	1.38	0.45	13.00	20.00	21.50	13.00
Control	1.44	0.36	8.25	17.75	16.75	6.75
SEm ±	0.017	0.015	0.325	0.347	0.374	0.348
CD at 5%	0.047	0.042	0.908	0.962	1.042	0.962

23.7 WAY FORWARD

It is evident that there is enough scope to provide the nutrients required for sugarcane through inorganic sources, along with obtaining the improvement of soil's physical, chemical, and biological properties. Present information reveals that the nutrients at 207.5 kg/ha nitrogen, 131 kg/ha phosphorus, and 143 kg/ha potassium can be supplied by using different sources of organic manures (Table 23.10). Additionally, there is also scope to further enhance the supplementation of nutrients by inculcating other sources of organics, namely, fish/poultry/bone meals, and concentrates of oilseed/cotton.

TABLE 23.10 Supply of Organic Sources of Nutrients to Sugarcane (kg/ha).

Treatment	N	P	K	Ca	Mg	S	Fe	Zn	Cu	Bo
Vermicompost @5 t/ha)	70	80	100	20	20	2.5	1.5	1.0	0.05	0.05
Trash @7.5 t/ha	37.5	11	43	–	–	–	1.5	–	–	–
P-solubilizers & VAM	–	40	–	–	–	–	–	–	–	–
S-oxidizer	–	–	–	–	–	20	–	–	–	–
Green manure	70	–	–	–	–	–	–	–	–	–
Intercropping	30	–	–	–	–	–	–	–	–	–
Total	**207.5**	**131**	**143**	**20**	**20**	**22.5**	3.0	1.0	0.05	0.05

The above results indicate that organic farming in sugarcane is possible. However, it needs to build up the organic base of the soil and biota to enhance the nutrient load, for obtaining increased cane yields on sustainable basis. Therefore, it is ideal that during initial stage, addition of organics along with 50% chemical fertilizers followed by 100% organic plus application of microbes during the second year and 100% organic plus microbes plus release of earthworms during the third year be adopted. One-hundred percent organic farming can be resorted to from the fifth year onward.

23.8 ECONOMICS

The highest return and benefit: cost ratio (B:C) were recorded in autumn when the cane is planted with FYM at 20 t/ha *Trichoderma viride* + lentil intercropping. In spring-planted cane, the highest profit and cost ratio (B:C) was observed with FYM 20 t/ha + *Trichoderma viride* + mung bean intercropping. These were followed by the autumn and spring planted cane with SPM 10 t/ha + FYM 10 t/ha (Srivastava et al., 2008). Kshirsagar (2008) studied the influence of organic farming on economics of sugarcane cultivation in Maharashtra. Report on the primary information gathered from two districts covering 142 farmers in which 72 growing OS while 70 were growing inorganic sugarcane (IS). The results concluded that OS cultivation enhances human labor employment by 16.9% and cost of cultivation is also lower by 14.2% compared to IS farming. Although the yield from OS was 6.79% lesser compared to the conventional crop, it is compensated by the higher premium price received and yield stability observed on OS farms. Overall, OS farming gave 15.63% higher profits than IS farms (Table 23.11).

TABLE 23.11 Economics of Organic Sugarcane Production.

Particulars	Organic sugarcane	Inorganic sugarcane	Percent over inorganic
Total cost of cultivation (Rs./ha)[a]	37,017.38	43,163.81	−14.24
Sugarcane yield (tonnes/ha)	96.63	103.56	−6.79
CV of sugarcane yield (%)	28.11	40.72	−12.61
Cost of production (Rs./tonne)	383.50	416.96	−8.02
Gross value of production (Rs./ha)	116,711.38	112,087.84	4.12
Gross profit (Rs./ha)	79,694.00	68,924.04	15.63
CV of gross profit (%)	39.76	45.68	−5.92
GVP/GCC	3.15	2.60	21.41

[a]This does not include the cost of harvesting, transport, and marketing.

23.9 CONCLUSION

Organic farming in sugarcane is possible in a phased manner which can be gained by integrating the available organic residues on the farm including that of the residues sugarcane and by-products generated by industry. Though yields are not higher during the initial period, this is known to enhance in the long run.

To maintain the higher sugarcane yields on a long-term basis, it is essential to evolve a system whereby adequate supplies of bulky manures coupled with biofertilizers are ensured. The addition of bulky manures and biofertilizers is inevitable in the nutrient management program for achieving sustainable sugarcane production. Application of FYM, cane trash, press mud, biocompost, green manuring, and vermicompost are essential for organic farming in sugarcane. Besides increasing the soil properties, it improves the profit of cane production. Intercropping and incorporation of green manures such as *dhaincha* and sunhemp are helpful in increasing the efficiency of nitrogen, enhancing the cane yields, and improving the different soil properties. It is necessary to ensure a better price for sugarcane grown on organics to compensate for the lower cane yields during initial periods.

KEYWORDS

- **organic manures**
- **farmyard manure**
- **green manure**
- **crop rotation**
- **vermicompost**
- **biocompost**
- **nontraditional additives**

REFERENCES

Feng, Y. X. Superficial View to the Integrated Control of ugarcane Diseases and Pests in Zhangjiang. *Sugarcane Canesugar* **2003,** *2*, 28–29.

Gana, A. K.; Busari, L. D. Contribution of Green and Farmyard Manure in the Nitrogen Nutrition of Sugarcane. *Sugar Technol.* **2006,** *8* (2 & 3), 175–179.

Kshirsagar, K. G. *Impact of Organic Farming on Economics of Sugarcane Cultivation in Maharashtra*; Gokhale Institute of Politics and Economics: Pune, 2008; p 15.

Kumar, V.; Sagwal, O. P. Effect of Sugarcane Trash and Fertilizer Nitrogen Applications on the Nutrient Status and Soil Physical Properties in Haryana Soil. *Indian Sugar* **1998,** *47*, 967–971.

Mumtaz, A. S.; Dur-e-Nayab; Iqbal, M. J.; Shinwari, Z. K. Probing Genetic Diversity to Characterize Red Rot Resistance in Sugarcane. *Pak. J. Bot.* **2011,** *43* (5), 2513–2517.

Radder, G. D. Prospects of Organic Farming in Sugarcane Cultivation. In *Proceedings of Seminar on Integrated Nutrient Management in Sugarcane*; Sameerwadi, Karnataka, 1998; pp 103–111.

Singh, K. P.; Srivastava, T. K. Sugarcane Productivity and Soil Fertility in Plant–Ratoon System under Integrated and Organic Nutrient Management in Sub-Tropics. *Indian J. Sugarcane Technol.* **2011,** *26* (10), 10–13.

Srivastava, T. K.; Singh, K. P.; Menhi, L.; Archna, S.; Pradip, K. Productivity and Profitability of Sugarcane (*Saccharum* spp Complex Hybrid) in Relation to Organic Nutrition under Different Cropping Systems. *Indian J. Agron.* **2008,** *53* (4), 310–313.

Vedprakash; Ram, M.; Lal, K. Effect of Continuous Application of Organic Manure and Inorganic Fertilizers on Yield and Quality Attributes of Sugarcane. *Coop. Sugar* **2009,** *40* (7), 65–67.

Wei, J. M. Damage and Control of Sugarcane Borers. *Hunan Agric. Sci.* **1996** (Suppl.), 34–35.

Yadahalli, K. B. Influence of Organic Amendments on Sugarcane Set Rot Development. *Int. J. Plant Sci.* **2008,** *3* (2), 556–557.

CHAPTER 24

Prospects of Endophytic Association of *Beauveria bassiana* in Pest Management Under Organic Farming

VIBHA PANDEY[1*] and RAKESH PANDEY[2]

[1]*Plant Pathology, Department of Plant Physiology, JNKVV, Jabalpur 482004, Madhya Pradesh, India*

[2]*Department of Entomology, BUAT, Banda 210001, Uttar Pradesh, India*

[*]*Corresponding author. E-mail: vibhapandey93@gmail.com*

ABSTRACT

Fungal endophytes are a rich source of novel compounds with potent biological activities in agriculture. Fungal endophyte, B. bassiana, has been found effective against aphids, thrips, lepidopterous larvae, and other insects as they produce toxic substances or signal molecules to modify the host response to such infestation. Artificial inoculation of such endophytes and their establishment in new environment are necessary under organic crop protection to minimize the attack of the wide range of insect pests. In addition, endophytes also ensure protection of crop losses from plant-parasitic nematodes and harmful pathogens. Besides, they secrete plant growth-promoting hormones, nutrient solubilizing substances, stress mitigating chemical compounds, and other resistance compounds against herbivores. Therefore, the establishment of endophytes after artificial inoculation is the primary objective to harness its beneficial impact on crops like plant growth promotion, nutrient solubilization, tolerances to biotic and abiotic stresses. Though the underlying mechanisms of plant-endophyte interactions are far

Organic Crop Production Management: Focus on India, with Global Implications, D. P. Singh, PhD, H. G. Prakash, PhD, M. Swapna, PhD, & S. Solomon, PhD (Eds.)

from clear, several studies have raised the hope of their potential application in agriculture, especially in mitigating abiotic and biotic stresses.

24.1 INTRODUCTION

Much attention is required for producing enough food without exploitation and degradation of earth's limited resources under agricultural systems using strategies that are also capable to cope with changing climatic conditions (Affaires et al., 2014). Organic farming can serve as a possible option to meet the demand as it works on the principle of eco-functional intensification, which includes making optimal use of natural resources to secure and improve agricultural productivity by minimizing negative environmental impacts such as loss of biodiversity, nutrient leakage, and soil degradation. The major aim of the organic farming in agricultural production is completing nutrient cycles via returning plant residues or manure from livestock besides incorporating perennial and leguminous plants to the soil. It also restricts the use of chemical-based synthetic fertilizers or synthetic pesticides. The success of any agricultural system is highly dependent on the microbial community and its composition as they are involved in different processes and functions of agriculture.

Asymptomatically and frequently, association of fungi within the plant tissues or plant harboring the fungi within its tissues are called endophytes. Moreover, the association of endophytes with plants does not result in manifestation of any harmful effects like chlorosis, lesion formation, growth retardation, and others on the host because such association is part of the life of pathogen for its survival (Hardoim et al., 2015; Puri et al., 2016). Hence, symptoms of host–pathogenic interactions cannot be seen on different parts, namely, roots, leaves, stems, branches, flowers, and fruits after endophytic colonization on the plant (Saikkonen et al., 2006). These endophytes have the capability to form confederacy with different groups of organisms belonging to plant kingdom to ensure generally indirect defense of plants from insect-pests (Hartley and Gange, 2009). Increasing risk from chemical pesticide to the ecosystem and its dweller have pushed the agriculture community to use these insect pathogenic fungal endophytes to manage lepidopterous larvae, thrips, aphids, and other insects that are threat to agriculture worldwide. They are specific to their hosts from an infection point of view but not too beneficial or nontarget insects (Akutse et al., 2014). Production of several toxic (Saikkonen et al., 2010) or signaling molecules by endophytic fungi either kill or modify the genetic response of host toward pest through triggering resistance mechanism.

Beauveria bassiana (Balsamo-Crivelli) Vuillemin has been found to occur endophytically in several plant species and has been inoculated artificially for establishment in various important crop plants (Vega, 2008; Vega et al., 2009; Vidal and Jaber, 2015). The Association of *B. bassiana* for suppression of deadly muscardine disease of silk was discovered in 1835 by Agostinio (Steinhaus, 1956) that opened the new way for management of insects through endophytic fungi. Thereafter, several mycoacaricides and mycoinsecticides have been developed worldwide since 1960 to manage insects. However, commercial mycoinsecticide product developed from *B. bassiana* was found more stable against lepidopteran group than other biological insect controlling agents (Thakur et al., 2011; Rana et al., 2008). Although *B. bassiana* endophyte has strong biocontrol potential but limitations associated with the change in its efficiency due to location and mode of action on different insects raise a big question mark on its use at wider level. Therefore, gathering knowledge about its functional role with the crop as an endophyte and biocontrol activity against insect-pest. Appropriate time of application as bioagent will help in determining its effect on host–plant interactions and also on associated insect-pest. Understanding the effect of ecological variations is necessary to justify the report of the neural and negative effect of *B. bassiana* on insect-pest. The virulence and pathogenesis of fungus is being decided by several bioactive secondary metabolites like peptides (beauveriolides, bassianolies, and beauvericin), polykitides (tenellin, oosporein, and bassianin), and acid (oxalic acids) that can also be further utilized under agriculture insect-pest management apart from pharmaceutical and industrial purposes (Xu et al.,2009).

24.2 *B. BASSIANA:* AN ENTOMOPATHOGENIC DEUTEROMYCETE

B. bassiana belongs to the Cordyciptacea family and order Hypocreales of Sordariomycetes class under Ascomycota division. Interesting features of many *Beauveria* isolates are high adaptability and high host specificity. Wider occurrence and diversity of *Beauveria* species have been found in association with various insects of the world that have increased its significance not only in agriculture but also in medical science. They parasitize the vectors, for example, tsetse fly (*Glossina morsitans*), sandfly (*Phlebotomus* sp.), and bugs of genera *Triatoma* and *Rhodnius* are agents of tropical infectious diseases, that justify its significance in medical sciences. Furthermore, higher survival of entomopathogenic fungi in the host population (i.e., insect population) and greater adaptability to different environmental conditions are the reasons for including them in long-term pest management strategies. In China, *B. bassiana* is used against several insects, namely, the pine caterpillar, green hopper, European corn borer, *Dendrolimus* spp., *Nephotettix* spp., and *Ostrinia nubilalis*. A commercial by-product named Boverin has been developed by the Soviet Union with *B. bassiana* to control pests like *Laspeyresia pomonella*, *Leptinotarsa decemlineata*, and *Cydiapom onella* (Pinnamaneni and Potineni, 2010).

24.3 MECHANISM OF ACTION

It is already documented by researchers that defense against insect depends on the production of different metabolites and organic volatile substances that can attract natural enemies of insects (Pineda et al., 2010). Synthesis and release of these metabolites is regulated through a network of regulatory mechanisms mediated through specific regulatory elements like mitogen-activated protein kinases, reactive oxygen species, jasmonic acid (JA), and ethylene transcription (ET) (Howe and Jander, 2008). Colonization of plant with beneficial endophytes is initially recognized as invades that trigger immunity response in the host. Later, hormone biosynthesis and an array of signaling molecules of defense system initiate due to the induced response of several JA and ET marker genes (Vos et al., 2015). There are several reports that prove endophytic association enhances plant growth (Waller et al., 2005) that in return help the plant to overcome loss caused by herbivore in terms of biomass. Colonization of *B. bassiana*on *Phoenix dactylifera* leaves has helped the plant to mitigate the stress response through secretion of protein compound that acted as a signaling molecule for induction of defense or stress response (Gomez-Vidal et al., 2009). However, there are few reports that support the induction of plant defense systems against insect in presence of *B. bassiana.* Moreover, the production of fungus-driven secondary metabolites in plants could be another method to sustain the negative effect of insect-pest attack. Metabolites like nonribosomal peptide-synthetase (Xiao et al., 2012), bassianolide (Kanaoka et al., 1978), beauvericin (Grove and Pople, 1980; Wang and Xu, 2012), bassiacridin (Quesada-Moraga and Vey, 2004), dipicolinic acid, and oxalic acid (Bidochka and Khachatourians, 1991) having insecticidal properties are produced by *B. bassiana.* Expected function of these metabolites during the infection process has been conceptualized and proved, while the release of these compounds during colonization or mode of action in insect body after its ingestion is yet not clear. Identification of specific fungal metabolite and its derivatives, produced by them among range of other compounds, can be determined through mass spectrometry (Meca et al., 2009), high-pressure liquid chromatography (Jirakkakul et al., 2008), and liquid chromatography-tandem mass spectrometry (LC-MS/MS) (Jestoi et al., 2009).

24.3.1 CONIDIA-CUTICLE ATTACHMENT AND FORMATION OF INFECTION STRUCTURE

Outer layer of dry spores of *B. bassiana* possesses intricately bonded fascicles of hydrophobic rodlets. No other vegetative cell of this fungus is reported to possess these rodlets layer except in conidial stage. Nonspecific hydrolytic force applied by these rodlets could be the reason for adhesion of dry spores on the outer cuticle layer of the insect body (Boucias and Pendland, 1988). The outer epicuticle and inner procuticle are the two layers of cuticle. The outer layer is thin and contains phenol-stablize protein, lipids, fatty acids, and sterols but lack chitin (Hackman, 1984). The inner layer constitutes the major part of cuticle which is composed of chitin fibrils embedded into protein matrix along with lipids and quinones (Neville, 1984). Arrangement of chitin in cuticle is in a helical manner that provides it laminate shape. However, glycoprotein-binding carbohydrate called lectins has also been found on conidial surface of *B bassiana* which is also known to facilitate the attachment of conidia to insect cuticle. Exact mechanism involved in host–pathogen (insect–fungus) interaction in terms of spore adhesion on insect body has yet not been elaborated (Latge and Monsigny, 1988). Penetration of host cuticle by entomopathogenic fungi takes place through direct penetration. After penetration, the conidia of *B. bassiana* germinate to form appresorium, a conidial cell differentiated into infection structure possessing infection pegs, on the host surface. Host–pathogenic relationship establishes after the formation of appressorium. Involvement of secondary messengers such as Ca^{2+} and cyclic (cAMP) between the cells apart from the structure of host surface and induction of biochemical changes has been reported to influence the appresorium (St Leger and Roberts,1991) formation.

24.3.2 CUTICULAR PENETRATION AND TOXIN PRODUCTION

The entomopathogenic fungi penetrate through the cuticle into the insect body to obtain nutrients for their survival. The penetration process into the host involves degradation of cuticle through enzymes as well as physical separation of lamellae by exerting the mechanical pressure caused by hyphae. Various extracellular enzymes, for example, chitinases, lipases, esterases along with four different classes of proteases are responsible for degradation of major components of insect cuticle. These extracellular enzymes play a major role during the fungal pathogenesis. The first enzymes produced in the cuticle are endoproteases (termed PR1 and PR2) and aminopeptidases and; these enzymes are produced at time of the formation of appressoria. The complex structure of insect cuticle suggests that synergistic action of different enzymes would require for penetration; however, much attention has given on the cuticle-active endoproteases. It is suggested that prior to hyphae penetration, the cellular disruption takes place by the cytotoxins. The mycosed insects exhibit behavioral symptoms, for example, partial or general paralysis, sluggishness, and decreased irritability, which are consistent with the action of neuromuscular toxins. The significant amount of toxic compounds is produced by *B. bassiana* within their hosts. The toxins beauvericin, beauverolides, bassianolide, and isarolides have been isolated from the hosts infected with *B. bassiana* (Hamill and Sullivan, 1969; Elsworth and Grove, 1977). The toxins produced by *B. bassiana* have shown diverse effects on different tissues of host.

24.3.3 FUNCTIONS OF CUTICLE-DEGRADING ENZYMES DURING FUNGAL INFECTION PROCESSES

The production of proteases during fungal infection processes implies an important role for cuticle degrading enzymes. Besides, proteolytic degradation of the cuticle, the utilization of host proteins in terms of nutrition, destruction of antifungal proteins of the host, and production of amine by the release of amino acids are the other possible roles which could raise the pH to make congenial growth conditions. The other effects may be indirect, for example, the proteolytic activation of toxin precursors. The production of enzymes and penetration of host cuticle in pathogenic process take place only when it is required to establish a nutritional relationship with the host. The synthesis of proteolytic enzymes and phospholipases in host cuticle culture may probably occur when the pathogen infects living insects.

24.4 MOLECULAR STUDIES

Characterization of organisms by implementing PCR-based tools has greatly advanced the understanding of the phylogenies and species in entomopathogenic fungi particularly in *B. bassiana*. Due to their potentiality as biological control agents of insect-pests, entomopathogenic fungi gained a lot of attention. A high diversity in *B. bassiana* was observed following molecular analyses through the use of isozyme systems from which genotypic classes could not be grouped in relation to host or geographical origins (St Leger and Frank, 1992). Various nonspecific DNA-based methods have been employed especially in *Beauveria* (Glare et al., 2008). Random amplified polymorphic DNA (RAPD) has been widely used in several studies. It is based on the use of short random primers that anneal to unspecified regions in the template DNA; however, universally primed PCR is based on longer general primers and a higher annealing temperature which makes it more robust in terms of reproducibility (Bulat et al., 1998, 2000; Lubeck et al., 1999; Sahu et al., 2011). Universally primed PCR was used in Denmark to separate sympatric isolates of *Beauveria* and to place isolates in genetic groups (Meyling and Eilenberg, 2006). A distinct relationship between the population structure of *B. bassiana* and two different host species (the genus *Ostrinia* under family Pyralidae and the genus *Sitona* under family Curculionidae) following Restriction Fragment Length Polymorphism (RFLP) probes and RAPD technique (Maurer et al., 1997) has been reported. The identification of similar genotypes suggested that these *B. bassiana* strains could have a clonal population structure maintained by the effective operation of heterokaryon incompatibility. Forty-eight isolates of indigenous strains of *B. bassiana* were collected from Central India by using protease zymography and RAPD analysis (Thakur and Sandhu, 2010). Another distinct group of strains from insect hosts of Lepidoptera and Coleoptera was observed with high genetic and biochemical diversities. In addition to RAPDs and RFLPs, amplified fragment length polymorphisms (AFLPs) technique has also been used for analysis. Digestion of PCR products of specific DNA regions, such as genes or ITS (internal transcribed spacer), with restriction enzymes, yields fragments of variable sizes. Bidochka et al. (2001) studied *Beauveria* species' characterization by using these RFLPs. de Muro et al. (2003, 2005) characterized *B. bassiana* by using AFLP, intersimple sequence repeats, simple sequence repeats, or microsatellites. Fungal systematic have widely been studied by following internal transcribed spacer sequences (Driver et al., 2000; Bowman et al., 1992).

Vegetative compatibility groups (VCGs) are defined as subspecific groupings within the same fungal species from which isolates can fuse hyphae to form heterokaryons and thus exchange genetic information. VCGs have been employed to study the diversity and population structure in several fungi (Glass and Kuldau, 1992). Among individuals of fungal species that have no sexual cycle; VCGs are the only known means of genetic exchange. Recently, the development of microsatellite markers has certainly provided the insight into the population ecology of *B. bassiana* (Rehner and Buckley, 2003; Enkerli et al., 2005). Arnold and Lewis (2005) revealed that *B. bassiana* has been linked to plants as an endophytic fungus. Genetic diversity was estimated by studying polymorphisms developed by a telomeric probe, an important tool used for the characterization of *B. bassiana* within and among VCGs (Viaud et al., 1996). cDNA libraries have been used in EST (expressed sequence tag) analysis of entomopathogenic fungus *Beauveria* (*Cordyceps*) *bassiana* (Cho et al., 2006) and is he most reliable method that demonstrates a protein is involved in pathogenicity of genetically engineered mutants. The multiplicity of cuticle degrading enzymes provides a major challenge in the establishment of particular enzymes that function in pathogenicity or in adapting new environment. The development of genetically engineered strains suitable for insect-pest control became possible after an increased understanding of the factors responsible for virulence. The complementary nitrate nonutilizing mutants (nit) are commonly used to identify compatible isolates because compatible and incompatible reactions between isolates are not visually detectable (Glass and Kundu, 1992; Puhalla, 1984; Leslie, 1993). Previously, the occurrence of VCGs in *Beauveria* has not been examined. Presently, the analysis of different data followed by interpretation in context to the origin of the strains is being undertaken to know the role of *B. bassiana* as endophyte.

24.5 ROLE OF ENDOPHYTIC B. BASSIANA

24.5.1 INSECT-PEST MANAGEMENT THROUGH ENDOPHYTIC B. BASSIANA

The detrimental effects of entomopathogenic fungal endophytes on the insect-pests feeding are being attained through colonization of plants by these endophytic pathogens and have been reported by several researchers (Quesada-Moraga et al., 2009; Gurulingappa et al., 2011; Gange et al., 2012;

Gathage et al., 2016; Resquín-Romero et al., 2016; Sànchez-Rodríguez et al., 2018). The reduction of insect herbivory by endophytic fungi may be due to a reduction in the development rate of insect (Akello and Sikora, 2012; Akutse et al., 2013). Besides, feeding deterrence (McGee, 2002; Vega, 2008), retardation of insect growth, reduced survival, and oviposition (Lacey and Neven, 2006; Martinuz et al., 2012) are other factors. The crop plants treated with endophytic entomopathogenic fungi exhibit a reduction in plant damage caused by insect-pests. Quesada-Moraga et al. (2009) reported the reduction in damage caused by poppy stem gall wasp (*Iraellalutipes*) in opium poppy; tunneling by larvae of European corn borer, (*Ostrinia nubilalis* Hübner*)* and *Sesamia calamistis* Hampson in maize treated with *B. bassiana* (Bing and Lewis, 1991; Lewis et al., 2001; Cherry et al., 2004). A reduction in the damage caused by *Helicoverpa zea* in tomato following treatment with *B. bassiana* was observed by Leckie (2002). It was also reported that *B. bassiana* and *Purpureocillium lilacinum* reduced the population of *H. zea* in cotton (Lopez and Sword, 2015). Fifty percent mortality of all larval instars and reduced longevity of *Tuta absoluta* (Meyrick) was observed when larvae fed with tomato leaves colonized by *B. bassiana* (Klieber and Reineke, 2016). Similarly, Qyyum et al. (2015) observed a reduction in damage caused by *Helicoverpa armigera* in tomato plants by artificial inoculation of *B. bassiana*. Reduced damage was reported against coffee berry borer (*Hypothenemus hampei*) in coffee plant treated with *B. bassiana* (Posada et al., 2007). Akello et al. (2008a, 2008b) observed reduced larval duration of banana weevil (*Cosmopolites sordidus*) with least plant damage in banana. The reduction in damage caused by the cotton aphid, *Aphis gossypii* Glover in cotton, and white jute stem weevil (*Apion corchori*) in white jute was reported by Gurulingappa et al. (2010) and Biswas et al. (2013). The accumulated mycotoxins in plant tissues resulted reduction in damage caused by different insect-pests (Gurulingappa et al., 2011). The beneficial effects of endophytic fungi attributed the reduced insect herbivory and it was mainly due to the production of fungal metabolites (Clay and Schardl, 2002).

24.5.2 DISEASE MANAGEMENT THROUGH ENDOPHYTIC B. BASSIANA

The potentiality of endophytic entomopathogenic fungi to control plant pathogens has also been reported by various workers besides insect-pest management (Ownley et al., 2004, 2008a; Griffin et al., 2006; Kim et al.,

2007, 2010; Vega, 2008, Vega et al., 2009; Jaber, 2015). The tomato and cotton seedlings were protected through seed soaking in *B. bassiana* conidia against plant pathogenic fungi *Rhizoctonia solani* and *Phytium myriotylum* (Ownley et al., 2008b). Roberti (1993) reported that the basal rot of onion (*Fusarium oxysporum* f. sp. *cepae*) was significantly reduced following treatment of onion bulbs with *B. bassiana*. Similar reports were observed by Jaber (2015) that the disease incidence and severity of downy mildew, caused by *Plasmopara viticola* (Berk. and Curt.) Berl. and de Toni, was significantly reduced in grapevine following colonization of leaf tissues by *B. bassiana*. There is also a report to offer protection against viral pathogens through *B. bassiana*. Jaber and Salem (2014) observed the reduction in disease incidence and severity of Zucchini yellow mosaic virus in *B. bassisana* inoculated squash plants.

24.5.3 OTHER ROLES OF ENDOPHYTIC B. BASSIANA IN AGRICULTURAL SYSTEM

In addition to insect-pests and disease management, the endophytic fungi provide other benefits to colonized hosts including increased plant growth (Lopez and Sword, 2015; Jaber and Enkerli, 2016), plant development, and nutrient uptake (nitrogen and phosphorus) into plants (Behie et al., 2012; Behie and Bidochka, 2014), improvement in plant hardiness (Khan et al., 2012) by creating a barrier that prevents other pathogenic organisms to colonize the same host (Martinuz et al., 2012) thus controlling phytopathogenic diseases (Moy et al., 2000). Endophytes are important sources of bioactive secondary metabolites, namely, alkaloids, phenolic acids, saponins, tannins, steroids, quinones, and terpenoids act as insect antagonist, exhibit antimicrobial, and many other properties enable the plant with strong defense mechanism (Gouda et al., 2016). Different growth parameters, namely, plant height, weight, etc., are also improved after plants inoculated with fungal endophytes. An improvement was reported in the height, fresh weight of shoots, and roots of *Vicia faba* plants artificial inoculation of *Beauveria brongniartii* and *B. bassiana* (Jaber and Enkerli, 2017). In another study conducted by Lopez and Sword (2015), it has also been observed that increased growth and dry biomass of colonized cotton plants by *B. bassiana* and *P. lilacinum*. Several other researchers have also documented the improved plant growth after treatment with endophytic entomopathogenic fungi (Kabaluk and Ericsson, 2007; Elena et al., 2011; Sasan and Bidochka,

2012; Liao et al., 2014; Jaber and Enkerli, 2016; Jaber and Araj, 2017). Chemicals induced by endophytes impede the growth and development of other competitors, including pathogenic organisms (Clark et al., 1989) that help plants to tolerate biotic stresses, for example, infestation of nematodes and other root-feeding activity by insects (Cosme et al., 2016). In addition, colonized plants also tolerate abiotic stresses, namely, drought, heat, salt, etc. (Khan et al., 2012). Endophytic fungi facilitate indirect enhancing seed dispersal through ants. The seeds of fescue (*Festuca arundinacea*) infected with *Neotyphodium coenophialum* were protected against seed harvesting ants, namely, *Pogonomyrmexrugosus* and *P. occidentalis* (Knoch et al., 1993).

24.6 CONCLUSION

It is well established that fungal endophytes initiate an array of resistance reactions inside the host plant to provide protection against insects' attack. Fungal endophyte, *B. bassiana*, has been found effective against aphids, thrips, lepidopterous larvae, and other insects as they produce toxic substances or signal molecules to modify the host response to such infestation. Artificial inoculation of such endophytes and their establishment in new environment are necessary under organic crop protection to minimize the attack of the wide range of insect pests. In addition, these endophytes also ensure protection of crop losses from plant-parasitic nematodes and harmful pathogens. Besides, they secrete plant growth-promoting hormones, nutrient solubilizing substances, stress mitigating chemical compounds, and other resistance compounds against herbivores. Therefore, the establishment of endophytes after artificial inoculation is the primary objective to harness its beneficial impact on crops like plant growth promotion, nutrient solubilization, tolerances to biotic and abiotic stresses. There is a need to enrich the knowledge about nutrient uptake by plant in presence of endophytes so as to optimize the doses of organic and inorganic fertilizers in artificially endophyte inoculated soil. Moreover, these endophytes also help plants to sustain other stresses, namely, drought, salt, and heat. It opens new dimensions to the knowledge and application of host–plant interactions in presence of endophytes. Secondary metabolites like peptides, polykitides, and acids secreted by endophytes open wider commercial utility of these substances in pharmaceutical and industrial usage apart from agricultural use. Hence, they can easily replace entomopathogenic fungi that are being sprayed over insect population under short-term pest management strategy. Successful establishment of endophytes in plants with entomopathogenic characters can offer their long-term utilization in insect-pest and disease management and can be utilized better in perspectives of organic farming. Though several research works have been carried out on the utilization of fungal endophytes in agricultural systems yet the possibilities of further research on its utilization are immense.

KEYWORDS

- **infection structure**
- **cuticular penetration**
- **insect pest management**

REFERENCES

Affaires, P. D.; Val, L. E.; Quentin, S. T.; Bretonneux, V. L. E. *Agriculture at a Crossroads* (Syntheis Report). International Assessment of Agricultural Knowledge, Science and Technology for Development (IAASTD), 2014.

Akello, J.; Dubois, T.; Coyne, D.; Kyamanywa, S. Effect of Endophytic *Beauveria bassiana* on Populations of the Banana Weevil, *Cosmopolites sordidus*, and Their Damage in Tissue-cultured Banana Plants. *Entomol. Exp. Appl.* 2008a, *129*, 157–165. DOI: 10.1111/j.1570-7458.2008.00759.x.

Akello, J.; Dubois, T.; Coyne, D.; Kyamanywa, S. Endophytic *Beauveria bassiana* in Banana (*Musa* spp.) Reduces Banana weevil (*Cosmopolites sordidus*) Fitness and Damage. *Crop Prot.* 2008b, *27*, 1437–1441. DOI: 10.1016/j.cropro. 2008.07.003.

Akello, J.; Sikora, R. Systemic Acropedal Influence of Endophyte Seed Treatment on *Acyrthosiphon pisum* and *Aphis fabae* Offspring Development and Reproductive Fitness. *Biol. Control* 2012, *61*, 215–221. DOI: 10.1016/j.biocontrol.2012.02.007.

Akutse, K.; Maniania, N.; Fiaboe, K.; Van Den Berg, J.; Ekesi, S. Endophytic Colonization of *Vicia faba* and *Phaseolus vulgaris* (Fabaceae) by Fungal Pathogens and Their Effects on the Life-history Parameters of *Liriomyza huidobrensis* (Diptera: Agromyzidae). *Fungal Ecol.* 2013, *6*, 293–301. DOI: 10.1016/j.funeco.2013.01.003.

Akutse, K. S.; Fiaboe, K. K.; Van Den Berg, J.; Ekesi, S.; Maniania, N. K. Effects of endophyte colonization of *Vicia faba* (Fabaceae) plants on the life–history of leaf miner parasitoids *Phaedrotoma scabriventris* (Hymenoptera: Braconidae) and *Diglyphus isaea* (Hymenoptera: Eulophidae). *PLoS One* 2014, *9*, e109965. DOI: 10.1371/journal. pone.0109965.

Arnold, A. E.; Lewis, L. C. Ecology and Evolution of Fungal Endophytes and Their Roles against Insects. In *Insect-Fungal Associations: Ecology and Evolution*; Vega, F. E., Blackwell, M., Eds.; Oxford University Press: Oxford, 2005; pp 74–96.

Behie, S. W.; Bidochka, M. J. Ubiquity of Insect-Derived Nitrogen Transfer to Plants by Endophytic Insect-pathogenic Fungi: An Additional Branch of the Soil Nitrogen Cycle. *Appl. Environ. Microbiol.* 2014, *80*, 1553–1560. DOI: 10.1128/AEM.03338-13.

Behie, S.; Zelisko, P.; Bidochka, M. Endophytic Insect-parasitic Fungi Translocate Nitrogen Directly from Insects to Plants. *Science* 2012, *336*, 1576–1577. DOI: 10.1126/science.1222289.

Bidochka, M. J.; Khachatourians, G. G. The Implication of Metabolic Acids Produced by *Beauveria Bassiana* in Pathogenesis of the Migratory Grasshopper, *Melanoplus sanguinipes*. *J. Invertebr. Pathol.* 1991, *58*, 106–117.

Bidochka, M. J.; Kamp, A. M.; Lavender, T. M.; Dekoning, J.; De Croos, J. N. A. Habitat Association in Two Genetic Groups of the Insect-Pathogenic Fungus *Metarhizium Anisopliae*: Uncovering Cryptic Species? *Appl. Environ. Microbiol.* 2001, *67* (3), 1335–1342.

Bing, L. A.; Lewis, L. C. Suppression of *Ostrinia nubilalis* (Hübner) (Lepidoptera: Pyralidae) by Endophytic *Beauveria bassiana* (Balsamo) Vuillemin. *Environ. Entomol.* 1991, *20*, 1207–1211. DOI: 10.1093/ee/20.4.1207.

Biswas, C.; Dey, P.; Satpathy, S.; Satya, P.; Mahapatra, B. Endophytic Colonization of White Jute (*Corchorus capsularis*) Plants by Different *Beauveria bassiana* Strains for Managing Stem Weevil (*Apion corchori*). *Phytoparasitica* 2013, *41*, 17–21. DOI: 10.1007/s12600-012-0257-x.

Boucias, D. G.; Pendland, J. C. Nonspecific Factors Involved in the Attachment of Entomopathogenic Deuteromycetes to Host Insect Cuticle. *Appl. Environ. Microbiol.* 1988, *54* (7), 1795–1805.

Bowman, B. H.; Taylor, J. W.; Brownlee, A. G.; Lee, J.; Lu, S. D.; White, T. J. Molecular Evolution of the Fungi: relationship of the *Basidiomycetes*, *Ascomycetes* and *Chytridiomycetes*. *Mol. Biol. Evol.* 1992, *9* (2), 285–296.

Bulat, S. A.; L¨ubeck, M.; Mironenko, N.; Jensen, D. F.; L¨ubeck, P. S. UP-PCR Analysis and ITS1 Ribotyping of Strains of *Trichoderma* and *Gliocladium*. *Mycol. Res.* 1998, *102* (8), 933–943.

Bulat, S. A.; Lubeck, M.; Alekhina, I. A.; Jensen, D. F.; Knudsen, I. M. B.; Lubeck, P. S. Identification of a Universally Primed-PCR-derived Sequence-characterized Amplified Region Marker for an Antagonistic Strain of *Clonostachys rosea* and Development of a Strain-specific PCR Detection Assay. *Appl. Environ. Microbiol.* 2000, *66* (11), 4758–4763.

Cherry, A. J.; Banito, A.; Djegui, D.; Lomer, C. Suppression of the Stem-borer *Sesamia calamistis* (Lepidoptera; Noctuidae) in Maize Following Seed Dressing, Topical Application and Stem Injection with African Isolates of *Beauveria Bassiana*. *Int. J. Pest Manag.* 2004, *50*, 67–73. DOI: 10.1080/ 09670870310001637426.

Cho, E. M.; Liu, L.; Farmerie, W.; Keyhani, N. O. EST Analysis of cDNA Libraries from the Entomopathogenic Fungus *Beauveria (Cordyceps) bassiana*. I. Evidence for Stage-specific Gene Expression in Aerial Conidia, in vitro Blastospores and Submerged Conidia. *Microbiology* 2006, *152*, 2843–2854.

Clark, C. L.; Miller, J. D.; Whitney, N. J. Toxicity of Conifer Needle Endophytes to Spruce Budworm. *Mycol. Res.* 1989, *93*, 508–512. DOI: 10.1016/j. phytochem.2010.01.015.

Clay, K.; Schardl, C. Evolutionary Origins and Ecological Consequences of Endophyte Symbiosis with Grasses. *Am. Nat.* 2002, *160*, S99–S127. DOI: 10.1086/342161.

Cosme, M.; Lu, J.; Erb, M.; Stout, M. J.; Franken, P.; Wurst, S. A. Fungal Endophyte Helps Plants to Tolerate Root Herbivory Through Changes in Gibberellins and Jasmonate Signaling. *New Phytol.* 2016, *211*, 1065–1076. DOI: 10.1111/nph.13957.

de Muro, M. A.; Mehta, S.; Moore, D. The Use of Amplified Fragment Length Polymorphism for Molecular Analysis of *Beauveria bassiana* Isolates from Kenya and Other Countries, and Their Correlation with Host and Geographical Origin. *FEMS Microbiol. Lett.* 2003, *229*, (2), 249–257.

de Muro, M. A.; Elliott, S.; Moore, D. Molecular Characterization of *Beauveria bassiana* Isolates Obtained from Overwintering Sites of Sunn Pests (*Eurygaster* and *Aelia* species). *Mycol. Res.* 2005, *109* (3), 294–306.

Driver, F.; Milner, R. J.; Trueman, J. W. H. A Taxonomic Revision of *Metarhizium b*ased on a Phylogenetic Analysis of rDNA Sequence Data. *Mycol. Res.* 2000, *104* (2), 134–150.

Elena, G. J.; Beatriz, P. J.; Alejandro, P.; Lecuona, R. *Metarhizium anisopliae* (Metschnikoff) Sorokin Promotes Growth and has Endophytic Activity in Tomato Plants. *Adv. Biol. Res.* 2011, *5*, 22–27.

Elsworth, J. F.; Grove, J. F. Cyclodepsipeptides from *Beauveria bassiana* Bals. Part 1. Beauverolides H and I. J. *Chem. Soc.*1977, [Perkin 1] *3*, 270–273.

Enkerli, J.; K¨olliker, R.; Keller, S.; Widmer, F. Isolation and Characterization of Microsatellite Markers from the Entomopathogenic Fungus *Metarhizium anisopliae. Mol. Ecol. Notes* 2005, *5* (2), 384–386.

Gange, A. C.; Eschen, R.; Wearn, J. A.; Thawer, A.; Sutton, B. C. Differential Effects of Foliar Endophytic Fungi on Insect Herbivores Attacking a Herbaceous Plant. *Oecologia* 2012, *168*, 1023–1031. DOI: 10.1007/s00442-011-2151-5.

Gathage, J. W.; Lagat, Z. O.; Fiaboe, K. K. M.; Akutse, K. S.; Ekesi, S.; Maniania, N. K. Prospects of fungal Endophytes in the Control of *Liriomyza* Leafminer Flies in Common Bean *Phaseolus vulgaris* under Field Conditions. *Biocontrol* 2016, *61*, 741–753. DOI: 10.1007/s10526-016-9761-0.

Glare, T. R.; Reay, S. D.; Nelson, T. L.; Moore, R. *Beauveria caledonica* is a Naturally Occurring Pathogen of Forest Beetles. *Mycol. Res.* 2008, *112* (3), 352–360.

Glass, N. L.; Kuldau, G .A. Mating type and Vegetative Compatibility in Filamentous Ascomycetes. *Annu. Rev. Phytopathol.* 1992, *30*, 201–224.

Gomez-Vidal, S.; Salinas, J.; Tena, M.; Vicente Lopez-Llorca, L. Proteomic Analysis of Date Palm (*Phoenix dactylifera* L.) Responses to Endophytic Colonization by Entomopathogenic Fungi. *Electrophoresis* 2009, *30*, 2996–3005.

Gouda, S.; Das, G.; Sen, S. K.; Shin, H. S.; Patra, J. K. Endophytes: A Treasure House of Bioactive Compounds of Medicinal Importance. *Front. Microbiol.* 2016, *7*, 1538. DOI: 10.3389/fmicb.2016.01538.

Griffin, M.; Ownley, B.; Klingeman, W.; Pereira, R. Evidence of Induced Systemic Resistance with *Beauveria bassiana* against *Xanthomonas* in Cotton. *Phytopathology* 2006, *96*, S42.

Grove, J. F.; Pople, M. The Insecticidal Activity of Beauvericin and the Enniatin Complex. *Mycopathology* 1980, *70*, 103–105.

Gurulingappa, P.; Sword, G. A.; Murdoch, G.; Mcgee, P. A. Colonization of Crop Plants by Fungal Entomopathogens and their Effects on Two Insect Pests when in Planta. *Biol. Control* 2010, *55*, 34–41. DOI: 10.1016/j.biocontrol. 2010.06.011.

Gurulingappa, P.; Mcgee, P. A.; Sword, G. Endophytic *Lecanicillium lecanii* and *Beauveria bassiana* Reduce the Survival and Fecundity of *Aphis Gossypii* Following Contact with Conidia and Secondary Metabolites. *Crop Prot.* 2011, *30*, 349–353. DOI: 10.1016/j. cropro.2010.11.017.

Hackman, R. H. Cuticle: Biochemistry. In *Biology of the Integument*; Bereiter-Hahn, J., Matolts, A. G., Richards, K. S., Eds.; Springer Verlag: Berlin, 1984; pp 626–637.

Hamill, R. L.; Sullivan, H. R. Determination of Pyrrolnitrin and Derivatives by Gas-liquid Chromatography. *Appl. Microbiol.* 1969, *18* (3), 310–312.

HarDOIm, P. R.; Van Overbeek, L. S.; Berg, G.; Pirttilä, A. M.; Compant, S.; Campisano, A. The Hidden World within Plants: Ecological and Evolutionary Considerations for Defining Functioning of Microbial Endophytes. *Microbiol. Mol. Biol. Rev.* 2015, *79*, 293–320. DOI: 10.1128/MMBR.00050-14.

Hartley, S. E.; Gange, A. C. Impacts of Plant Symbiotic Fungi on Insect Herbivores: Mutualism in a Multitrophic Context. *Annu. Rev. Entomol.* 2009, *54*, 323–342. DOI: 10.1146/annurev.ento.54.110807.090614.

Howe, G. A.; Jander, G. Plant Immunity to Insect Herbivores. *Annu Rev Plant Biol.* 2008, *59*, 41–66.

Jaber, L. R. Grapevine Leaf Tissue Colonization by the Fungal Entomopathogen *Beauveria bassiana* and its Effect against Downy Mildew. *Biocontrol* 2015, *60*, 103–112. DOI: 10.1007/s10526-014-9618-3.

Jaber, L. R.; Araj, S. E. Interactions Among Endophytic Fungal Entomopathogens (Ascomycota: Hypocreales), the Green Peach Aphid *Myzus persicae* Sulzer (Homoptera: Aphididae), and the Aphid Endoparasitoid *Aphidius colemani* Viereck (Hymenoptera: Braconidae). *Biol. Control* 2017, *116*, 53–61. DOI: 10.1016/j.biocontrol.2017.04.005.

Jaber, L. R.; Enkerli, J. Effect of Seed Treatment Duration on Growth and Colonization of *Vicia faba* by Endophytic *Beauveria bassiana* and *Metarhizium brunneum*. *Biol. Control* 2016, *103*, 187–195. DOI: 10.1016/j.biocontrol.2016.09.008.

Jaber, L. R.; Enkerli, J. Fungal Entomopathogens as Endophytes: Can They Promote Plant Growth? *Biocontrol Sci. Technol.* 2017, *27*, 28–41. DOI: 10.1080/09583157.2016.1243227.

Jaber, L. R.; Salem, N. M. Endophytic Colonisation of Squash by the Fungal Entomopathogen *Beauveria bassiana* (Ascomycota: Hypocreales) for Managing Zucchini Yellow Mosaic Virus in Cucurbits. *Biocontrol Sci. Technol.* 2014, *24*, 1096–1109. DOI: 10.1080/09583157.2014.923379.

Jestoi, M.; Rokka, M.; Jarvenpaa, E.; Peltonen, K. Determination of *Fusarium* Mycotoxins Beauvericin and Enniatins (A, A1, B, B1) in Eggs of Laying Hens Using Liquid Chromatography– Tandem Mass Spectrometry (LC–MS/MS). *Food Chem.* 2009, *115*, 1120–1127.

Jirakkakul, J.; Punya, J.; Pongpattanakitshote, S.; Paungmoung, P.; Vorapreeda, N.; Tachaleat, A.; Klomnara, C.; Tanticharoen, M.; Cheevadhanarak, S. Identification of the Nonribosomal Peptide Synthetase Gene Responsible for Bassianolide Synthesis in Wood-Decaying Fungus *Xylaria* sp. BCC1067. *Microbiology*, 2008, *154*, 995–1006.

Kabaluk, J. T.; Ericsson, J. D. Seed Treatment Increases Yield of Field Corn When Applied for Wireworm Control. *Agron. J.* 2007, 99: 1377–1381. DOI: 10.2134/agronj2007.0017N.

Kanaoka, M.; Isoga, A.; Murakosh, S. I.; Ichjnoe, M.; Suzuki, A.; Tamura, S. Bassianolide, A New Insecticidal Cyclodepsipeptide from *Beauveria bassiana* and *Verticillium lecanii*. *Agric. Biol. Chem.* 1978, *42*, 629–635.

Khan, A. L.; Hamayun, M.; Khan, S. A.; Kang, S.-M.; Shinwari, Z. K.; Kamran, M. Pure Culture of *Metarhizium anisopliae* LHL07 Reprograms Soybean to Higher Growth and Mitigates Salt Stress. *World J. Microbiol. Biotechnol.* 2012, *28*, 1483–1494. DOI: 10.1007/s11274-011-0950-9.

Kim, J. J.; Goettel, M. S.; Gillespie, D. R. Potential of *Lecanicillium* species for Dual Microbial Control of Aphids and the Cucumber Powdery Mildew Fungus, *Sphaerotheca fuliginea*. *Biol. Control* 2007, *40*, 327–332. DOI: 10.1016/j.biocontrol.2006.12.002

Kim, J. J.; Goettel, M. S.; Gillespie, D. R. Evaluation of *Lecanicillium longisporum*, VertalecR against the Cotton Aphid, *Aphis gossypii*, and Cucumber Powdery Mildew, *Sphaerotheca fuliginea* in a Greenhouse Environment. *Crop Prot.* 2010, *29*, 540–544. DOI: 10.1016/j.cropro.2009.12.011.

Klieber, J.; Reineke, A. The Entomopathogen *Beauveria bassiana* has Epiphytic and Endophytic Activity against the Tomato Leaf Miner *Tuta absoluta*. *J. Appl. Entomol.* 2016, *140*, 580–589. DOI: 10.1111/jen.12287.

Knoch, T. R.; Faeth, S. H.; Arnott, D. L. Endophytic Fungi Alter Foraging and Dispersal by Desert Seed-harvesting Ants. *Oecologia* 1993, *95*, 470–473. DOI: 10.1007/BF00317429.

Lubeck, M.; Alekhina, I. A.; L¨ubecks, P. S.; Jensen, D. F.; Bulat, S. A. Delineation of *Trichoderma harziannum* into Two Different Genotypic Groups by a Highly Robust Fingerprinting Method, UP-PCR, and UP-PCR Product Cross-Hybridization. *Mycol. Res.* 1999, *103* (3), 289–298.

Lacey, L. A.; Neven, L. G. The Potential of the Fungus, *Muscodor albus*, as a Microbial Control Agent of Potato Tuber Moth (Lepidoptera: Gelechiidae) in Stored Potatoes. *J. Invertebr. Pathol.* 2006, *91*, 195–198. DOI: 10.1016/j.jip.2006.01.002.

Latge, J. P.; Monsigny, M. Visualization of Exocellular Lectins in the Entomopathogenic Fungus *Conidiobolus obscurus*. *J. Histochem. Cytochem.* 1988, *36*, 1419–1424.

Leckie, B. M. Effects of *Beauveria bassiana* Mycelia and Metabolites Incorporated into Synthetic Diet and Fed to Larval *Helicoverpa Zea*; and Detection of Endophytic *Beauveria Bassiana* in Tomato Plants Using PCR and ITS Primers. *Master Thesis*, University of Tennessee, Knoxville, TN, 2002.

Leslie, J. F. Fungal Vegetative Compatibility. *Annu. Rev. Phytopathol* 1993, *31*, 127–150.

Lewis, L. C.; Bruck, D. J.; Gunnarson, R. D.; Bidne, K. G. Assessment of Plant Pathogenicity of Endophytic *Beauveria bassiana* in *Bt* Transgenic and Non-Transgenic Corn. *Crop Sci.* 2001, *41*, 1395–1400. DOI: 10.2135/cropsci2001.4151395x.

Liao, X.; O'brien, T. R.; Fang, W.; Leger, R. J. S. The Plant Beneficial Effects of *Metarhizium* Species Correlate with their Association with Roots. *Appl. Microbiol. Biotechnol.* 2014, *98*, 7089–7096. DOI: 10.1007/s00253-014-5788-2.

Lopez, D. C.; Sword, G. A. The Endophytic Fungal Entomopathogens *Beauveria bassiana* and *Purpureocillium Lilacinum* Enhance the Growth of Cultivated Cotton (*Gossypium Hirsutum*) and Negatively Affect Survival of the Cotton Bollworm (*Helicoverpa zea*). *Biol. Control* 2015, *89*, 53–60. DOI: 10.1016/j. biocontrol.2015.03.010.

Martinuz, A.; Schouten, A.; Sikora, R. Systemically Induced Resistance and Microbial Competitive Exclusion: Implications on Biological Control. *Phytopathology* 2012, *102*, 260–266. DOI: 10.1094/PHYTO-04-11-0120.

Maurer P.; Couteaudier Y.; Girard P. A.; Bridge P. D.; Riba, G. Genetic Diversity of *Beauveria bassiana* (Bals.) Vuill. and Relatedness to Host Insect Range. *Mycol. Res.* 1997, *101*, 159–164.

McGee, P. Reduced Growth and Deterrence from Feeding of the Insect Pest *Helicoverpa Armigera* Associated with Fungal Endophytes from Cotton. *Aust. J. Exp. Agric.* 2002, *42*, 995–999. DOI: 10.1071/EA01124.

Meca, A.; Sepulveda, B.; Ogona, J. C.; Grados, N.; Moret, A.; Morgan, M.; Tume, P. *In vitro* Pathogenicity of Northern Peru Native Bacteria on *Phyllocnistis citrella* Stainton (Gracillariidae: Phyllocnistinae), on Predator Insects (*Hippodamia convergens* and *Chrisoperna externa*), on *Citrus aurantifolia* Swingle and White Rats. *Span J. Agric. Res.* 2009, *7*, 137–145.

Meyling, N. V.; Eilenberg, J. Isolation and Characterization of *Beauveria bassiana* Isolates from Phylloplanes of Hedgerow Vegetation. *Mycol. Res.* 2006, *110* (2), 188–195.

Moy, M.; Belanger, F.; Duncan, R.; Freehoff, A.; Leary, C.; Meyer, W. Identification of Epiphyllous Mycelial Nets on Leaves of Grasses Infected by Clavicipitaceous Endophytes. *Symbiosis* 2000, *28*, 291–302.

Neville, A. C. Cuticle: Organization. In *Biology of the Integument*; Bereiter-Hahn, J., Matolts, A. G., Richards, K. S., Eds.; Springer Verlag: Berlin, 1984; pp 611–625.

Ownley, B. H.; Dee, M. M.; Gwinn, K. Effect of Conidial Seed Treatment Rate of Entomopathogenic *Beauveria bassiana* 11-98 on Endophytic Colonization of Tomato Seedlings and Control of *Rhizoctonia* Disease. *Phytopathology* 2008a, *98*, S118.

Ownley, B. H.; Griffin, M. R.; Klingeman, W. E.; Gwinn, K. D.; Moulton, J. K.; Pereira, R. M. *Beauveria bassiana*: Endophytic Colonization and Plant Disease Control. *J. Invertebr. Pathol.* 2008b, 98: 267–270. DOI: 10.1016/j.jip.2008.01.010.

Ownley, B. H.; Pereira, R. M.; Klingeman, W. E.; Quigley, N. B.; Leckie, B. M. *Beauveria bassiana*, A Dual Purpose Biological Control with Activity against Insect Pests and Plant Pathogens. In *Emerging Concepts in Plant Health Management*; Lartey, R. T., Caesar, A. J., Eds.; Research Signpost: Trivendrum, 2004; pp 255–269.

Pineda, A.; Zheng, S. J.; van Loon, J. J. A.; Pieterse, C. M. J.; Dicke, M. Helping Plants to Deal with Insects: The Role of Beneficial Soil-Borne Microbes. *Trends Plant Sci.* 2010, *15*, 507–514.

Pinnamaneni, R.; Potineni, K. Mechanisms Involved in the Entomopathogenesis of *Beauveria bassiana. Asian J. Environ. Sci.* 2010, *5*, 65–74.

Posada, F.; Aime, M. C.; Peterson, S. W.; Rehner, S. A.; Vega, F. E. Inoculation of Coffee Plants with the Fungal Entomopathogen *Beauveria bassiana* (Ascomycota: Hypocreales). *Mycol. Res*. 2007, *111*, 748–757. DOI: 10.1016/j. mycres.2007.03.006.

Puhalla, J. E. A Visual Indicator of Heterokaryosis in *Fusarium oxysporum* from Celery. *Can. J. Bot.* 1984, *62*, 540–545.

Puri, A.; Padda, K. P.; Chanway, C. P. Evidence of Nitrogen Fixation and Growth Promotion in Canola (B*rassica napus* l.) By an Endophytic Diazotroph *Paenibacillus polymyxa* P2b-2R. *Biol. Fertil. Soils* 2016, *52*, 119–125. DOI: 10.1007/ s00374-015-1051-y.

Quesada-Moraga, E.; Vey, A. Bassiacridin, a Protein Toxic for Locusts Secreted by the Entomopathogenic Fungus *Beauveria bassiana. Mycol. Res.* 2004, *108*, 441–452.

Quesada-Moraga, E.; Munoz-Ledesma, F.; Santiago-Alvarez, C. Systemic Protection of *Papaver somniferum* L. against *Iraella luteipes* (Hymenoptera: Cynipidae) by an Endophytic Strain of *Beauveria bassiana* (Ascomycota: Hypocreales). *Environ. Entomol.* 2009, *38*, 723–730. DOI: 10.1603/022. 038.0324.

Rana, I. S.; Kanojiya, A.; Sandhu, S. S. Mosquito larvicidal potential of fungi isolated from larval mosquito habitats against *Aedes aegypti. J. Biol. Control* 2008, *22*, 179–183.

Rehner, S. A.; Buckley, E. P. Isolation and Characterization of Microsatellite Loci from the Entomopathogenic Fungus *Beauveria bassiana* (Ascomycota: Hypocreales). *Mol. Ecol. Notes* 2003, *3*, 409–411.

Resquín-Romero, G.; Garrido-Jurado, I.; Delso, C.; Ríos-Moreno, A.; Quesada-Moraga, E. Transient Endophytic Colonizations of Plants Improve the Outcome of Foliar Applications of Mycoinsecticides against Chewing Insects. *J. Invertebr. Pathol.* 2016, *136*, 23–31. DOI: 10.1016/j.jip.2016.03.003.

Saikkonen, K.; Lehtonen, P.; Helander, M.; Koricheva, J.; Faeth, S. H. Model Systems in Ecology: Dissecting the Endophyte–Grass Literature. *Trends Plant Sci.* 2006, *11*, 428–433. DOI: 10.1016/j.tplants.2006.07.001.

Saikkonen, K.; Saari, S.; Helander, M. Defensive Mutualism between Plants and Endophytic Fungi? *Fungal Divers* 2010, *41*, 101–113.

Sasan, R. K.; Bidochka, M. J. The Insect-pathogenic Fungus *Metarhizium robertsii* (Clavicipitaceae) is Also an Endophyte that Stimulates Plant Root Development. *Am. J. Bot.* 2012, *99*, 101–107. DOI: 10.3732/ajb.1100136.

St Leger, R. J.; Frank, D. C. Molecular Cloning and Regulatory Analysis of the Cuticle-Degrading- Protease Structural Gene from the Entomopathogenic Fungus *Metarhizium anisopliae*. *Eur. J. Biochem.* 1992, *204* (3), 991–1001.

St Leger, R. J.; Roberts, D. W. A Model to Explain Differentiation of Appressoria by Germlings of *Metarhizium anisopliae*. *J. Invertebr. Pathol.* 1991, *57*, 299–310.

Steinhaus, E. A. Fungal Infections. In *Principles of Insect Pathology*,1st ed.; 1956; p 318.

Thakur, R.; Sandhu, S. S. Distribution, Occurrence and Natural Invertebrate Hosts of Indigenous Entomopathogenic Fungi of Central India. *Indian J. Microbiol.* 2010, *50* (1), 89–96.

Thakur, R.; Jain, N.; Pathak, R.; Sandhu, S. S. Practices in Wound Healing Studies of Plants. Evidence Based Complementary and Alternative Medicine, 2011; p 17.

Vega, F. E. Insect Pathology and Fungal Endophytes. *J. Invertebr. Pathol.* 2008, *98*, 277–279. DOI: 10.1016/j.jip.2008.01.008.

Vega, F. E.; Goettel, M. S.; Blackwell, M.; Chandler, D.; Jackson, M. A.; Keller, S. Fungal Entomopathogens: New Insights on their Ecology. *Fungal Ecol.* 2009, *2*, 149–159. DOI: 10.1016/j.funeco.2009.05.001.

Viaud, M.; Couteaudier, Y.; Riba, G. Genomic Organisation of *Beauveria bassiana*: Electrophoretic Karyotyping, Gene Mapping and Telomeric Fingerprinting. *Fungal Genet. Biol.* 1996, *20*, 175–183.

Vidal, S.; Jaber, L. R. Entomopathogenic Fungi as Endophytes: Plant-Endophyte-Herbivore Interactions and Prospects for Use in Biological Control. *Curr. Sci.* 2015, *109*, 46–54.

Vos, C. M. F.; De Cremer, K.; Cammue, B. P. A.; De Coninck, B. The Toolbox of spp. in the Biocontrol of Disease. *Mol. Plant Pathol.* 2015, *16* (4), 400–412.

Wang, Q.; Xu, L. Beauvericin, A Bioactive Compound Produced by Fungi: A Short Review. *Molecules* 2012, *17* (12): 2367–2377.

Xiao, G.; Ying, S. H.; Zheng, P.; Wang, Z. L.; Zhang, S.; Xie, X. Q.; Shang, Y.; St Leger, R. J.; Zhao, G. P.; Wang, C. Genomic Perspectives on the Evolution of Fungal Entomopathogenicity in *Beauveria bassiana*. *Sci. Rep.* 2012, *2*, 483.

Xu, Y.; Orozco, R.; Kithsiri Wijeratne, E. M.; Espinosa-Artiles, P.; Leslie Gunatilaka, A. A.; Patricia Stock, S.; Molnár, I. *Biosynthesis of the Cyclooligomer Depsipeptide Bassianolide, an Insecticidal Virulence Factor of Beauveria bassiana. Fungal Genet. Biol.* 2009, *46*,353–364.

CHAPTER 25

Earthworms: An Important Ingredient for Organic Farming

R. A. SINGH*

Organic Farming, C.S. Azad University of Agriculture & Technology, Kanpur 208002, Uttar Pradesh, India

**Corresponding author. E-mail: rasinghcsau@gmail.com*

ABSTRACT

Earthworms have always been regarded as friends of the farmers. Earthworms have a very positive effect on the physical, chemical and biological parameters of the soils. The earthworm is a soil biotechnologist and a solid waste manager. Earthworms are known to consume large quantities of organic litter or waste and convert them into manure, which is used as valuable compost, known as 'vermicompost'. Their use has been found to improve the pod yield of peanut and vegetable pea by using organic components like FYM and vermiculture. Peanut and vegetable pea could successfully be raised under sequential cropping by the use of FYM at 100¦q/ha + vermiculture to peanut and 100¦q/ha FYM to vegetable pea for completing the starter dose 40¦kg N + 60¦kg P2O5 + 40¦kg K2O/ha, besides better management of natural resources and residue for higher productivity and monetary return. It is beneficial for the farm families residing in the riverine tract of Uttar Pradesh.

25.1 INTRODUCTION

Earthworms are important ingredients for recycling of decomposable wastes back to the soil in agriculture, horticulture, and forestry, and other areas.

Organic Crop Production Management: Focus on India, with Global Implications, D. P. Singh, PhD, H. G. Prakash, PhD, M. Swapna, PhD, & S. Solomon, PhD (Eds.)

Earthworms are beneficial basically due to their life activities in soils and their importance in vermiculture and vermin-composting. It is a well-known fact that sustainable agriculture involves the successful management of all available resources for agriculture to satisfy the changing human needs while maintaining and enhancing the quality of environment and conserving resources. The intensive agriculture with modern agricultural technologies ushered in *Green Revolution* which is attributed to the use of high-yielding varieties, more of inputs like fertilizers, pesticides, weedicides, and better irrigation facilities, besides efficient management. This resulted in increased production of wheat and rice but affected sustainability of the Indian agriculture in various ways. Problems that emerged are varied, notably, heavy dependence on fertilizer inputs, increased micronutrients deficiency, reduced cultivation acreage for pulses and others, depleted subsoil water table, increased dependence to pesticides and weedicides, and deterioration of plant substrate—the soil. Therefore, without conservation and management of soil fertility or its sustainability, continued intensive farming would turn green cultivable lands into deserts. The continued irrigation would degrade soil to kallor/salty/leechy and sandy without damage management.

Excessive use of chemical fertilizer also has ecological and economic implications. These are well known to affect soil chemistry, depletion of soil micronutrients, and may cause water pollution. Some of these as chain reaction affecting human health, namely, higher nitrogen application in some leaf vegetables leading to accumulation of nitrites in leaves which affect various body functions in humans. The utilization of earthworms' natural activities may reduce these problems (Bhatnagar and Palta, 1996). The use of chemical fertilizers and pesticides may cause soil erosion, resulting in the loss of humus which supports microbes like algae, fungi, soil bacteria, and microbes like earthworms and insects that are responsible for the fertility of the soil. Earthworms help the natural maintenance of all these beneficial factors toward soil management.

Indiscriminate, rather injudicious, use of pesticides and weedicides also causes environment threat, killing the beneficial nontargets organisms like earthworms that are essential for maintaining ecological balance.

25.2 SIGNIFICANCE OF EARTHWORMS IN ORGANIC FARMING

The principle of organic farming is to produce food of good quality and quantity by using eco-friendly technologies, which can coexist with nature.

Such practices exclude the use of chemical fertilizer, pesticides, and weedicides. The system depends upon the use of leguminous plants and microbial inoculations for nitrogen fixation, crop rotation, organic manures, vermiculture and vermicompost (organic waste), and biological control methods. To all these, vermiculture and its utilization is an important component. The scientific recommendation is that 1 kg earthworms decompose by 4–5 kg of organic waste every 24 h.

25.3 BENEFICIAL ACTIVITIES OF EARTHWORMS

The important activities of earthworms are given below.

Feeding—It results in breakdown of soil particles and mixing of soil nutrients and bacteria in the digestive process as well with deposit of casts, which is responsible for automatic conversion of organic wastes. These micronized soil particles, lead to an increase of particle surface area which increases moisture absorption, holding capacity, and air circulation.

Burrowing—Burrowing of earthworms brings about tillage of soil. Interestingly in most conventional forms of tillage, it is up to depths of 30 cm, while earthworms carry out tillage up to 3 m, without adversely affecting plants in any manner.

Microbial action—Increase microbial action. Casts form fine quality biofertilizer having up to 1000 times more microbes than the surrounding soil.

Promotes porosity—Feeding on soil particles with burrowing promotes porosity, which increase percolation and infiltration of water.

Check soil erosion—Proper drainage is promoted and in many soil, erosion is checked as with casts.

Particle binding—There is a degree of particle binding which promotes the growth of vegetation.

Soil porosity—Increase soil porosity that leads to increase in soil aeration.

Soil temperature—Reduce the severity of soil temperature fluctuation essential for plant growth.

Soil fertility and structure of soil—Earthworms incorporate plant residues, organic waste, and dung with soil from surface. These materials are often deposited inside burrows where these are left for decomposition. Thus, organic matter is mixed in different soil layers. The organic matter decomposition, that is, humification is hastened by earthworm activities. This is brought about by the passage of organic matter through gut of earthworms.

Microorganisms within gut pass out with the casts and thereby increase the soil microbial content. Thus, there is a dissemination of several types of microorganisms. Further, activities like feeding, burrowing, and excretion of earthworms on soil favor the growth of soil microorganisms that synthesize polymers present in humus substances.

Humification enhances the water holding capacity of soil, improves soil structure, and increases ionic activities. All these further add to the fertility and fitness of soil for agriculture. In other words, we can say that soil structure change improves with the activity of the earthworms.

Nutrient availability—More than 15–30% of phosphorus and more than 6% of nitrogen are made available to plants by worm activities in soil. Similarly, the availability of trace plant nutrients like calcium, magnesium, potassium, phosphorus, and molybdenum to plants is improved through worm casts.

Soil aeration—Earthworms improve soil aeration by making pores with its burrowing, feeding, and casting activities. Aerating pores made by activities of earthworms vary from large ones (2.0–11.0 mm) to small ones (namely, feeding and casting).

These pores have important roles in the total decomposition process. This is because with appropriate oxygen supply, growth of aerobic microorganisms dominates anaerobic decomposition facilitating complete degradation. This is faster and reduces foul smell. In reduced aeration, oxygen supply is less which results in the dominance of anaerobic microorganisms and decomposition where breakdown is only partial. Aeration is, therefore, essential requirement in decomposition process. In conventional composting, this is brought about through a labor-intensive process by periodic turning over and mixing. While in vermicomposting, all processes related to aeration, mixing, and turning over is done with earthworm activities.

Not only live earthworms are beneficial but also benefits can be accrued even after the death of worms in natural conditions. Nitrogen content in the soil is increased with decay of dead earthworms. It has been estimated that in an earthworm species, dead individual can yield up to 10 mg of nitrates. Even after death, a population of 3.75 million/ha earthworms would yield approximate 217 kg of nitrate of soda/ha.

25.4 VERMICULTURE

The following point to be considered for the planning of vermiculture:

a. Site selection.
b. Availability of decomposable organic waste, its daily quantity and quality, alternate organic material, transportation, and stocking.
c. Marketable outlet and requirement with future scope.
d. Collection and study of knowhow on earthworms to be cultured.
e. Collection and procurement of suitable species.
f. Testing of the suitability of species on performance in available organic waste.
g. Maintenance of seed culture for eventual large-scale culture.

In addition to these, numerous allied points should also be considered, these are as follows:

- Funding.
- Availability of labor.
- Quality of water.
- Packaging.

Among the aforementioned points of vermiculture, the first is selection of suitable species for culture and eventual vermicomposting.

25.5 SELECTION OF SUITABLE SPECIES

There are three types of earthworms that can be used in vermiculture. These are (1) epiges, (2) endoges, and (3) aneciques.

25.5.1 EPIGES

They are small sized with uniform body color, live on surface litter or dung, tolerate disturbance, have active gizzard, but have limited period of activity. These are phytophagous. Most of these species are good biodegrading agents, so are good nutrient releasers, but do not redistribute nutrients. These have insignificant role in humus formation and are not good for use in field conditions for soil reclamation, unless careful manipulations are done. These have a short life cycle with high reproduction and regeneration rate.

25.5.2 ENDOGES

These are small or large-sized worms with weak pigmentation, found in topsoil layer of organic and mineralized matter. They burrow branchings that are horizontal and moderately tolerate disturbance. According to one expert, these show some preference in the selection of feed substrate (buffered and predictable) conditions. These are geophagous and their life cycles are of intermediate duration but have the potential in soil improvements due to high efficiency in energy utilization from poor soils, so can be used in field with some manipulations.

25.5.3 ANECIQUES

These are large-sized worms with pigmentation only at anterior and posterior ends. These are largely nocturnal, deep burrowing, and pull leaves or litter matter into soil (burrows), and are phytogeophagous. Tolerance to disturbance is poor and reproduction rate as evidenced from cocoon production is low. Since they emerge from deep burrow to the surface for casting, they play a useful role in mixing nutrients.

The species have a high metabolic rate with high conversion rates, small body size, high reproductive rate, high productivity, and are early colonizers of new environments (thrive on high organic humus). Epigeic and endogeic earthworms, therefore, are most suitable for vermicompostings. The species have a large body size, with long life span and low metabolic rates and have lesser movement outside burrows.

25.6 CHARACTERISTIC OF SUITABLE SPECIES FOR VERMICOMPOSTING

The following characters of the worms need to be considered:

1. Worms should be efficient convertor of plant or animal biomass to body proteins, so that its growth rates are high.
2. These should have high consumption, digestion, and assimilation rate (composting qualities).
3. Worms should have wide adaptability (tolerance) to environmental factors (capability to live in varying temperature conditions).

4. These should have feeding preference as well as adaptability for a wide range of organic material (high and rich organic matter).
5. Worms should produce large numbers of cocoons that should not have long hatching time, so that fast multiplication and organic matter conversion may occur.
6. Growth rate, maturity from young one to adult stage should be fast.
7. Worms should have compatibility or tolerance with other worms (as with possibility of mixture of species by amateurs).
8. Worms should be disease-resistant.

25.7 VERMICOMPOSTING

Vermiculture and vermicomposting are two interlinked and interdependent processes that when conjoined can be referred as vermitechnology. Vermiculture can only be done on compostable or decomposable organic matter. Composting is the outcome of earthworm activities. So both the processes can be brought about simultaneously.

25.8 ADVANTAGES OF VERMICOMPOSTING

1. Huge quantities of domestic, agricultural, and rural industrial organic wastes can be recycled for various usages.
2. Reduces pollution.
3. Vermicompost substitution with fertilizer input will reduce economic input.
4. Vermicompost can be produced nearest to the site, where it will be used.
5. Vermicompost with or without worms can be marketed, in case of surplus production, for generating extra income.

25.9 VERMICOMPOSTING MATERIALS

Commonly used composting materials are given below:

Animal dung—Cattle dung, sheep dung, horse dung, and goat dung are used in vermicomposting. Except cattle dung, use of other animal dungs requires various necessary preliminary testing and precautions for pathogens and responses to earthworms. The tetanus virus is common in horse dung, which is lethal to human beings. In sheep dung, the initial growth of earthworms is found to be poor.

Agricultural waste—Select all the items discarded after harvesting and threshing, that is, stem, leaves, husk (except paddy husk), peels, vegetable waste, orchard leaf litter, processed food wastes, cane trash and baggase, banana leaves, and processing wastes.

Forestry waste—Shavings, peels, sawdust, pulp, and forest leaf litter.

City leaf litter—Road site planted leaves.

Waste paper and cotton cloth—These are decomposable organic waste that can be used for vermincomposting.

City refuge—Kitchen waste with little manipulation can be used for vermicomposting.

Biogas slurry—After recovery of biogas, if not required for agricultural use, namely, in conventional composting may be use for vermicomposting.

25.10 PREPARATION OF VERMICOMPOST

After necessary pretests and selection of suitable species, vermicomposting bed is prepared. This is in fact the most important aspects of the whole program. The steps described for schemes are as follows:

- Available container or pit is to be cleaned for removing unwanted chemical or other material, if any present. The 2 × 2 × 1 M size pit can be prepared under the tree shade at an elevated place or use the same size container of wood or plastic with 10–12 holes in the bottom to drainage water.
- At the bottom, an 11 cm thick layer of brick or stone is laid. Over this, a 2-cm thick layer of sand or *maurang* should be layered. In the third layer, 15 cm thick layer of productive soil is laid, and moist this layer with the spraying of water. Over this layer, 5–7 cm thick layer of partially rotten powdered cow dung is put, thereafter, 1 kg live earthworms are gently released over it.
- After the release of earthworms, 5–10 cm thick layer of kitchen waste, crop residue, *Jalkumbhi*, tree leaves, and others are put over it. Spray the filled material with water till 20–25 days. Thereafter, a thick layer of 5–10 cm rottable material should be laid twice in a week till the complete filling of pit or container. The whole pit is covered with hessian cloth or locally available broad leaves. Maintain 50% moisture in filled material with spraying of water.

- After 6–7 week, the vermicompost is prepared. When vermicompost and vermicasting are ready for collection, top layer appears somewhat dark brown, granular as if used dry tea leaves have been spread over the layer. Watering should be stopped for 2–3 days and compost should be scrapped gently from top layers to a depth, so that it appears vermicomposted. This should then be removed aside and left undisturbed for 6–24 h. If there are adult worm present these would go down or move away from to composted material.
- The prepared compost is heaped in the tree shade and dried, sieve out the cocoons from vermicompost. The mesh size for removing cocoons is 2 mm galvanized mesh, some cocoons invariably go along with the compost and would lead to natural disposal of earthworms.

The 20–25% moisture should be maintained in the prepared compost. For commercialization, vermicompost should be packed in plastic bags or hessian bags of marketable quality, that is, in small or big packing.

25.11 APPLICATION OF VERMICOMPOST

The technique of using vermicompost is the same as FYM in the field. For various usages, vermicompost application can be done in the following manner.

In agriculture—Utilization of vermicompost in agriculture has so far been limited due to low-level production of vermicompost. However, the first application mixed with FYM in equal proportion at seeding stage may be useful.

The quantity of vermicompost for different crops is given below:

- In cereals and pulses apply 5 t/ha vermicompost before seeding.
- In oilseed crops apply 7.5 t/ha vermicompost before seeding.
- In spices and vegetable crops use vermicompost at 10 t/ha before seeding.
- In commercial cultivation of flowers apply vermincompost at 12.5 t/ha before sowing.

In horticulture—Vermicompost application is preferred at the time of plantation of horticultural plants. It is applied by mixing 5 kg vermicompost with equal quantity of cow dung manure/farmyard manure per plant. In old plantation, the method involves the preparation of a ring around plant base of 0.5–1 feet depth and 1–2' feet wide. In this ring, mixture of vermicompost at 5 kg/plant and equal quantity of FYM or cow dung manure is filled. Over this, a thin layer of soil is put and finally covered with organic matter comprising dry leaves, weeds (minus seeds), husk, coir, or even old hessian. Watering should be done to complete the process.

25.12 CASE STUDY: USING VERMICULTURE TO IMPROVE THE PRODUCTION OF PEANUT AND VEGETABLE PEA GREEN POD

Organic farming is gaining importance in India, but it faces the problem of nonavailability of sufficient organic manures due to use of cow dung as fuel. However, a large amount of residue is available at farms that may be utilized for composting and effectively recycled in the system itself as organic manure. These residues can quickly be converted into good compost by using the technology that converts organic materials into compost at a faster rate. The earthworms in general are beneficial to agriculture. They are good friends of the farmers, as they continuously plough and make the soil friable. The importance of earthworms as a super composter has been proved beyond doubt (Venkataramani, 1995). These creatures help mix materials, aerate the heap, and hasten the decay of organic matter. Composting through earthworms is advantageous in preventing the leaching of nutrients and even in conserving nutrients, bacteria, enzymes, vitamins, and moisture contents.

The lack of adequate facilities for soil testing is the main constraint, especially, in the peanut-growing area of southwestern semiarid zone IV, which is mostly deficient in soil organic matter. Therefore, the groundnut (peanut) cultivation is not possible here without sufficient addition of organic matter. To improve the peanut and vegetable pea green pod production under peanut–vegetable pea cropping system, the present experiment was planned and executed under location base study.

25.12.1 METHODOLOGY

A field experiment was conducted during 1996–1997 and 1997–1998 at the Regional Research Station of Chandra Shekhar Azad University of Agriculture & Technology, Mainpuri. The soil was sandy loam, having pH 8.5 organic carbon 0.45%, total nitrogen 0.04%, available phosphorus 10 kg/ha, and available potassium 278 kg/ha. Thus, the fertility status of the experimental soil was low. The treatments comprised conventional system

(N_{20} + P_{30} + K_{45} kg/ha), FYM at 100 q/ha, vermiculture at 60,000/ha, FYM 100 q/ha + vermiculture at 60,000/ha, *neem* leaf powder (NLP) at 100 kg/ha, FYM at 100 q/ha + NLP at 100 kg/ha, vermiculture 60,000/ha + NLP at 100 kg/ha and FYM at 100 q/ha + vermiculture at 60,000/ha + NLP at 100 kg/ha. FYM was applied to peanut 1 month before sowing, whereas NPK and botanical pesticides were applied at the time of sowing as per treatment. The vermiculture was released in furrows at peanut planting before planking. The peanut cv. *Amber* was planted in rows 45 cm apart, using 75 kg kernel/ha in the first fortnight of July, harvested after 115 days in the first fortnight of November, during both the experimental years.

The succeeding crops of vegetable pea cv. *Azad P-1* were planted in rows 30 cm apart, using 100 kg seed/ha in the second fortnight of November and green pods were harvested between the periods of 85 and 90 days after seeding in the first fortnight of February. Vegetable pea was raised on the residue of natural resources. The starter dose of 40 kg N + 60 kg P_2O_5 + 40 kg K_2O was supplied through the application of 100 q/ha FYM/ha and residue of nutrients applied to peanut crops. The irrigations were given to vegetable pea as and when required. The experiment was carried out in RBD with three replications.

25.12.2 RESULTS

25.12.2.1 PEANUT: EFFECT ON GROWTH CHARACTERS

Significantly higher plant stand was noted in the conventional system (246,295 plants/ha), whereas minimum stand was observed with the use of NLP at 100 q/ha (124,483 plants/ha). A special point was that the application of NLP at 100 kg/ha alone or in combination with other ingredients significantly reduced the plant stand of peanut in pooled results of 2 years because its allelopathic effect adversely affected the germination of peanut kernels as NLP was applied along with peanut in the same furrows. The plants affected with significantly higher bud necrosis disease were also counted in the conventional system (30.85%) and the least was noted in vermiculture + NLP at 100 kg/ha (12.43%). The application of FYM, vermiculture, and NLP alone, or in combination, reduced the incidence of bud necrosis disease (Table 25.1).

TABLE 25.1 Effects of Different Treatments on Growth, Yield Parameters, and Yield of Peanut (Pooled Data of 2 Years).

Treatment		Plant stand/ha	Plants affected with BND (%)	Pods/ plant	Pod yield (q/ha)
T_1	Conventional system	246,295	30.85	15.71	9.65
T_2	FYM @100 q/ha	226,334	23.70	17.77	11.53
T_3	Vermiculture @60,000/ha	233,742	18.75	20.93	12.16
T_4	FYM @100 q/ha + vermiculture @60,000/ha	245,265	15.63	22.66	13.20
T_5	*Neem* leaf powder (NLP) @100 kg/ha	124,483	15.29	23.55	10.37
T_6	FYM @100 q/ha + NLP @100 kg/ha	150,101	13.40	24.72	10.50
T_7	Vermiculture @60,000/ha + NLP @100 kg/ha	152,363	12.43	25.10	11.73
T_8	FYM @100 q/ha + vermiculture @60,000/ha + NLP @100 kg/ha	157,507	15.10	24.99	11.77
CD 5%		30,811	9.20	6.18	1.19

25.12.2.2 EFFECTS ON YIELD ATTRIBUTES OF PEANUT

The minimum pods/plant were recorded in conventional system (15.71 pods/plant), and the highest in vermiculture at 60,000/ha + NLP at 100 kg/ha (Table 25.1).

The different treatments could not influence significantly the pod weight/plant, kernel weight/plant, and 100 kernel weight in pooled results of 2 years. However, these yield attributes were found highest in the combination of vermiculture at 60,000/ha and FYM at 100 q/ha. The release of earthworms in combination with FYM provided better environment for pods and kernel development due to more organic casting of earthworms.

25.12.2.3 EFFECTS ON POD YIELD OF PEANUT

The release of earthworm as a vermiculture at 60,000/ha (6/m^2) gave higher pod yield when compared with conventional system (9.65 q/ha) and use of NLP at 100 kg/ha (10.37 q/ha) and FYM at100 q/ha + NLP at 100 kg/

ha (10.50 q/ha). The efficiency of vermiculture increased by 8.50% when released in amended field of FYM at100 q/ha (13.20 q/ha).

Among all the treatments, application of FYM at 100 q/ha + vermiculture at 60,000/ha recorded higher pod yield compared with other ingredient combinations. These results confirm the findings of Agasimani et al. (1994).

The application of NLP alone or in combination with other ingredients reduced the germination of peanut kernels, causing reduction in pod yield under these treatments. This may be due to allelopathic effect of the highest dose of NLP.

25.12.2.4 VEGETABLE PEA: EFFECTS ON GROWTH, YIELD ATTRIBUTING TRAITS, AND GREEN POD YIELD OF SUCCEEDING CROP

Vegetable pea was sown after peanut to utilize the benefit of natural resources applied to peanut during the rainy season.

The residue of FYM at 100 q/ha, in combination with vermiculture at 60,000/ha with starter dose of NPK supplied through FYM at 100 q/ha maximized the growth and yield attributing traits of vegetable pea, that is, pods/plant, closely followed by residue of vermiculture @60,000/ha with starter dose of NPK supplied through 100 q/ha FYM (Table 25.2).

The release of earthworms in peanut during the rainy season and FYM at 100 q/ha applied to vegetable pea at sowing to fulfill the starter dose of NPK increased the green pod yield of succeeding crop of vegetable pea (123.62 q/ha) sown in winter season over all the treatments, except that in the treatment of residue of FYM at 100 q/ha applied in the association of vermiculture and starter dose of NPK supplied through FYM at 100 q/ha used in vegetable pea (124.10 q/ha). The better growth and yield traits under residue of FYM at 100 q/ha in association with vermiculture at 60,000/ha and in residue of vermiculture alone with starter dose of NPK supplied through FYM at 100 q/ha applied to vegetable pea increased the green pod yield significantly in comparison with other tested treatments.

TABLE 25.2 Effecs of Different Treatments on Growth, Yield Parameters, and Yield of Peanut (Pooled Data of 2 Years).

Treatment[a]		Pods/ plant	Kernels/ plant	Green pods weight/plant (g)	Yield of green pods/ha
T_1	Conventional system	9.60	45.16	40.50	118.89

TABLE 25.2 *(Continued)*

Treatment[a]		Pods/ plant	Kernels/ plant	Green pods weight/plant (g)	Yield of green pods/ha
T_2	FYM @100 q/ha	9.65	53.16	40.95	120.45
T_3	Vermiculture @60,000/ha	9.88	59.33	42.05	123.62
T_4	FYM @100 q/ ha + vermiculture @60,000/ha	9.92	63.27	42.20	124.10
T_5	NLP @100 kg/ha	9.45	45.00	40.15	118.16
T_6	FYM @100 q/ha + NLP @100 kg/ha	9.52	47.38	40.45	119.01
T_7	Vermiculture @60,000/ha + NLP @100 kg/ha	9.56	48.33	40.65	119.56
T_8	FYM @100 q/ ha + vermiculture @60,000/ha + NLP @100 kg/ha	9.58	48.22	40.70	119.74
CD 5%		N.S.	4.04	N.S.	3.41

[a]FYM at 100 q/ha was given under all the treatments to fulfill the starter dose of NPK for vegetable pea.

25.12.3 OBSERVATIONS RECORDED ON ACTIVITIES OF EARTHWORMS

It is observed that farmyard manure should never be fed directly to the plants as manure, as it mostly serves as the food for the living organisms present in the soil like earthworms and bacteria. In the present situation, the released earthworms made the soil rich and aerable. The earthworm casting was noted more at pegging stage and as these were rich in nitrogen (2.5%), sulfur (2.9%), and potash (1.4%), they helped in easy pegging into the soil and retained more moisture, leading to better pod development in the combination of FYM and earthworms. Vermicompost acted as a good medium for the growth and development of microbes in the soil and made the nutrients available for plant uptake (Kale et al., 1987) and thus increased the pod yield of peanut. Earthworms multiplied very rapidly from July to September and ate organic debris continuously and excreted digested material at frequent intervals on the soil surface. They come to the soil surface about 8–10 times in a day for excretion. In this process, they mixed various soil compounds and improved the soil surface. The population of earthworms was recorded to be higher in soil having FYM. The population varied from 50,000 to 400,000/ha earthworms, depending upon organic material amended. Earthworms casting was estimated to be 25–36 tons/ha, which was also dependent upon the population counted. The maximum casting was found with the integration of FYM. This casting also helped in improving the growth and yield of vegetable pea-green pods.

25.13 SUMMARY AND CONCLUSION

During 1996–1997 and 1997–1998, a study was undertaken at the Regional Research Station of Chandra Shekhar Azad University of Agriculture & Technology, Mainpuri, to improve the pod yield of peanut and vegetable pea by using organic components like FYM and vermiculture. Release of earthworms as vermiculture at 60,000/ha (6/m^2) in peanut was found significantly better, gave higher pod yield (12.16 q/ha) than all other ingredient combinations except FYM at 100 q/ha + vermiculture at 60,000/ha (13.20 q/ha). Release of earthworms in peanut significantly increased the yield of succeeding crop of vegetable pea (123.62 q/ha green pods) over all the treatments, except the treatment with FYM at 100 q/ha in combination with vermiculture at 60,000/ha (124.10 q/ha green pods). Thus, the peanut and vegetable pea could successfully be raised under sequential cropping by the use of FYM at 100 q/ha + vermiculture to peanut and 100 q/ha FYM to vegetable pea for completing the starter dose 40 kg N + 60 kg P_2O_5 + 40 kg K_2O/ha, besides better management of natural resources and residue for higher productivity and monetary return.

The farm families residing in the riverine tract of Uttar Pradesh may be apprised of the use of vermicompost for improving the production of peanut and green pods of vegetable pea under peanut–vegetable pea cropping system.

KEYWORDS

- **earthworms**
- **vermiculture**
- **epiges**
- **endoges**
- **aneciques**
- **vermicompost**
- **peanut**

REFERENCES

Agasimani, C. S.; Patil, R. K.; Rabishanker, G.; Manikeri, I. M. Effect of Manures and Fertilizer on Activity of Earthworm and Productivity of *Kharif* Groundnut. *Groundnut News* **1994,** *6* (2), 5–6.

Bhatnagar, R. K.; Palta, R. K. *Earthworm Vermiculture and Vermi-composting*; Kalyani Publishers: Ludhiana, 1996; pp 1–105.

Kale, K. D.; Baro, K.; Sreenivas, M. N.; Bhagyaraj, D. J. Influence of Worm Casting on Growth and Mycrorrhizal Colonization of Two Ornamental Plants. *South Indian Hort.* **1987,** *35*, 433–437.

Venkataramani, G. Ecological Farming Available Option for Future. *The Hindu Survey of Indian Agriculture* **1995,** 23–31.

CHAPTER 26

Issues and Challenges in Marketing of Organic Produce in India

S. R. SINGH

CCS National Institute of Agricultural Marketing, Jaipur 302033, Rajasthan, India
E-mail: sattramsingh@gmail.com

ABSTRACT

The growing health consciousness among the consumer and increasing awareness about organic food has led numerous opportunities for organic producers. Along with opportunities, there also arise challenges, due to lack of awareness and low level of education, most of the information of market, technology, certification, and source of demand are beyond the reach of growers, which is a major bottleneck in developing the organic food market. Moreover, certification processes are quite tedious and difficult to understand and most of the literature on organic cultivation is not in farmer-friendly language. Some of the common marketing challenges faced by the farmers, are lack of warehousing facility, lack of price information, inadequate demand for crop, costly transportation, market price variations, and lack of government support. These aspects need to be looked into, so that the farmers are benefitted to a greater extent.

26.1 INTRODUCTION

It is well documented that after independence, India was facing severe problem of food insecurity and it was difficult to provide food for all households.

Organic Crop Production Management: Focus on India, with Global Implications, D. P. Singh, PhD, H. G. Prakash, PhD, M. Swapna, PhD, & S. Solomon, PhD (Eds.)

With Green Revolution coming to our rescue, food security could be ensured in the country. The team of scientists worked day and night to make it a success to pull out the country from the crisis. During this period, along with the high-yielding crop varieties used, different synthetic agrochemicals such as fertilizers and pesticides were utilized by the farmers to enhance the productivity of crops. The various externalities have had an adverse effect on the natural resources and human health, as well as on agriculture itself, leading to environmental and health problems.

The increasing environmental and health consciousness has given rise to a major shift in consumer preference toward quality of the produce and food products, in India and also particularly in the developed countries (Makadia and Patel, 2015). Globally, there is an increasing and robust demand for organic food that is considered safe and hazard free, with a fast growth of the sector.

The varied agroclimatic conditions in India offer a lot of scope for organic cultivation of a variety of crops. The traditional organic farming being practiced for a long time in several parts of the country, that is, the northeastern states, hilly regions, and Sikkim, is an added advantage. This can serve as an excellent opportunity for the agricultural growers and farmers to tap the market for organic produce, which is growing steadily in the domestic as well as export arena. In the recent past, this trend has increased manifold due to better consumer awareness.

26.2 ORGANIC PRODUCTS

Organic products are grown in an environmentally and socially responsible way, without the use of chemical fertilizers and pesticides. This method involves sound soil management, preserving the reproductive and regenerative capacity of the soil, and ensuring good plant nutrition. This takes care of soil and human health, influences the environment in a positive manner, and produces nutritious food rich in energy.

The statistics of 2013 reveal that India occupies 15th rank globally, with respect to organic agriculture (Source: FIBL & IFOAM Year Book 2015). The total area under organic certification is 5.71 million ha, including 26% cultivable area (1.49 million ha) and the rest 74% (4.22 million ha) being forest and wild area for collection of minor forest produce (2015–2016). By default, a sizeable quantity of minor forest produce and herbs from the jungle is organically produced in our country.

For supporting organic production movement in the country, National Programme for Organic Production (NPOP) was initiated in 2000 and National Project on Organic Farming (NPOF) in 2004, by the Government of India. The NPOF scheme takes care of capacity building through service providers, financial support to biofertilizer production units, fruit and vegetable waste compost, and vermihatchery units, apart from human resource development through training, certification, and inspection. Basic infrastructure and support have been created through these initiatives to boost up the production of crops organically. Various projects were initiated under NPOP programs for organic cultivation countrywide, to improve productivity during transition from inorganic to organic farming. These supply organically grown certified seeds, biopesticides, biofertilizers, and organic manures at subsidized price and also facilitate the growers to make their own inputs like compost, vermicompost, biopesticides, and biofertilizers. A few lead/progressive farmers are trained to supply the organic inputs at subsidized rates, to benefit the growers. An established network of agricultural centers with trained staff in the field of surveillance, disease, and pest forecasting can successfully carry out the disease and pest management, especially during the transition phase, which is otherwise very challenging. The pesticides used in organic farming are either of natural origin or simple chemical products, with rapid breakdown of the active ingredients when exposed to sunlight.

The Union Government's determined approach coupled with the focus of various state governments have helped many states to achieve a remarkable progress in organic farming. In January 2016, Sikkim became the first Indian State to go fully organic, with the state committed to provide chemical-free agricultural produce. Sikkim Organic Mission is supporting production of crops and 50,000 ha of land has been transformed for organic farming in the state. This has inspired other states to follow suit and to announce detailed policies for organic farming. The other northeast states like Manipur are also trying to achieve the organic status as the land, and environment is similar to that of Sikkim.

The area of organic cultivation in various states are given in Table 26.1:

TABLE 26.1 Area of Cultivation of Organic Crops in Major States in India During the Year 2014–2015.

S. No.	States	Area under Organic Crops (ha)
1	Madhya Pradesh	461,775
2	Maharashtra	198,352

TABLE 26.1 *(Continued)*

S. No.	States	Area under Organic Crops (ha)
3	Rajasthan	155,020
4	Telangana	103,556
5	Odisha	95,897
6	Karnataka	93,963
7	Gujarat	76,813
8	Sikkim	75,851
9	Uttar Pradesh	61,082
10	Uttarakhand	37,221
11	Jharkhand	30,364
12	Assam	28,433
13	Kerala	25,899
14	Jammu & Kashmir	25,515
15	Andhra Pradesh	18,252
16	West Bengal	17,890

Source: APEDA. Apeda.gov.in/apedawebsite/organic/index.htm

26.3 POTENTIAL BENEFITS OF ORGANIC FARMING

The poor productivity of our soils is due to the low organic matter content, arising from indiscriminate use of agrochemicals and fertilizers is a major problem in our country. The organic matter content of the soil can be enhanced using green manure, vermicompost, humus, and rotten farmyard manure. The natural nutrients from these organics help in a build-up of organic content in the soil, thereby increasing the fertility of soil and enhanced the availability of nutrients, mainly due to increase in microbial activity. There are several indirect benefits also from organic farming to the growers as well as the consumers.

Organic farming is an alternative way to counter the problems of sustainability, global warming, and food security. Specific standards are formulated with detailed guidelines, aiming at social and ecological sustainability of agrosystems, along with efficient food production. Since the early 1990s, the term organic agriculture has become legally defined and has originated from strategies like biodynamic agriculture, regenerative agriculture, nature farming, and others developed in different countries.

26.4 ORGANIC PRODUCE MARKET IN INDIA

The organic markets in India are largely spread across various sectors like food and beverages, health and wellness, beauty and personal care, and textile industries. Various essential oils produced organically are being used for health and personal care to cure body ailments. The highest growth is observed in the organic food segment, followed by textile, beauty, and personal care. The current Indian domestic market is estimated at Rs. 40,000 million that is likely to increase by Rs. 100,000–120,000 million by 2020, with a similar increase in exports of various products like cereals, pulses, sugarcane, spices, fruits and vegetables, and essential oils.

Organic food market, packaged food, and beverages are the emerging niche markets in India. More focus should be given to capture the market as its primary consumers are high-income group primarily living in urban areas. In 2016, the total market size for organic packaged food in India was Rs. 533 million. This has grown at 17% by 2015, and is expected to reach manifold in future. India's exports of organic products increased by 17% between 2015–2016 and 2016–2017. In India, the maximum demand for organic food comes from metro cities, malls, and hotels located in the metros. Various market players and companies, such as Conscious Foods, Sresta, Eco Farms, Organic India, Navdanya, and Morarka Organic Foods are entering this sector due to increasing share of profit in this market segment. Market shares of various commodities have been presented in Table 26.2.

TABLE 26.2 Export of Organic Produce During Financial Year 2015–2016.

S. No.	Commodity	Share (%)
1	Oilseeds	50
2	Processed food	25
3	Cereals and millets	17
4	Pulses	2
5	Tea	2
6	Spices	1
7	Dry fruits	1
8	Others	2

Source: APEDA.

Organic export market is flourishing, along with the growing domestic market. India is the second-largest exporter of organic products in Asia after

China, with the major export destinations being the United States, European Union, Canada, and New Zealand. The increasing export market and the Government support have contributed toward the success of organic cultivation in India. Various schemes and policies are launched by the Government of India with the help of NCOP and NCOF to support organic farming in India. The remaining volume of organic produce after export is sold for domestic consumption in local markets. In other words, the sale of this share of organic produce is uncategorized. Oilseeds comprised half of India's overall organic food export, followed by processed food products at 25%. India produced around 1.35 million tonnes of certified organic products including food products in 2015–2016, like sugarcane, oilseeds, cereals and millets, cotton, pulses, medicinal plants, tea, fruits, spices, dry fruits, vegetables, coffee, and essential oils. The production extends also to the organic cotton fiber, functional food products, etc.

26.5 UPCOMING TRENDS

The current market size for organic food products is US$ 533 million with an increasing trend in the subsequent years. According to ASSOCHAM, in the metropolitan cities, the demand has increased by 95% in the last 5 years. This is a result of the awareness generated among the consumers of metro cities who are health cautious and want to choose safe food for consumption.

26.6 MAKE THE MARKETING POSSIBLE

Many organic food companies are adopting the online route to expand their consumer base. The organic stores are mainly located in metro and mini metro cities of India, where sizeable proportion of rich consumers are living. These companies are reaching out to the rest of the consumers through online channels. Some of the players that have established their own online website include Farm2Kitchen, Organic Shop, Naturally Yours, and Organic India. Premium food retail chains such as Godrej Nature's Basket and many more also sell organic food brands such as Navdanya and 24 Mantra online. Multinational and national companies are adopting different methods to capture targeted consumer segment to sell their organic produce.

26.7 MARKETING BY DIFFERENT WAYS- ORGANIC FOOD RESTAURANTS AND CAFES

With an increasing incidence of health problems such as diabetes, anxiety, and stress plaguing urban India, many entrepreneurs are venturing into the area of organic cafe and marketplace. For example, Devang House is a fully organic cafe located in New Delhi, which organizes the Organic Living Festival every fortnight that creates awareness among growers and increases the consumer base of organic produce. It also sells herbal lifestyle products like dental powder and digestive tonics and uses natural and organic produce in its menu. Various restaurants in metro cities are serving organic menu for food lovers and have a direct contact with organic growers who supply fresh vegetables and other items to the cafes. This trend is increasing in the metros and second-line cities too.

26.8 BOOST THE MARKET OF ORGANIC FOODS

Awareness creation among stakeholders, growers, suppliers, food lovers, restaurants, and cafe are necessary to make them aware and enlightened about the health benefits of organic foods. Many events are taking place throughout the country to generate awareness about the benefits of organic farming. Despite all the government efforts to boost production, there are several challenges that remain and these are mostly related to value addition and marketing of organic products. The companies need to bear the cost of collecting the produce from small farmers and transportation and handling costs; bear losses on account of perishability, quality, and rejections, and maintain a buffer margin for quality variations and disaggregation to reach out to retail points. The warehousing protocols and product manufacturing protocols are even stringent in terms of fumigation during storage, use of preservatives and added ingredients while manufacturing. The packaging requirement for organic produce is also stringent, wherein natural packing materials are to be used, which increases the cost of packaging. Despite the awareness about the benefit of organic produce, the higher-priced organic food could not reach to various consumers as they are not in a position to buy organic produce from the market. The value chain of organic produce is very fragmented because the essential commodity act could be invoked any time in any commodity rendering large-scale investments in storage and infra-structure, which otherwise is unavailable. Therefore, innovative solutions to optimize scale and maintain profitability for organic products are required. A comprehensive plan for the development of organic sector value chain can cut down operation cost and make the organic market profitable and scalable by bringing down the cost of cultivation simultaneously (Narayanan, 2005).

26.9 EXPORT OF ORGANIC PRODUCE

The total volume of export during 2015–2016 was 263,687 MT with a realization of around 298 million USD. The major destinations are European Union, the United States, Canada, Switzerland, Korea, Australia, New Zealand, the southeast Asian countries, the Middle East, and South Africa.

26.10 MARKETS FOR ORGANIC FOOD—NEED OF THE DAY

The major weakness in the Indian organic food industry is the lack of established marketing channels or green markets, making it difficult for small and marginal farmers to sell the produce. Lack of awareness about market information, cultivation, export practices, certification of crops, grading, packaging, and existing market channels in the locality is another challenge. Inadequate information about the procedures also is a reason for the slowdown of the growth of the market in India. There is an urgent need for quality improvement of organic products, and related activities like packaging, logistic infrastructure, and technical support to the stakeholders.

26.11 FOCUS ON QUALITY MANAGEMENT IN PRODUCTION AND PROCESSING

Even with the government measures to make organic products popular in the domestic market, the consumers anticipate a price on par with nonorganic products. A lowering of price can be expected only after the completion of conversion stage when there is also an increase in production, with the simultaneous influence of other factors like rising demand and economies of scale. For the Indian organic market to be successful, implementation of high-quality standards is a must. All stakeholders need to be well-aware of the principles and standards of organic agriculture.

26.12 ORGANIC MARKET AS A BUYER/CONSUMER-DRIVEN MARKET

Simultaneous producer- and market-driven approaches exist in parallel in the organic products trade in the developing countries. International trade has increased with the growing demands for the products in the international markets in the past 15 years. Due to the low awareness about organic food and its benefits, the Indian organic food market is buyer/consumer driven rather than producer/supply driven. Even though at present, the affordability of organic food is low among the common man, with increasing health awareness, the demand has also started increasing. The producers/suppliers need to create more awareness among population about organic produce for a better market.

26.13 LACK OF STRATEGY FOR DEVELOPMENT OF ORGANIC MARKET

The Indian organic market is typical in the pregrowth-phase market, where consumers are already aware of the possible hazards and quality of produce. This is advantageous and favors the fast growth of the organic market.

26.14 SCATTERED PRODUCERS, PROCESSORS, AND TRADERS

For the success of any business sector, networking and collaboration among trade partners are essential and this is one area where Indian organic produce markets are yet to pick up. The Indian organic market is characterized by small and scattered producers, processors, and traders, with very minimum networking among the players. Small-scale producers have a crucial role in expanding the organic export sector also. Being labor intensive in nature, it is compatible with traditional peasant practices. The development of organic supply chains and networking with the small and marginal producers is a viable solution. This also involves linking the different factors, from farm to fork to achieve a more effective and market-oriented sustainable flow of goods. A sound knowledge to develop a workable structure and assurance of its sustainability, along with partnership and integration are the key factors of success of these supply chains.

26.15 ADULTERATION AND POOR QUALITY OF ORGANIC INPUTS

Poor quality and adulteration of organic inputs due to the absence of a proper marketing channel has been a major problem in India. Conventional/modern input dealers and retailers are less interested in organic produce due to erratic supplies of the produce, low demand due to low awareness, and inadequate distribution network. However, recently the government has formulated product standards and specifications for inputs like organic/biofertilizers. Application of poor and adulterated organic inputs in the market due to improper inspection and regulations leads to poor performance and in turn, to a lack of trust in the entire practice of organic farming. The government needs to establish organic input marketing channels and to put necessary market regulations in place in the country to ensure the timely availability of quality inputs and to instill confidence among the growers.

26.16 SMALL FARMS WITH WEAK ORGANIZATIONAL STRUCTURE

Neither domestic nor export market for organic products is yet developed. As the lion share of the organic produce come from small farmers; wholesalers/traders account for 60% of the distribution system of organic products. Large organized producers distribute their products through supermarkets as well as through self-owned stalls. Considering the profile of existing consumers of organic products, supermarkets, and restaurants are the major marketing channels for organic products. While certification is mandatory for exports, products for domestic consumption are mostly uncertified due to the existence of small or marginal farmers, small cooperatives, or trade fair companies. This also leads to an incomplete product range that is mostly available as a small or local brand, In contrast, to that in developed countries like the United States and Europe. Therefore, a shift to organized retailing and marketing from the present unorganized pattern is inevitable if the organic produce markets are to be successful (Azam et al, 2019).

26.17 PRICE STRUCTURE FOR ORGANIC PRODUCTS

Awareness about the scope of markets and price structure for self-claimed and certified organic produce is necessary among the growers also. There are often unrealistic expectations regarding the prices for organic produce, where the price demand can be 100–400% more than that for conventionally produced products. A drop in the yields is often cited as the reason for this high demand. But the export markets for the certified products are well organized, with the premium up to around 50%. If the grower owns the certificate and sells the produce to an exporter, the premium is around 25–30%. With no certificate, the premium is between 15 and 25% (NCOF, Ghaziabad).

26.18 CREATION OF SEPARATE "GREEN CHANNELS" FOR ORGANIC FOODS

Most of the farmers sell their organic products in the local conventional market with no separate marketing channels for the organic produce, thereby forcing the farmers toward distress sales. A typical example is the case of sugarcane farmers who produce organic sugarcane in Uttar Pradesh and Maharashtra, with no option other than sugar factories for the sale of the organically grown crop. The farmers do not get the premium prices and also there is no benefit to the society. There needs to be a separate "Green marketing channel" exclusively for the organic products. The Public–Private-Partnership model of marketing of organic products at the district level is expected to boost marketing of organic sector in the country.

26.19 PREMIUM PRICES FOR ORGANIC PRODUCTS

Lack of premium prices in the market for organic produce leads to a risky situation for the organic growers. This, coupled with lower productivity reduces the per area returns compared with conventional farming. Fixing premium prices for at least the staple food crops like paddy, wheat, *jowar, bajra,* and others is necessary for the organic markets to thrive.

26.20 CREATION OF DEMAND THROUGH MORE AWARENESS PROGRAMS

Long-term experiments conducted all over the world have shown that organic farming increases the crop productivity while sustaining the ecosystem. Hence, governments need to be proactive in motivating the farmers toward organic farming through awareness programs and field demonstrations. These programs should also sensitize the end users about the benefits of organic food and other produce.

26.21 INPUT/CONVERSION SUBSIDIES FOR ORGANIC FARMING

A shift from inorganic to organic farming demands some time for adjustment of the soil to the accompanying changes. There might also be the problem of lower yields initially, as compared with conventional farming. An interval of 2–3 years is essential for attaining competitive yields in organic farming and during these initial years, there should be provision of input/conversion subsidies by the government to organic growers, to overcome these losses.

26.22 EXTENDING MORE R& D INVESTMENTS AND TECHNICAL SUPPORT TO FARMERS

Compared with that in the developing countries, in India, investments on organic research and development are very low. There is a need for a specific extension and technical support division exclusively for addressing the problems of organic farmers. Thus, an active involvement of state agricultural universities and agricultural departments is necessary for rapid popularization and expansion of organic farming in the country.

26.23 FARMER-FRIENDLY CERTIFICATION PROCESS

The procedures and high costs involved in the organic certification process is another major impediment for the expansion of organic certified area in the country. The complex and high cost of certification process is a burden to many of the small and marginal growers during conversion phase. Development of an innovative, cost effective, and simple certification process would help in attracting more growers and also in expanding the area under organic farming.

26.24 AVAILABILITY OF HIGH-QUALITY CERTIFIED ORGANIC INPUTS

In India, many times, the absence of quality organic inputs in markets, leads to the use of adulterated inputs, leading to yield losses to farmers, and also loss of faith in the practice. So, the development of organic input marketing channels in the country with assured availability of quality inputs is the need of the hour. It will not only improve the productivity and efficiency of organic farming in the country but also will help in increasing awareness about the safety and quality of foods and long-term sustainability of the system. Organic farming has emerged as an alternative system of farming, which not only addresses the quality and sustainability concerns, but also ensures a debt free, profitable livelihood option. Within a short span of 5 years, it has grown into a mainstream agricultural practice with an admirable growth rate. Institutional mechanisms and governmental support have been a major factor for this enormous growth especially during the 11th Five-Year Plan. But more efforts are needed for a proper marketing strategy.

26.25 CONCLUSION

Due to lack of awareness and low level of education, most of the information of market, technology, certification, and source of demand are beyond the reach of growers, which is a major bottleneck in developing the organic food market. Moreover, certification processes are quite tedious and difficult to understand and most of the literature on organic cultivation is not in farmer-friendly language. These aspects need to be looked into, so that the farmers are benefitted to a greater extent. There is an immediate need to develop a separate green market exclusively for organic produce (as in states like Sikkim and Kerala), with proper transportation and logistics and also to improve the present marketing system of APMC, enabling them to handle the organic produce cultivated in and around. A cluster approach to develop a favorable momentum toward the practice is beneficial. The farmers should be made aware of the Government schemes and subsidies for the production of organic crops in those areas that are suitable for cultivation and they should be also be urged to make maximum use of such facilities.

ACKNOWLEDGMENT

This chapter is based on the review of secondary data, official reports, and previous studies relevant to the organic food industry in India. The majority of reviewed content is based on studies conducted in India and most of the data collected is sourced from various ministries of the Government of India and national and international organizations, News Papers, articles, and Journals. The author is highly thankful to one and all.

KEYWORD

- **organic products**
- **marketing**
- **potential benefits**
- **organic food restaurants**
- **price structure**

REFERENCES

APEDA. http://apeda.gov.in/apedawebsite/organic/Organic_Products.htm

Makadia, J. J.; Patel, K. S. Prospects, Status and Marketing of Organic Products in India-A Review. *Agri. Rev.* **2015,** *36* (1), 73–76.

Azam, M. D.; Shaheen, M.; Narbariya, S. Marketing Challenges and Organic Farming in India—Does Farm Size Matter? *Int. J. Nonprofit Voluntary Sector Market.* **2019,** *24* (8) .DOI: 10.1002/nvsm.1654.2019

Narayanan, S. Organic Farming in India: Relevance, Problems and Constraints. https://www.nabard.org/demo/auth/writereaddata/File/OC2038.pdf. 2005

CHAPTER 27

Perspectives and Potentials of Organic Farming in Sugarcane Cultivation

GOVIND. P. RAO*

Department of Plant Pathology, ICAR-Indian Agricultural Research Institute, New Delhi 110012, India

**Corresponding author. E-mail: gprao_gor@rediffmail.com*

ABSTRACT

Sugarcane is an important commercial and industrial crop in India which has contributed significantly to the growth of Indian Agriculture and National GDP. In sugarcane it emphasizes management practices involving substantial use of organic manures and bio-fertilizers, residue recycling and management of insect-pests and diseases through the use bio-pesticides and bioagents . The organic farming is very remunerative and beneficial for its social benefits in terms of resources, improved human health, and environment. Thus, it is recommended to properly measure and quantify the social benefits of organic sugarcane farming for having an idea about the extent of subsidy that could be justified for promotion of organic farming in India.

27.1 INTRODUCTION

Organic production is a holistic system specially designed for optimizing the yield and fitness of diverse communities within the agroecosystem inclusive of microorganisms of soil, plants, people, and animals. Organic production aims to develop sustainable and environmentally friendly enterprises.

Organic Crop Production Management: Focus on India, with Global Implications, D. P. Singh, PhD, H. G. Prakash, PhD, M. Swapna, PhD, & S. Solomon, PhD (Eds.)

Sugarcane (*Saccharum* species hybrids) is being cultivated in India since time immemorial. Agroclimatic conditions of India are conducive for sugarcane production. Sugarcane crop contributes significantly in the agroindustrial development of the country. The performance of Indian sugar industries depends upon the domestic sugarcane production. Sugarcane is the primary and major source for production of sugar and contributes to nearly 78% of the global sugar production. In India, it occupies 3.5% (about 5 million ha) of the total cropped area. Indian sugar industry is the largest agro-based industry after cotton textiles industry. About 524 sugar mills located in India produces 20.14 million tonnes of sugar (Solomon, 2016).

Sugarcane is the primary source of production of white sugar in India (63%). Sugarcane is also used for making *gur* and *khandsari* (25%), and 12% canes are used in seed, feed, and chewing purpose. Molasses, press mud, and bagasse are important by-product of the sugar industry. Alcohol is produced from molasses. Press mud is used in the field as a organic matter and for other purpose, bagasse is used in cogeneration plants to generate electricity and other work. Sugarcane, a long-duration crop requiring more amount of water and nutrients, is grown under subtropical as well as tropical conditions of India. Sugarcane needs 750–1200 mm of water during its entire life cycle for its optimum growth (Solomon, 2016). However, it needs 1500–1700 mm water (rainfall+ irrigation) from planting to harvesting in subtropical region. Well drained with a neutral pH (6.0–7.0) and optimum depth (>60 cm) alluvial to medium black cotton soils, are suitable for sugarcane production. Sugarcane planted in sandy to sandy-loam soils, with almost neutral pH under assured irrigated conditions in the north India provides the average yield. The crop takes about one to one and a half years to mature in different agroclimatic zones.

With the basic objective of achieving socially and ecologically sustainable agroecosystems, organic production systems are specifically prepared standards for food production. The systems take care of reducing the use of external chemical inputs through judicious and efficient use of on-farm resources in comparison to chemical fertilizers and pesticides use, thereby, avoiding the use of chemical fertilizers and insecticides. Organic farming systems have the answer to several challenges faced by the agriculture sector during the last two decades (Kshirsagar, 2006). Environmental security, conservation of nonrenewable resources along with better food quality are the major advantages of adopting organic farming (Kshirsagar, 2006). There is an immense potential in India for producing various organic products due to its varied agroclimatic conditions. The organic farming being practiced in various parts of the country holds promise to tap the untapped potential for the organic products, the demand for which is growing fast in the domestic market as compared to export market (Bhattacharya and Chakraborty, 2005).

In the present chapter, an updated scenario of organic practices adopted for organic sugarcane (OS) cultivation has been discussed on the basis of the results obtained from previous studies.

27.2 PRACTICES SUGGESTED FOR OS FARMING

27.2.1 PRECULTIVATION PRACTICES

To make the soil loose and fluffy, two shallow plowing, at right angles to each other, are required but deep ploughing should be avoided. Compost/vermicompost at 2.5 tonnes/acre with 100 kg rock phosphate and 4.0 kg PSB should be applied with the first ploughing.

Depending upon the availability of crop stubbles, sprinkle 200 L/acre of biodecomposer consisting of 2 kg *Trichoderma viride* over the residue before its incorporation into the soil, using a tractor driven rotavator. Twin or paired row planting method is recommended for providing enough space for the intercrops. Ridges and furrows should be made in the North-South direction for facilitating enough solar harvesting. Including leguminous crops in the sugarcane-based cropping system has been found quite advantageous. Only matured pods of the leguminous crops should be picked up and the crop residue should be incorporated in the soil for its enrichment. At the time of planting, use of 2 q *neem* leaf/seed manure and 5 q concentrated manure in the furrows helps in good germination and yield (Kshirsagar 2006; Sundara, 2011).

27.2.2 BIOFERTILIZERS

Due to the heavy drawing of nutrients from the soil by sugarcane, there is a need for its regular corresponding replenishment. With the increasing crop residue and farm waste. Chances of searching the alternative uses for supplementing the soil are quite remote with the increased crop residue and farm waste. Application of chemical fertilizer and insecticides for the long run leads to soil quality deterioration along with the accumulation of heavy metals in the soil and ultimately cause concern of reduction in the yield level along with adverse impact on human health. Progressive reductions of yield in successive crops due to stubble decline is major challenge on profitability of the Indian sugarcane industry, which can be attributed to poor soil aeration, weed competition, drainage, and *Pythium* root rot. Application of biofertilizers is suggested for promoting sustainable organic farming to obtain more acceptable product in the global market. The productivity of sugarcane is enhanced significantly by the use of various biofertilizers like *Trichoderma, Pseudomonas, Azotobacter, Azospirillum, Gliocladium* sp. AM fungi (*Glomus intraradices, Glomus fasciculatum,* and Gigaspora albida) and Seaweeds (*Sargassum wightii, Ulva lactuca, Caulerpa racemosa, Gelidiella acerosa,* and *Gracilaria edulis*) (Kumari et al., 2014). Biofertilizers improve different physical properties of soil like aeration, porosity, and water infiltration by forming and stabilizing soil aggregates. Use of composted organic materials for increasing the production and productivity of sugarcane also suppresses root rot disease. The research experiments conducted at various locations have clearly revealed that a comparable sugarcane yield is achievable by meeting the nutrient demand entirely through organics,

The compost or FYM (farmyard manure) or well-decomposed press mud may be used at 80 t/ha either in the furrows before planting or before last ploughing. The quantity of organic manure may be adjusted in such a fashion for supplying 280 kg N/ha through one or more sources of organic matter such as compost, FYM, or decomposed press mud, depending upon the content of their nitrogen. Setts should be collected from 8 to 10 months old disease-free nursery crop. Two budded setts should be preferred to three-budded setts. It is also recommended to use the seed material collected from organically grown sugarcane crop. By adopting 120 cm spacing in paired row by trench planting method, about 80,000 two-budded setts (75–80 q) are needed for planting 1-hectare area. After planting sugarcane crop, green manure crops like sunhemp or *dhaincha* may be sown on one side of the ridges as an intercrop on the third or fourth day after planting of sugarcane. After around 45 days of the sowing, the intercrop will be ready to be used as green manure. After harvesting the same, incorporate the whole biomass in situ. At 30, 60, and 90 days after planting, weeds may be removed by hand hoeing and hand weeding. For controlling the weeds, only nonchemical weed management technologies like hand weeding and mechanical weed control methods may be used. On 30 and 60 days after planting of sugarcane, 5 kg each of *Azospirillum* and phosphobacteria per hectare, respectively, may be used. The biofertilizers may be mixed thoroughly with 5 tonne of farmyard manure/ha to enhance the bulkiness for its application in the field. Give a light earthing up and Irrigate immediately after a light earthing up. The dried and senescent leaves should be removed at fifth and seventh months for its use as mulch in alternate furrows.

27.2.3 SEED-SETT/BUD SELECTION FOR PLANTING

About 8–10-month-old, disease-free, healthy cane should be selected for seed. Only two buds setts may be selected as per organic farming practices. Generally, the lowest two internodes should be avoided for their use as seed material. By trench method planting in paired row (row to row distance 120 cm) around 80,000 two budded setts are required for 1 ha. The individual buds are also an important alternative for significantly improving germination and saving a large quantity of cane for crushing purpose. In 1-ha area, about 8750–10,000 buds/seedlings should be planted. Although, sugarcane is cultivated by planting the two-budded setts directly in the field, but sometimes, buds or seed setts can be planted in polythene bags to raise a nursery in case of short supply of water or occupancy of the field by some other crop. Polythene bags of 10 × 15 cm, filled with soil and compost mixture (in 1:1 ratio) are closely used for raising the nursery in the field. Having a water source nearby. In these polythene bags, scooped buds are sown at 1–2 cm depth. Cut pieces of sugarcane with only one bud can also be used for raising the nursery. These bags are irrigated twice a week. Generally, saplings are ready for transplanting in 2 months period.

For sett treatment, *Azospirillum, Azotobacter,* and PSB biofertilizer may be used instead of chemical fertilizer and pesticides. The setts are dipped in suspension for 30 min. There is another option for dipping freshly cut seed setts in cow dung and cow urine slurry for a period of 10–15 min, followed by dipping in *Azospirillum* and *Azotobacter* solution for 30 min (Anonymous, 2006).

27.2.4 INTERCROPPING

Although sugarcane is generally cultivated as a monocrop under the traditional system, but intercropping of sugarcane with a wide variety of legume and nonlegume crops is preferred under organic farming to maintain nutrient balance and diversity. Generally, mungbean, urdbean, groundnut, cowpea, and chickpea are commonly grown as intercrops but organic farmers in various parts of India have developed their own plantation patterns, depending upon the crop duration, sowing time, and availability of water (Geetha et al., 2015).

Manure should be applied and field may be ploughed after the harvesting of the rabi crop or the previous sugarcane crop. Two feet wide furrows may be prepared and residue of the leguminous or nonleguminous crop may be incorporated in the field. A furrow may be prepared after every fifth row with the help of a bullock drawn plough. The treated sugarcane setts are planted in these furrows, leaving three furrows blank in between two sugarcane rows. After that, the setts are covered with soil and mulch with legume residue. The leguminous seeds are sown in the rows on either side of the sugarcane in two rows, one row on each slope of the ridge. Thus, there are four rows of legumes between two sugarcane rows. The legume pods are harvested and residue is used as mulch. Mixed seeds of chickpea+ coriander + mustard seeds (10:1:0.25 kg) are sown in one row in the mungbean furrows. All the three rows between two sugarcane rows (no irrigation in sugarcane furrow) are irrigated in the field. Intercrops will also be ready for harvesting by February. In the mungbean furrows, the plough should run between two sugarcane rows for earthing up of soil over growing sugarcane plants. The sugarcane ridge is mulched with crop residue. Different combination of legumes, coriander, mustard, and vegetables can be grown in furrows in-between sugarcane rows for higher economic returns (Kanwar et al., 1990; Shinde et al., 2009).

27.2.5 RATOON MANAGEMENT

Generally, it has been observed that yield of ratoon crop remains comparatively lower than that of plant crop. But in case of good management of the ratoon crop with all the recommended package of practice, the cane yield from the ratoon crop may almost be the same or marginally lower (around 5%) compared with that of the last plant/ratoon crop. One ratoon after the plant crop is a common practice for organic farming. Residue should not be burned after the harvest as the residue is converted into compost by decomposer, which can be applied in the same field. The field may be ploughed lightly between two paired rows. One furrow should be made between two rows of sugarcane for enhancing sprouting. The intercrops can be sown as described earlier. The ridges can be covered with the last crop residue and other farm wastes for acting as a mulch. The sugarcane roots are allowed to sprout along with the growing intercrops with biofertilizers.

27.2.6 BIOCONTROL OF SUGARCANE DISEASES AND PESTS

The costs incurred on biopesticides on control of pests and diseases are comparatively low under organic farming due of use of homemade biopesticides by organic farmers or microbial extracts purchased from other sources. In organic farming, no inorganic plant protection measures are adopted. For integrated management of major fungal diseases such as red rot, wilt, sett rot, and seedling rot, use of biological control is an effective tool in OS farming practices. Further, to control the pathogens under field conditions infecting stalk tissue with deep seated fungi, limited amount of fungicides can be used. Although sincere efforts are being made for developing disease resistant varieties of sugarcane having durable resistance but frequent development of new strains of red rot pathogen makes the varieties susceptible in the field (Viswanathan et al., 2019). Under this scenario, diseases like red rot, wilt, smut, sett rot and seedling rot are managed through the natural enemies or biocontrol agents which is an important component of OS farming. The different environmental factors influence occurrence and severity of soil-borne diseases such as wilt and sett rot. In spite of certain limitations of biological management, significant number of researchers clearly expressed some benefits in adopting biocontrol measures for the control of sugarcane diseases (Mohanraj et al., 2002). An investment in biocontrol results in improved sugarcane yield. The biocontrol agents are self sustainable. However, a by-product of sugar industry, the "press mud" can be used as a suitable carrier for formulating biocontrol agents to mass production. Apart from identifying efficient biocontrol agents, delivering them at right time and dosage is very important in the controlling diseases. Under disease-endemic locations, wilt and red rot diseases can be managed by the sett treatment with PGPR/*Trichoderma* followed by soil application of the biocontrol agents formulated in talc/press mud is recommended (Viswanathan and Rao, 2011).

Several efficient fungal and bacterial antagonists have been identified for management of diseases with proven efficacy under in vivo and in vitro conditions. Fungal bioagents like *Trichoderma, Chaetomium,* and bacterial antagonists individually and the combination of *Pseudomonas fluorescens* and *Bacillus* species have been found quite effective for management of red rot incidence. Delivery of the antagonists has been standardized for field application through sett treatment. Under field conditions, press mud formulation of *Trichoderma* has been found effective against wilt. Application of *Trichoderma* formulation has also been observed very efficient against seedling rot caused by *Pythium* spp and therefore, it is recommended to use the same in seedling trays to manage the disease (Viswanathan and Rao, 2011).

Borers and white woolly aphids of sugarcane are also very crucial insect pests. Among them, the white woolly aphid *(Ceravacuna lanigera)*, is a major insect. The early shoot borer can be efficiently managed by release of *Trichogramma chilonis* at 50,000 per ha, 45 days after planting. Tricho cards pasted with 0.2 cc eggs of *Trichogramma chilonis* parasite can be obtained from the parasite breeding laboratories. These cards may be stapled in the field at 25 cards/ha. These cards may be equally distributed in 25 places once in 15 days at the 4–11 months old stage of the crop. Alternatively, pheromone traps may be set up in the field at 25/ha spaced at 20 m apart at the 5 month old crop stage. The male moths of internode borer may be trapped and killed. The pheromone vials in the traps may be replaced in seventh and ninth months.

The white wooly aphids can also be controlled by inundative release of *Chrysoperla* at 2500–5000 eggs/ha. For effective management of white wooly aphids, thrips, white flies, mealy bugs, and mites, the following organic sprays have been recommended. A total of 2 kg custard apple leaves may be mixed with 2 kg *Pongamia* leaves and 2 kg *Ipomaea* leaves with 500 mL *neem* oil in an earthen pot along with 8–10 L of hot water and 200 g soap powder. This solution is mixed thoroughly in 15 L of water for its use as a foliar spray. Similarly, mixture of onions in cow urine and sour buttermilk are used as effective spray for the control of wooly aphids (Anonymous, 2006).

27.2.7 WATER CONSERVATION IN ORGANIC FARMING

When sugarcane crop is 3–4-month old and 150 cm tall, all the dried leaves should be removed for using them as a mulch for reducing evaporation. Planting of one or two rows of maize fence around the sugarcane field has also been found quite effective in reducing wind speed and ensuring low evaporation. For reducing irrigation water requirement, pipe-based irrigation system or drip irrigation should be adopted. Evaporation and loss of water can also be reduced by profuse intercropping and mulching.

27.3 ECONOMICS OF SUGARCANE CULTIVATION UNDER ORGANIC AND INORGANIC FARMING

Several studies undertaken earlier have revealed that the per acre cost of sugarcane cultivation on organic farms (Rs. 45,974.50) was less in comparison to the costs involved on inorganic farms (Rs.54,331.82) (Rajendran et al., 2000; Shivanaikar et al. 2014). This difference in costs can be attributed to higher cost incurred on chemical fertilizers, cost on more quantity of setts used with higher plant density and more human labor employed in inorganic sugarcane farms. With a positive net return on organic as well as inorganic category of the farms, the per acre gross returns were higher (Rs. 82,328) on organic farms in comparison to inorganic farms (Rs. 81,360). The net return on inorganic farm was Rs. 27,028.18 against Rs. 36,353.90 obtained in organic farms. Higher benefit: cost ratio was recorded on organic farms (1.79) in comparison to 1.50 in inorganic farms (Thakur and Sharma, 2005; Shivanaikar et al. 2014). Hence, organic cultivation of sugarcane is economically better as compared with inorganic cultivation besides economic gain it also improves soil health. No use of chemical fertilizer and lower cost on planting material, irrigation and plant protection chemicals were responsible for the lower cost in OS farming. Apart from the cost reduction, OS farming was also more cost efficient due to lower cost of production per tonne (Rs 334) than inorganic sugarcane farming (Rs 366) (Table 27.1).

TABLE 27.1 Cost of Cultivation of Sugarcane under Organic and Traditional Farming (in Rs/ha).

Operations charges	**Organic**	**Inorganic**	**Per cent**
Land preparation	5838	5307	10.01
Seed and planting	5372	6974	−22.97
Manures and manuring	10,534	5242	100.95
Chemical fertilizers	0.00	8980	–
Weeding and interculture	5157	4959	3.99
Irrigation	5986	7587	−21.00
Plant protection	781	1274	−38.70
Others	1964	1792	9.60
Total*	35,632	42,115	−15.39

*This cost is exclusive of the cost on harvesting, transport, and marketing.

Source: Kshirsagar (2006).

Variable costs in respect of organic farms primarily consists of costs incurred on human labor, organic manures (FYM, green manuring, vermicompost, biofertilizers, and biopesticides), setts, and land preparation. The most important cost constituent in total cost of cultivation on organic farms is the expenditure on organic manures. While the other constituents of variable cost like interest on working capital and cost of machine and labour contributes to only 5 and 4.89%, respectively, to the total costs incurred on sugarcane cultivation in the organic farms (Sujatha et al., 2006). Similarly, the expenditure on chemical fertilizers has been the most significant cost item in the total cost of sugarcane cultivation on inorganic farms. The fixed cost constitutes nearly 22 and 18.5% to the gross cost of sugarcane cultivation on organic and inorganic farms, respectively. The net returns recorded from organic and inorganic sugarcane farms were Rs. 36,353.90 and 27,028.18, respectively. The return per rupee of cost of cultivation of sugarcane was more (1.8) in organic farming in comparison to inorganic farming (1.50). Total human labor cost was also found lower on organic farms against inorganic farms. More number of times of application of chemical fertilizer on more number of times, hand weeding, and irrigation increase the labor cost involved in inorganic farming of sugarcane. In OS cultivation, majority of the organic farmers followed wider spacing with the requirement of fewer setts. Whereas the farmers followed narrow spacing in respect of inorganic sugarcane cultivation so they required slightly more quantity of setts.

Although per acre average yield of sugarcane was lower in organic farms (43.56 tonnes) in comparison to inorganic farms (45.2 tonnes) as it takes 4–5 years to regain the soil fertility for higher yield as during transition from inorganic to organic farming. Among different break-up costs, the costs on planting and irrigation were slightly higher under organic farming than traditional farming. While the costs incurred on fertilizer application and plant protection chemicals were significantly higher under traditional farming (Waykar et al. 2006; Lampkin and Padel, 1994).

27.4 IMPACT OF OS CULTIVATION

Although the conventional high input intensive sugarcane cultivation is treated as unsustainable and detrimental to the environment, the viability of OS farming as an alternative agricultural system is also questioned. Therefore, there is a need to compare the impact of OS farming on input use, costs, yields, and returns with respect to conventional inorganic sugarcane farming for cost-benefit analysis. There are several issues of concern in the OS farming, which needs immediate redressal. For example, the characteristics and economic status of sugarcane farmers producing organically and the impact of OS farming on input use pattern, yields, costs, returns, and profits from the sugarcane crop may also be studied in-depth.

Sugarcane is one of the major employment generating sectors, providing employment to over 7.50% of total rural population in India. The data also revealed that sugarcane cultivation, especially the organic cultivation requires large number of human labor days, which can be attributed to enhanced requirement of laborers for carrying out preparatory tillage, manuring, green manuring, and managing the pests and diseases operations on organic farms. Apart from it, the intercropping adopted on organic farms, with crops having different time of sowing/planting and harvesting schedules can adjust the labor demand for stabilizing employment, which means that organic farming helps rural masses in availing number of opportunities for their sustainable employment throughout the year.

Sugarcane being a long duration and water juggling crop, requires water in enormous quantity for its successful cultivation is a cause of great concern for the farmers, scientists, and policy makers. This problem can be addressed by conservation and judicious use of water as water resources have endangered the stability and sustainability. The alternative organic approaches like trash mulching should be initiated as a common practice to preserve the soil moisture. Other irrigation practices like drip irrigation, fertigation, and irrigations in alternate trench/furrow, and others should be prioritized for saving groundwater.

Sharing their experiences, the farmers recalled that they faced the most difficult time during the conversion period from conventional farming to organic farming, which can be attributed to lack of knowledge about the organic farming principles, several significant changes in agricultural practices during the conversion period, conversion period of minimum 36 months for successfully completing the conversion, reduction in cane productivity right from the beginning of the conversion period, nonreceipt of premium prices, reduced farmers income during the conversion period, and noncooperation from neighboring farmers practicing conventional agriculture. These were the major constraints in the adoption and spread of organic farming. Thus, it is suggested that apart from training, the beginners should also get support for certification of organic produce and its marketing during this period. The beginners would like to shift to organic farming in stages, rather than trying to convert all the landholding at the once stroke, if possible.

Use of organic manures, organic fertilizers, biofertilizers, vermicompost, and biopesticides is very high in organic farming in comparison to conventional farming as chemical fertilizers and pesticides are substituted with these organic inputs. With the expansion of area under organic farming, the demand for the critical organic inputs is likely to be enhanced (Kumari et al., 2014). Thus, it is most desirable to ensure the smooth flow of these inputs, otherwise it will not be able to expand the area under organic farming in different States. Involvement of farmers for the production of certified biofertilizers and vermicompost along with the biopesticides would be of prime importance in this context. Therefore, it is recommended to launch specific schemes by the government and sugar factories for involvement of self-help groups (SHG) in production of biofertilizer and other inputs needed for OS farming. Such cooperative schemes for production of certified organic inputs should consist of major components of training, financial assistance, facilities for distribution, and marketing. This will be helpful to organic farmers in smooth supply of quality organic inputs at a cheaper and reasonable price.

27.5 CONCLUSION

The organic farming is very remunerative and beneficial for its social benefits in terms of resources, improved human health, and environment. Thus, it is recommended to properly measure and quantify the social benefits of OS farming for having an idea about the extent of subsidy that could be justified for promotion of OS farming in India. The central government should take immediate initiatives: to start awareness program for organic farming for sustainable agriculture, setting up of vermicomposting, biofertilizer production and, liquid biofertilizers units for availability of organic matter, provide training for application of bioplastics and green manure crops to retaining soil moisture, train farmers for the enrichment of soil fertility through trash mulching. In this context, a high-level committee comprising of representative of all the stakeholders may be constituted by the Government for identification of the high potential regions along with the high potential crops, formulating and prioritizing the policies and strategies for promoting the organic farming to reap the benefits of a rapidly growing national and international market for organic products.

KEYWORDS

- **precultivation practices**
- **biofertilizers**
- **bud selection**
- **intercropping**
- **water conservation**

REFERENCES

Anonymous. *Package of Organic Practices from Maharashtra for Cotton, Rice, Redgram, Sugarcane and Wheat*; Maharashtra Organic Farming Federation (MOFF): Maharashtra, 2006; p 186.

Bhattacharya, P.; Chakraborty, G. Current Status of Organic Farming in India and Other Countries. *Indian J. Fertilizers* **2005,** *1* (9), 111–123.

Geetha, P.; Sivaraman, K.; Tayade, A. S.; Dhanapal, R. Sugarcane Based Intercropping System and its Effect on Cane Yield. *J. Sugarcane Res.* **2015,** *5* (2), 1–10.

Kanwar, R. S.; Mehta, S. P.; Sharma, K. K.; Singh, S.; Bains, B. S.; Singh, N. Studies on Autumn Planted Sugarcane Based Cropping Systems. *Bharatiya Sugar* **1990,** *165* (2), 33–35.

Kshirsagar, K. G. Organic Sugarcane Farming for Development of Sustainable Agriculture in Maharashtra. *Agric. Econ. Res. Rev.* **2006,** *19*, 145–153.

Kumari, R.; Nautiyal, A.; Dhaka, N. D.; Bhatnagar, S. Potential of Organic Farming in Production of Environment Friendly Sugarcane. *Curr. Nutr. Food Sci.* **2014,** *10* (3), 173–180.

Lampkin, L. H.; Padel, S., Ed. *The Economics of Organic Farming—An International Perspective*; CAB International Publishers, 1994.

Mohanraj, D.; Padbanaman, P.; Viswantahan, R. Biological Control of Major Diseases of Sugarcane. In: *Biological Control of Major Crop Diseases*, Gnanamanickam, S. S., Ed.; Marcel Dekker Inc.: New York, 2002; pp 161–178.

Rajendran, T. P.; Venugopalan, M. V.; Tarhalkar, P. P. *Organic Cotton Farming in India*; Technical Bulletin No. 1; Central Institute for Cotton Research: Nagpur, 2000.

Shinde, N.; Patil, B. L.; Murthy, C.; Mamledesai, N. R. Profitability Analysis of Sugarcane Based Intercropping Systems in Belgaum District of Karnataka. *Karnataka J Agric. Sci.* **2009,** *22* (4), 820–823.

Shivanaikar, M.; Guledagudda, S. S.; Mokashi, P. Economics of Sugarcane Cultivation under Organic and Inorganic Farming in Bagalkot District of Karnataka. *Int. J. Commun. Bus. Manage* **2014,** *7* (1), 84–87.

Solomon, S. Sugarcane Production and Development of Sugar Industry in India. *Sugar Technol.* **2016,** *18* (6), 588–602.

Sujatha, R. V.; Eswara Prasad, Y.; Suhasini, K. Comparative Analysis of Efficiency of Organic Farming *vs* Inorganic Farming- A Case Study in Karimnagar District of Andhra Pradesh. *Agric. Econ. Res. Rev.* **2006,** *19* (2), 232.

Sundara, B. Agrotechnologies to Enhance Sugarcane Productivity in India. *Sugar Technol.* **2011,** *13* (4), 281–298.

Thakur, D. S.; Sharma, K. D. Organic Farming for Sustainable Agriculture and Meeting the Challenges of Food Security in 21st Century: An Economic Analysis. *Indian J. Agric. Econ.* **2005,** *60* (2), 205–219.

Viswanathan, R.; Malathi, P. P. Biocontrol Strategies to Manage Fungal Diseases in Sugarcane. *Sugar Technol.* **2019,** *21* (2), 202–212.

Viswanathan, R.; Rao, G. P. Disease Scenario and Management of Major Sugarcane Diseases in India. *Sugar Technol.* **2011,** *13* (4), 336–353.

Waykar, K. R.; Yadav, D. B.; Shendage, P. N.; Sale, Y. C. Economics of Grape Production under Organic and Inorganic Farming in the Nasik District of Maharashtra State. *Agric. Econ. Res. Rev.* **2006,** *19* (2), 240.

CHAPTER 28

Organic Farming for Sustainable Agriculture and Livelihood Security under Changing Climatic Conditions

D. K. SINGH*, SHILPI GUPTA, and Y. SHARMA

Department of Agronomy, College of Agriculture, G.B.P.U.A. & T., Pantnagar, U.S. Nagar 263145, Uttarakhand, India

Corresponding author. E-mail: dhananjayrahul@rediffmail.com

ABSTRACT

Different studies emphasize on the need for a shift to a sustainable model that promotes good yields with better resource use efficiency and ecosystem services, with no further expansion of agriculture into natural forests. Many innovative technologies and agricultural models have emerged in the past half century with a potential to improve the agriculture of developing countries but, that has not been reflected in creating a major poverty reduction. It is because resource poor farmers need readily available and low-cost technologies suitable for their farm situations that can increase the local food production and farm income . Only those resource conserving agricultural models that aim for long term sustainability of rural livelihoods provide an opportunity to resolve the issues of feeding the world population through small family farms. Organic farming is holistic production management system promotes and enhances agro-ecosystem health, including biodiversity, biological cycles, and soil biological activity. It emphasizes the use of management practices with reference to the use of off-farm inputs, taking into account those conditions required by locally

Organic Crop Production Management: Focus on India, with Global Implications, D. P. Singh, PhD, H. G. Prakash, PhD, M. Swapna, PhD, & S. Solomon, PhD (Eds.)

adopted systems. This is accomplished by using, where possible, agronomic, biological, and mechanical methods as opposed to using synthetic materials to fulfill any specific function within the system.

28.1 INTRODUCTION

The decrease in food quality, soil fertility, degradation of cultivated lands, water, and air are the main outcomes of rapidly increasing crop output as a result of indiscriminate use of high amount of fertilizers and pesticides which, in turn, is threatening food security. Thus, this indiscriminate exploitation of natural resources and nonjudicious use of agriculture inputs without considering the carrying capacity of soil is ultimately threatening the sustainability of exhaustive cropping systems predominantly rice–wheat. The soils under these exhaustive cropping systems are showing signs of fatigue and are not increasing productivity even with increased use of inputs. The stagnant level of crop productivity can be attributed to diminishing nutrient supplying capacity of the soil, imbalanced nutrition, and less or no application of organic manures. Therefore, in order to address all these concerns, there is a need to develop a system, which is not only productive and low cost but also resource conserving and sustainable for upcoming decades. To develop this system and function properly various strategies, namely, enrichment of soil, use of self-reliance inputs, maintenance of natural cycles and life forms, rainwater conservation, temperature management, integration of animals' minimum dependence on renewable energy sources like solar and animal power need to be considered and implemented in our system.

Organic farming systems are based on the philosophy of "feed the soil to feed the plant." The basic precept is implemented through a series of approves practices which ultimately increases soil organic matter, biological activity, and nutrient availability. In order to convert toward organic, one must have the know-how of the system and long-term strategies must be developed first in accordance with the area. FAO/WHO guidelines have defined organic agriculture as follows:

> "A holistic production management system promotes and enhances agro-ecosystem health, including biodiversity, biological cycles, and soil biological activity. It emphasizes the use of management practices with reference to the use of off-farm inputs, taking into account those conditions required by locally adopted systems. This is accomplished by using, where possible, agronomic, biological, and mechanical methods

as opposed to using synthetic materials to fulfill any specific function within the system" (FAO/WHO Codex Alimentarius Guidelines)".

Due to requirement of long conversion period of 2–3 years, organic farming is practiced hardly. Data on the comparative productivity of crops under organic farming versus intensive farming are not available. Limited organic farming being practiced in India can be attributed to the contract farming. Nutrient management in organic farming is the most challenging task due to bulky nature of organic sources and their slow rate of release of nutrients and is mostly met through inclusion of legumes in cropping systems, green manure, farmyard manure, and others. Green manure applied in organic farming provides an economical alternative to escalating prices of nitrogen fertilizer and has become an integral part of effective technology to sustain the economics of agriculture production system, ensuring productive capacity of soil without causing any impoverishment and combating many ecological and environmental problems (Bana and Pant, 2000). Under organic farming, the addition of organic manures promotes the growth and activity of useful microorganisms in soils which in turn can benefit the crop and organic matter binds soil particles into structural units thus improving soil structure and maintaining favorable conditions for aeration and permeability. Application of soil organic matter is helpful in enhancing the rate of infiltration and percolation of water and water holding capacity. In this context, an experiment was taken up to evaluate the organic farming practices on high-value cropping sequence, especially, basmati rice–based cropping sequence with the aim of finding out the impact of organic, inorganic, and integrated nutrient management practices on productivity of basmati rice–based cropping system and soil health, including the soil physicochemical properties in terms of bulk density, organic carbon, available nitrogen, phosphorus, potassium, sulfur, and other micronutrients.

28.2 ORGANIC FARMING'S POTENTIAL TO MITIGATE CLIMATE CHANGE

Global food production is severely altered by climate change. It has been reported that agriculture is not affected by climate change but also contributes to it. Ten to twelve percent of global greenhouse gas (GHG) emission is due to human food production. The changes in land use (deforestation, overgrazing, soil degradation, etc.) play significant contributions to the global CO_2 emissions. In order to adapt to more unpredictable and extreme weather conditions (drought and floods), reduce GHGs emissions by food production and reverse carbon losses in soils, the capacity of agriculture production must be boosted and therefore, sustainable agriculture and food supply systems are the need of the hour. Therefore, organic farming is treated as one of the most sustainable approaches in agricultural production by emphasizing recycling techniques with low external inputs along with high output strategies which are based on enhancing soil fertility and diversity at all levels and make soils less susceptible to erosion. In this chapter, organic farming systems are evaluated in the context of changing climatic scenarios. As climate change is a complex and global problem, there is a need to develop strong framework for recommendations for future development and research requirements in organic farming. Organic farming is an adaptation strategy and systematic approach to address issues of climate change and variability. Thus, organic farming is a strong and sustainable option and has additional potential as mitigation strategy through production management, minimizing energy, randomization of nonrenewable resources, and carbon sequestration. Generally, the global warming potential of organic farming systems in temperate climates is comparatively smaller (when calculated per land area) than that of chemical and integrated systems. But, this difference declines, when calculated per product unit (Badgley et al. 2007). But under dry or water constraints, organic farming may outperform chemical farming, both in respect of per crop area and per harvested crop unit.

Emissions avoidance and carbon sequestration are the major components of mitigation strategy of organic farming. For achieving emission avoidance, lower emissions of N_2O are a good technique due to lower nitrogen input, as per the assumption that 1–2% of the N used in cropping systems is emitted as N_2O, irrespective of the form of the nitrogen input. IPCC uses this default value as 1.25%, but the recently concluded studies reveal significantly lower values as in the case of semiarid areas (Barton et al., 2008). Second, there is less erosion in organic farming systems than conventional ones generally through less CO_2 emissions through erosion due to better structure of soil and more plant cover and, finally, by lowering CO_2 emissions from farming system inputs (insecticides/pesticides and fertilizer produced using fossil fuel).

The highest mitigation of carbon in organic farming is achieved through carbon sequestration in soils. Soil carbon sequestration is a natural as well as deliberate process. Under this process, carbon dioxide is either removed from the atmosphere or diverted from emission sources and stored in the oceans, terrestrial environments (vegetation, soils, and sediments). Carbon sequestration is greatly enhanced through farm management or agronomical practices such as enhanced use of organic manures, conservation tillage, cover crops, nutrient management, irrigation, restoring degraded soils, pasture management, use of intercrops, and green manures, increase the carbon sequestration by enhancing the content of soil organic matter and improved structure of the soil. This increasing soil organic content in farming systems has also been identified as a major carbon mitigation option.

28.3 ORGANIC AGRICULTURE REDUCES EMISSIONS OF NITROUS OXIDE AND METHANE

Emission of nitrous oxide is directly associated with the concentration of easily available mineral nitrogen in soils. After fertilization, there is high emission of nitrous oxide which is highly variable. According to IPCC, nitrous oxide is emitted by 1.6% of applied nitrogen fertilizer. In organic farming, the ban of mineral nitrogen and the reduced livestock units per hectare considerably reduce the concentration of easily available mineral nitrogen and, thus, the N_2O emissions in soils. In addition to this, the following factors in organic farming also contribute to lower emissions of nitrous oxide: First, diversified crop rotations with green manure improve soil structure and diminish emissions of nitrous oxide and secondly, soils managed organically are more aerated and have significantly lower mobile nitrogen concentrations. Thus, in both cases, there is reduced emission of nitrous oxide.

The nitrous oxide emission rates for conventional farming were found to be higher in comparison to organic farming in the case of five European countries (Petersen et al., 2006). Flessa et al. (2002) also observed decreased emission rates of nitrous oxide in the organic farm, although yield-related emissions were not reduced in a long-term study in southern Germany.

Nearly 14% of the GHG emissions are contributed by methane (Barker et al., 2007). Two-thirds of this is of anthropogenic origin and mainly from agriculture. Enteric fermentation and manure management are responsible for methane emissions to a large extent form which in turn are directly proportional to livestock numbers. Organic agriculture has an impact on reduction of emissions, as there are a limited number of livestock in organic farms (Kotschi & Müller-Sämann, 2004; Olesen et al., 2006; Weiske et al., 2006). There is limited availability of data on methane emissions from livestock particularly with respect to the reduction of GHG emissions from ruminants and manure heaps.

28.4 ORGANIC FARMING SEQUESTERS CO_2 IN THE SOIL

Arable cropland and permanent pastures lose soil carbon through mineralization, overgrazing and soil erosion. Loss of 12 million ha/year in global arable land loss has been estimated which is 0.8% of the world cropland area (1513 million ha) (Pimentel et al., 1995). Annual carbon dioxide emissions from intensively cropped soils were equivalent to 8% of national industrial CO_2 emissions. If farm practices are not changed, the typical arable soils will continue to lose organic carbon content. The uses of improved farm practices particularly organic farming and conservation tillage are helpful in reducing soil erosion (Bellamy et al., 2005) and carbon loss is converted into gains, resulting thereby in removal of considerable amount of CO_2 from the atmosphere. Organic land management are also helpful in reducing soil erosion and converting carbon losses into gains (Reganold et al., 1987), especially due to the use of green and animal manure, soil fertility-conserving crop rotations with intercropping and cover cropping and composting techniques.

28.5 RESEARCH EXPERIENCES OF ORGANIC FARMING AT PANTNAGAR, UTTARAKHAND

Field experiments are being conducted at Breeder's Seed Production Centre of G.B. Pant University of Agriculture and Technology, Pantnagar, under the Indian Council of Agriculture Research funded the Network Project on Organic Farming for the period of 1 decade during 2004–2013. The experimental soil was silty loam in texture (45.8% silt, 31.40% sand, and 22.8% clay), medium in organic carbon (0.65%), low in available N (238 kg/ha), medium in P (16.7 kg/ha), K (156 kg/ha), and high in available sulfur (29.3 kg/ha). pH and electrical conductivity were 7.4 and 0.34 dS/m, respectively. Three management practices, 100% organic, 100% inorganic, and integrated (50% organic and 50% inorganic) as horizontal strip along with four cropping systems, namely, *basmati* rice–wheat, *basmati* rice–lentil, basmati rice–vegetable pea, and *basmati* rice—*Brassica napus* as vertical strips, were tested in strip plot design.

28.6 PRODUCTION SUSTAINABILITY

Basmati rice grain yield trend during the 10 years of experiment showed an increasing trend ranging from 22.66 q/ha in the first year (2004) to 35.03

q/ha in the 10th year (2013) with an increase of 54.59% over initial grain yield (Fig. 28.1). This increasing trend in the grain productivity obtained during the study period revealed that organic mode after 3 years of conversion period was helpful in increasing yields and during the last 3 years (2011–2013), it registered the highest basmati rice grain yield than other two modes of nutrient management. Similar results of gradual increase in seed productivity of rice with the use of organics over a period of time were also observed (Surekha, 2007). The highest yield under organic treatments was mainly due to buildup of organic matter, improvement in physicochemical properties of soil, and accumulation of nutrients over the years. Mankotia (2007) also reported similar trend where higher productivity of rice was obtained due to in situ green manuring with *Sesbania* and application of FYM. Although almost similar transition in seed productivity was observed in the plots which received 50% nutrients through organic and 50% through inorganic sources, the increase was not much higher as compared to organic showing only 38.1% increase in grain yield of basmati rice after 10 years. However, 100% inorganic plots showed the highest grain yield of rice up to 3 years but thereafter, the organic and integrated modes expressed their superiority over it. This declining trend in productivity under inorganic mode can be attributed to higher bulk density of soil that does not favor proper root growth of crops.

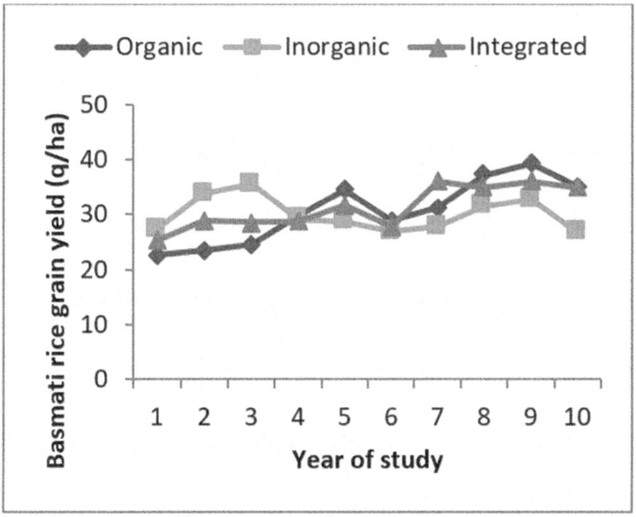

FIGURE 28.1 Productivity of basmati rice under organic, inorganic, and integrated farming during the 10 years of experimentation (2004–2013).

The average data over 9 years of organic farming revealed that that yield gap in the case of leguminous crops between the 100% inorganic fertilizer treatment and 100% organic treatments was bridged up year after year. Also, lower grain yields in the plots with 100% organic can be attributed to the less readily available nutrients in the initial years of transition as nutrient cycling processes in first-year organic systems converge from inorganic N fertilization to organic source and slow release rates of organic materials. It was observed that wheat and *B. napus* seed productivity revealed almost same trend over the years, that is, in both the crops, grain yield was significantly higher in 100% inorganic treated plots followed by integrated plot and minimum was recorded in case of 100% organic treatments. However, in case of lentil, seed yield was significantly higher in 100% organic treatment in the first year and the trend shifted to 100% inorganic treated plots during the last years. A reverse trend was observed in respect of vegetable pea, where significantly lower productivity was recorded in 100% organic treatments during first year, followed by 100% inorganic plots and finally from the fifth year onwards significantly higher, grain yield was recorded with 100% organic treatments (Fig. 28.2).

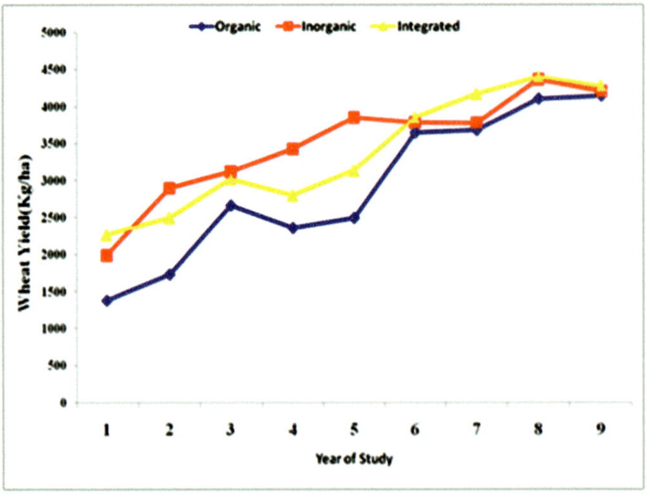

FIGURE 28.2 Wheat grain yields over 10 years of experimentation (2004–2013).

The mean data of rice grain equivalent yield over 10 years of study (Fig. 28.3) revealed that higher rice grain equivalent yield (44.8 q/ha) was obtained in integrated plots followed by inorganic (43.2 q/ha) and organic modes (42.8

q/ha). After 10 years of study, RGEY in organic mode was enhanced to about 59.1% over initial RGEY in comparison to inorganic mode where increase in yield was 25.9% over initial year.

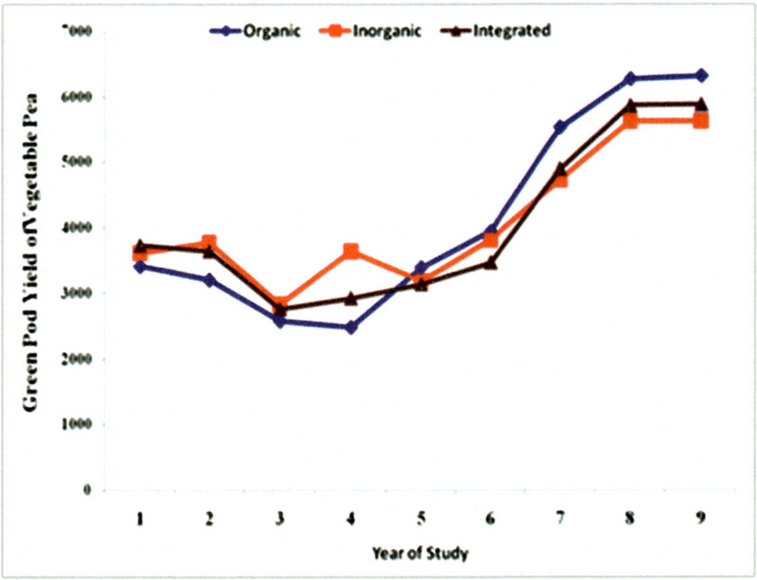

FIGURE 28.3 Green pod yield of vegetable pea over 10 years of experimentation (2004–2013).

28.7 ECONOMIC SUSTAINABILITY

Organic farming development in Uttarakhand could have wide-ranging ramifications in rural employment, ecological sustainability, and remunerative agriculture. Basmati rice with vegetable pea—*Sesbania* green manure or Basmati rice—chickpea with *Sesbania* green manure proved to be more remunerative under organic production system in foot hill and *tarai* areas of Uttarakhand. Since the last decade, researches on organic crop production revealed that the average net realization under organic mode of cultivation over 10 years was recorded Rs. 44,456/ha, while in inorganic mode, it was Rs. 40,824 /ha (Fig. 28.4). Net returns were found higher with organic mode of management in spite of slightly higher cost of cultivation under organic mode, as compared to chemical one after 3 years of conversion and charging premium price. Therefore, an economic empowerment was observed when basmati rice–based cropping systems were managed organically.

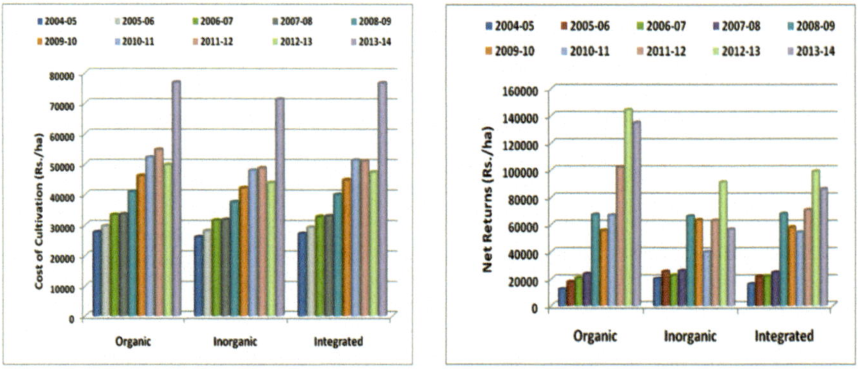

FIGURE 28.4 Total cost of cultivation and net return from basmati rice–based cropping system.

28.8 SOIL SUSTAINABILITY

28.8.1 SOIL BULK DENSITY, ORGANIC CARBON, AND ORGANIC CARBON STOCK

It is evident from data (Table 28.1) that bulk density of soil decreased appreciably (by 8.03% from 1.37 to 1.24 g/cc) after 10 years of experiment under organic mode followed by integrated mode (1.33 g/cc) while under inorganic mode, the value of bulk density increased to 1.44 g/cc contrary to other tow modes. The lowering of bulk density under organic and integrated modes may be ascribed to increased organic carbon, good soil aggregation, and increased pore space. Francis et al. (2006) also reported similar results. Organic carbon content in 0–15, 15–30, 30–45, and 45–60 cm soil depth was increased by 41, 49, 56, and 59%, respectively, under organic nutrient management. These organic sources besides adding organic carbon to the soil also enhance root growth due to reduced bulk density. Results reported by Tripathi et al. (2007) also confirm the same. Soil organic carbon stock decreased with increasing soil depth from 0 to 60 cm with higher soil organic carbon (SOC) stock in organic mode in each soil depth. In organic mode, the SOC stocks were found 24.18, 23.42, 21.58, and 20.22 t/ha in 0–15, 15–30, 30–45, and 45–60 cm depth, respectively, which was 21.69, 25.40, 28.96, and 31.65% higher, respectively, than inorganic mode. Thus, application of higher amount of organic manure under organic mode led to higher organic carbon content in soil leading to higher carbon stock in soil in comparison to inorganic one where no organic matter was applied.

TABLE 28.1 Depth Wise Bulk Density (t/m^3), Organic Carbon (%), and Soil Organic Stock (t/ha) after 10 Years of Experimentation.

	Bulk density (t/m^3)	Organic carbon (%)	SOC stock (t/ha)
		Depth (0–15 cm)	
Organic	1.24	1.3	24.18
Inorganic	1.44	0.92	19.87
Integrated	1.33	1.14	22.74
		Depth (15–30 cm)	
Organic	1.28	1.22	23.42
Inorganic	1.42	0.82	17.47
Integrated	1.38	1.06	21.94
		Depth (30–45 cm)	
Organic	1.32	1.09	21.58
Inorganic	1.46	0.7	15.33
Integrated	1.40	0.99	20.79
		Depth (45–60 cm)	
Organic	1.39	0.97	20.22
Inorganic	1.51	0.61	13.82
Integrated	1.50	0.89	20.03

28.8.2 TRANSITION IN AVAILABLE NUTRIENT STATUS OF SOIL

Data (Table 28.2) revealed that available nitrogen and sulfur increased to a great extent in organic mode as compared to inorganic and integrated modes. The maximum increment in available nitrogen and sulfur percentages after 10 years over initial year of experimentation was recorded 58.4 and 31.1%, respectively. The values of available phosphorus were almost the same under inorganic (214%) and integrated modes (218%) as compared to 204% in organic mode of cultivation. Likewise, available micronutrients viz., Fe, Cu, and Zn in soil after a decade of continuous crop cycles under organic mode of cultivation increased to the extent of 85.0, 18.7, and 61.9%, respectively, over their initial values. The percent increment in available Mn in soil was found higher under integrated mode (398.0%) in comparison to organic (295%) and inorganic modes (193%).

TABLE 28.2 Changes in Available N, P, K, S, Zn, Cu, Fe, and Mn Status of Soil after a Decade of Study Over Initial.

		N (kg/ha)	P (kg/ha)	K (kg/ha)	S (kg/ha)	Zn (mg/kg)	Cu (mg/kg)	Fe (mg/kg)	Mn (mg/kg)
100% Organic	Value	377	50.8	244	38.4	1.36	3.69	55.94	12.35
	% change	58.4	204	56.4	31.1	61.9	18.7	85.0	295
100% Inorganic	Value	354	52.4	250	33.2	0.97	2.55	35.43	9.16
	% change	48.7	214	60.3	13.3	15.5	−15.0	17.2	193
Integrated (Org. + inorg.)	Value	369	53.1	243	38.1	1.02	3.33	51.35	15.59
	% change	55.0	218	55.8	30.0	21.9	11.0	69.8	398
Initial		238	16.7	156	29.3	0.84	3.00	30.24	3.13

28.9 ENVIRONMENTAL SUSTAINABILITY

28.9.1 METHANE AND NITROUS OXIDE EMISSIONS FROM BASMATI RICE FIELDS

Neither CH_4 emissions (Fig. 28.5) nor N_2O emissions (Fig. 28.6) differed significantly between plots under inorganic, integrated, and organic nutrient management. Results also showed that plots that received the same amount of N through different rates of organic amendments did not significantly differ in their N_2O emissions compared to purely urea fertilized plots (Fig. 28.6). Although CH_4 emissions were not significantly different, plots fertilized with organic amendments (integrated, organic) tended to emit less CH_4 than to the plots under inorganic nutrient management (Fig. 28.5), which contradicts the results of earlier studies. In a study conducted over several seasons, Wassmann et al. (1996) found that rice fields, which received 120 kg N/ha through rice straw (high organic input), had higher CH_4 emissions compared to the fields which received the same amount of N through urea (low organic input). In addition to it, CH_4 emissions was enhanced when substituted with organic amendments (*Sesbania* GM, 60-kg N/ha; urea, 90-kg N/ha) in comparison to urea treatments (150 kg N/ha) (Wassmann et al., 2000).

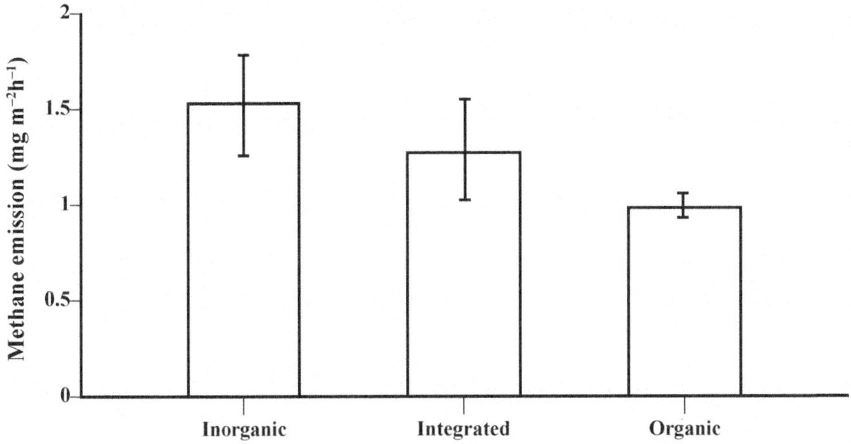

FIGURE 28.5 Mean CH_4 emissions (mg/m^2/h) from plots under inorganic, integrated, and organic nutrient management.

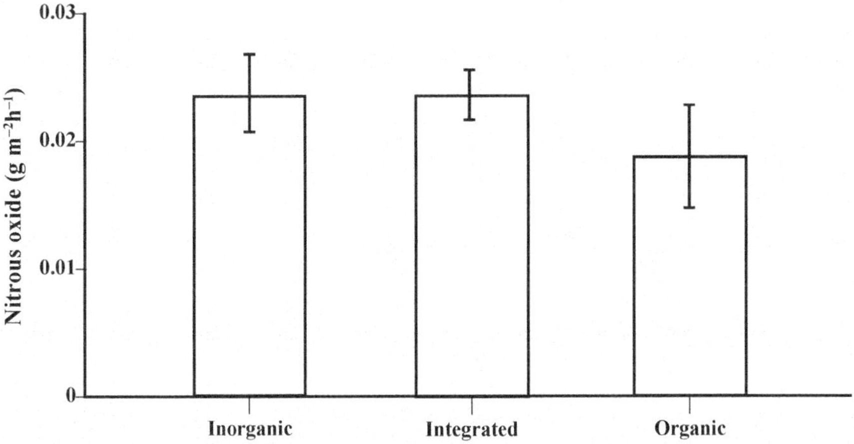

FIGURE 28.6 N_2O emissions (g/m^2/h) from plots under inorganic, integrated, and organic nutrient management.

ACKNOWLEDGMENT

The authors express their gratitude to the Director, Experiment Station, Gobind Ballabh Pant University of Agriculture and Technology, Pantnagar (India), for providing necessary facilities; Indian Council of Agricultural

Research, India Research Programme under Network Project on Organic Farming and HELVETAS Swiss Inter-cooperation, India for providing financial assistance.

KEYWORDS

- **climate change**
- **nitrous oxide**
- **methane**
- **production sustainability**
- **economic sustainability**
- **soil sustainability**
- **environmental sustainability**

REFERENCES

Bana, O. P. S.; Pant, K. Green Manuring for Ecologically Sound Crop Production. *Indian Farmers Digest* **2000,** *33* (2), 19–20.

Badgley, C.; Moghtader, J.; Quintero, E.; Zakem, E.; Jahi Chappell, M.; Avilés-Vázquez, K.; Samulon, A.; Perfecto, I. Organic Agriculture and the Global Food Supply. *Renew. Agric. Food Syst.* **2007,** *22* (2), 86–108.

Barker, T.; Bashmakov, I.; Bernstein; Bogner, L. J. E.; Bosch, P. R.; Dave, R.; Davidson, O. R.; Fisher, B. S.; Gupta, S.; Halsnæs, K.; Heij, G. J.; Kahn Ribeiro, S.; Kobayashi, S.; Levine, M. D.; Martino, D. L.; Masera, O.; Metz, B.; Meyer, L. A.; Nabuurs, G. J.; Najam, A.; Nakicenovic, N.; Rogner, H. H.; Roy, J.; Sathaye, J.; Schock, R.; Shukla, P.; Sims, R. E. H.; Smith, P.; Tirpak, D. A.; Urge-Vorsatz, D.; Zhou, D. *Technical Summary on Contribution of Working Group III to the Fourth Assessment Report of the Intergovernmental Panel on Climate Change*. In *Climate Change 2007: Mitigation*; Metz, B., Davidson, O. R., Bosch, P. R., Dave, R., Meyer, L. A., Eds.; Cambridge University Press: Cambridge and New York, 2007. http://www.mnp.nl/ipcc/pages_media/FAR4docs/final_pdfs_ar4/TS.pdf

Barton, L.; Kiese, R.; Gatter, D.; Butterbach Bahl, K.; Buck, R.; Hinz, C.; Murphy, D. Nitrous Oxide Emissions from Cropped Soil in Semi-Arid Climate. *Global Change Biol.* **2008,** *14*, 177–192.

Bellamy, P. H.; Loveland, P. J.; Bradley, R. I.; Lark, R. M.; Kirk, G. J. D. Carbon Losses from All Soils Across England and Wales 1978–2003. *Nature* **2005,** *437*, 245–248.

Francis, Z.; Bobbi, L. H.; Francis, J. L.; Henry, J.; Olaacken, O. A.; Barry, M. O. Predicting Phosphorus Availability from Soil Applying Compost and Non-composted Cattle Feedlot Manure. *J. Environ. Qual.* **2006,** *35*, 928–937.

FAOSTAT FAO. *Statistical Database Domain on Fertilizers: Resource STAT- Fertilizers*; Food Agriculture Organization of the United Nations (FAO): Rome, Italy, 2009. http://faostat.fao.org/site/575/default aspx (accessed Dec 10, 2017).

Flessa, H.; Ruser, R.; Dörsch, P.; Kamp, T.; Jimenez, M. A.; Munch, J. C.; Beese, F. Integrated Evaluation of Greenhouse Gas Emissions (CO_2, CH_4, N_2O) from Two Farming Systems in Southern Germany. *Agric. Ecosyst. Environ.* **2002,** *91*, 175–189.

Mankotia, B. S. Effect of Fertilizer Application with Farmyard Manure and In Situ Green Manures in Standing Rice–Wheat Cropping System. *Indian J. Agric. Sci.* **2007,** *77* (8), 512–514.

Kotschi, J.; Müller-Sämann, K. *The Role of Organic Agriculture in Mitigating Climate Change*; International Federation of Organic Agriculture Movements (IFOAM): Bonn, 2004.

Petersen, S. O.; Regina, K.; Pöllinger, A.; Rigler, E., Valli, L., Yamulki, S., Esala, M., Fabbri, C., Syväsalo, E., Vinther, F. P. Nitrous Oxide Emissions from Organic and Conventional Crop Rotations in Five European Countries. *Agric. Ecosyst. Environ.* **2006,** *112*, 200–206.

Pimentel, D.; Harvey, C.; Resosudarmo, P.; Sinclair, K.; Kurz, D.; McNair, M.; Crist, S.; Shpritz, L.; Fitton, L.; Saffouri, R.; Blair, R. Environmental and Economic Costs of Soil Erosion and Conservation Benefits. *Science* **1995,** *267*, 1117–1123.

Reganold, J.; Elliott, L.; Unger, Y. Long-term Effects of Organic and Conventional Farming on Soil Erosion. *Nature* **1987,** *330*, 370–372.

Surekha, K. Nitrogen Release Pattern from Organic Sources of Different C: N Ratios and Lignin Content and Their Contribution to Irrigated Rice (*Oryza sativa*). *Indian J. Agron.* **2007,** *52* (3), 220–224.

Tripathi, H. P.; Mauriya, A. K.; Kumar, A. Effect of Integrated Nutrient Management on Rice-Wheat Cropping System in Eastern Plain Zone of Uttar Pradesh. *J. Farm. Syst. Res. Dev.* **2007,** *13* (2), 198–203.

Wassmann, R.; Buendia, L. V.; Lantin, R. S.; Bueno, C. S.; Lubigan, L. A.; Umali, A.; Neue, H. U. Mechanisms of Crop Management Impact on Methane Emissions from Rice Fields in Los Banos, Philippines. *Nutr. Cycl. Agroecosyst.* **2000,** *58* (1–3), 107–119.

Wassmann, R.; Neue, H. U.; Alberto, M. C. R.; Lantin, R. S.; Bueno, C.; Llenaresas, D.; Rennenberg, H. Fluxes and Pools of Methane in Wetland Rice Soils with Varying Organic Inputs. *Environ. Monitor. Assess.* **1996,** *42* (1–2), 163–173.

Weiske, A.; Vabitsch, A.; Olesen, J. E.; Schelde, K.; Michel, J.; Friedrich, R.; Kaltschmitt, M. Mitigation of Greenhouse Gas Emission in European Conventional and Organic Dairy Farming. *Agric., Ecosyst. Environ.* **2006,** *112*, 221–232.

CHAPTER 29

Prospects of Organic Farming in Uttar Pradesh

H. G. PRAKASH*, R. K. PANDEY, and D. P. SINGH

Chandra Shekhar Azad University of Agriculture & Technology, Kanpur 208002, Uttar Pradesh, India

**Corresponding author. E-mail: drhp1962@gmail.com*

ABSTRACT

The Ganga basin region of Northern India, which includes Uttar Pradesh and Uttarakhand, is also known as the food bowl of India. A major part of the country's rice and wheat production happens here. Intensive farming, however, has resulted in severe pressure on land, water and other natural resources. Over the years, river Ganga has become heavily polluted due to runoff from fertilizers and other industrial waste.

Organic farming is a productive and sustainable system and there is a consensus among all the stakeholders on its eco-friendly nature and inherent ability to protect human health and agro-biodiversity. In Uttar Pradesh, labor is quite abundant and relatively cheaper, there is large opportunity to employ the labor on this sector which will provide local employment and restrict labor migration. Organic farming has great potential in the state to solve the problem caused by the intensive agriculture and overuse of chemicals. Efforts have been made by the Government of Uttar Pradesh to promote organic farming in the state. Recently, government is encouraging organic farming within a radius of 10 km on both the banks in 27 districts through which the Ganga passes in the state.

Organic Crop Production Management: Focus on India, with Global Implications, D. P. Singh, PhD, H. G. Prakash, PhD, M. Swapna, PhD, & S. Solomon, PhD (Eds.)

29.1 INTRODUCTION

Uttar Pradesh covers 2.41 lakh km^2, accounting for 7.3% of the total geographical area of India with 16.16% of the country's population as per 2011 Census, that is, 19.95 crore with 79.2% residing in rural areas. Located in the Indo-Gangetic plain and intersected by rivers, Uttar Pradesh's economy is basically agrarian in nature with more than 60% of its population depending on agriculture in one or the other way for their livelihood. Being the largest food grains producing state of the country, Uttar Pradesh offers a diverse agroclimatic condition, most conducive for farm production. The state has the distinction of being the largest sugarcane- and sugar-producing state of the country. The state also offers excellent investment opportunities for industrial development. The most of the villages in Uttar Pradesh are small with an average population of around 2500 per *panchayat*. In economic terms, it is divided into four major regions, western, central, eastern, and Bundelkhand regions and also into nine agroclimatic zones, western plains, central-western plains, southwestern plains, central plains, Bhabhar and *Tarai* regions, northeastern plains, eastern plains, Vindhyan region, and Bundelkhand.

The state has contributed significantly to the Green Revolution of the country. Uttar Pradesh is the largest producer of wheat, potato, sugarcane, and milk and ranks third in paddy production. Agriculture still constitutes the backbone of the state economy, providing livelihood security to about two-thirds population of the state. The state is endowed with ample alluvial soil along with diverse agroclimatic profile, suitable for farming of a large number of crops. Due to large cultivated area, its share in national agricultural production is quite impressive but low crop productivity has hindered the realization of ultimate potential.

29.2 RELEVANCE OF ORGANIC FARMING

The agriculture sector will play an important role in economic security of the country in the future also. As the largest private enterprise (~138 million farm families) of the country, about 14% is contributed to the national GDP by agriculture and engages more than 52% of the workforce. Therefore, the growth in agriculture and allied sectors remains a prerequisite condition for inclusive growth. India is predominated by small farming agriculture. By ignoring the ecology and environment, the present pattern of economic development cannot be helpful in sustaining the needs of man without substantial degradation of the factors essentially required for supporting the lives of all the living things on this planet called "'Earth." Negative impacts of excessive use of the chemical fertilizers, insecticides, pesticides, herbicides, and fungicides in conventional agriculture system, affecting the soil and human health, highlight the importance of an eco-friendly alternative farming system. Residues of pesticides, fertilizers, and other chemicals used during the process of crop production, which are highly injurious to human health are being found in the food. The Green Revolution, a major milestone of India's agricultural achievements, which has transformed the country from the stage of food deficiency to self-sufficiency, encouraged the use of fertilizers and pesticides for high yielding varieties. As a result, production has increased from 50.82 million tons (1950–1951) to 265.57 million tons (2013–2014), but indiscriminate and excessive use of chemicals has put forth a big question mark on sustainability of agriculture in the long run. Although the traditional methods of farming in ancient India were by default "organic," there is greater awareness among the farmers for the adverse impacts of chemical inputs on the soil, environment, and human health. To fulfill and address social, ecological, and economic issues together, organic farming could play a vital role.

Organic manures are not only treated as a source of nutrients but they are also helpful in enhancing biodiversity and activity of the microbial population in soil, improving soil structure, and number of other related physiological chemical and biological processes of the soil. In order to growing more to feed, we have taken the wrong way of sustainability. All stakeholders have begun to think of various alternative farming systems, which can protect environment and beneficial for human health in various ways. Organic agriculture is treated as the best among all of them due to its scientific approach and its wider acceptance by all the stakeholders. This system has several benefits in comparison to the conventional one, in addition to the protection of the environment through improved soil fertility, better water quality, and prevention of soil erosion and human health by generating number of rural employments, etc.

29.3 PHYSICAL FEATURES

Uttar Pradesh is a border state of the country with its northern frontiers adjoining Nepal which earlier extended up to Tibet before the creation of

Uttarakhand with the Shivalik ranges near Tibet border. Uttar Pradesh's boundaries touch Delhi, Rajasthan, and Haryana in the west, Madhya Pradesh in the south, and Bihar in the east. In geophysical terms, the Shivalik range of the Himalayas in the North, the river Yamuna and the Vindhyas in the West, the south-west, and the South and the Gandak rivers in the east demarcate the terrain called Uttar Pradesh.

29.3.1 NATURAL RESOURCE MANAGEMENT

Land is a basic natural resource where the history has recorded the development of human beings with other living beings and, along with water and plants, has taken place from the beginning of the creation. Ineffective management of natural resources affected biodiversity, agriculture productivity, and ecological balance. To ensure planned development and to achieve required production of food grains, fodder, and biofuel, it is necessary to implement soil and water conservation programs in problematic areas on priority basis. Schemes implemented also provide local employment to the agriculture laborers, small and marginal farmers. Nonscientific use of land creates numerous problems like land degradation, ravine, and water logging. In Uttar Pradesh, most of the holdings (75.4%) are marginal, accounting 33.7% of the total area and 2.7% of holdings having more than 4 ha of land and accounting for 19.2% of the total area clearly reveal severe inequities in the ownership of landholding. Despite all odds, the state contributes 32% wheat, 40% sugarcane, and 26% potato produced in the country.

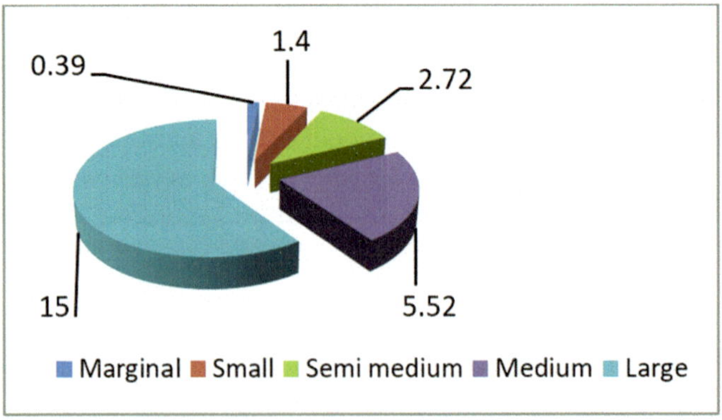

FIGURE 29.1 Size of holding in ha.
Source: Agricultural Statistics at a Glance 2012, Department of Agriculture and Cooperation, Ministry of Agriculture, Govt. of India.

There is 241.70-lakh ha total geographical area in Uttar Pradesh, out of which 165.46-lakh ha area is under cultivation. Gross cropped area is 258.65-lakh ha with the cropping intensity of 156%. More than half (93.49 lakh ha) of the cultivated area is sown more than once. The hare of Uttar Pradesh in total geographical area, net sown area, and gross cropped area of the country are 7.90, 11.84, and 13.28%, respectively. In Uttar Pradesh, the average size of holding is around 0.83 ha and per capita land area of the state is 0.14 ha as compared to 0.33 ha for the country as a whole. The marginal and small peasants own 64.77% of the total land area. The average size of holding of marginal farmers is only 0.40 ha. Thus, the state has to support more people with lesser land (2013–2014).

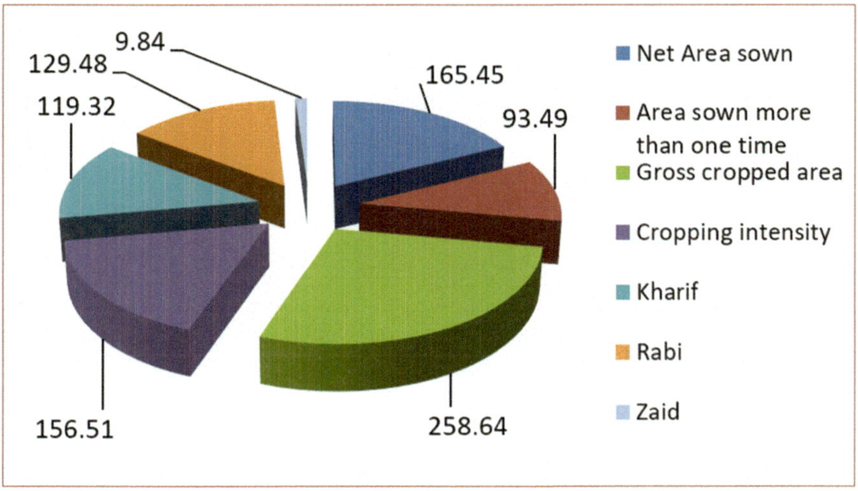

FIGURE 29.2 Land use pattern (lakh ha).
Source: Directorate of Agriculture, Govt. of Uttar Pradesh 2013–2014.

29.3.2 SOILS

The soils of western Uttar Pradesh comprising the districts of Meerut, Muzaffarnagar, Saharanpur, Bijnor, Moradabad, Bareilly, and Pilibhit are generally deep brown and loamy in few pockets, also mixed with sand. The soil is shallow, gravely, and full of stones—being generally acidic. The soils of the western plains comprising the districts of Meerut, Muzaffarnagar, and Saharanpur) are deeper and fertile. Moving east toward the districts of Bijnor, Moradabad, Bareilly, and Pilibhit, the soil gets to be loamy. When we further move toward Pilibhit, some of the soils show acidic while other reveals alkaline properties. The soil in the central region of the state comprising the districts of Lakhimpur Kheri, Sitapur, Lucknow, Barabanki, Hardoi, Kanpur, and Azamgarh is loamy and sandy loams. In the east Uttar Pradesh comprising the districts of Gorakhpur, Basti, Mahrajganj, Siddarth-nagar, and Gonda, there are two types of soil., "*Bhat*" and "*Banjar*." The soil in the north-western Uttar Pradesh contains less phosphate. The soils of Jaunpur, Azamgarh, and Mau districts lacks in potash and the drier areas are known as "*Usar*" and "*Reh*." The soils of Aligarh, Mainpuri, Kanpur, Etah, Etawah, Sitapur, Unnao, Raibareilly, and Lucknow are salt affected which is locally called "*Usar*" and "*Reh*" soils. There is mixed red and black soil in Jhansi Division of the Uttar Pradesh. In Mirzapur and Sonebhadra as well as the Karchhana and Meja tehsils of Allahabad besides Chakia and Varanasi districts, the black soil is sticky, calcareous, and fertile. It expands as it soaks moisture and contracts on drying up. There are two types of red soil, namely,, "*Parwa*" and "*Rackar*" in the upper plateau of the abovementioned districts.

29.3.3 AGROCLIMATIC ZONES

The state is divided into nine agroclimatic zones, Bhabar and *Tarai*, western Plains, central-western plains, south-western plains, central plain, Bundelkhand, northeastern Plain, eastern plains, and Vindhyan region as shown in Table 29.1.

TABLE 29.1 Agroclimatic Zones of Uttar Pradesh.

S. no	Agroclimatic zones	Geographical area in ha	Districts	Soil type
1	*Tarai* and Bhabar	1,847,319	Saharanpur (58%), Muzaffarnagar (10%), Bijnaur (79%), Moradabad (21%), Rampur (40%), Bareilly (19%), Pilibhit (75%), Shahjahanpur (6%), Kheri (39%), Bahraich (47%), Shravasti (71%)	Alluvial, least to medium phosphorus, medium-to-high potassium, and highly carbonized matter
2	Western plains	1,637,424	Saharanpur (42%), Muzaffarnagar (90%), Meerut, Baghpat, Ghaziabad, Gautam Budh Nagar, Bulandshahar	Alluvial pH value normal to sodic and carbonic matter from least to medium

TABLE 29.1 *(Continued)*

S. no	Agroclimatic zones	Geographical area in ha	Districts	Soil type
3	Mid-western	1,697,125	Bareilly (81%), Badaun, Pilibhit (25%), Moradabad (79%), J. P. Nagar, Rampur (60%), Bijnor (21%)	Almost alluvial, normal to slight sodic, and contains medium corbonic matters
4	South-western semidry	2,234,222	Agra, Firozabad, Aligarh, Hathras, Mathura, Mainpuri, Etah	Alluvial and Arawali
5	Mid-plain/central	5,647,307	Shahjahanpur (94%), Kanpur Nagar, Kanpur Dehat, Etawah, Auraiya, Farrukhabad, Kannauj, Lucknow, Unnao, Raebareli, Hardoi, Kheri (61%), Sitapur, Fatehpur, Allahabad (58%), and Kaushambi	Alluvial, pH normal to sodic and containing carbonic matter from least to medium quantity
6	Bundelkhand	2,961,006	Lalitpur, Jhansi, Jalaun, Hamirpur, Mahoba, Banda, and Chitrakoot	*Rakar, Parwa, Kabar,* and *Maar*
7	Northeastern	2,955,485	Gorakhpur, Maharajganj, Deoria, Kushinagar, Basti, Sant Kabir Nagar, Siddharthnagar, Gonda, Bahraich (63%), Balrampur, and Shrawasti (29%)	Alluvial, calcarius
8	Eastern plains	3,808,718	Azamgarh, Mau, Ballia, Pratapgarh, Faizabad, Ambedkar Nagar, Barabanki, Sultanpur, Varanasi, Chandauli, Jaunpur, Ghazipur, and Sant Ravidas Nagar (86%)	Alluvial, sodic, and *Diara* soil
9	Vindhyan	1,381,840	Allahabad (42%), Sant Ravidas Nagar (14%), Mirzapur, and Sonbhadra	*Kali, Bhari* red granules, and alluvial soil in-plane area

Source: Department of Land Development and Water Resources, Government of UP.

29.3.4 IRRIGATION

The net irrigated area in the state is 140.26 lakh ha. The canals and private tube wells are the main sources of irrigation accounting for approximately 18.23 and 71.96% of the net irrigated area, respectively. Over three-fourths of the sown area is irrigated. The state has fairly large canal network, which account for about 22% of the irrigated area. Ground water is easily tappable and accounts for about 78% of irrigated area.

29.3.5 CLIMATE

The climate of the state is generally subtropical and semiarid type. The month of May is the hottest month varying in maximum temperature from 45 to 47°C and the coldest month is January, temperature varies from 5 to 8°C whereas annual normal rainfall of the state is 947 mm varying from 594 to 1400 mm and relative humidity varies from 25 to 95%. Generally, maximum rainfall occurs during monsoon season (June to July) by the south-west monsoon.

29.4 PERFORMANCE OF AGRICULTURAL SECTOR

Agro-allied sectors happen to be the key sector in Uttar Pradesh that engages more than 65% of workforce, most of whom are below the poverty line. However, the performance of agricultural sector has remained far from satisfactory. From the sixth Five-Year Plan onwards, it can be observed that agricultural growth rate of the state economy declined continuously under various Five-Year Plans and the same was less than the national average during the eighth and ninth Plans because of fractions in size of holding and poor harvest plus. Thus, the pressure of population and abysmal poverty has exerted pressure on the state, ultimately resulting in fragmentation of landholdings. Uttar Pradesh is one of the major wheat-, rice-, chickpea-, pigeonpea-, field pea-, lentil-, mungbean-, urdbean-, mustard-, groundnut-, sesame-, and linseed-producing states. Sugarcane is the most important cash crop and area under this crop has increased significantly with the expansion of irrigated area. The performance of major crops has been discussed in the following sections.

29.4.1 AREA UNDER MAJOR CROPS

In terms of total area cultivated, food grains dominate as nearly about four-fifth of the gross cropped area is occupied by food grains while the national level share of food grain cultivated area is about 38%. Within food grains, cereals and pulses occupy about 69 and 10% of the total cropped area, respectively. Major cereals (rice and wheat) are sown in 61% of the area and area under coarse cereals is nearly 10% of total area. Maize and pearl millet (*bajra*) are the most important coarse cereals accounting for three-fourth of the total area under coarse cereals. Nonfood grain crops of the state, which include oilseeds, vegetables, fruits, spices, flowers, sugarcane, cotton, tobacco, and others, are sown in about 20% area. Sugarcane is the main nonfood grain and commercial crop of the state which is cultivated in about 8.6% of the gross cropped area. Oilseeds such as rapeseed and mustard, sesame, groundnut, linseed, sunflower, and soybean are being cultivated in about 4.45% area during the year 2013–2014 (Fig. 29.3).

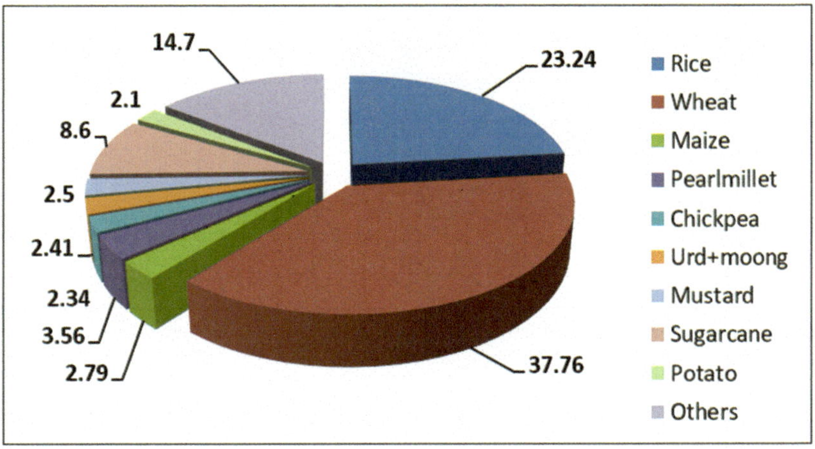

FIGURE 29.3 Percent area under major crops.
Source: Directorate of Agriculture, Uttar Pradesh 2013–2014.

29.4.2 CONTRIBUTION OF MAJOR CROPS

Uttar Pradesh is the largest producer of sugarcane which contributes about 40.26% followed by wheat about 31.7% to national food basket, whereas the state is the second-largest producer of rice, which accounts for about

13.78% of the national production. During 2013–2014 in Uttar Pradesh, the mean wheat yield was 31.10 q/ha while the average productivity of barley, rice, *bajra*, maize, and *jowar* was 29.06, 24.44, 20.59, 18.50, and 9.78 q/ha, respectively. The mean yield of pigeonpea, chickpea, urdbean, and mungbean was 9.26, 4.56, 4.42, and 3.04 q/ha, respectively, during 2013–2014. The average productivity of mustard was 10.03 q followed by 7.61 q/ha for groundnut during 2013–2014 in Uttar Pradesh. The average productivity of sugarcane and potato during 2013–2014 was 637.78 and 214.40 q/ha, respectively, and in same year, cropping intensity was 156.51% which was higher than previous years (Fig. 29.4).

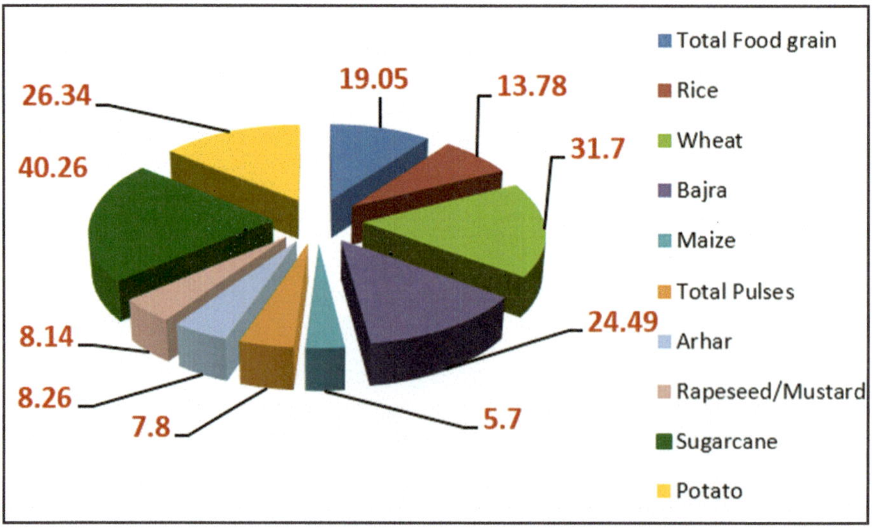

FIGURE 29.4 Contribution of major crops in the national food basket.
Source: Directorate of Agriculture, Uttar Pradesh.

29.4.3 PRODUCTIVITY STATUS OF AGROCLIMATIC ZONES

The productive status of different crops under various agroclimatic zones has been described below:

Food Grains: The average productivity of food grains is 25.11 q/ha while cereals' production is 27.60 q/ha in the state. The maximum (33.84 q/ha) average food grain productivity was observed in the western plain zone while minimum (12.28 q/ha) in Bundelkhand zone. In the case of cereals, again western plain zone leads with 34.63 q/ha followed by Bundelkhand zone with 20.32 q/ha average productivity in the year 2013–2014. The average total cereal production of Vindhyan, eastern plains, and Bundelkhand zone is observed lower than the state and national average.

Pulses: Average total pulses productivity of western and Bundelkhand zone is lower than that of the state average which is 6.36 q/ha. The pulses productivity varies from 5.18 (Bundelkhand zone) to 11.97 q/ha (south-western semiarid zone).

Oilseeds: The average productivity of oilseeds is 7.32 q/ha in the state. The maximum oilseed's productivity (14.34 q/ha) was recorded in the south-western semiarid zone whereas lower oilseed productivity was observed in Bundelkhand zone (3.56 q/ha). Average oilseed productivity of Bundelkhand, eastern, and Vindhyan zone is lower than the state productivity.

Sugarcane: The maximum sugarcane productivity (702.15 q/ha) was recorded in western zone which is higher than the state productivity, whereas lower sugarcane productivity was observed in Bundelkhand zone (439.07 q/ha) with state average of 637.78 q/ha which is at par with Vindhyan zone (source: Directorate of Agriculture, Uttar Pradesh, 2013–2014).

Potato: The maximum potato productivity (255.93 q/ha) was recorded in the south-western semiarid zone, whereas lower potato productivity was observed in eastern zone (150.48 q/ha) with the state average of 214.40 q/ha which is at par with that of Vindhyan and Bundelkhand zones. The potato productivity of most of the zone is lower than that of the state and national average, except the south-western semiarid and mid-western zone (Figs. 29.5 and 29.6).

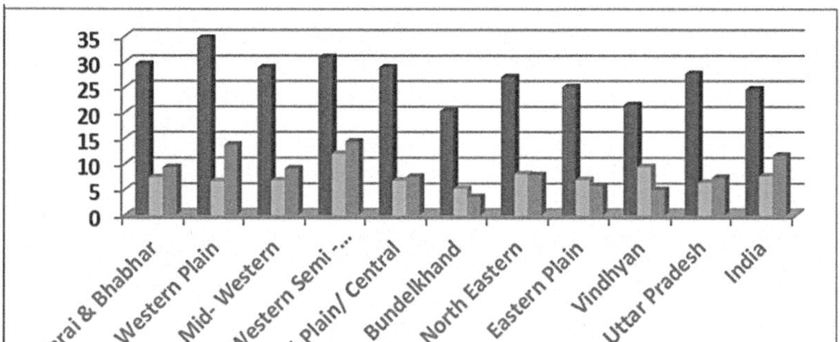

FIGURE 29.5 Productivity (q/ha) of cereals, pulses, and oilseed crops.
Source: Directorate of Agriculture, Uttar Pradesh.

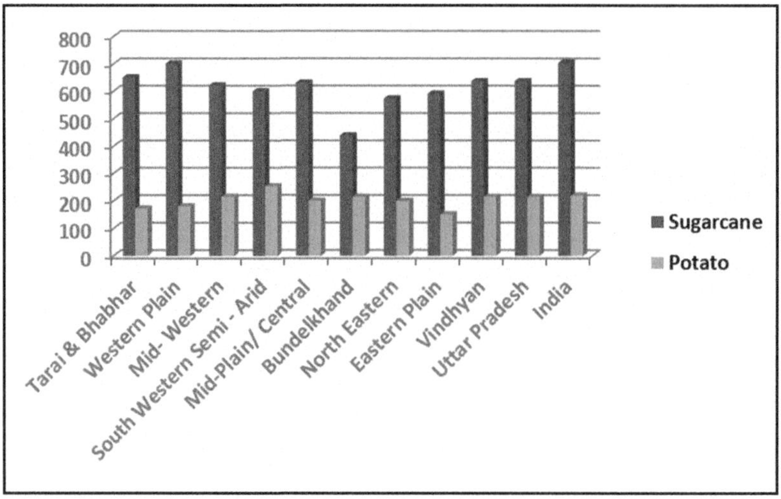

FIGURE 29.6 Productivity (q/ha) of sugarcane and potato.
Source: Directorate of Agriculture, Uttar Pradesh.

Horticulture: Uttar Pradesh holds a leading position for horticultural crops in the country and the diverse climate of the state is suitable for producing all kind of horticultural crops, that is, fruits, vegetables, flowers, medicinal and aromatic plants, root and tuber crops and spices, besides bee-keeping and mushroom cultivation, as subsidiary enterprises along with their processing and value addition. Varied agroclimate of the state allows growing a large number of these crops throughout the year, ensuring that it is available on regular basis. These crops have inherent benefit of achieving higher yield per unit area of land as compared to other crops, leading to higher income and employment generation for the farmers in rural areas. Horticulture crops contribute in a major way to the family income, employment, and nutrition per unit of area of small farmers. Area under fruit crops is 4.25 lakh ha with the production of 88.91 lakh MT. However, area under vegetable production is about 11.45 lakh ha with 235.72 lakh MT and spices crops cover 0.71 lakh ha with the production of 2.73 lakh MT during 2013–2014.

Livestock: Uttar Pradesh is the highest milk producing state (251.98 lakh MT) in India and holds a share of more than 18% in the total milk production of the country in which 69.46% was accounted for by buffaloes followed by 25.38 and 5.16% by cows and goats, respectively. Apart from being largest milk producer, the state also has the largest number of cows and buffaloes. Livestock Census 2012 depicted that the total population of livestock was 68,715,147 lakh in the state of which, buffaloes accounted for 44.57%

followed by 28.46, 22.68, and 1.97% of cattle, goats, and sheep, respectively. Out of a total population of bovines of 50,182,401 lakh, the buffaloes accounted for 61.03% followed by 38.97% of cattle. The livestock resources of Uttar Pradesh also include 145.94 lakh goats, 13.71 lakh sheep, 19.73 lakh pigs, and 105.79 lakh poultry. The state is also well known for *Kherigarh, Ponwar, Gangatiri*, and *Kenkatha* breeds of cow; *Bhadwari* breed of buffalo; *Jamunapari* and *barbari* breeds of goats and *Jalauni* breed of sheep. The meat production in Uttar Pradesh was only 956,000 tons in 2011–2012 which increased to 1,221,000 tons in 2013–2014 showing 27.72% increase over the period. The wool production has also increased by 3.73% during corresponding period (Fig. 29.7).

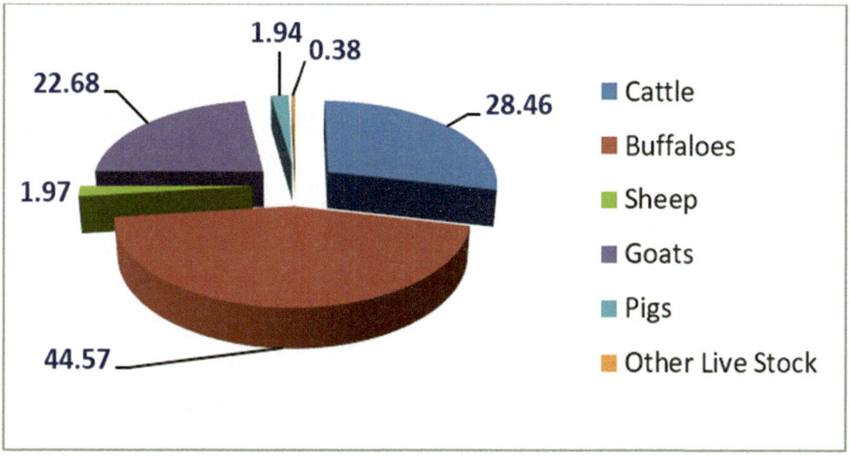

FIGURE 29.7 Percent livestock in Uttar Pradesh.
Source: Department of Animal Husbandry, Government of Uttar Pradesh.

29.5 PRINCIPLES OF ORGANIC FARMING

Organic farming does not permit the use of any synthetic inputs, namely, chemical fertilizers, fungicides, insecticides, pesticides, hormones, food additives, etc. and allows the use of crop rotations, crop residues, animal manures, off-farm organic waste, mineral-grade rock additives, and biological system of nutrient mobilization and plant protection. It is not just based on replacing inorganic inputs with organic matter rather it is based on the four principles as advocated by International Federation of Agriculture Movement (Deshmukh and Babar, 2015):

- Organic farming should be based on living ecological system and cycles, promoting working with them, emulating them and helpful in sustaining them.
- Organic farming should build on relationships for ensuring fairness in respect to the common environment and life opportunities.
- Organic farming should be adopted as a precautionary manner for protecting the health and well-beings of present and future generations along with environment.
- Sustainable organic agriculture should be helpful in enhancing the health of soil, plant, animal, human, and planet as one and indivisible.

29.6 ADVANTAGES OF ORGANIC FARMING

Although there are a number of benefits of switching over from conventional farming techniques to organic farming, yet all the benefits may not be feasible for all. Therefore, it is important to highlight some of the following benefits of organic agriculture:

29.6.1 SOIL QUALITY

Organic farming is based on the quality of the soil. The organic content of the soil is built up by the natural plant nutrients obtained from farmyard manures, composts, green manures, and plant residues. Soil under organic farming conditions has lower bulk density, higher water holding capacity, higher microbial biomass carbon, and nitrogen and higher soil respiration activities in comparison to the soils under conventional farming. It reveals that microbial activity under organic farming is enhanced due to the availability of sufficient amount of nutrients to the crops.

29.6.2 HIGH PREMIUM

As the organic food products fetch 20–30% higher prices than conventional food products, there are ample opportunities for farmers to get a high premium price which improves their standard of living by providing a fair chance to flourish.

29.6.3 LOW INVESTMENT

The organic agriculture requires low capital investment in comparison to the conventional farming where a number of chemical inputs are used. Apart from it, the production of organic manures is so simple that it does not require any sophisticated techniques for its production. Further, the costs incurred on organic manures and pesticides are very low due to its local production. Since farming is influenced by a number of external factors such as climate (namely, rainfall), pests, diseases, in cases of natural calamity, and irregular rainfall, small farmers practicing organic farming have to suffer less even in case of crop failure due to their low investments.

29.6.4 LOW DEPENDENCE ON LOANS FROM MONEY LENDERS

It is very unfortunate to note that several farmers commit suicide due to their inability to pay the debts taken by them. As costly chemical fertilizers, insecticides, and pesticides are not used in organic farming, there is no need for the farmers to take the loans from the local money lenders. Therefore, the farmers are not compelled to take such an extreme step due to crop failure.

29.6.5 EMPLOYMENT OPPORTUNITIES

Number of studies undertaken in India on organic farming reveals its labor-intensive nature in comparison to the conventional farming system. Thus, organic farming could be an attractive option for the state like Uttar Pradesh having large percentage of labor unemployment and underemployment. Apart from it, the problem of periodical unemployment can also be reduced due to diversification of the crops with their different planting and harvesting schedules, requiring more number of laborers.

29.6.6 CONSUMER AWARENESS AND DEMAND FOR HEALTHY FOOD

In urban areas of the country, as per capita income is increasing, change in lifestyle and food habits are visible. Since literacy is on the rise and media is making consumers more aware regarding their healthy food habits, demand for organic products is growing in domestic market. There has been a trend in the last decade for products associated with lifestyle choices and process quality which ultimately justify premium price of organic products. This may enhance the domestic consumption of organic foods.

29.6.7 PROTECTING AND ENHANCING BIODIVERSITY

Intensification of agriculture reflects that the sector enters into more and direct competition for scarce land, water, and other natural resources resulting degradation of biodiversity. Chemicals have destroyed many beneficial insect species and have caused environmental degradation. Organic farming will be beneficial for improving degraded biodiversity in areas where intensive agriculture persists and it will also be helpful for maintaining the biodiversity in other traditional farming-dominated area.

29.6.8 SYNERGY WITH LIFE FORMS

A synergy has been developed between various plant and animal life forms by the organic farming which can be easily understandable by the marginal and small farmers for its adoption. The traditional knowledge that the farmers have can be an added advantage in organic farming so as to get fruitful outcomes.

29.6.9 SOCIAL IMPACT

Organic production has significant social impact on rural society by substantially reducing pollution, conserving energy, soil nutrients, reducing cost for production, and ensuring the supply of food for future generations. Besides, farmers may have better standards of living, can get good product prices; low unemployment, reduced labor emigration, and reduced health risks are some other benefits of organic farming.

29.7 CONCLUSION

Organic farming is a productive and sustainable system and there is a consensus among all the stakeholders on its eco-friendly nature and inherent ability to protect human health and agro-biodiversity. In Uttar Pradesh, labor is quite abundant and relatively cheaper, there is large opportunity to employ the labor on this sector which will provide local employment and restrict labor migration. Organic farming has great potential to solve the problem caused by the intensive agriculture and overuse of chemicals. Efforts have been made by the Government of Uttar Pradesh to promote organic farming in the state. In addition, the Government has set-up different organizations for the marketing of the organic produce, produced in the State. As per capita income is increasing, changes in lifestyle and food habits are visible in urban and semiurban areas and demand for organic and processed agro-allied products has been increased resulting in a development of the Indian organic food industries, having the immense potential to boost the Indian economy and rural prosperity.

KEYWORDS

- **organic farming**
- **physical features**
- **natural resource management**
- **soils**
- **agroclimatic zones**
- **irrigation**
- **climate**
- **employment opportunities**
- **consumer awareness**
- **social impact**

REFERENCES

Agricultural Statistics at a Glance. Government of India, Ministry of Agriculture & Farmers Welfare, Department of Agriculture, Cooperation and Farmers Welfare, Directorate of Economics and Statistics, New Delhi, 2015.

Deshmukh, M. S.; Babar, N. Present Status and Prospects of Organic Farming in India. *Eur. Acad. Res.* 2015, *3* (4), 4271–4287.

GOUP. *Statistical Abstract*, Economics & Statistics Division, State Planning Institute, UP, 2013.

GOUP. *Statistical Diary*, Economics & Statistics Division, State Planning Institute, UP, 2013.

GOUP. *Statistical Diary*, Economics & Statistics Division, State Planning Institute, UP, 2014.

Vision 2025. IARI Perspective Plan. Indian Agricultural Research Institute, New Delhi.

Vision 2030. Indian Council of Agricultural Research, New Delhi.

CHAPTER 30

Bio-Organic Approach: Success Story of Organic Farming at Bafna Farms, Pune

SANTOSH B. CHAVAN and PRAKASH M. BAFNA*

Jay Research and Biotech India Private Limited, Bafna Group, 111 Tower 1, World Trade Centre, Kharadi, Pune 411014, Maharashtra, India

Corresponding author. E-mail: prakash@bafnagroup.com

ABSTRACT

A multitude of experiments done at Bafna Farm indicated that if the microbial activity of the soil is increased, it can improve soil health, plant health and develop disease and insect resistance in the plant. Use of organic farming can reduce chemical fertilizer and pesticides to the extent of 25% due to bio-organic products applications. As chemical uses reduced, the population of friendly insects increased in the Bafna Farm. Contrary to the national average the organic carbon at Bafna Farms ranges from 2 to 2.5% for most of the plots. This was mainly due to recycling of crop residues, green manuring and use of bioagents in controlling diseases and pests.

30.1 INTRODUCTION

A bio-organic approach is a method of farming that avoids the use of pesticides, fertilizers, genetically modified organisms, antibiotics, and growth hormones in crop cultivation. It is a holistic approach to optimize the productivity and fitness of diverse communities within the agroecosystem, including soil organisms, plants, livestock, and people. The principal goal of

Organic Crop Production Management: Focus on India, with Global Implications, D. P. Singh, PhD, H. G. Prakash, PhD, M. Swapna, PhD, & S. Solomon, PhD (Eds.)

organic production is to develop enterprises that are sustainable and harmonious with the environment.

The general principles of bio-organic production are as follows:

- It should protect the environment, minimize soil degradation and erosion, decrease pollution, optimize biological productivity, and promote a sound state of health.
- It should maintain long-term soil fertility by optimizing conditions for biological activities within the soil.
- It should maintain biological diversity within the system.
- It should recycle materials and resources to the greatest extent possible within the enterprise.
- It should provide attentive care that promotes health and meets the behavioral needs of livestock.
- It should prepare organic products, emphasizing careful processing, and handling methods in order to maintain the organic integrity and vital qualities of the products at all stages of production.
- It should rely on renewable resources in locally organized agricultural systems.

A bio-organic approach promotes the use of crop rotations and cover crops and encourages balanced host/predator relationships. Organic residues and nutrients produced on the farm are recycled back to the soil. Cover crops and composted manure are used to maintain the soil's organic matter and fertility. Preventative insect- and disease-control methods are practiced, including crop rotation, improved genetics, and resistant varieties. Integrated pest and weed management, and soil conservation systems are valuable tools on an organic farm.

The bio-organic standards generally prohibit products of genetic engineering and animal cloning, synthetic pesticides, synthetic fertilizers, sewage sludge, synthetic drugs, synthetic food processing aids and ingredients, and ionizing radiation. Prohibited products and practices must not be used on certified organic farms for at least 3 years prior to harvesting the certified organic products. In bio-organic production, farmers choose not to use some of the convenient chemical tools available to other farmers. The design and management of the production system are critical to the success of the farm. In this chapter, we present a success story of Bafna Farms at Pune, where Jay Research and Biotech India Private Ltd. (JRABIPL) has produced table grapes, pomegranate, and sugarcane using bio-organic approach.

30.2 HISTORY OF BAFNA FARMS AND ESTABLISHMENT OF JAY RESEARCH AND BIOTECH INDIA PVT. LIMITED

In 1988–1989, the Bafna Group purchased 100-acre land for putting up a paper mill project. Due to some unavoidable circumstances, the paper mill project was canceled and the land has been converted into an agriculture field without any background or knowledge. Agricultural production has been during 1990–1994 with 70-acre land for the cultivation of pomegranate and 30 acres for grapes. Till 1999, Bafna Farms was incurring losses due to the lack of knowledge and they were planning to sell the land, as they felt this is not their "baby." In 1999, Shri Prakash Bafna attended a seminar on "Importance of Bio-organic Products for Agriculture" organized by honorable Shri Sharad Chandraji Pawar. The take-home message of the seminar was "Bio will be the future of Agriculture." By the year 2000, the demand for bio-organic products was on the rise as the use of chemicals (fertilizer and pesticide) for table grape and pomegranate production was reduced. From 2000 to 2004, bio-organic products were purchased from different bio-product companies. Sometimes, results were good and sometimes were not very encouraging. To understand the benefits and strength of bio-organic farming, Mr. Bafna, the founding director of Bafna Farms, visited National Chemical Laboratory, Pune. To overcome the problem of low yield, a few scientists suggested Mr. Bafna to set up a small laboratory for his own farm. In 2005, Bafna Group started a partnership firm in the name of Jay Biotech and started producing bio-organic preparations with the help of technical experts for their own farm. Now the company has become a private limited in the name of JRABIPL and has the following R&D activities:

1. They started a collaboration with Dadelos and AINIA, Valencia, Spain, under Indo-Spain Collaboration in Biotechnology Programme of Department of Biotechnology, New Delhi, and Centre for the Development of Industrial Technology, Spain. The theme of this project was "Bio-sanitary and bio-stimulants products based on the combination of microorganisms and organically stabilized bio-nutrients."
2. They started collaborations with RMIT University, Melbourne, Australia, on "Novel Eco-friendly Nano-formulations along with bio-organisms for the control of powdery/downy mildew in grapes."
3. They imparted R&D training to BCIL trainees under BITP since 2010–2011. Till date, 50 students have been trained.
4. They started a collaboration with Bharati Vidyapeeth for vegetable crops management.

5. They started a collaboration with NCL, Pune, for the control of mealybugs in grapes, DBT Funded, successfully completed.

30.3 ISSUES IN TABLE GRAPES PRODUCTION AND THEIR MANAGEMENT USING BIO-ORGANIC APPROACH

In India, table grapes production suffers due to different microbial pathogens such as downy mildew (Gessler et al., 2011), powdery mildew (Miazzi et al., 2003), and anthracnose (Thind et al., 2004). Table grape production is also affected by insect pests such as thrips (Roditakis, 2007), flea beetle (Mani et al., 2014), jassids (Mani et al., 2014), mealy bug (Daane et al., 2012), stem borer (Sunitha et al., 2017), and mites (Duffner et al., 2001). Several chemical insecticides (Krewer et al., 2002), fungicides (Qin et al., 2010, Mohapatra et al., 2010), and acaricides (Karabhantanal et al., 2012) are being used to control the abovementioned pathogens and pests. To control the above problems, various qualities of chemical pesticides were used, which has created the problem of resistance to pathogens and pests. Since CIB-registered chemical insecticides are available in limited quantities, the same chemicals are used continuously with increased doses. This also leads to the development of resistance in insect pests resulting in the increased levels of pesticide residue in grapes which is harmful to humans and also to the environment. Thus, in Bafna Farm, the focus was laid on bio-organic approach that involves the use of less chemicals and more bio-organic products for table grape production. In this chapter, we are focusing on the use of bio-organic products that were developed at the R&D unit of Jay Research and Biotech India Private Limited, Pune, located within the premises of Bafna Farms. The list of all the bio-organic products along with their application has been provided in Table 30.1. As the focus was given on the bio-organic approach, the use of chemicals was reduced slowly and steadily every year and now in 2017–2018 season, the use of chemical pesticides and fertilizers was reduced to 70–75% to that of earlier the year 2000.

TABLE 30.1 Bio-organic Products Developed by JRABIPL, Pune, for Sustainable Agriculture.

Brand name	Application
These products should be given at regular interval for any agricultural crop	
Jay Bio-Multi	Soil health, disease resistance, pesticide degradation, and nutrient uptake
Jay Bio-Gold	White root development

TABLE 30.1 *(Continued)*

Brand name	Application
Jay Bio-Amrut	White root development and disease resistance
Jay Anti-stress	For sunburn and insect control
Jay Comp	Decomposition of green manure
Jay Azo	Nitrogen availability
Jay PSB	Phosphorous availability
Jay KSB	Potash availability
These products need to be given as per the requirement	
Jay Bio-Killer	Caterpillar, white grub, and borer control
Jay Bio-Ramban	Thrips, jassids, mealy bug, and mite control
Jay Verti	White fly, mealy bug, and mite control
Jay Bio-Force	Nematode control
Jay Bio-Prahar	Fungal pathogen control
Jay Bio-Shakti	Fungal pathogen control
Jay Bio-Special	Mite control

30.4 SOIL AND WATER QUALITY AT BAFNA FARMS

Earlier, the soil quality of Bafna Farms was too bad (Fig. 30.1a). The level of soil sodium was >1000 ppm, calcium carbonate ($CaCO_3$) level was 12–20%, pH of the soil was 7.5–8.3. The water quality available for Bafna Farms is still bad (Fig. 30.1b) and becoming bad every year. For example, sodium level in water for table grape cultivation should be less than 2 meq/L, while it was found 4.3 meq/L in river Mula–Mutha, 8.68 meq/L for river Bhima, 13.73 meq/L for one of the wells in the farm, and 16.48 meq/L for one of the tube well in the farm (Fig. 30.1c). Daily irrigation with such poor-quality water would have deteriorated the soil if the bio-organic approach would not have been used. The above-mentioned data can be correlated with earlier scientific studies done. The analysis depicted heavy pollution of the Mula–Mutha waters as BOD was high with lower or the absence of DO at some sites. The nitrates, nitrites, sulfates, phosphates, and sodium chloride also showed high values at highly polluted sites (Vinaya and Mane, 2007). Similarly, high values of BOD, total dissolved solids, nitrates, and chlorides have been reported in the water of Bhima River (Jagtap, 2014). Surprisingly, the total dissolved solute level of water is 1200–1600 ppm, while it should be less than 500 ppm for any agricultural crop (Stagnari et al., 2016).

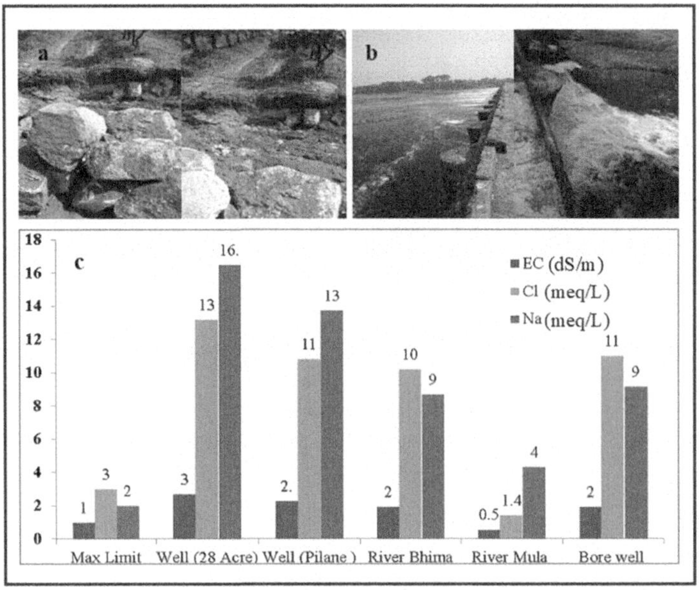

FIGURE 30.1 (a) Soil in Bafna Farms; (b) Water source for Bafna Farms rivers Mula–Mutha and Bhima; (c) Sodium level in irrigation water for grape production should be less than 2 meq/L; however, in Bafna Farm, it was from 4 to 16 meq/L. Similarly, chloride level should be less than 3 meq/L and in Bafna Farm, it was from 10 to 13 meq/L except river Mula (1.5 meq/L).

We have no option but to use the available water quality; thus, the initial concern was to improve the soil quality, specifically organic carbon, by focusing on the use of bio-organic products along with cow-dung manure at least two times a year and the application of cow-dung slurry at regular interval. In addition to this, the use of microbial products started in 2000, it was 10–15% and now, in 2017, it is 70–75%. The global use of bio-organic products is below 5%. The use of chemicals was reduced from 90% to 25–30%. Importantly, earlier (2006–2008), the production of table grapes was 3–5 tons/acre and now it has reached 10–12 tons/acre with excellent quality grapes (Fig. 30.2) by using bio-organic way.

30.5 FACILITIES AVAILABLE AT BAFNA FARMS

Sand and disc filters for drip irrigation and RO system for water purification have also been installed at Bafna Farms.

Sand filters for drip irrigation
RO system for water purification

FIGURE 30.2 Sand and disc filters for drip irrigation, and RO system for water purification at Bafna Farms.

30.6 ACHIEVEMENTS OF BAFNA FARMS USING BIO-ORGANIC PRODUCTS

30.6.1 GRAPE PRODUCTION USING THE TECHNOLOGY DEVELOPED AT JRABIPL

1. Producing export-quality grapes in low-quality soil and bad-quality water.
2. A 100% export of grapes to the United Kingdom and Europe for the last 17 years.
3. No pesticide residue problem till today due to the uses of microbial products.
4. In 2010, all the Indian grapes except those from Bafna Farms, exported to Europe, were rejected due to the residue of chlormequat chloride (CCC); however, Bafna Farms' grapes qualified in Indian as well as European laboratories and all the samples had below MRL value even after using CCC in fruiting season.
5. Developed a technology to grow two crops in a year for grapes (13th crop in 6.5 years of variety Manjari Navin).
6. Reduced the use of chemical fertilizers and pesticides by 50–75%.
7. Successfully grown "zero chemical nitrogen" grapes without using chemical nitrogen for 3 years on 60 acres by substituting chemicals with bio-products.
8. Under scientific trials, successfully grew grapes without chemical pesticides for 4 years. The yield of the grapes was 70% to that of the commercial production practices. Based on these results, more focus was given to the use of bio-organic products (Fig. 30.3).

FIGURE 30.3 Different varieties of grapes in Bafna Farms: (a) Fantasy; (b) Thomson; (c) Crimson Red; (d) Autumn royal; (e) Manjari Navin; (f) Quality check by European buyer; (g) Packing of grapes in pack house; (h) Special weighing balance designed by Mr. Bafna for table grape; (i) Precooling facility check; (j) Cold store facility check.

30.6.2 GLOBAL ISSUE: PESTICIDE RESIDUE IN AGRICULTURAL CROPS

Since 2000, table grapes for export purposes are being produced in Bafna Farms. Till date, none of the grape's samples failed in Indian as well as the UK/European laboratory, and nor recommended to be reanalyzed. More than 75% of samples are found below level of quantification (BLQ) and the rest of samples are well below the allowed maximum residue level (MRL) by European Union (EU) (Europa and APEDA). Table 30.2 presents pesticide MRL data for table grapes in different countries. All pesticide residue levels are below 15% to that of the allowed MRL (Fig. 30.4). Such results for pesticide residue reduction are due to the continuous use of bio-organic preparations esp., microbial preparation at the laboratory of Bafna Farms. Figure 30.4 shows the pesticide residue level allowed by EU and detection in Bafna Farms' table grapes.

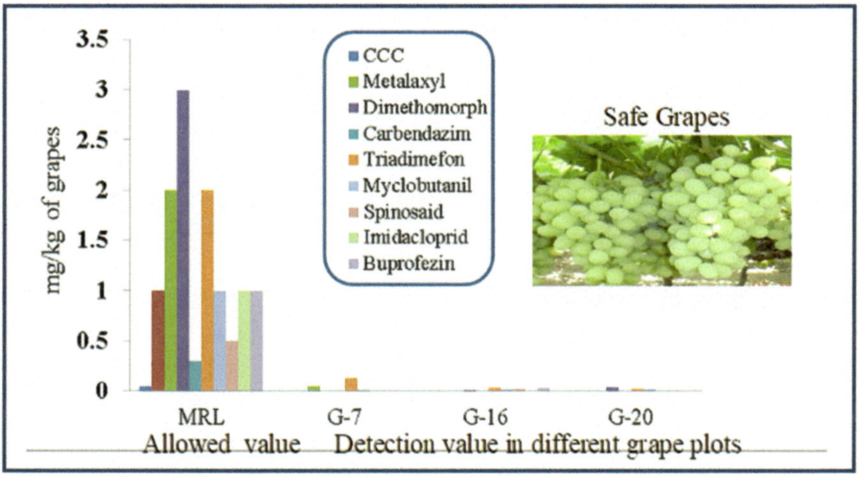

FIGURE 30.4 Allowed maximum pesticide residue level by EU and detection in Bafna Farms' table grape; G-7 is Thomson variety, while G-16 is Fantasy and G-20 is Manjari Navin.

30.6.3 POMEGRANATE PRODUCTION USING THE TECHNOLOGY DEVELOPED AT JRABIPL

1. Pomegranate plants were grown with "zero" chemical nitrogen from planting to the beginning of the first crop.
2. Successfully grew pomegranates with less chemical nitrogen (15 kg/acre/year). Generally, most pomegranate growers use up to 100 kg/acre/year.
3. Grew pomegranate (3-year plant) in less water (9.8-L water/plant/day) due to bio-organic fertilizer use and subsurface drip irrigation technology during the summer period (ambe bahar). Generally, 50–60 L of water is recommended/plant/day on an average by the pomegranate plant.
4. No antibiotic use, *streptomycin,* for *Xanthomonas* blight (Tellya: oily spot) control from planting till date. Normally, 15–20 sprays are recommended per year.

TABLE 30.2 Pesticide MRL Data for Table Grapes in Different Countries and Detected in Bafna Farms during Last 9 Years.

G 10	EU	Russia	Indonesia	China	2016–2017	2015–2016	2014–2015	2013–2014	2012–2013	2011–2012	2010–2011	2009–2010	2008–2009
Pesticide	MRL	MRL	MRL	MRL									
Cymoxanil (20%)	0.3	0.1	N/A	N/A	0.012	BLQ	BLQ	BLQ	BLQ	BLQ	BLQ	BLQ	BLQ
Dimethomorph (50)	3.0	N/A	2.0	N/A	0.011	0.048	BLQ	BLQ	BLQ	BLQ	BLQ	BLQ	0.06
Mancozeb (75)	5.0	0.1	N/A	N/A	BLQ	BLQ	BLQ	BLQ	BLQ	BLQ	BLQ	BLQ	BLQ
Metalaxyl (25%)	2.0	N/A	1.0	1.0	BLQ	0.008	BLQ	0.022	BLQ	BLQ	BLQ	BLQ	BLQ
Hexaconazole (5%)[a]	0.01	N/A	N/A	N/A	BLQ	BLQ	BLQ	BLQ	BLQ	BLQ	BLQ	BLQ	BLQ
Myclobutanil (10%)	1.0	N/A	N/A	N/A	0.135	0.101	BLQ	0.158	0.028	BLQ	BLQ	BLQ	0.03
Penconazole (25)	0.4	0.3	0.2	N/A	BLQ	BLQ	BLQ	BLQ	BLQ	BLQ	BLQ	BLQ	BLQ
Propiconazole (10%)	0.3	N/A	N/A	N/A	BLQ	BLQ	BLQ	BLQ	BLQ	BLQ	BLQ	BLQ	BLQ
Sulfur (80%)	N/A	N/A	N/A	N/A	BLQ	BLQ	BLQ	BLQ	BLQ	BLQ	BLQ	BLQ	BLQ
Tebuconazole	0.5	N/A	N/A	N/A	BLQ	BLQ	BLQ	BLQ	BLQ	BLQ	BLQ	BLQ	BLQ
Triadimefon (25)[a]	0.3	N/A	N/A	N/A	BLQ	0.091	BLQ	BLQ	BLQ	BLQ	BLQ	BLQ	BLQ
Azoxystrobin	2	N/A	N/A	N/A	BLQ	BLQ	BLQ	BLQ	BLQ	BLQ	BLQ	0.06	0.1
Carbendazim (46%)[a]	0.3	N/A	N/A	N/A	BLQ	BLQ	BLQ	BLQ	BLQ	BLQ	BLQ	BLQ	BLQ
Clothianidin (50%)	0.7	N/A	N/A	N/A	BLQ	BLQ	BLQ	BLQ	BLQ	BLQ	BLQ	BLQ	BLQ
Iprodione	10	N/A	N/A	N/A	BLQ	BLQ	BLQ	BLQ	BLQ	0.025	BLQ	BLQ	BLQ
$CuSO_4$	50	N/A	N/A	N/A	BLQ	BLQ	BLQ	BLQ	0.66	0.883	BLQ	BLQ	BLQ

TABLE 30.2 *(Continued)*

G 10	EU	Russia	Indonesia	China	2016–2017	2015–2016	2014–2015	2013–2014	2012–2013	2011–2012	2010–2011	2009–2010	2008–2009
Pesticide	**MRL**	**MRL**	**MRL**	**MRL**									
Abamectin (1.8%)	0.01	N/A	N/A	N/A	BLQ	BLQ	BLQ	BLQ	BLQ	BLQ	BLQ	BLQ	BLQ
Buprofezin (25%)	1.0	N/A	N/A	N/A	0.056	0.027	BLQ	BLQ	BLQ	BLQ	BLQ	BLQ	BLQ
Dichlorvos (76%)[a]	0.01	N/A	N/A	N/A	BLQ	BLQ	BLQ	BLQ	BLQ	BLQ	BLQ	BLQ	BLQ
Fipronil (5%)[a]	0.005	N/A	N/A	N/A	BLQ	BLQ	BLQ	BLQ	BLQ	32.02%	BLQ	BLQ	BLQ
Imidacloprid (18%)	1.0	N/A	N/A	N/A	0.007	0.007	BLQ	0.012	BLQ	BLQ	BLQ	BLQ	BLQ
Lambda-cyhalothrin (5)	0.2	0.2	N/A	N/A	BLQ	BLQ	BLQ	BLQ	BLQ	BLQ	BLQ	BLQ	BLQ
Thiamethoxam	0.4	N/A	N/A	N/A	BLQ	BLQ	BLQ	BLQ	BLQ	BLQ	BLQ	BLQ	BLQ
Emamectin benzoate (5%)	0.1	N/A	N/A	N/A	BLQ	BLQ	BLQ	BLQ	BLQ	BLQ	BLQ	BLQ	BLQ
Spinosad (45)	0.5	N/A	N/A	N/A	0.022	0.104	BLQ	BLQ	BLQ	0.045	BLQ	0.05	BLQ
6 BA	0.01	N/A	N/A	N/A	BLQ	BLQ	BLQ	BLQ	BLQ	BLQ	BLQ	BLQ	BLQ
GA	N/A	N/A	N/A	N/A	BLQ	BLQ	BLQ	BLQ	BLQ	BLQ	BLQ	BLQ	BLQ
NAA	0.1	N/A	N/A	N/A	BLQ	BLQ	BLQ	BLQ	BLQ	BLQ	BLQ	BLQ	BLQ
CCC	0.05	N/A	N/A	N/A	BLQ	0.014	BLQ	0.02	0.022	0.011	BLQ	0.026	BLQ

MRL, Maximum residue level.

[a]Not Approved by European Union; N/A: Not available.

5. According to the observations at Jay Research Biotech India Ltd., the indiscriminate use of chemical nitrogen, water management, heavy preventive antibiotic, etc. is the main cause for pomegranate pests and pathogens such as nematodes, *Ceratocystis/Fusarium* wilt, and bacterial blight are harmful. As per the observation at the lab for nematode control, *Paecilomyces or Trichoderma* are the best solutions, while *Pseudomonas/Bacillus/Trichoderma* is the best solution along with respective chemicals for *Ceratocystis/Fusarium or Xanthomonas*.

30.6.4 SUGARCANE PRODUCTION USING THE TECHNOLOGY DEVELOPED AT JRABIPL

JRABIPL produced 100-tonne (1000 q) sugarcane per acre using drip irrigation (variety: CO 86032; 16 months) in similar low-quality soil using less water (13,000–15,000 L/acre/day, whereas normal recommendation is 30,000 L/acre/day) and less N P K use (85:82:81 kg/acre/season). The average farmer's yield of sugarcane in this area is 30–40 tonnes per acre.

More than 95% of sugarcane production in India is using flood irrigation methods. Under such a system, water consumption is more than 50,000 L/day/acre. Increased TDS and salt levels are matters concerned for sugarcane production. In most of the regions in India, the TDS of water is higher than 1000 ppm, and in some of the sugarcane-growing regions near Pune, it is 2000 ppm. If 50,000-L water is given/acre/day, it is 50–100-kg salts being dumped every day in an acre of soil. Thus, the sugarcane production in and nearby Pune reduced to 30–40 tons. The situation in a few of the plots is too bad with yields as low as 20 tons/acre.

30.7 SOIL PREPARATION AND DRIP INSTALLATION TECHNOLOGIES IN BAFNA FARMS FOR LESS WATER CONSUMPTION

The soil was having total rock at 1-ft depth. All the rock stones were removed from the soil using JCB and pock land machines, up to 3-feet depth from the soil surface. From a 5-acre plot, 350 tractor trolleys full of stones were removed (Fig. 30.5a). The system to remove excess water to avoid the lodging of water near the planting area was done (Fig. 30.5b). The concept of subsurface drip irrigation was adopted from sugarcane plantation. In sugarcane, a subsurface drip line was put at 10 in. below the soil surface

(Fig. 30.5e and f). Similar methodologies were used for pomegranate plantation. In pomegranate, total three drip lines were installed with three different subsurfaces: two subsurface in-line drips and one online surface drip. The irrigation pattern for three drip lines is on every alternate day, that is, first irrigation is given using subsurface drip line from the eastern side, on the alternate day the second irrigation is done using surface drip, and the third irrigation is given using subsurface installed in the West side. Thus, watering is given in subsurface drip line every sixth day and for surface drip-line watering, it is given every fourth day. In this way, continuous watering is avoided. The subsurface in-line pressure compensate drip was inserted at 18" (inches) from the soil surface (Fig. 30.5g) and at a 3-feet distance to the eastern and the western sides of the plant. The drip discharge is 1.6 L/h and the drip distance is 40 cm. Additional surface online drip was laid near the plant for initial growth. The dripper with rubber tubing was installed at a 3-feet distance from the plant (North/South). Initially, rubber tubing water discharge was near the plant for initial growth and once the plant grew up, the tubing water discharged into a pit at a 3-feet distance (north/south) into a 500-mL plastic bottle with holes in it for water discharge to the cow-dung manure added in the pit (Fig. 30.5h). In all, the water availability to the pomegranate plants was kept at a 3-feet distance to keep water away from the trunk also as mentioned above irrigation is done only using one drip line at a time and thus continuous moisture is not there. This avoids nematode development, fungal wilt, and subsequently, *Xanthomonas* blight. Excess water is one of the main reasons for all the abovementioned problems in pomegranate. Now in grapes, one subsurface drip line was added at a 5-feet distance, exactly in the center of two lines for the development of root systems in the center. One surface in-line drip line was kept near the plant at a 1-feet distance. Usually, in agriculture, a single drip line is used for most of the crop productions. Due to the single line, water is applied at a single point and it affects the uniform root development.

30.8 ACTIVITIES FOR ORGANIC CARBON DEVELOPMENT AND SALT LEVEL REDUCTION FROM SOIL USING BIO-ORGANIC APPROACH AT BAFNA FARMS

As the soil quality is poor, the soil was dug to a depth of 3 feet using JCB and Pock land machines. Pits were made using a machine till the depth of 3 feet for manure application (Fig. 30.5i). Initially, cow-dung slurry 10 L was added to the pit along with bio-organic preparations (Fig. 30.5j and k) such as Jay Bio-Multi, 5 mL/L; Jay Bio-Force, 5 mL/L; Jay Bio-Killer, 5 mL/L; Jay Bio-Prahar, 5 mL/L; Jay Bio-Gold, 3 mL/L; and Jay Bio-Amrut, 3 mL/L. After slurry application, cow dung was added in the pit (Fig. 30.5l–n). As cow-dung manure is enriched with the above bio-organic preparation and that too is added in a pit, it acts like a natural fermenter in the field and slowly and steadily microbes multiply near the root zone of the plants and once it is established, it starts developing in-built resistance toward pest and pathogen attack. Afterward, it is necessary to give bio-organic products at regular intervals, that is, every 10 days till the first 6 months. Then the duration can be increased to 15 days and finally 1 month depending upon the stage of the crops. Microbial preparations that are necessary to be given at regular intervals are given in Table 30.1. The application of microbial preparation in this way is like an investment for the future as these microbes are isolated from soil and given back to the soil with enhanced virulence and fertility activities. If these microbes are enriched with organic nutrients, they multiply in the soil and in the future, the requirement of chemicals as well as microbial preparations will be reduced. In Bafna Farm, chemical pesticides/fertilizers were reduced by 70–75%. Importantly, 60-acre grapes were grown without chemical nitrogen for 3 years. This was achieved due to regular soil and petiole analysis. Initially, 120 kg/acre nitrogen was used, slowly it came down to 22 kg and for 3 consecutive years, it was "zero." This was possible due to the initial regular application of Jay Azo, and the doses slowly reduced to only two times in a year, that is, once at the time of backward pruning and other at the time of forwarding pruning in table grapes. It was also noticed that the problems of insect pests and pathogens can be increased if chemical nitrogen use is increased. From the experience of grape production, pomegranate plants were also grown without chemical nitrogen, that is, "zero nitrogen" from the date of planting till the starting of the first crop, and the purpose behind this is to grow pomegranate slowly and naturally.

The uses of microbial preparation were found to be unlimited and still search for newer ones continues (Baez-Rogelio et al., 2017). Other important observations from Bafna Farms are as follows:

Reduction in soil sodium level from 1000 to 520 ppm (importantly, regularly sodium is added in the soil through drip irrigation water) chemical pesticide degradation, quality of agricultural produce is increased.

The advantages of microbial preparations over chemicals are that they are eco-friendly, and their doses get reduced slowly as they multiply in the soil, while doses of chemical fertilizers go on increasing (Belen et al., 2016; Stockmann et al., 2013; Hassan et al., 2010). Now, most of the pests do not die even under higher doses due to the development of resistance in insect pests and microbial pathogens. Microbial preparation works using different mechanisms such as the production of antimicrobial metabolites, enzymes/proteins, carbohydrates, acids, morphological structures, and unknown molecules. Recently (September 2017), the spraying of hazardous chemicals was the main cause of death of 35 farmers in Vidarbha, Maharashtra. This type of incidence happens due to the lack of safety measures in using more doses of chemicals to kill the pest instantly and the lack of knowledge toward harm to human health due to the chemicals. In farming, precautions need to be taken from the date of sowing till the harvest. Thus, there should be a schedule for each and every activity with notifications of atmospheric conditions for the respective crops. Following are the special practices carried out in Bafna Farms, Pune, for table grapes, pomegranate, and sugarcane management.

FIGURE 30.5 Soil preparation, drip installation, and activities for organic carbon development. (a) Removal of stones from soil using pock land machine; (b) Trenches in the plot to remove excess water to avoid lodging; (c) Drip pipe line; (d) Surface and subsurface drip lines; (e) Subsurface drip installation; (f) Subsurface drip check in sugarcane plot; (g) Subsurface drip check in pomegranate plot; (h) Online surface drip in pomegranate with extension tubing and bottle; (i)–(n) Pit preparation for cow-dung slurry application and cow-dung/sugar factory manure application.

30.9 DATA MAINTENANCE AND PRACTICES AT BAFNA FARMS SINCE 25 YEARS

The following data were collected at Bafna Farms for the management of grapes, pomegranate, and sugarcane crops:

1. pesticide application and detection details regarding residue level;
2. soil/water/petiole/leaf analysis for nutrient level periodically;
3. plot-wise farm activity details;
4. plot-wise spraying details;
5. plot-wise drip application details;
6. plot-wise water application details;
7. farm weather updates like temperature, humidity, wind direction, rainfall, and sunlight availability;
8. farm input inward/outward details;
9. pest and pathogen attack recording as per plots;
10. manpower requirement per activity;
11. the availability of electricity for the farm from different locations;
12. scientific testing of each and every material before use in the farm;
13. trial confirmation of each and every product for 3 years in scientific trial plots before commercial use;
14. use of reverse osmosis "RO" water for the spraying of biological or chemical products;
15. eco-friendly approach for pest and pathogen management (Fig. 30.6a–i); and
16. regular farmers' meet for recent updates in agriculture (Fig. 30.6j).

Every year, more than 5000 farmers, buyers from the United Kingdom, Europe, scientists, and industrialists visit Bafna Farm to understand agricultural developments due to the use of bio-organic products. Also teams from JRABIPL and Bafna Farm visit different places to share the developments at Bafna Farm due to the regular use of bio-organic products (Fig. 30.7).

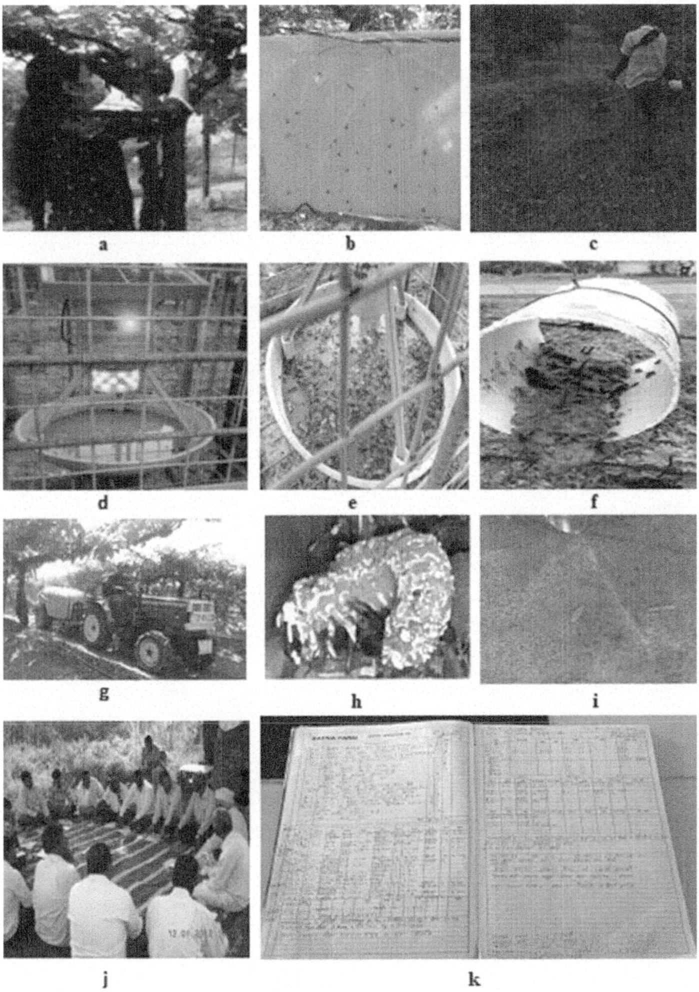

FIGURE 30.6 (a) Daily observation in the plot for pest/pathogen attack or abnormalities in crop; (b) Tagging of yellow sticky trap in the field for obtaining the level of sucking insect pest in a plot; (c) Weed removal; (d) and (e) Solar light trap for insect control; (f) Food trap for fruit fly control; (g) Timely spraying to avoid pest and pathogen attack; (h) Use of microbial products to reduce chemical pesticides (entomopathogenic activity of *Metarhizium anisopliae* on white grub of sugarcane and further development of mycosis on the larvae); (i) As chemical uses is less in Bafna Farms, friendly insect population increased in the farm (ladybug beetle larvae reducing the level of powdery mildew infection by feeding activity); (j) Regular farmers' meet either in Bafna Farms or at different farmers' field to exchange the knowledge/information in agriculture; (k) Most importantly, maintenance of data of all the activities related to agriculture.

FIGURE 30.7 Sharing of knowledge for sustainable agriculture.

30.10 CONCLUSIONS

Overall, it has already been proved in Bafna Farm that if the microbial activity of the soil is increased, it can improve soil health, plant health and develop disease and insect resistance in the plant. It also helps chemical pesticide residue management and the production of "Safe and Healthy produce." The use of chemical fertilizer and pesticides was reduced from 90 to 25% due to bio-organic products applications. As chemical uses reduced, the population of friendly insects increased in the Bafna Farm. The doses of microbial products were also reduced as they multiply naturally due to the excellent organic carbon levels developed in the soil. In

India, the organic carbon level in most places is less than 0.5%. However, in Bafna Farms, it is 2–2.5% for most of the plots. It was also observed that the results were excellent if microbial products are used fresh as it is a living system. Importantly, in Bafna Farms, RO water is used for the spraying of bio-organic or chemical products. The TDS/pH of water due to salt level affects the efficiency of the product.

In Jay Biotech, important microbial technologies have been developed like maintaining virulence and fertility power of the microbial strains. Due to hard work, tireless efforts, and passion for more knowledge toward sustainable agriculture of a farmer (Mr. Prakash Bafna), the bio-organic technology of Jay Biotech is benefitting. In this way, "a non-agricultural land was converted into a fertile land with export quality production." The ultimate vision is to develop the Bafna Farm model at a global level with the slogan "For the Farmer by the Farmer."

ACKNOWLEDGMENT

We are thankful to NHM, Government of India, DBT, New Delhi & BCIL, New Delhi for their financial assistance and NCL, Pune, for their scientific collaborations. We are also thankful to CSA University of Agriculture and Technology, Kanpur, and NCOF, Govt. of India for the publication of this chapter. The technical and scientific help of Dr. Shadab Khan is greatly acknowledged in writing this chapter.

KEYWORDS

- **table grapes**
- **soil and water quality**
- **pesticides residue**
- **soil preparation**
- **organic carbon development**
- **data maintenance**

REFERENCES

Baez-Rogelio, A.; Morales-Garc; Quintero-Hernandez, Y. E. V.; Munoz-Rojas, J. Next Generation of Microbial Inoculants for Agriculture and Bioremediation. *Microbial. Biotechnol.* **2017,** *10*, 19–21.

Belen, M. A.; Martínez-Cuenca, M. R.; Bermejo, A.; Legaz, F.; Quiñones, A. Liquid Organic Fertilizers for Sustainable Agriculture: Nutrient Uptake of Organic *versus* Mineral Fertilizers in Citrus Trees. *PLoS One* **2016,** *11*, e0161619.

Daane, K. M.; Almeida Rodrigo, P. P.; Bell, V. A.; Walker, J. T. S.; Botton, M.; Fallahzadeh, F. et al. Biology and Management of Mealy bugs in Vineyards. *Arthropod Manage. Vineyards* **2012,** 271–307.

Duffner, K.; Schruft, G.; Guggenheim, R. Passive dispersal of the grape rust mite Calepitrimerusvitis (Nalepa) 1905: (Acari, Eriophyoidea) in vineyards. *J. Pest. Sci.* **2001,** *74*, 1–6.

Gessler, C.; Pertot I.; Perazzolli, M. *Plasmopara viticola*: A Review of Knowledge on Downy Mildew of Grapevine and Effective Disease Management. *Phytopathol. Mediterr* **2011,** *50*, 3–44.

Hassan, H. S. A.; Hagag, L. F.; Rawash, M. A.; El-Wakeel, H.; Abdel-Galel, A. Response of Klamata Olive Young Trees to Mineral, Organic Nitrogen Fertilization and Some Other Treatments. *Nat. Sci.* **2010,** *8*, 59–65.

Jagtap, P. Study of Water Quality of Bhima River Water at Tulapur Dist. Pune. Maharashtra (India). *Int. J. Res. Biosci. Agric. Technol.* **2014,** *2*, 155–163.

Karabhantanal, S. S.; Udikeri, S. S.; Vastrad, S. M.; Wali, S. Y. Bio-efficacy of Different Acaricides Against Red Spider Mite, Tetranychusurticae on Grapes. *Pest Manage. Hort. Ecosyst.* **2012,** *18*, 94–97.

Krewer, G.; Dutcher, J. D.; Chang, C. J. Imidacloprid Insecticide Slows Development of Pierce's Disease in Bunch Grapes. *J. Entomol. Sci.* **2002,** *37*, 101–112.

Mani, M.; Shivaraju, C.; Rao, M. S. Pests of Grapevine: A Worldwide List. *Pest Manage. Hort. Ecosyst.* **2014,** *20*, 170–216.

Miazzi, M.; Hajjeh, H.; Faretra, F. Observations on the Population Biology of the Grape Powdery Mildew Fungus. *Uncinulanecator. J. Plant Pathol.* **2003,** *85*, 123–129.

Mohapatra, S.; Ahuja, A. K.; Deepa, M.; Jagadish, G. K.; Prakash, G. S.; Kumar, S. Behaviour of Trifloxystrobin and Tebuconazole on Grapes under Semi-arid Tropical Climatic Conditions. *Pest Manage. Sci.* **2010,** *66*, 910–915.

Qin, G.; Zheng, Zong, Y.; Chen, Q.; Hua, D. Inhibitory Effect of Boron against *Botrytis cinerea* on Table Grapes and Its Possible Mechanisms of Action. *Int. J. Food Microbiol.* **2010,** *138*, 45–150.

Roditakis, N. E. Assessment of the Damage Potential of Three Thrips Species on White Variety Table Grapes—*In-vitro* Experiments. *Crop Protection* **2007,** *26*, 476–483.

Stagnari, F.; Jan, S.; Angelica, G.; Michele, P. Sustainable Agricultural Practices for Water Quality Protection. In *Water Stress and Crop Plants: A Sustainable Approach*; Ahmad, P. Ed.; John Wiley & Sons, Ltd.: Chichester, 2016. DOI: 10.1002/9781119054450.ch6.

Stockmann, U.; Adams, M. A.; Crawford, J. W.; Field, D. J.; Henakaarchchi, N.; Jenkins, M. The Knowns, Known Unknowns and Unknowns of Sequestration of Soil Organic Carbon. *Agric. Ecosyst. Environ.* **2013,** *164*, 80–99.

Sunitha, N. D.; Khan, K. H.; Giraddi, R. S. Studies on Yield Loss by Grape Stem Borer *Celosternas cabrator* Fabr. (Cerambycidae: Coleoptera). *J. Entomol. Zool. Stud.* **2017,** *5*, 1352–1355.

Thind, T. S.; Arora J. K.; Mohan, C.; Raj, P. Epidemiology of Powdery Mildew, Downy Mildew and Anthracnose Diseases of Grapevine. *Diseases Fruits Vegetables* **2004,** *1*, 621–638.

Vinaya, V.; Mane, T. T. Studies on Water Pollution of Mula, Mutha and Pawana Rivers in Summer Season in the Pune City Region. *Nat. Environ. Pollut. Technol.* **2007,** *6*, 499–506.

CHAPTER 31

Organic Farming of Leafy Vegetables for Mitigating Hunger

RAJIV

Vegetable Research Station, C.S. Azad University of Agriculture and Technology, Kalyanpur, Kanpur 208024, Uttar Pradesh, India
Corresponding author. E-mail: rajiv.agro69@gmail.com

ABSTRACT

Vegetable production with organic farming has been practiced by many farmers in India. However, one of the problems is the yield lower than that by chemical farming. It has been noticed that the continuous use of chemical fertilizers has resulted in nutritional imbalance in soil, depletion of soil organic matter, contamination of food, and water, thereby affecting human health, as well as an adverse effect on biodiversity. The supply of nutrients through organic sources can be adopted for avoiding the hazardous effects of fertilizers and maintaining sustainability. Organic inputs improve the quality of produce and are eco-friendly. Leafy vegetables in organic fertilization treatments tend to grew better and resulted in a final higher total yield than those in chemical fertilization treatments, which was attributed to the high nutrient sustainability of organic fertilizer and the improved biological properties of the soil. The dry matter, protein, vitamin C content, etc. increased in organically grown leafy vegetables. Organic produce have better color, smoothness, and texture and also contain more minerals, enzymes, trace elements, and antioxidants. It can be concluded that the organic production of leafy vegetables is feasible and economic, and organic farming offers an opportunity to achieve the objectives of nutritional security.

Organic Crop Production Management: Focus on India, with Global Implications, D. P. Singh, PhD, H. G. Prakash, PhD, M. Swapna, PhD, & S. Solomon, PhD (Eds.)

31.1 INTRODUCTION

Globally, India ranks second in vegetable production, contributing to 15.8 and 14% of the global vegetable area and production, respectively. With 9.8 million-ha area under vegetables and 171 million tonnes of production, the average productivity is 17.4 tonnes/ha (Anonymous, 2016). India has exported 699,600.34 MT of fresh vegetables other than onion worth Rs. 2119.50 crores during the year 2015–2016 (Singh et al., 2017). The major markets for Indian vegetables are UAE, the United Kingdom, Saudi Arabia, the Netherland, Sri Lanka, Bangladesh, Malaysia, Nepal, Qatar, and Pakistan.

Vegetables are an excellent source of highly digestible carbohydrates, nutritionally complete protein, essential nutrients, antioxidants, glucosinolates, fiber, and vitamins, particularly niacin, riboflavin, thiamin, and vitamins A and C. Generally, vegetables are consumed for nutrition, maintenance of health, and prevention of diseases. International Food Policy Research Institute (IFPRI), Washington 2016, revealed that 15.2% of the Indian citizens are malnourished and deprived of enough food (quantity and quality). In India, per capita, land resources are 0.12 ha, which is going to shrink further in the future. Thus, attaining higher productivity in a sustainable manner is the need of the hour.

Continuous use of chemical fertilizers has resulted in nutritional imbalance, depletion of soil organic matter, contamination of food and water, and adverse effect on biodiversity as well as on human health. Considering these factors along with the higher cost of chemical fertilizers, it is necessary to use some alternatives, which besides sustaining the productivity with improving quality of produce will also be friendly to the environment. Supplying nutrients through organic sources can be opted for avoiding the hazardous effects of fertilizers and maintaining sustainability. Organic manures like FYM or vermicompost and bio-fertilizers like *Azotobacter* and *Phosphobacteria* may play a major role in supplementing the crop nutrients through their direct addition, improvement in soil condition, nitrogen fixation, and the solubilization of fixed forms of phosphorus in soil (Bhardwaj and Gaur, 1970). Organic farming is based on the principle of sustainable farming and it is a viable alternative to the consumption of contaminated agricultural produce and the bad impact of agricultural practices on the environment.

FAO suggested that "Organic agriculture is a unique production management system which promotes and enhances agro-ecosystem health, including biodiversity, biological cycles, and soil biological activity, and this is accomplished by using on-farm agronomic, biological and mechanical methods in exclusion of all synthetic off-farm inputs". In simple words, organic agriculture is the farming system without the addition of artificial chemicals

(Tomar et al., 2017). India is now one of the countries where organic farming regulations are fully implemented and the area under organic farming practices is increasing rapidly in all the states across the country. In fact, states like Uttarakhand and Sikkim have been declared as fully organic states in agricultural development strategies (Singh and Choudhary, 2017).

Organic products are nutritive, safe, eco-friendly, and often fetch a premium price. Thus, organic farming may play an important role in livelihood security. Organic food has high demand and it gets a high premium over conventionally grown food (Bhattacharya and Chakraborty, 2005). Organic farming is based on the maximum use of on-farm inputs and thus, the average cost of cultivation in an organic system is reduced due to the use of farm-derived inputs. The loss of nutrients in organic manure is less due to its slow release. Besides primary and secondary nutrients, the continuous addition of organic manure assures a regular supply of micronutrients also, and there is a positive correlation between organic matter and micronutrient availability. Ramesh et al. (2009) have reported a similar increase in soil health by the addition of organic manures in organic farming.

According to the Agricultural & Processed Food Products Export Development Authority (APEDA), by the end of March 2017, India has brought more than 4.45 million-ha area under organic certification, comprising 1.44 million ha (32.35%) under cultivation and 3.0 million ha (67.6%) under wild harvest collection. The overall status of the area under the organic certification process in different states of India has been presented in Table 31.1.

TABLE 31.1 Area under Organic Certification Process (2016–2017).

S. no.	States	Organic area (ha)	In conversion area (ha)	Total farm area (ha)	Wild harvest area (WH) (ha)	Total WH + farm
1	Andhra Pradesh	9812.8	7871.0	17,683.8	155,099.1	172,783.0
2	Arunachal Pradesh	21.4	3989.7	4011.2	68300.0	72,311.2
3	Assam	2544.0	21,326.3	23,870.3	60.0	23,930.3
4	Bihar	0.0	1.2	1.2	678.0	679.2
5	Chhattisgarh	2339.7	10,372.3	12,712.1	167,040.0	179,752.1
6	Goa	14,116.7	1645.6	15762.4	0.0	15,762.4
7	Gujarat	36,034.2	28,206.8	64,241.0	6253.9	70,495.0
8	Haryana	4482.7	528.8	5011.5	20.1	5031.7
9	Himachal Pradesh	5903.7	6473.0	12,376.7	2000.0	14,376.7

TABLE 31.1 *(Continued)*

S. no.	States	Organic area (ha)	In conversion area (ha)	Total farm area (ha)	Wild harvest area (WH) (ha)	Total WH + farm
10	J & K	9550.0	13,058.2	22,608.3	159,000.0	181,608.3
11	Jharkhand	88.4	26,725.4	26,813.9	10,000.0	36,813.9
12	Karnataka	22,478.2	58,610.8	81089.0	859.7	81,948.8
13	Kerala	13,809.2	11,003.4	24,812.7	18,889.1	43,701.8
14	Lakshadweep	895.5	0.0	895.5	0.0	895.5
15	Madhya Pradesh	213,968.1	250,891.2	464,859.4	1,827,837.9	2,292,697.3
16	Maharashtra	84,338.7	139,668.7	224,007.5	68,384.2	292,391.7
17	Manipur	0.0	241.4	241.4	0.0	241.4
18	Meghalaya	1414.8	8214.7	9629.5	0.0	9629.5
19	Mizoram	0.0	210.0	210.0	0.0	210.0
20	Nagaland	1508.6	3191.2	4699.9	0.0	4699.9
21	New Delhi	9.2	0.0	9.2	0.0	9.2
22	Odisha	36,710.3	55,479.7	92,190.1	7546.0	99,736.1
23	Pondicherry	2.8	0.0	2.8	0.0	2.8
24	Punjab	434.5	598.0	1032.5	16,616.0	17,648.5
25	Rajasthan	46,088.7	105,521.1	151,609.9	387,912.2	539,522.1
26	Sikkim	72,145.4	3072.8	75,218.2	0.0	75,218.2
27	Tamil Nadu	2058.2	3654.5	5712.7	5062.8	10,775.6
28	Telangana	4457.4	5230.3	9687.8	0.0	9687.8
29	Tripura	203.5	0.0	203.5	0.0	203.5
30	Uttar Pradesh	39,929.4	16,319.8	56,249.3	45,210.5	101,459.9
31	Uttarakhand	18,510.3	12,397.0	30,907.4	62679.0	93,586.4
32	West Bengal	4759.8	416.1	5176.0	0.0	5176.0
Total		648,617	794,920	1,443,538	3,009,449	4,452,987

Source: Yadav (2017).

In India, coriander (*Coriandrum sativum* L.), amaranth (*Amaranthus* sp.), spinach beet (*Beta vulgaris*), Malabar/Chinese spinach (*Basella alba/rubra*), *bathua* (*Chenopodium album*), fenugreek (*Trigonella foenum-graecum*), cabbage (*Brassica oleracea* var. *capitata*), etc. are the commonly consumed leafy vegetables. Leafy vegetables may play a vital role not only in preventing hidden hunger but also in health maintenance. The importance of leafy vegetables is increasing because they are rich in phytochemicals and antioxidants. Leafy vegetables help in lowering the risk of cancer, cardiovascular disease, and other age-related disorders. This group of vegetables has great potential in curing vitamin A deficiency.

31.2 NUTRITIVE VALUE OF SOME MAJOR LEAFY VEGETABLES

31.2.1 CORIANDER

One-hundred grams of fresh leaves of coriander contain 87.9% moisture, 3.5% protein, 0.6% fat, 6.5% carbohydrates, 1.7% mineral matter, 0.14% calcium, 0.06% phosphorus, and vitamin A 10,460–12,600 IU (Singh et al., 2004). One hundred grams of dry seeds contain 11.2% moisture, 14.1% protein, 16.1% fat, 21.6% carbohydrates, 32.6% fiber, 4.4% mineral matter, 0.63% calcium, 0.37% phosphorus, and 17.9-mg iron (Singh et al., 2004). Coriander leaves are an important source of vitamins and per 100-g edible portion contains 6918-μg carotene, 135-mg vitamin C, 0.05-mg thiamine (B_1), 0.06-mg riboflavin (B_2), and 0.8-mg niacin (Anonymous, 2004). Coriander dry seeds are rich in dietary fiber, which binds to bile salts (produced from cholesterol) and decreases their reabsorption in the colon and thus helps lower serum LDL cholesterol levels. The seeds are an excellent source of minerals like iron, copper, calcium, potassium, manganese, zinc, and magnesium (Fig. 31.1).

FIGURE 31.1 Coriander.

31.2.2 MALABAR SPINACH/CHINESE SPINACH/INDIAN SPINACH/POI (BASELLA)

It is a wholesome leafy vegetable and a good source of vitamins and minerals, particularly calcium (109 mg), iron (10 mg), vitamin A (8000 IU), vitamin C (102 mg), and folic acid (140 μg) per 100 g of fresh weight (Anonymous, 2016). Being a rich source of folic acid, it is also recommended for pregnant women. It has been used successfully to check malnutrition in children. The essential amino acids, namely, arginine, leucine, lysine, threonine, isoleucine, and tryptophan, are present in this vegetable.

31.2.3 BATHUA (CHENOPODIUM)

As a nutritious leafy vegetable, *bathua* is rich in protein (3.7 g), calcium (150 mg), iron (4.2 mg), β carotene (1740 μg), niacin (0.60 mg), and phosphorus (80 mg) per 100 g of fresh weight (Anonymous, 2016). It is useful in abdominal pain and eye diseases and acts as a laxative, diuretic, anthelmintic, and tonic for boosting stamina.

31.2.4 AMARANTH (CHAULAI)

Amaranth leaves and succulent stems are very good sources of iron and vitamin A. Per 100 g of fresh leaves contain 85.7% moisture, 2500–11,000 IU of vitamin A, 130–173 mg of vitamin C, and 100–130 IU of vitamin B (Singh et al., 2004). It also contains 397-mg calcium, 83-mg phosphorus, 3.49-mg iron, 341-mg potassium, 122-mg magnesium, etc. while its fresh leaves have trace elements, namely, copper, manganese, zinc, sulfur, and molybdenum. It is used as a blood purifier and in dysentery (Anonymous, 2004).

31.2.5 SPINACH BEET (PALAK)

Hundred grams of spinach leaves contain 92.7% moisture, 2.3-g protein, 0.7-g fat, 10.9-mg iron, 80-mg vitamin C, 25,000-IU vitamin A, 380-mg calcium, and 30-mg phosphorus (Singh et al., 2004; Anonymous, 2004).

Being a rich source of vitamin A, calcium, and iron, it is recommended for pregnant women. It has a mild laxative and diuretic effect also.

31.2.6 FENUGREEK

One hundred grams of fresh leaves of fenugreek contain 81.8% moisture, 4.9% protein, 0.9% fat, 1.6% mineral matter, 1.0% fiber, 9.8% carbohydrates, 0.47% calcium, 0.05% phosphorus, vitamin A 3900 IU, and vitamin C 14 mg (Singh et al., 2004). Its fresh tender leaves, pods, and tender stems are consumed as vegetables alone or in combination with potatoes. It removes indigestion, prevents constipation, and stimulates the liver. It is a diuretic.

31.2.7 CABBAGE

Its 100-g edible portion contains 91.9% moisture, 1.8-g protein, 0.1-g fat, 1.0-g fiber, 0.5-mg iron, 49-mg calcium, 44-mg phosphorus, 31.7-mg magnesium, 120-μg carotene, 50-mg vitamin C, etc. (Anonymous, 2004). It is commonly used as cooked vegetable and *salad*. It has an anticancer property and a curative effect on scurvy disease (Fig. 31.2).

FIGURE 31.2 Cabbage.

31.2.8 CELERY

It has important medicinal properties with very low calorific values. Celery is an ideal food for people who are dieting. It contains 6.3-g protein, 0.6-g fat, 1.4-g fiber, 0.4-mg iron, 230-mg calcium, 140-mg phosphorus, 52-mg magnesium, 3990-μg carotene, and 62-mg vitamin C per 100-g edible portion (Anonymous, 2004).

31.3 OBJECTIVES OF ORGANIC LEAFY VEGETABLES PRODUCTION

The objectives of organic leafy vegetables' production (Upadhayay, 2013) are given below:

1. the production of leafy vegetables of high nutritional quality in sufficient quantity,
2. sustainable use of local resources,
3. work with natural systems,
4. encourage and enhance the biological cycle within the farming system,
5. maintain and increase the soil fertility and productivity,
6. use renewable resources in locally organized agricultural systems,
7. adequate returns and safe working environment,
8. avoid all forms of pollution,
9. ensure the basic biological functions of soil–water–nutrients–human continuum, and
10. widen the social and ecological impact of leafy vegetable cultivation.

31.4 CONSTRAINTS OF ORGANIC LEAFY VEGETABLES' PRODUCTION

1. Poor yield during the conversion period.
2. Shortage of organic inputs, namely, FYM, vermicompost, compost, poultry manure, and cakes.
3. Insufficiency of bioagents and biopesticides.
4. Scarcity of organic seeds.
5. Higher cost of organic inputs.
6. Lack of awareness for organic certification of produce.
7. Lack of consumer awareness for organic vegetables and their products.

8. Higher price of organic produce affects the consumer's choice in the market.

31.5 APPROACHES TO PRODUCE ORGANIC LEAFY VEGETABLES

Different approaches (Malhotra, 2010; Upadhayay, 2013) available to be used alone or in combination to produce organic coriander are given below:

31.5.1 ORGANIC AMENDMENT

Manures used in organic nutrition management are as mentioned below:

1. ***Bulky Organic Manures:*** FYM, vermicompost, poultry manure, rural/urban compost, Nadep compost, sewage sludge, crop residues, etc. are bulky organic manures and good organic sources of plant nutrients.
2. ***Green Manures:*** *Dhaincha*, sunhemp, cowpea, urdbean, mungbean, and others are used for green manuring.
3. ***Concentrated Organic Manures:*** *Neem* cake, groundnut cake, caster cake, etc. are concentrated organic manures and may be used as organic inputs.
4. ***Biodynamic:*** Cow horn manure, horn silica, compost preparations BD 502–508, etc. are indigenous and biodynamic preparations that can also be used in organic farming.

31.5.2 BIOFERTILIZERS

Azotobacter, Azospirillum, phosphorus solubilizing bacteria (PSB), potassium-mobilizing bacteria, mycorrhizae, and microbial consortium are natural fertilizers containing carrier-based microorganisms that can help in sustainable vegetable production in organic farming.

31.5.3 BIOAGENTS AND BIOPESTICIDES

Trichoderma viride, Pseudomonas fluorescens, Verticillium lecanii, Bacillus thuringiensis, Beauveria bassiana, yellow sticky traps, *neem* seed kernel extract, *neem* oil, *neem* cake, onion extract, bio-plants' growth promoters, biological agents (protozoa and other natural enemies), and others. are some important components of organic farming which can be used for effective management of diseases and insects in leafy vegetables.

31.5.4 CULTURAL PRACTICES

Summer plowing, soil solarization, use of disease-free seeds, avoiding excessive irrigation, adopting field sanitation measures, following long-term crop rotation, uprooting and burning all the infected plants, timely sowing, growing resistant varieties, and weed control through crop rotation and manual weeding practices should also be adopted in organic farming.

31.5.5 FOLIAR FEEDING

Seaweed, filtered solutions of manures, fish emulsion, and others may be used for foliar feeding. Seaweed is an excellent source for growth hormones like gibberellins, auxins, and cytokinins as well as for trace elements.

31.5.6 MINERAL FERTILIZERS

The mineral fertilizers, which are allowed in organic agriculture, are based on ground natural rock. Rock phosphate, plant ashes, stone powder, and others are mineral fertilizers and they may only be used as a supplement to organic manures.

31.6 TECHNOLOGIES AND MANAGEMENT PRACTICES FOR ORGANIC LEAFY VEGETABLES PRODUCTION

For organic leafy vegetables' production, minimum of 2–3 years are required as a conversion period in organic plots, and an isolation distance of at least 25 m for the organically cultivated plots from the neighboring plots is to be maintained to avoid contamination (Malhotra, 2010). It is most essential that all the crops in the organically cultivated plots must be grown through organic methods of production. Organically produced seed material should be used in organic cultivation but if it is not available, initially conventional seed material can also be used. Crop rotation involving legume crops should be followed and regular surveillance is necessary for effective management of diseases and insects in organic cultivation. The strategies for organic leafy vegetables' production are given below:

31.6.1 SITE SELECTION AND SOIL

The field for organic vegetable production should be free from pests and diseases infestations. The soil and climate should be suitable for the crop. The regular supply of good-quality irrigation water is essential for organic cultivation. Leafy vegetable crops prefer well-drained loam or clay–loam soil containing sufficient organic carbon with a pH range of 6.0–7.5. The soil should also be tested for fertilizers and pesticide residues and contamination with heavy metals before planting the crop.

31.6.2 MANAGEMENT OF SOIL FERTILITY

Soil fertility is the primary step in any organic farming system. The organic inputs, namely, farmyard manure (FYM), **compost, poultry manure,** vermicompost, **green manures, and microbial fertilizers,** may be applied as an alternative to chemical fertilizers to enhance soil fertility.

Well-rotten FYM at 20–25 t/ha at 15–20 days before sowing or vermicompost at 5.0–7.5 t/ha or **poultry manure at** 5.0–7.5 t/ha or *neem* cake at 2.0 t/ha can be applied depending upon fertility status of the soil. The application of a mixture of different organic inputs such as FYM at 10 t/ha + vermicompost at 3.5 t/ha or FYM at 10 t/ha + poultry manure at 2.5 t/ha or vermicompost at 3.5 t/ha + poultry manure at 2.5 t/ha along with bio-fertilizers-produced yield equal to the conventional inorganic system in vegetable crops (Anonymous, 2016). Compost is also a good source of plant nutrients in organic vegetable production, which improves the soil structure and stimulates beneficial microorganisms. The C:N ratio should be about 20:1 in compost. The organic manures should be applied as a basal. The nutrient contents of organic manures have been presented in Tables 31.2 and 31.3.

Green manuring is a major component of organic farming. Green manure crops are raised and then turned into the soil for decomposition just before flowering (about 40–45 days after sowing). It should be incorporated about 20–25-cm deep into the soil. The major green manure crops are *Sesbania aculeata, Setosphaeria rostrata, Crotalaria juncea,* green gram, and black gram. The incorporated green manure crops increase the activity of soil organisms, build up organic matter in the soil, and improve soil structure and fertility. The nutrient potential of green manure crops has been presented in Table 31.4. The time gap between turning the green manure crop and planting the next crop should be about 2–3 weeks (Fig. 31.3).

FIGURE 31.3 Green manuring.

The application of biofertilizers such as nitrogen-fixing bio-fertilizers and phosphorous-solubilizing biofertilizers enhance atmospheric nitrogen fixation and increase soil health, nitrogen uptake, and the availability of phosphorous. Generally, *Azospirillum* and PSB at 5 kg each/ha are recommended for leafy vegetable crops for direct soil application. In this method, 5 kg of bio-fertilizer is mixed in 100 kg of compost or FYM and thereafter, this mixture is broadcast in 1 ha of land and plowed in at the time of sowing or 24 h before sowing. Microbial consortium at 12.5 kg/ha can also be applied instead of *Azospirillum* and PSB. The microbial consortium in conjunction with organic sources of nutrients increases the productivity of leafy vegetables and can be used for organic production. Microbial consortium contains nitrogen-fixing, phosphorus- and zinc-solubilizing, and plant growth–promoting microbes in a single formation, and the application of microbial consortium increases the nutrient use efficiency as well as yield (Rajiv and Singh, 2017).

Organic manures provide organic matter as well as all the nutrients that are required by plants (Chandra, 2005; Chandrasekaran et al., 2010; Gaur, 2003; FAI, 2004). They improve the physical, chemical, and biological properties as well as the structure and texture of the soil. Organic manures help in restoring effective microorganisms in the soil and higher microbial activity increased the availability of nutrients to the plant. Thus, organic manures increase the fertility and productivity of the soil. The application of poultry manure or FYM or vermicompost at 7.5 t/ha or 20–30 t/ha or 7.5–10 t/ha, respectively, can ensure 28–35% higher yield compared to conventional system (Anonymous, 2016).

TABLE 31.2 List of Important Organic Manures and Their Nutrient Contents.

Manure	Nutrient contents (in percentage)		
	Nitrogen (N)	Phosphorus (P_2O_5)	Potash (K_2O)
Farmyard manure	0.80	0.41	0.74
Nadep compost	0.5–0.8	0.5–0.6	1.2–1.4
Vermicompost	3.00	1.00	1.50
Poultry manure	3.03	2.63	1.40
Rural compost	1.22	1.08	1.47
Sewage sludge	1.5–3.5	0.75–4.00	0.3–0.6
Neem cake	5.2–5.6	1.10	1.50
Castor cake	4.0–4.4	1.9	1.4
Groundnut cake	6.5–7.5	1.3	1.5
Sesame cake	4.7–6.2	2.1	1.3
Cowpea[a]	0.71	0.15	0.58
Green gram[a]	0.72	0.18	0.53
Sunnhemp[a]	0.75	0.12	0.51
Black gram[a]	0.85	0.18	0.53

[a]Green manures, fresh.

TABLE 31.3 Average Secondary and Micronutrient in Vermicompost and FYM.

S. N.	Nutrient	Vermicompost	FYM
1.	Calcium	0.44%	0.91%
2.	Magnesium	0.15%	0.91%
3.	Iron	175.2 ppm	146.5 ppm
4.	Manganese	96.51 ppm	69.0 ppm
5.	Zinc	24.43 ppm	14.5 ppm
6.	Copper	4.89 ppm	2.6 ppm

TABLE 31.4 Nutrient Potential of Green Manures.

S. no.	Green manure	Biomass (tonnes)	N accumulation (kg/ha)
1.	*Sesbania aculeate*	22.50	145.00
2.	*Setosphaeria rostrata*	20.06	146.00
3.	*Crotalaria juncea*	18.40	113.00
4.	*Tephrosia purpurea*	6.80	6.00
5.	Green gram	6.50	60.20
6.	Black gram	5.12	51.20
7.	Cowpea	7.12	63.30

Organic farming is based on the minimum use of off-farm inputs and the maximum use of on-farm inputs. Generally, vegetable crops produce a huge amount of biomass, which can be recycled to produce Nadep compost. Nadep compost is prepared in the aerobic tank made of bricks and cement. Farm wastes (both dry and green) along with a small amount of cattle dung or biogas slurry mixed with almost an equivalent amount of soil are composted for 3–4 months. About 2–3 tonnes of compost is obtained from a tank of Nadep compost and it can be filled up 3–4 times in a year. Thus, in a year about 6–8-tonnes compost can be produced from one tank. The Nadep compost on an average contains 0.5–1.0% nitrogen, 0.5–1.0% phosphorus, and 1.2–1.4% potassium (Gaur, 2003). Well-matured Nadep compost at 15–20 t/ha can be applied. It should be mixed in the soil about 2 weeks before sowing of the crop (Figs. 31.4 and 31.5).

FIGURE 31.4 Nadep compost tank.

FIGURE 31.5 Vermicompost unit.

31.6.3 CROP AND VARIETY

Organic seeds of recommended variety should be used for sowing. The variety should also be high yielding, early in vigor, pest, and disease-resistant and popular among the vegetable growers.

31.6.4 WEED MANAGEMENT

Deep summer plowing and hand weeding are the most common methods to reduce weed pressure in leafy vegetables. Mulching with black polythene film is also an effective means for suppressing weeds in organic vegetable production by inhibiting germination and growth of weeds. Organic mulches include plant or by-products like sawdust, straw, leaves, and sugarcane bagasse. Soil solarization is also an eco-friendly method of weed control. In this method, the soil is covered with transparent polythene film that traps the heat inside and raised the soil temperature to a level lethal to weed species. Crop rotation in the organic field should also be adopted.

31.6.5 PEST MANAGEMENT THROUGH ORGANIC METHODS

The mechanical practices, namely, crop rotation, following optimum planting time and use of healthy planting material, should be followed to avoid or reduce the risk of losses from insects. The aphids and leaf-eating caterpillars are serious insect-pests in leafy vegetable crops. Pest monitoring once a week should be done to keep a close watch on the appearance of insect pests. Populations of the sucking pests is counted on three leaves, that is, top, middle, and lower per plant, whereas percent damage assessment is made through counting the total number of plants and affected plants for cutworms and defoliators. Aphids/whitefly can be monitored by setting up yellow pan/ sticky traps at 10/ha.

Trap Crops: Trap crop protects the main crop from certain pests. Row intercropping and border trap cropping are two methods of trap crop planting. Indian mustard as a trap crop can be utilized for the management of diamondback moth, aphid, and leaf webber on cabbage. Mustard attracts more than 80% of the cabbage pests.

Intercropping: Intercropping with other crops limits the infestation from the major pest. Coriander intercropped with cabbage reduces the infestation of aphids because of the diverse nature of the intercrop, which checks the multiplication of major pests.

Sticky Traps: Sticky traps can be incorporated into the management practices of various pests like whiteflies, leaf miners, and thrips. Yellow and blue sticky traps are used to monitor whitefly, leaf miner, and thrips infestation, respectively, and have been recommended in IPM practices.

Natural or Biopesticides: Biopesticides are products of microbial and plant origin. The nymphs and adults of aphids (*Hyadaphis coriandri* Das.) suck sap from the tender leaves and reduce the yield and quality of leaves. Aphids can be controlled by clipping off the heavily infested parts, pressurized water spray, and foliar spray in the crop with *neem* seed kernel extract at 3% or *neem* oil at 1% or *V. lecanii* at 3 mL/L of water, which may be repeated three times at 15-days interval. Leaf-eating caterpillars can be controlled by foliar spray of *B. thuringiensis* at 1 kg/ha or NPV at 250 LE/ha. Mites (*Petrobia latens*) can be controlled by foliar spray of *neem* oil at 1%. All soil insects can be managed by soil application of *B. bassiana* at 4 kg/ha + 80 kg FYM or *neem* cake at 400 kg/ha. Garlic extract is a natural product used as broad-spectrum pesticide. *Neem*, *Sabadilla*, and *Pyrethrum* extracts are also used as pesticides (Table 31.5).

TABLE 31.5 Natural or Biopesticides.

S. no.	Botanical pesticide	Source	Nature of the product	Mode of action
1	Allicin	Garlic	Broad-spectrum pesticide	Acts as antibacterial and antifungal biopesticide
2	Nicotine sulfate	Tobacco	Insecticides	Aphids, thrips, spider, mites, and other sucking insects
3	Sabadilla	Sabadilla lily	Insecticides	Caterpillars, leaf hoppers, thrips, sink bug, and squash bugs
4	Nemacide	*Neem* tree	Insecticides	Grass hopper, moth, etc.

Source: Singh and Choudhary (2017).

Biological Control: In the biological control method, natural enemies manage the pest population. Diamondback moth in cabbage can be managed by the release of adult larval parasitoids (*Cotesia plutellae, Diadegma semiclausum*) at 15,000/ha at weekly intervals during the initiation of larval damage.

31.6.6 DISEASE MANAGEMENT THROUGH ORGANIC METHODS

Wilt (*Fusarium oxysporum*) is a serious disease in leafy vegetable crops. In this disease, the affected plants show yellowing and drooping of leaves, thereafter drying and leading finally to the death of the plant. It can be effectively controlled through integrated approaches of soil solarization, use of disease-free seed, crop rotation, and seed treatment with *T. viride* at 4 g/kg seed followed by the soil application of *T. viride* at 4 kg/ha + 80-kg FYM or seed treatment with *P. fluorescens* at 10 g/kg seed followed by the soil application of *P. fluorescens* at 5 kg/ha + 100-kg FYM or seed treatment with *P. fluorescens* at 10 g/kg seed followed by the soil application of neem cake at 150 kg/ha. Powdery mildew (*Erysiphe polygoni* D.C.) is an important disease in leafy vegetable crops and a white powdery mass appears on the leaves, branches, and fruits in affected plants. Affected leaves lose their chlorophyll and dry up. It can be controlled by avoiding the late sown crop, foliar spray of onion extract at 5%, and spraying the crop with wettable sulfur at 1.0 kg/ha.

Blight (*Alternaria poonensis*) is also an important disease in leafy vegetable crops and dark brown spots appear on the stem and leaves on the affected plants. Blight can be managed through the use of disease-free seeds, avoiding excessive irrigation, and adopting field sanitation measures. Stem gall (*Protomyces macrosporus Unger*) is one of the serious diseases, its symptoms appear on the stem, leaf stalk, leaf veins, and peduncles as galls/swellings, and seeds are deformed. It can be managed through soil solarization, following long-term crop rotation and the use of disease-free seeds. The infected plants should be uprooted and burnt. Growing resistant variety is also an effective method for the management of diseases in organic vegetable production (Table 31.6). Sulfur- and copper-based products are the only labeled fungicides allowed in organic certification. Sulfur is labeled for the control of powdery mildew and copper for anthracnose, bacterial spot, early blight, late blight, etc.

Biofertilizers like *Azotobacter*, *Azospirillum*, PSM, and phosphorus-mobilizing microbes have antifungal activities without any residual or toxic effect, resulting in sustainable and quality vegetable production (Singh and Choudhary, 2017).

TABLE 31.6 Varieties of Vegetable Crops Tolerant/Resistant to Disease.

S. no.	Crop	Disease	Varieties
1.	Cabbage	Black rot	Pusa Mukta
		Black Leg	Pusa Drum Head
2.	Coriander	Stem gall	Pant Haritima (Pant D-1), Hisar Sugandh

31.7 QUALITY OF ORGANIC LEAFY VEGETABLES

There is a high demand for organic vegetables in domestic and international markets because of their quality. The dry matter, protein, vitamin C content, etc. increase in organically grown leafy vegetables. Organic produce have better color, smoothness, and texture and also contain more minerals, enzymes, trace elements, and antioxidants. In organically produced cabbage, vitamin C content increases by 17% (Anonymous, 2016). The use of organics in the cultivation of leafy vegetables had a significant influence on the quality of fresh leaves in terms of oxalate content.

According to studies at ICAR-IIVR, Varanasi, vitamin C content in organically grown spinach, fenugreek, and cabbage increased by 51.12, 25.76, and 41.31%, respectively, and the total phenolic compounds and

TABLE 31.7 Quality of Leafy Vegetables Grown Under Organic Farming.

Particular	Spinach		Fenugreek		Cabbage	
	Organic (vermicompost at 10 t/ha)	Inorganic (recommended dose of inorganic fertilizer)	Organic (vermicompost at 10 t/ha)	Inorganic (recommended dose of inorganic fertilizer)	Organic (vermicompost at 10 t/ha)	Inorganic (recommended dose of inorganic fertilizer)
Ascorbic acid (g/100 g)	0.65	0.43	1.66	1.32	234.31	165.81
Total phenol (g/100 g)	10.48	6.52	12.97	9.35	48.85	34.33
Antioxidant (%)	42.73	30.80	93.02	65.63	14.35	12.54

Source: Singh et al., 2017

peroxidase activity improved by 44 and 38%, respectively, in organically produced cabbage. Thus, organic farming offers an opportunity to achieve the objectives of nutritional security. Qualities of some leafy vegetables grown under organic farming have been presented in Table 31.7.

31.8 INITIATIVES AT CSAUA&T, KANPUR

Scientific studies on the organic production of leafy vegetables to ascertain the productivity potential and economic feasibility are undertaken at Kalyanpur Vegetable Research Station of C.S. Azad University of Agriculture and Technology, Kanpur, as mentioned below.

31.8.1 ORGANIC FARMING IN CORIANDER (CORIANDRUM SATIVUM L.)–RADISH (RAPHANUS SATIVUS L.) CROPPING SEQUENCE

Coriander (leaf crop)–radish crop sequence during the *Rabi* season is popular in many vegetable-growing areas. In this sequence, farmers sow coriander crops in the month of September after *Kharif* crop and radish crop in the month of December. Thus, coriander and radish both crops are taken in sequence during the same *Rabi* season, and the coriander crop is grown for fresh green leaf purposes.

A field study was conducted on organic farming in coriander cv. Pant Haritima–radish cv. Japanese White sequence during *Rabi* season of 2015–2016 and 2016–2017. Organic inputs applied to supply plant nutrients to the vegetable crops include FYM, vermicompost, and microbial consortium. Plant protection measures were adopted through organic methods under organic treatments. Results of the study revealed that the application of 100% recommended dose of nitrogen through vermicompost + microbial consortium at 12.5 kg/ha (PP with organic methods) recorded the highest marketable green leaves yield of 92.62 q/ha in coriander and root yield of 363.25 q/ha in radish (Tables 31.8 and 31.9). It was followed by conventional practices (recommended FYM + fertilizer + PP with chemicals) + microbial consortium at 12.5 kg/ha. However, both of these treatments were statistically at par in terms of yield. The return per rupee invested of Rs 3.69 (Table 31.10) of the sequence is also highest under the treatment of the application of 100% recommended dose of nitrogen through vermicompost + microbial consortium at 12.5 kg/ha (PP with organic methods). Results of the trial indicated that organic farming in coriander (leaf crop)–radish sequence is feasible and profitable. Thus, on the basis of the results of the study, an organic package of practice with 100% recommended dose of nitrogen through vermicompost + microbial consortium at 12.5 kg/ha for coriander (leaf crop)–radish sequence has been recommended for the agroclimatic condition of Zone-IV in 35th Group Meeting of AICRP on vegetable crops held at ICAR-IIHR, Bangalore, during June 24–27, 2017.

TABLE 31.8 Effect of Treatments on Fresh Green Leaves Yield of Coriander.

Treatment	Fresh green leaves yield (q/ha)			Increase/decrease in yield over conventional practices	
	2015–2016	2016–2017	Pooled	q/ha	%
T_1—Conventional practices (recommended FYM + fertilizer + PP with chemicals)	78.36	88.89	83.63	–	–
T_2—100% RDN through vermicompost (PP with organic methods)	82.86	86.96	84.91	(+) 1.28	(+) 1.53
T_3—100% RDN through FYM (PP with organic methods)	71.10	73.69	72.40	(−) 11.23	(−) 13.43
T_4—Conventional practices (recommended FYM + fertilizer + PP with chemicals) + microbial consortium at 12.5 kg/ha)	83.18	89.50	86.34	(+) 2.71	(+) 3.24
T_5—100% RDN through vermicompost + microbial consortium at12.5 kg/ha (PP with organic methods)	89.86	95.37	92.62	(+) 8.99	(+) 10.75
T_6—100% RDN through FYM + microbial consortium at 12.5 kg/ha (PP with organic methods)	77.36	78.88	78.12	(−) 5.51	(−) 6.59
T_7—Recommended FYM + fertilizer + microbial consortium at 12.5 kg/ha (PP with organic methods)	82.12	89.12	85.62	(+) 1.99	(+) 2.38
SEm±	3.17	3.09	2.95	–	–
CD (P=0.05)	9.76	9.51	8.41	–	–

RDN, Recommended dose of nitrogen.

TABLE 31.9 Effect of Treatments on Root Yield of Radish.

Treatment	**Root yield (q/ha)**			**Increase/decrease in yield over conventional practices**	
	2015–2016	**2016–2017**	**Pooled**	**q/ha**	**%**
T_1—Conventional practices (recommended FYM + fertilizer + PP with chemicals)	302.43	323.84	313.14	–	–
T_2—100% RDN through vermicompost (PP with organic methods)	318.32	325.85	322.08	(+) 8.94	(+) 2.85
T_3—100% RDN through FYM (PP with organic methods)	272.18	277.93	275.05	(−) 38.09	(−) 12.16
T_4—Conventional practices (recommended FYM + fertilizer + PP with chemicals) + microbial consortium at 12.5 kg/ha)	319.83	354.48	337.15	(+) 24.01	(+) 7.67
T_5—100% RDN through vermicompost + microbial consortium at 12.5 kg/ha (PP with organic methods)	336.38	390.12	363.25	(+) 50.11	(+) 16.00
T_6—100% RDN through FYM + microbial consortium at 12.5 kg/ha (PP with organic methods)	292.12	296.14	294.13	(−) 19.01	(−) 6.07
T_7—Recommended FYM + fertilizer + microbial consortium at 12.5 kg/ha (PP with organic methods)	316.30	349.92	333.11	(+) 19.97	(+) 6.38
SEm±	11.25	12.62	11.29	–	–
CD (P=0.05)	34.65	38.87	32.09	–	–

RDN, Recommended dose of nitrogen.

TABLE 31.10 Effect of Treatments on Economic Parameters of Coriander (Leaf Crop)–Radish Sequence (₹/ha).

Treatment	Coriander (leaf crop)–radish sequence				
	Cost of cultivation (₹/ha)	Gross return (₹/ha)	Net return (₹/ha)	Return per rupee invested (₹)	Per day return (₹/ha)
T_1—Conventional practices (recommended FYM + fertilizer + PP with chemicals)	84,352	255,846	171,494	3.03	1491.25
T_2—100% RDN through vermicompost (PP with organic methods)	75,928	262,049	186,121	3.45	1618.44
T_3—100% RDN through FYM (PP with organic methods)	68,616	223,642	155,026	3.26	1348.05
T_4—Conventional practices (recommended FYM + fertilizer + PP with chemicals) + microbial consortium at 12.5 kg/ha)	87,554	271,778	184,224	3.10	1601.95
T_5—100% RDN through vermicompost + microbial consortium at 12.5 kg/ha (PP with organic methods)	79,132	292,376	213,244	3.69	1854.29
T_6—100% RDN through FYM + microbial consortium at 12.5 kg/ha (PP with organic methods)	71,810	239,907	168,097	3.34	1461.71
T_7—Recommended FYM + fertilizer + microbial consortium at 12.5 kg/ha (PP with organic methods)	88,456	268,815	180,359	3.04	1568.34

RDN, Recommended dose of nitrogen.

31.8.2 ORGANIC PRODUCTION OF AMARANTH

A study on the organic production of amaranth was conducted during the *Zaid* season of 2015 and 2016. Organic inputs applied to supply plant nutrients to the crop included vermicompost, FYM, and *neem* cake alone and also, along bio-inoculants (PSB and *Azospirillum*). Amaranth variety "*Local Green*" was used in the experiment. The results of the study revealed that the application of vermicompost at 5 t/ha + PSB + *Azospirillum* at 5 kg/ha each recorded the highest green leaves yield (177.30 q/ha) followed by vermicompost at 5 t/ha (167.92 q/ha). Yield under inorganic fertilizer treatment (149.53 q/ha) was less compared to vermicompost alone and along with PSB + *Azospirillum* (Table 31.11). The highest C:B ratio of 2.76 was observed under the application of vermicompost at 5 t/ha + PSB + *Azospirillum* at 5 kg/ha each, which was higher than inorganic fertilizer treatment (2.50). Hence, the organic production of amaranth is feasible and economic. On the basis of the results of the study, the application of vermicompost at 5 t/ha + PSB + *Azospirillum* at 5 kg/ha each has been recommended for the organic production of amaranth for the agroclimatic condition of Zone-IV in the 34th Group Meeting of AICRP on vegetable crops held at ICAR-IARI, New Delhi, during May 10–13, 2016.

TABLE 31.11 Effect of Organic Sources of Nutrients on Green Leaves Yield of Amaranth.

Treatment	Green leaves yield (q/ha)			Gross income (₹/ha)	Cost of cultivation (₹/ha)	Net income (₹/ha)	C:B ratio
	2015	2016	Mean				
T_1—Recommended dose of NPK at 100:50:50 kg/ha	148.36	150.70	149.53	139,416.25	39,843.00	99,573.25	2.50
T_2—Vermicompost at 5 t/ha	164.13	171.70	167.92	156,514.75	42,753.00	113,761.75	2.66
T_3—FYM at 20 t/ha	153.53	158.18	155.85	145294.35	41582.00	103,712.35	2.49
T_4—Neem cake at 2 t/ha	111.43	114.52	112.98	105,322.40	59,939.00	45,383.40	0.76
T_5—Vermicompost at 5 t/ha + PSB + *Azospirillum* at 5 kg/ha each	172.63	181.96	177.30	165,246.20	44,006.00	121,240.20	2.76
T_6—FYM at 20 t/ha + PSB + *Azospirillum* at 5 kg/ha each	161.12	168.22	164.67	153,492.65	42,793.00	110,699.65	2.59

TABLE 31.11 *(Continued)*

Treatment	Green leaves yield (q/ha)			Gross income (₹/ha)	Cost of cultivation (₹/ha)	Net income (₹/ha)	C:B ratio
	2015	2016	Mean				
T_7—Neem cake at 2 t/ha + PSB + *Azospirillum* at 5 kg/ha each	118.47	122.00	120.23	112,088.50	61,189.00	50,899.50	0.83
CD at 5%	16.01	16.92	–	–	–		–
CV	6.12	6.23	–	–	–		–

31.8.3 ORGANIC PRODUCTION OF SPINACH BEET

A field study on the organic production of spinach beet was conducted during the *Rabi* season of 2017–2018. In the study, different organic inputs, namely, vermicompost, FYM, and *neem* cake, were applied alone and along with bio-inoculants (PSB and *Azospirillum*) to supply plant nutrients to the crop. Spinach beet popular variety '*All Green*' was used in the experiment. Results of the study revealed that the application of 100% RDN through vermicompost alone and along with PSB + *Azospirillum* at 5 kg each/ha recorded green leaves yield of 123.69 q/ha and 114.12 q/ha (Table 31.12), respectively, which was higher than inorganic fertilizer treatment (100.07 q/ha).

TABLE 31.12 Effect of Treatments on Green Leaves Yield of Spinach Beet.

Treatment	Green leaves yield (q/ha)
T_1—Recommended NPK at 100:50:50 kg/ha	100.07
T_2—100% RDN through vermicompost	114.12
T_3—100% RDN through FYM	104.00
T_4—100% RDN through neem cake	87.03
T_5—100% RDN through vermicompost + PSB + *Azospirillum* at 5 kg/ha each	123.69
T_6—100% RDN through FYM + PSB + *Azospirillum* at 5 kg/ha each	112.86
T_7—100% RDN through neem cake + PSB + *Azospirillum* at 5 kg/ha each	92.24
CD at 5%	13.30
CV	7.13

RDN, Recommended dose of nitrogen.

31.9 CONCLUSION

Continuous use of chemical fertilizers has resulted in nutritional imbalance in soil, depletion of soil organic matter, contamination of food, and water, thereby affecting human health, as well as an adverse effect on biodiversity. The supply of nutrients through organic sources can be adopted for avoiding the hazardous effects of fertilizers and maintaining sustainability. Organic inputs improve the quality of produce and are eco-friendly. These ensure sustainability in production and soil health along with a pollution-free environment. The dry matter, protein, vitamin C content, etc. increased in organically grown leafy vegetables. Organic produce have better color, smoothness, and texture and also contain more minerals, enzymes, trace elements, and antioxidants. Finally, it may be concluded that the organic production of leafy vegetables is feasible and economic, and organic farming offers an opportunity to achieve the objectives of nutritional security. Organic leafy vegetables may play a vital role not only in preventing hidden hunger but also in health maintenance.

KEYWORDS

- **leafy vegetables**
- **organic amendment**
- **biofertilizers**
- **bioagents**
- **biopesticides**
- **cultural practices**
- **soil fertility**
- **green manures**
- **weed management**

REFERENCES

Anonymous. *Leaf Vegetables*; Agrotech Publishing Academy: Udaipur (Rajasthan), 2004; p 296.

Anonymous. *Indian Horticulture* (January–February); ICAR, Krishi Anusandhan Bhavan-I, Pusa, New Delhi, **2016,** *61* (1), 84.

Bhardwaj, K. K. R.; Gaur, A. C. The Effect of Humic and Fulvic Acids on the Growth and Efficiency of Nitrogen Fixation of *Azotobacter Chroococcum*. *Folia Microbiol.* **1970,** *15*, 364–367.

Chandra, K. *Organic Manures*. Regional Centre of Organic Farming: Hebbal, Banglaore, 2005.

Chandrasekaran, B.; Annadurai, K.; Somasundaram, E. *A Text Book of Agronomy. Nutrient Management*; New Age International (P) Limited, New Delhi, 2010; pp 432–443.

FAI. *Fertilizer Statistics 2003–04*; The Fertilizer Association of India: New Delhi, 2004.

Gaur, A. C. *Report on Organic Manures*; U.P. Council of Agricultural Research: Lucknow, 2003; p 67.

Malhotra, S. K. Organic Production of Seed Spices. In *Organic Horticulture: Principles, Practices and Technologies*; Westville Publishing House: New Delhi, 2010, pp 83–119.

Rajiv; Singh, R. Organic Coriander Production for Sustainable Agriculture and Livelihood Security. In *Souvenir of National Conference on Organic Farming for Sustainable Agriculture and Livelihood Security under Changing Climatic Conditions*, C. S. Azad University of Agriculture & Technology, Kanpur, Dec 12–13, 2017, pp 124–132.

Ramesh, P.; Panwar, N. R.; Singh, A. B.; Ramana, S.; Subba, Rao A. Impact of Organic Manure Combinations on Productivity and Soil Quality in Different Cropping Systems in Central India. *J. Plant Nutr. Soil Sci.* **2009,** *172*, 577–585.

Singh, B.; Choudhary, S. Organic Cultivation of Vegetables under Protected Conditions. In *Souvenir of National Conference on Organic Farming for Sustainable Agriculture and Livelihood Security under Changing Climatic Conditions*, C. S. Azad University of Agriculture & Technology, Kanpur, Dec 12–13, 2017, pp 88–92.

Singh, B.; Singh, S. K.; Singh, S. Organic Vegetable Production for Sustainable Growth and Improved Livelihood under Changing Climatic Conditions. In *Souvenir of National Conference on Organic Farming for Sustainable Agriculture and Livelihood Security under Changing Climatic Conditions*, C. S. Azad University of Agriculture & Technology, Kanpur, Dec 12–13, 2017, 93–98.

Singh, N. P.; Bhardwaj, A. K.; Kumar, A.; Singh, K. M. *Modern Technology on Vegetable Production*; International Book Distributing Co.: Lucknow, 2004; p 353.

Tomar, B. S., Jat, G. S.; Singh, J. *Organic Vegetable Production: Need, Challenges and Strategies. Souvenir of National Conference on Organic Farming for Sustainable Agriculture and Livelihood Security under Changing Climatic Conditions*, C. S. Azad University of Agriculture & Technology, Kanpur, Dec 12–13, 2017; pp 99–106.

Upadhayay, N. C. Organic Potato Production. In *Compendium of Summer School on Advances in Quality Potato Production and Post Harvest Management,* Central Potato Research Institute, Shimla, July 16–Aug 5, 2013; pp 59–66.

Yadav, A. K. Organic Agriculture in India: An Overview. *Souvenir of National Conference on Organic Farming for Sustainable Agriculture and Livelihood Security under Changing Climatic Conditions*, C. S. Azad University of Agriculture & Technology, Kanpur, Dec 12–13, 2017; pp 01–08.

CHAPTER 32

Organic Production of Basmati Rice

RITESH SHARMA[1], VIVEK YADAV[2], and VIJAY KUMAR YADAV[3*]

[1]*Basmati Export Development, Modipuram, Meerut 250110, Uttar Pradesh, India*

[2]*Rice Research Station (SVPUA&T, Meerut), Nagina, Bijnor 246762, Uttar Pradesh, India*

[3]*C.S. Azad University of Agriculture and Technology, Kanpur 208002, Uttar Pradesh, India*

[*]*Corresponding author. E-mail: vkyadu@gmail.com*

ABSTRACT

India is the major producer and exporter of Basmati rice. It is the major rice variety in demand, with two third of its production exported every year from India. With rise in awareness of the benefits of organic products, the demands for such products have been increased in the national as well as international market. In order to meet the international standards, India has switched to organic cultivation of basmati rice. Lack of knowledge, high input cost, low yield were the early challenges for the farmers adopting organic farming for the first time. Currently used organic techniques include the use of organic manure, biofertilizers, crop rotation, mixed farming and integrated pest management. This quality of Basmati rice is derived from the intermediate amylose content in the rice and the intermediate GT (gelatinization temperature). The most important characteristic of them all is the distinctive aroma. Incidentally, the aroma in basmati arises from a cocktail of over 100 compounds—hydrocarbons, alcohols, aldehydes, and esters.

Organic Crop Production Management: Focus on India, with Global Implications, D. P. Singh, PhD, H. G. Prakash, PhD, M. Swapna, PhD, & S. Solomon, PhD (Eds.)

2-Acetyl-1-pyrroline is a molecule specifically to be noted in this regard. Basmati rice is now covered under GI and can be grown only in seven states of India: Punjab, Haryana, Uttarakhand, Himachal Pradesh, Delhi, Jammu and Kashmir (Jammu, Samba, and Kathua districts) and 30 districts of Uttar Pradesh

32.1 INTRODUCTION

The word '*Basmati*' is derived from two Sanskrit roots (*vas* = *aroma*) and (*mayup* = ingrained or present from the beginning). While combining *mayup* with *vas*, it changed to *mati* making *vasmati.* Generally, people pronounce it as '*basmati*'. The great Himalayan foothills of the Indian subcontinent is the treasurer of basmati rice which is long cylindrical scented rice. It is grown for centuries with its unique characteristics like high aroma, unmatched taste, and the ability to elongate twice to their original size after cooking. Basmati rice is unique among other aromatic long-grain rice varieties. Basmati rice is characterized by unique cooking and eating quality characteristics with a distinct aroma. Superfine long slender grain (>6.61 mm long), pleasant aroma, extreme elongation on cooking (>1.8 times), and soft texture of cooked rice are some of the distinctive features of basmati rice. The milled rice is translucent depending on the degree of milling, color may vary from creamy to pearly white during cooking, and a high-volume expansion is achieved by the linear elongation of the cooked kernel with minimum breadth-wise swelling. After cooking, the rice remains fluffy (nonsticky, tender, and moist) without bursting or splitting. Cooked rice has a longer shelf-life and does not turn stale easily if properly covered and kept at even room temperature. It is known for its remarkable taste, mouthfeel, and after-feel.

The basmati rice has elongation post-cooking and no other rice in the world has this character in combination. Basmati rice is available in the market as brown rice (only husk removed) and brown parboiled steamed. Brown basmati rice is also available but the most commonly used is white basmati. The grain of basmati is long and the texture is firm and tender without splitting, and it is nonsticky. This quality is derived from the intermediate amylose content in the rice and the intermediate GT (gelatinization temperature). The most important characteristic of them all is the distinctive aroma. Incidentally, the aroma in basmati arises from a cocktail of over 100 compounds—hydrocarbons, alcohols, aldehydes, and esters. 2-Acetyl-1-pyrroline is a molecule specifically to be noted in this regard. the rice is fluffy and the grains stay separate because of the flavor and texture. It complements curries in an excellent way. It is also suited for *biryani* and *pulao* with saffron added to provide extra flavor and color.

As per the certificate issued from the Registrar of the Geographical Indication (GI), basmati is now covered under GI and can be grown only in seven states of India: Punjab, Haryana, Uttarakhand, Himachal Pradesh, Delhi, Jammu and Kashmir (Jammu, Samba, and Kathua districts) and 30 districts of Uttar Pradesh (Table 32.2). The state-wise basmati areas of seven basmati producing states are given in Table 32.1.

TABLE 32.1 State-Wise Basmati Area in India.

S. no	States	Area (000 ha)
1	Haryana	719.60
2	Punjab	615.60
3	Uttar Pradesh	266.20
4	Jammu and Kashmir	62.30
5	Uttarakhand	15.62
6	Himachal Pradesh	8.00
7	Delhi	1.50
	Total	1688.82

Source: APEDA (2017).

TABLE 32.2 Districts-Wise Basmati Area in Uttar Pradesh.

S. no	Districts	Area (000 ha)
1	Agra	0.9
2	Aligarh	27.6
3	Auraiya	2.0
4	Baghpat	3.9
5	Bareilly	13.1
6	Bijnor	6.9
7	Badaun	4.2
8	Bulandshahar	30.7
9	Etah + Kasganj	8.6
10	Farrukhabad	2.7
11	Firozabad	3.9

TABLE 32.2 *(Continued)*

S. no	Districts	Area (000 ha)
12	Etawah	9.0
13	Gautam Buddha Nagar	21.4
14	Ghaziabad + Hapur	10.0
15	Hathras	7.3
16	Mathura	29.4
17	Mainpuri	18.7
18	Meerut	7.3
19	Moradabad	3.4
20	J.P Nagar	4.1
21	Kannauj	2.2
22	Muzaffarnagar + Shamli	11.1
23	Pilibhit	4.4
24	Rampur	2.0
25	Saharanpur	23.1
26	Shahjehanpur	4.4
27	Sambhal	3.7
	Total	266.2

Source: APEDA (2017).

Due to health concerns, the demand is increasing for organically grown food products worldwide. India has great potential to become a leading exporter of organic food in the international market. Basmati rice can be a very good option for organic cultivation and the government is making every effort to produce and export it around the world.

32.2 EXPORT

India is leading from the front as far as export of the Basmati rice in the global market is concerned. The export of basmati rice has been growing steadily; from 7.71 lakh metric tonnes in 2003, its production has reached an estimated 4.05 million metric tons, worth Rs. 21,604.58 crores (or 3230.24 US$ Mill.) during the year 2016–2017 on the robust demand from the traditional markets in west Asia. Almost 132 countries import basmati from India every year, out of which, Saudi Arabia, Iran, United Arab Emirates, Iraq, and Kuwait are the major importers.

32.3 AGRO TECHNIQUES FOR THE ORGANIC PRODUCTION OF BASMATI RICE

Rice is a major crop that receives the maximum quantity of fertilizers (40%) and pesticides (17–18%). There are two major concerns in organic rice farming: nutrient management and pest management. How the sources of organic nutrients can be effectively and efficiently managed for achieving higher productivity and best quality basmati rice is described next.

32.3.1 SELECTION OF BASMATI VARIETIES

Basmati is categorized into two broad heads—Premium Indian basmati (traditional basmati) and Indian basmati or evolved basmati (crossed/hybrid).

32.3.1.1 PREMIUM INDIAN BASMATI (TRADITIONAL BASMATI)

Basmati 370 is grown widely in both India and Pakistan and is known for its best quality. The other famous categories, basmati 386, basmati 217, and type-3, have been evolved from Dehraduni basmati, Taraori basmati (HBC-19), and Ranbir basmati (IET-11348).

32.3.1.2 INDIAN BASMATI OR EVOLVED BASMATI (CROSSED/HYBRID)

The nontraditional, basmati rice high yielding varieties/hybrids as notified by the Seeds Act are Pusa basmati-1 (IET-10364), Punjab basmati-1 (Bauni basmati), Kasturi (IET-8580), Haryana basmati-1 (HKR-228/IET-10367), Mahi Sugandha, Improved Pusa basmati-1 (Pusa-1460), Pusa basmati-1121, Vallabh basmati-22, Pusa basmati-6 (Pusa-1401), Punjab basmati-2, Basmati CSR-30, Vallabh basmati-21 (IET-19493), Pusa basmati-1509 (IET-21960), Vallabh basmati-23, Vallabh basmati-24, Pant basmati-1 (IET-21665), Pant basmati-2 (IET-21953), Punjab basmati-3, and Pusa basmati-1637. Farmers prefer high-yielding Pusa-1509 in place of low-yielding Pusa basmati-1121 due to its shorter growing time period and early maturity trait. Pusa basmati-1 is continuously reducing in acreage over years and is now preferred only by farmers who cultivate it for their self-consumption; it still has good market potential in Europe and Saudi Arabia, which is about 1.5–2 lakh tonnes each.

32.3.2 CHARACTERISTICS OF GOOD SEED

Seed is the important key for the cultivation of a good and healthy nursery. The seed must be true to its type, that is, genetically pure, free from admixtures and should belong to the proper variety or strain of the crop and their duration, and should be according to agroclimatic and cropping system of the locality. Seed should be pure, viable, and vigorous with high-yielding potential. Seed should be free from seed-borne diseases and pest infections and should be clean; free from seeds of any weed or inert materials. Seed should contain the optimum amount of moisture (10–12%). It should have high germination percentage (more than 80%) and germinate rapidly and uniformly when sown (Fig. 32.1).

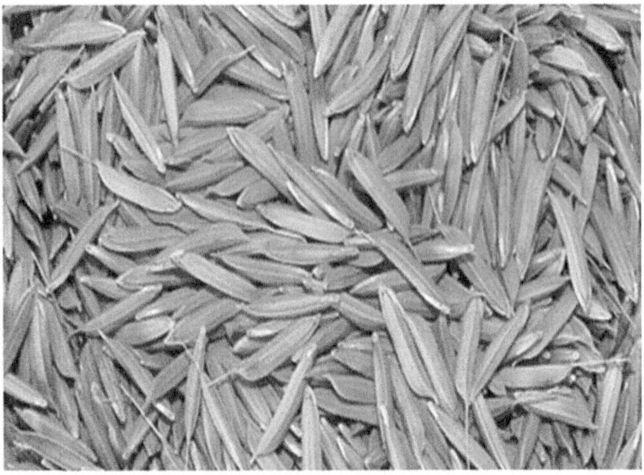

FIGURE 32.1 Seed of basmati rice.

32.3.3 NURSERY AREA AND SEED RATE

For sowing of 1-ha area, 15–20-kg good-quality seed is recommended. Raising a nursery for getting healthy seedlings is an important operation. In northern India, maximum farmers use 1000-m^2 area for sowing of 60-kg seed, which results in a very thin and weak nursery. The proper seed rate of 20–25 kg per 1000 m^2 area results in healthy, vigorous seedlings.

32.3.4 PURIFICATION OF SEED

For the success of any crop, the sowing of healthy and pure seeds is the prime requirement. Before sowing of seeds, the unfilled, infected, and lightweight seeds are separated. To select suitable seeds, dipping the seeds in a 10% common salt solution is a simple attest. After pouring the salt solution, stir the seed and the lighter seeds will float on the water surface. These should be discarded and the healthy seeds that are set down at the bottom of the container should be washed two to three times with clean water and used for sowing (Fig. 32.2).

FIGURE 32.2 Purification of seed.

32.3.5 SEED TREATMENT

The seed treatment helps to improve the resistance to pests and diseases, hardening against drought and environmental shocks. The paddy seeds can be pregerminated before sowing in the nursery bed. The common practice followed is by soaking the seeds in a vessel containing water for 24 h, treated with *Pseudomonas fluorescens* and *Trichoderma harzianum* at 10 g/kg, each. This is stored as a heap for another 24–36 h in shade by covering with gunny bags. Keep it moist by lightly sprinkling water, if required. Other method of seed treatments is soaking of seeds in a mixture of 1 kg each of cow dung, cow urine, 91-L water, and 100-g cow's ghee for 10 min. The seeds are dried in the shade for 6–10 h. When such soaked seeds are used for sowing, it results in healthy seedlings and high yields (Fig. 32.3).

FIGURE 32.3 Seed treatment.

32.3.6 FIELD PREPARATION

It is required to raise seedlings on specially prepared seedbeds and go for transplanting the paddy seedlings into lowland puddled soil. This helps in providing the seedlings a substantial head start avoiding the weeds. Select a fertile, well-drained upland field near the source of irrigation water. Paddy seeds require soaking in water and are pre-germinated to raise seedlings quickly in the field or seedbed. For better germination and good growth of seedlings, field should be prepared very well. Field preparation for nursery raising is the most effective and crucial for getting healthy seedlings. Good agriculture practices such as deep summer plowing to increase water holding capacity of the soil, the prevention of disease and pest due to solarization, green manuring for increasing organic matter content and to improve soil health also, plowing twice before growing nursery at 10-day interval after irrigation, etc. should be adopted to prepare filed for nursery cultivation. This will make the nursery area weed-free. Plow the field at 20–25 cm with a mold board plow in summer to expose the eggs of harmful insects, pests, and rhizomes of weeds. Keep it open 10–15 days for solarization (sunbath of nursery field) (Fig. 32.4).

FIGURE 32.4 Field preparation.

32.3.7 SEEDBED PREPARATION

After proper field preparation, prepare small and narrow seedbeds (1.5 × 20 m^2). Maintain 30–50-cm channel/bund/walking space in between two beds with proper leveling. A thorough bed preparation with a spade (*Fawda)* and proper leveling helps to destroy the weeds that grow in this seedbed and also for uniform irrigation. The application of 225-g urea or 500-g ammonium sulfate and 500-g single super phosphate per 10-m^2 nursery area can be done. Uniformly broadcast about two to three handfuls of fertilizer on the seedbed. Mix the fertilizer well with the soil by shallow digging and then irrigate the seedbed and level it with a small planker. After this, you can broadcast the germinated seed very carefully without damage. Broadcast the seed uniformly in the seedbed. Sowing of the nursery should be done in the evening to avoid damage due to sunlight (Figs. 32.5 and 32.6).

FIGURE 32.5 Seed bed preparation.

FIGURE 32.6 Puddling of field.

32.3.8 NURSERY RAISING

Apply well-decomposed farmyard manure (FYM) at the rate of 1.5–2.0 t/ha, vermicompost at 1 t/ha, biofertilizer (blue-green algae preparation) at 250 g/ha. The land should be fertile and free of soil problems like salinity. The seedbed area is plowed twice and then puddled by giving two or three more plowings. After 10 days, the field is again plowed twice and leveled. This process is done to decrease weed problems because most of the germinated weed seeds are destroyed by the plowing process. When the field is ready as fine soft puddled condition, raise the seedbeds (4–5-cm high) of 1.5–2.0 m width and of convenient length with 45 cm channel all around are constructed. Raised beds should be prepared only in the areas where waterlogging is a problem. The surface of the seedbed is leveled in a manner that there is a gradual proclivity toward both sides to facilitate the drainage of water (Figs. 32.7 and 32.8).

FIGURE 32.7 Presowing seed.

FIGURE 32.8 Sowing of seed.

Sow (homogenously broadcast) pre-germinated and treated seeds on a drained bed at the rate of 1-kg seed/40-m^2 area. A higher density of seed resulted in weaker seedlings. It will be also more difficult to pull out seedlings and there will be more chances of injury to the long roots of adjacent seedlings. Beds should be moist for the first few days and if possible, give light irrigation in the evening time. Never flood the beds, keep the beds submerged in a shallow layer of water when seedlings are about 2 cm in height.

32.3.9 UPROOTING OF SEEDLING

The optimum age of nursery for transplanting is 20–25 days after sowing, while in the case of SRI technology it should be 8–12 days old. The seedbed should be flooded that the uprooting of the seedling will be easy without damage. Uproot a few seedlings at a time by holding them from the base as low as possible and pulling sideways. Always handle seedlings with extra care. Seedlings that are handled gently during uprooting and transporting recover much more quickly when transplanted. To remove mud, wash the seedlings gently in water (Fig. 32.9).

FIGURE 32.9 Derooting of nursery.

32.3.10 SEEDLING TREATMENT

The seedlings can be dipped in a solution of biodynamic preparation 500 and cow pat pit (CPP) or a solution of *T. harzianum* and *P. fluorescens* 10 g each per liter water. It will help in protecting the crop from various fungal and bacterial diseases (Fig. 32.10).

FIGURE 32.10 Nursery bundle.

32.3.11 GREEN MANURE AND FIELD PREPARATION FOR TRANSPLANTING

The preparation of the main field is extremely vital for rice cultivation. The field should be well leveled, if necessary a laser land leveler can be used for leveling the field before sowing of green manure crop. For green manure, sesbania (*Dhaincha*), sun hemp (*Sanai*), cowpea, mungbean, and others can be used to curtail the nutrient requirement of rice crops. The sowing of sesbania or sun hemp can be done 40–45 days before transplanting (Fig. 32.11).

FIGURE 32.11 Field preparation.

The green manure crop should be incorporated in 2–4 in. of standing water through the rotavator, 1 or 2 days before transplanting. By this method, the mixing of green manure and puddling can be done together to save the cost and maximum uptake of the nutrients by the rice crop.

32.3.12 NUTRIENT MANAGEMENT

- Growing leguminous plants in the rotation would provide an additional crop to the farmer and would also balance the process of biological nitrogen fixation.
- Summer deep plowing is very helpful for the disinfection of nematodes, insect pests, and diseases.

- In new clearings, green manure crops and legumes like cowpea and horse gram can be grown for 2–3 years to build up soil fertility.
- Many sources, that is, FYM, compost, press mud, green manure, vermicompost, and biofertilizers are utilized for the fulfillment of the nutrient requirement. The application of organic manures/green manure/vermicompost should be based on soil testing.
- During final puddling, well-decomposed FYM at 10 t/ha should be added into the soil without leaving them in small piles in the field for a long period. Otherwise, the nutrients will be lost. The vermicompost may apply at 2–5 t/ha if required. The topdressing of vermicompost at tillering and booting stage may be done at 1 t/ha.
- Biofertilizers *Azospirillum* or PSB/PSM at 2 g/ha can be applied by adding 25-kg vermicompost or FYM. Blue-green algae at 10 kg/ha are also recommended for application in the main field.
- During various phases of cultivation of rice, any formulation, that is, biodynamic spray, liquid manure, vermi wash, *Amritpani*, cow's urine and dung brew (*Matka Khad*), *Panchagavya,* and *Amruthajalam*, can be adopted for maintaining a healthy crop.

32.3.13 TRANSPLANTING

- Use 20–25-day-old seedlings at three to four leaf stages for transplanting as old seedling can reduce your yield (Fig. 32.12).
- Always irrigate the field before uprooting the seedlings to reduce injuries and it will be very helpful for bakanae disease control.
- One-fourth upper portion of leaf clipping will protect the seedling from wind damage and will manage the stem borer attack in the main field in the early stage.
- The seedlings should be transplanted the same day after uprooting.
- Under normal conditions, two to three healthy seedlings per hill at 20 × 15-cm distance are transplanted.
- On average, 30 hills/m^2 are recommended to assure an adequate population in the rice field.
- If an aged seedling is used, increase the number of the seedling.
- Delayed transplanting leads to early flowering of main tillers, poor tillering, and reduction in yield.

- The seedlings should be transplanted at 2–3-cm depth, straight without any injury to the seedling, as deeper planting delays and reduced tillering.
- Yield can be maintained steady of each variety of rice if all the practices are properly followed.

FIGURE 32.12 Transplanting of nursery.

32.3.14 WATER MANAGEMENT

Assured and timely supply of irrigation water has a reflective influence on the yield of the crop. In the life cycle of rice plants, there are some critical stages wherein water requirement is high. Such critical crop growth stages are the initial seedling period (about 10–15 days), tillering, panicle initiation, flowering, and milking stage, where moisture stress significantly reduces yield. Irrigate before crack formation and there should be standing water of 2–5 cm during critical stages. If possible, drain the field from time to time to prevent the insect pest attack.

32.3.15 WEED MANAGEMENT

Weeds pose a major threat; however, in transplanted rice cultivation, weed problem is less due to green manure, puddling, and efficient water management. In the early stage of the crop, proper weed management is necessary to obtain a good yield. Water should be maintained 3–5 cm up to 20 days after transplanting to reduce the weed flora. Hand weeding is to be practiced. Rotary or cono weeders can be used based on availability.

Pata practice is also helpful to overcome the weed population. In this practice, 15–20 days after transplanting, one *pata* made of wood or a bundle of bamboos (approximate weight 10–15 kg), which can be pulled out by a person, is used in a standing crop. It should be ensured that this practice can be followed only in standing water (4–6 in.). By this practice, the soil will get disturbed resulting in a very good tillering and the overall development of the crop. This practice can be repeated at an interval of 2–5 days (Fig. 32.13).

FIGURE 32.13 Weed management.

32.3.16 MANAGEMENT OF INSECT-PESTS AND DISEASES

Organic farming systems should be carried out in such a way that the losses from pests, diseases, and weeds are minimized. Emphasis is placed on the use of a balanced fertilization program, use of crops and cultivars well adapted to the environment, fertile soils of high biological activity, adapted rotations, green manures, and so on.

The majority of pests and diseases in rice could be effectively kept below the economic threshold level by the judicious maintenance of microclimate. The pests should be controlled by the integration of physical, cultural, and biological methods, including plant-based preparations permitted under the National Standards for Organic Products.

Cultural activities like 3–5 deep plowing in summer destroyed the pathogens of different diseases and eggs of insect pests. Crop rotation, green manures, and adding plant residue in the soil are also effective to minimize the harm by insect pests and diseases.

32.3.17 FIELD PREPARATION

Apply FYM precolonized with *T. harzianum* or *P. fluorescens*. For the precolonization of FYM, *Trichoderma* and *Pseudomonas* are to be added at a monthly interval in the FYM pits at 100 g/pit. These pits should be covered with sugarcane leaves or rice straw. Water should be sprayed at regular intervals (at least once after bioagent application) and 7–15 days before FYM use to maintain moisture. At the turning of green manure, spray *T. harzianum* at 5 g/L just before plowing.

32.3.18 NURSERY SOWING

Seed treatment should be done as recommended. Set up one pheromone trap per 100-m^2 nursery area for the monitoring of stem borer. Release *Trichogramma japonicum* or *T. chilonis* (150,000 parasites /ha), if required.

32.3.19 TRANSPLANTING

Transplanting is done by drenching of *Pseudomonas fluorescence* (1 g/m^2) in nursery soil 1 day before uprooting of the seedling or dipping the roots of seedling in a solution of *T. harzianum* and *P. fluorescens* at 5 g each per liter water.

Clipping one-fourth of the leaf from the tip will help to eliminate the egg masses of the stem borer. It should be collected separately and should be destroyed outside the field.

Avoid close planting and maintain an optimal population (25 plants/m^2).

Avoid planting under full or partial shade to avoid bacterial leaf blight (BLB). Once BLB attacks plants in shade, these plants become the source of inoculums for the remaining fields.

The uprooting of seedling should be done under flood conditions, that is, the nursery area must be irrigated 1 day before or the same day of uprooting the seedling. It will help to reduce the infection of Bakanae disease.

32.3.20 AFTER TRANSPLANTING TILL MATURITY

Use pheromone traps with 5-mg lure per trap, 20 traps/ha on a spacing of 20×25 m for the management of stem borer and replace lure after 15 days.

Maintain a height of 50 cm for the trap at the early stage of the crop and raise the height to 20–30 cm above the crop.

Apply *Trichogramma japonicum* at 150,000 parasites per ha, whenever needed.

Keep the bunds clean by trimming them and remove the grassy weeds that serve as an alternate host for diseases, especially for the management of sheath blight and other insect pests.

Use rope pulling in the field for effective management of leaf folders and other insects. Through this practice, all larvae will drop in water and die.

Avoid water stagnation and keep the field wet except at critical stages. Drain the field as frequently as possible for managing sheath blight, BPH, and other insect pests.

Remove the infected plants and destroy them. The field should be clean and weed-free.

Set up light traps to monitor and manage pests. Establish bird perch for birds to sit. They destroy the egg masses of insect pests.

Many formulations are available for the management of insect pests and diseases and can be applied, whenever needed. Some of them are described here under the following:

Cow's Urine: To counter the attack of boring or sucking pests, cow urine serves as a good pest manager and tonic at 10% concentration. Diluted cow urine (10 L in 100 L of water) at the rate of 200–300 L/ha should be sprayed on the whole plant at regular intervals except on the panicle (grain portion).

Neem Leaf Extract: *Neem* leaf extract is sprayed to counter the attack of any pest. It is prepared by grinding and soaking fresh leaves for 24 h in water. Twenty-five liters of this extract are mixed with 100 L of water and sprayed on the plants at the rate of 300–500 L/ha.

Neem Oil Spray: *Neem* oil available at 2% can also be sprayed for repelling the pest. Any detergent powder at the rate of 40–50 g for 100 L of spray solution can be used as an emulsifier.

***Cow Dung Spray*:** Paddy blast and bacterial blight diseases of paddy can be minimized by the spraying of cow-dung slurry after proper straining. This spray can minimize disease development and the spread of bacterial blight. A suspension of 500-gm fresh cow dung per 10-L water strained through a muslin cloth can be sprayed through sprayer or broom up to grain filling stage on a weekly interval.

***Bael Leaf Extract*:** A decoction of *bael* leaves (100-g leaves boiled in 2-L water) cooled and added to 100-L water and adequate quantity of cow urine, when sprayed, control diseases of rice.

***Bioagents*:** Spray mixed formulation of compatible strains of *T. harzianum* and *P. fluorescens* (5 g/L) at panicle initiation against sheath blight, sheath rot, and neck blast. One to two sprays are to be given at weekly intervals, as per need.

***Neem Seed Kernel Extract (NSKE)*:** NSKE should be used as a prophylactic before the onset of pests.

***Asafoetida Spray*:** If sucking pest appears on the crop, they can be repelled by spraying a preparation of 10-g asafoetida, and 4-L cow urine in 10-L water.

***Chili–garlic Soup*:** Chili–garlic soup is effective as an insect repellent against most insects.

***Tobacco Tea*:** Tobacco tea is effective against most of the pests.

32.3.21 CONTAMINATION CONTROL

- All relevant measures should be taken to minimize contamination from outside and within the farm.
- A suitable buffer zone should be maintained between organic and conventional blocks to prevent possible contamination with chemicals.
- The by-products like leaves, stems, and cherry husk should be in the field for composting.
- Products based on polyethylene/polypropylene or polycarbonate should not be burnt or left out in farmland.
- The accumulation of heavy metals and other pollutants should be kept minimal.
- The use of polychloride-based products is prohibited.

32.3.22 HARVESTING/THRESHING

Harvesting can be done manually by sickle or by combined machines. When the grains turn to golden color and the moisture is 14–18%, the harvesting can be done. Generally, harvested rice crop is spread over the stables for drying for 2–3 days. The threshing is done manually by beating on wooden pallets or drums to separate the grains from the plant. This paddy should dry in shade after winnowing. Winnowing can be done by tractor-/power-driven fans (Fig. 32.14).

FIGURE 32.14 Harvesting of paddy.

32.4 ORGANIC CERTIFICATE

There are two systems of organic certification, first *Third Party* and *Participatory Guarantee System (PGS)*. There are seven companies, mostly foreign, doing organic certification in Uttar Pradesh. They charge Rs. 75,000/ visit in addition to their operational expenses. Minimum two visits for 3 years are required to complete the *Conversion period* before certification for *full organic* certification can be issued. No doubt, the high cost of certification had always been a matter of concern for resource-poor farmers. But the Government of India has started the PGS system in which groups of farmers certify each other's production and product. Under minimum supervision from the government, certificates are issued which are valid for the Indian market. *Green Certificate* is issued for the first 3 years, while a *blue logo* is issued after becoming full organic and regular.

KEYWORDS

- **basmati**
- **export**
- **nursery area**
- **green manure**
- **transplanting**
- **water management**
- **weed management**
- **contamination control**

REFERENCE

APEDA. *Basmati Survey Report (Volume 1)*; New Delhi, 2017.

Index

C

E

S

U

V